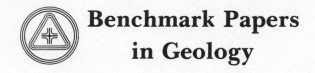 # Benchmark Papers in Geology

Series Editor: Rhodes W. Fairbridge
Columbia University

Published Volumes and Volumes in Preparation

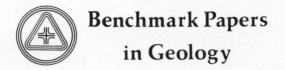

Benchmark Papers
in Geology

—— A *BENCHMARK* TM Books Series ——

GLACIAL ISOSTASY

Edited by
JOHN T. ANDREWS
University of Colorado

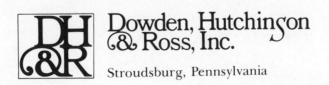

Dowden, Hutchinson & Ross, Inc.
Stroudsburg, Pennsylvania

Library of Congress Cataloging in Publication Data

Andrews, John T   comp.
    Glacial isostasy.

    (Benchmark papers in geology, v. 10)
    Bibliography: p.
    1. Isostasy--Addresses, essays, lectures.
2. Glacial epoch--Addresses, essays, lectures.
I. Title.
QE511.A6   551.1'3   73-12624
ISBN 0-87933-051-1

Exclusive distributor outside the United States and
Canada: John Wiley & Sons, Inc.

Acknowledgments
and Permissions

ACKNOWLEDGMENTS

American Geophysical Union—*Journal of Geophysical Research*
"Glacial Rebound and the Deformation of the Shorelines of Proglacial Lakes"

Cambridge University Press—*Geological Magazine*
"On the Cause of the Depression and Re-elevation of the Land During the Glacial Period"

Geological Survey of Norway—*Geology of Norway*
"General View of Shore Levels on the Coast and Fjord Area of Finnmark"

Iceland Glaciological Society—*Jökull*
"Late- and Post-glacial Rise in Iceland and Sub-crustal Viscosity"

International Association of Hydrogeological Sciences—*International Association of Hydrogeological Sciences Publication 79*
"The Extent of the Würm Glaciation in the European Arctic"

Macmillan (London and Basingstoke)—Selections from *The Quaternary Ice Age*

Universitetsforlaget, Publishers to the Norwegian Universities—*Det Norske Videnskaps-Akademi i Oslo*
"The Strandflat and Isostasy"

University of Chicago Press—*Journal of Geology*
"A Reconstruction of Water Planes of the Extinct Glacial Lakes in the Lake Michigan Basin"

PERMISSIONS

The following papers have been reprinted with the permission of the authors and the copyright owners.

Academia Scientiarum Fennica (Soumalainen Tiedeakatemia)—*Annales Academiae Scientiarum Fennicae*
"The Relation Between Raised Shores and Present Land Uplift in Finland During the Past 8000 Years"

American Geophysical Union—*Journal of Geophysical Research*
"Viscosity of the Mantle from Relaxation Time Spectra of Isostatic Adjustment"
"On Crustal Flexure"

The Arctic Institute of North America—*Arctic*
"Radiocarbon-dated Postglacial Delevelling in Northeast Greenland and Its Implications"

Edinburgh Geological Society—*Transactions of the Edinburgh Geological Society*
"A Re-interpretation of the Literature on Late-glacial Shorelines in Scotland with Particular Reference to the Forth Area"

Elsevier Publishing Company—*Palaeogeography, Palaeoclimatology, and Palaeoecology*
"Eustatic Changes During the Last 20,000 Years and a Method of Separating the Isostatic and Eustatic Factors in an Uplifted Area"

Information Canada—*Geographical Bulletin*
"Postglacial Emergence at the South End of Inugsuin Fiord, Baffin Island, N.W.T."

Institute of British Geographers—*Transactions and Papers 1966, Publication 39*
  "Late-glacial and Post-glacial Shorelines in South-east Scotland"

University of Manitoba Press—*Life, Land and Water*
  "Postglacial Uplift—A Review"
  "Geology of Glacial Lake Agassiz"

National Research Council of Canada—*Canadian Journal of Earth Sciences*
  "Isostatic Response to Loading of the Crust in Canada"

Regents of the University of Colorado—*Arctic and Alpine Research*
  "Differential Crustal Recovery and Glacial Chronology (6,700 to 0 BP), West Baffin Island, N.W.T.,
    Canada"

Societas Scientiarum Fennica—*Commentationes Physico-Mathematicae*
  "A Profile Across Fennoscandia of Late Weichselian and Flandrian Shore-lines"

M. Ters—*Études sur le Quaternaire dans le monde*
  "Marginal Subsidence of Glaciated Areas: United States, Baltic, and North Sea"

University of Washington, Seattle—*Quaternary Research*
  "Possible Causes of the Variability of Postglacial Uplift in North America"

American Journal of Science (Yale University)—*American Journal of Science*
  "Postglacial Uplift in North America"
  "Late-Pleistocene Fluctuations of Sealevel and Postglacial Crustal Rebound in Coastal Maine"

Yale University Press—Selections from *The Changing World of the Ice Age*

Series Editor's Preface

The philosophy behind the "Benchmark Papers in Geology" is one of collection, sifting, and rediffusion. Scientific literature today is so vast, so dispersed, and, in the case of old papers, so inaccessible for readers not in the immediate neighborhood of major libraries that much valuable information has been ignored by default. It has become just so difficult, or so time consuming, to search out the key papers in any basic area of research that one can hardly blame a busy man for skimping on some of his "homework."

This series of volumes has been devised, therefore, to make a practical contribution to this critical problem. The geologist, perhaps even more than any other scientist, often suffers from twin difficulties—isolation from central library resources and immensely diffused sources of material. New colleges and industrial libraries simply cannot afford to purchase complete runs of all the world's earth science literature. Specialists simply cannot locate reprints or copies of all their principal reference materials. So it is that we are now making a concerted effort to gather into single volumes the critical material needed to reconstruct the background of any and every major topic of our discipline.

We are interpreting "Geology" in its broadest sense: the fundamental science of the Planet Earth, its materials, its history, and its dynamics. Because of training and experience in "earthy" materials, we also take in astrogeology, the corresponding aspect of the planetary sciences. Besides the classical core disciplines such as mineralogy, petrology, structure, geomorphology, paleontology, and stratigraphy, we embrace the newer fields of geophysics and geochemistry, applied also to oceanography, geochronology, and paleoecology. We recognize the work of the mining geologists, the petroleum geologists, the hydrologists, the engineering and environmental geologists. Each specialist needs his working library. We are endeavoring to make his task a little easier.

Each volume in the series contains an Introduction prepared by a specialist (the volume editor)—a "state of the art" opening or a summary of the objects and content of the volume. The articles, usually some thirty to fifty reproduced either in their entirety or in significant extracts, are selected in an attempt to cover the field, from the key papers of the last century to fairly recent work. Where the original works are in foreign languages, we have endeavored to locate or commission translations. Geologists, because of their global subject, are often acutely aware of the oneness of our world. The selections cannot, therefore, be restricted to any one country, and whenever possible an attempt is made to scan the world literature.

To each article, or group of kindred articles, some sort of "Highlight Commentary" is usually supplied by the volume editor. This should serve to bring that article into historical perspective and to emphasize its particular role in the growth of the field. References, or citations, wherever possible, will be reproduced in their entirety —for by this means the observant reader can assess the background material available to that particular author, or, if he wishes, he too can double check the earlier sources.

A "benchmark," in surveyor's terminology, is an established point on the ground, recorded on our maps. It is usually anything that is a vantage point, from a modest hill to a mountain peak. From the historical viewpoint, these benchmarks are the bricks of our scientific edifice.

Rhodes W. Fairbridge

Preface

This collection of reprints on the field evidence for, and the geophysical interpretation of, the crustal deformation associated with glacio-isostatic loading and unloading of large portions of the earth's surface is intended to serve a variety of readers at a variety of levels. The importance of the study of glacio-isostasy is succinctly noted by Crittenden (1967: *Geophys. J. Roy. Astron. Soc.*, **14**, 261–279), who stated: "The response of the earth to transient loads of ice or water provides the most explicit information available about the nonelastic properties of the mantle." It is also an invaluable component of studies in the broad area of Quaternary geology. Not only are the warped shorelines and raised marine deposits the raw material which the Quaternary geologist must interpret for the use of geophysical modelers, but also the raised strandlines that are so characteristic of formerly glaciated areas serve as valuable time horizons in the landscape. Thus, many aspects of the chronology of the postglacial period can be interpreted by the interrelationships between the raised strandlines and other deposits. In the field of archeology, the slow glacio-isostatic recovery of land in North America and Fennoscandia has resulted in the location of camps of maritime hunting people being displaced a few to several tens of meters above the present sea level. In such areas, a knowledge of the changing rate of postglacial uplift can be used to narrow the search for a specific cultural occupancy.

In selecting these "benchmark" reprints of papers in the field of Glacial Isostasy, I am aware of the number of excellent and exciting papers that had to be omitted for reasons of space or considerations of coverage of selected topics. Nevertheless, I feel that the selections included in this volume bring together much of the flavor of the subject and highlight both past and present trends.

The volume will provide a comprehensive overview of studies in the features and processes that we term collectively "glacial isostasy." It will be a useful addition to the shelves of many Quaternary geologists and to solid earth geophysicists who are interested in using the field evidence for modeling the earth's response to loads over time scales of 10^3 to 10^4 years and who should be aware of the interpretative problems faced in field work and the accuracy of the resulting maps or graphs of crustal deformation through time (generally over the last 12,000 years). Certainly, I would hope that graduate students engaged in research in one of the aspects of the subject would find this volume useful. The combined bibliographies of all the papers and the general reference list will provide a comprehensive view of the literature. Undergraduate courses in glacial geology or associated topics will find the book useful as supplemental reading to such established texts as R. F. Flint's (1971) textbook, where glacial isostasy is covered in an informative chapter.

Specific acknowledgments for assistance in preparing this volume are given at the end of the introduction.

John T. Andrews

Contents

III. GEOPHYSICAL RESEARCH IN GLACIAL ISOSTASY

Contents by Author

Introduction

The effects of the Quaternary ice sheets on the surface of the world have been many and profound. Not only were they responsible for significant erosion and deposition but also they were involved in, were indeed the cause of, huge mass transfers. During the repeated waxing and waning of the Laurentide, Fennoscandia, Great Britain, and other ice sheets, water was taken from the oceans and deposited as snow on land during the development phase; during deglaciation this mass of water was once again returned to the oceans. These repeated mass exchanges resulted in a glacio-eustatic control of sea level that has considerable scientific interest in its own right. However, in this collection of reprints, we will examine only crustal aspects of the mass transfer process, whereby the addition and subtraction of water (in the form of snow and ice) imposed upon or removed considerable loads from the continents, a slow adjustment to which took place by depression and uplift, respectively, in an attempt to maintain isostatic equilibrium. Other effects of mass transfer on the globe, such as variations in the rate of rotation and shift of the poles of rotation to maintain inertial symmetry will not be treated in this collection.

The term *glacial-isostasy* is defined as the perturbations on regional isostatic equilibrium caused by the growth and retreat of ice sheets. In this volume we will deal exclusively with the Quaternary phenomena. *Isostasy* refers to the state of mass compensation that maintains at a certain depth in the earth equal hydrostatic forces. Satellite-derived gravity maps indicate that, for large areas of the world, observed gravity is close to that predicted by theory, but that large areas of mass deficiency and surplus do occur. The theory of isostasy implies a state of hydrodynamic equilibrium so that gravity anomalies will be reduced to zero in time. One of the contributions of glacial-isostasy has been to provide some measure of the speed of this reaction.

Other terms for glacio-isostatic recovery include postglacial uplift, postglacial recoil, postglacial rebound, postglacial deleveling, and postglacial emergence. The last two terms take into account the fact that in local areas the net effect of postglacial rebound and the postglacial rise of world sea level is a decrease in the apparent amount of rebound; submergence may occur in some places.

The study of glacial-isostasy has a broad spectrum of interest for diverse fields in both the earth and the life sciences. Geologists and geomorphologists, for example, have paid specific attention to the geometry of crustal deformation as seen by the warping and tilting of raised shorelines, or "strandlines," (*see* definition below). These isochronous surfaces have the properties of time-stratigraphic units and hence can be used to date such features as glacial advances (by cross-cutting relationships), paleoclimatic events, or specific prehistoric archaeological sites, to name but a few of the many uses to which the chronology of these surfaces have been put.

Geophysicists, for their part, are concerned with the response of the earth's crust to the ice loads, and in the study of raised marine and lake shores they see the potential for gaining some insight into the deep structure of the earth's mantle. The former Laurentide Ice Sheet had a radius of about 1800 km and a thickness at the ice centers of about 4000 m—such a load is compensated for by elastic and viscous responses at depths in excess of 200 km in the earth's mantle. Indeed, Cathles (1971, p. IX–18) suggests that the Fennoscandian data can be used to infer mantle properties (specifically, its viscosity) to a depth of 1500 km.

Thus the geophysicist is interested in glacial rebound primarily as a process whose characteristic response should be decipherable in terms of variations in mantle structure (*see*, for example, McConnell, 1968, this volume).

There are, therefore, two separate facets of studies of glacial-isostasy that are in some ways quite distinct, although, hopefully, we would expect a degree of interchange of ideas between the scientists studying the phenomenon from these two perspectives. The first of these facets is, of course, the field studies, which involve the finding and interpretation of the geological evidence for postglacial rebound and crustal deformation as a consequence of the ice loading/unloading cycle. This body of information is then made available to the geophysicists, who, in the second facet, analyze and manipulate the data to ascertain the answers to certain questions that are posed. This volume of reprints and associated commentary is accordingly organized into these two broad areas.

I will now review certain aspects of both the field geology and geophysical sections, mainly as a means for introducing the reader to some of the broader concepts and to some of the terminology because, as in many specialist areas today, a dictionary of the research subject is required, is, indeed, essential for an understanding of the actual reprints themselves. Many of the earliest scientific references are in one or other of the Fennoscandian languages and are not generally available. However, the importance of this work pervades even the modern Fennoscandian and English language literature (see reprints in this volume). However, this volume probably does not do full justice to the early Fennoscandian research.

Definitions

A problem in many areas of science is the variety of meanings that can be attached to a specific work or phrase. For example, in North America confusion has arisen over the use of "isobase." Definitions of certain words that are frequently used in this volume are therefore given:

Late glacial: used in a literal sense to refer to the period following deglaciation of a site.

Postglacial: used in a literal sense to refer to the period following deglaciation of a site. The length of postglacial time is measured from the present.

Marine limit (ML): the maximum elevation which the late- or postglacial sea reached at a coastal site. The elevation is measured with respect to present sea level. Over most of northern Canada, the marine limit was formed at the instant of deglaciation.

Eustatic sea level (E): world-wide sea level measured with respect to present sea level. Changes of the world's hydrological cycle are indicated by eustatic sea levels lower or higher than at present on "stable coasts."

Postglacial uplift (Up): the glacio-isostatic recovery of the earth's crust between the instant of deglaciation and the present. Because eustatic sea levels were lower than the present during the retreat of the Wisconsin ice sheet, the elevation of a marine limit does not equal the amount of postglacial uplift. Instead, postglacial uplift is defined as the sum of the marine-limit elevation and the change in eustatic sea level between the date of deglaciation and the present. Up = ML − E, where both ML and E are measured from present sea level. [In Arctic Canada, the equation in general is Up = ML − (−E).] Synonyms for postglacial uplift include postglacial rebound and postglacial recovery.

Postglacial emergence: the upward movement of land relative to sea level, by convention, considered positive (movement in the opposite direction is, of course, submergence, which is considered a negative movement).

Deleveling: a general term used by R. A. Daly (1934) and others to denote crustal movement, regardless of whether it was positive, negative, or a combination of the two.

Isobase: A line joining points of equal postglacial emergence (coastal sites) or equal postglacial uplift (glacial lake shorelines), where these elevations are the result of postglacial emergence or uplift operating over the same length of time. By convention, time is measured from the present.

Equidistant diagram: a diagram drawn in a plane that is orthogonal to the local system of isobases. On this graph, the *y* axis is elevation above present sea level and the *x* axis, distance. The projection of a strandline (see below) on the plane shows the extent and geometry of crustal deformation. Synonyms include strandline diagram.

Strandline: the trace of a former sea level or lake level as indicated by a variety of raised (or submerged) morphological/sedimentological features, such as beach terraces, deltas, and shingle ridges. Synonyms include raised beach, water plane, and lake or marine shoreline.

Relation diagram: a graph on which the x axis is the elevation above present sea level of a particular strandline (for instance, a strandline dated from 6500 BP), and the y axis is the elevation of other strandlines above this "reference level." Other strandlines are related to the reference strandline by an equation of the form $y = a + bx$.

Quaternary Studies on Glacial Isostasy

Early Research

Although Jamieson (1882) and Shaler (1874) are commonly regarded as the scientific fathers of glacial isostasy, it is clear from a reading of early explorations that there existed prior to the 1850s some understanding of the nature of the field evidence and some insight into the controlling processes. The Fennoscandian uplift was first commented on by the Swedish bishop of Abo in 1616 and was based on observations of harbors becoming dry and new islands emerging. His observations caused "water-marks" to be cut in bedrock for surveying purposes. The process of land uplift was consequently much discussed in Fennoscandia during the seventeenth and eighteenth centuries, and in 1838 Bravais (1840) made his benchmark studies on the inclined water planes in Finmark (see reference in Paper 11). In northwest Greenland, Elisha Kent Kane was led to remark:

> . . . but I may say that the opportunity which I had today of comparing the terrace and boulder lines of Mary River and Charlotte Wood Fiord enables me to assert positively the interesting fact of secular elevation of the crust . . . (Kane, 1856, Vol. 2, p. 81).

The secular rise of the shoreline is so rapid that the native peoples who live on the shores of the North American arctic seas were also well aware that the relative position of land and sea were changing sufficiently to be observed and discussed. Bell, in 1896, used Eskimo folklore to comment on the land emergence in Hudson Bay. Today, on the west side of Hudson Bay, the Cree Indians actively argue as to whether the land is rising or the sea falling!

The field basis for the physical reality of glacial isostasy is the presence along the shores of areas formerly covered by the Laurentide, Fennoscandia, and British ice caps of unequivocal evidence of once higher sea levels. The evidence is well preserved in Maine, the Canadian Maritime provinces, the Gulf of St. Lawrence, and coastal areas in Norway, Finland, Sweden, Scotland, Northern Ireland, Alaska, and Patagonia, to name a few localities that were studied and described at an early date.

The highest level that was reached by the sea in any area is called the marine limit. The marine limit may be mapped by a variety of different criteria (De Geer, 1892; Bird, 1954) but the most commonly used are (1) the highest deposit (actual or stratigraphic) containing marine fossils, most notably marine littoral-type bivalves; (2) the surface of a perched delta that grades laterally into an end/lateral moraine system marking a former glacier margin; (3) the highest raised shingle or sand beach deposit, with or without a "fossil cliff", scarp, or wave-cut terrace or "beach"; (4) the lowest limit of till or perched boulders or, conversely, the upper-most limit of washed bedrock; (5) the highest level of *in situ* marine silts or clays; and (6) lowest level of lateral drainage channels and other marginal supra-aquatic features. Of these six criteria, the second is the most reliable, especially as the delta bottomset, foreset, or topset beds are frequently fossiliferous and hence, since the late 1950s, can be radiocarbon-dated to provide an age for the glacier stand and associated relative sea level.

In 1892 De Geer (see this volume) was able to present a first working map of the elevations of marine limits for eastern North America based on the work of Low and others from Labrador and reports from the Gulf of St. Lawrence and coastal New England. This pioneer synthesis indicated that the marine limit increased northward toward the former ice centers in Labrador and Hudson Bay, thus suggesting that postglacial rebound near the areas of the former ice centers was greater than toward the periphery; that, in fact, there is a rough geographic correspondence between the outline of the Laurentide Ice Sheet and the outer limits of glacio-isostatic recovery. Such a correspondence is, of course, vital to the general theory of glacial isostasy, although it has been questioned by many researchers, notably Russian scientists, who point to the uplift of areas clearly never glaciated during the Wisconsin Glaciation (Nikolayev, 1967).

Although the recognition of the marine limit is fundamental to the field appraisal of postglacial crustal recovery, studies in Fennoscandia and Britain in the late nineteenth and early twentieth centuries successfully recognized lower, impor-tant raised beaches, terraces, or deltas that indicated that the sea level did not fall steadily since deglaciation, but rather that there had been significant sea level still-stands. Stratigraphically, it was also demonstrated (e.g., De Geer, 1893) that marine transgressions had interrupted the general fall of relative sea level caused by glacio-isostatic recovery. These observations indicated that a complete knowledge of crustal response to glacial unloading also demanded a knowledge of the world-wide eustatic sea level changes due to the addition of water to the oceans from the melt-ing of the ice caps. This had added perhaps 130 m to world sea level over the course of the last 15,000 years, although the addition was not linear with time but varied because of the deglacial history of the main and minor ice sheets.

A parallel development to the studies on raised marine deposits began in the late nineteenth century and was concerned with the study of the shorelines of the major glacial lakes in Scandinavia and in North America (e.g., Upham, 1896; Goldthwait, 1908) that once covered vast tracts of land, notably in the region of the present Great Lakes and the Canadian Prairies (Fig. 1). Of these, Glacial Lake Agassiz, in the vicinity of Winnipeg, covered 205,000 km² at its maximum extent (Elson, 1967), while in the basins of the modern Great Lakes, a complicated se-quence of lakes developed due to the interaction of glacio-isostatic uplift with the lower cols which were exposed as the ice retreated (Prest, 1970). The new Glacial Map of Canada (Prest et al., 1968) shows that vast tracts of southern Canada and western Canada east of the Rocky Mountains are mantled by glacio-lacustrine deposits (Fig. 1) laid down in numerous proglacial lakes.

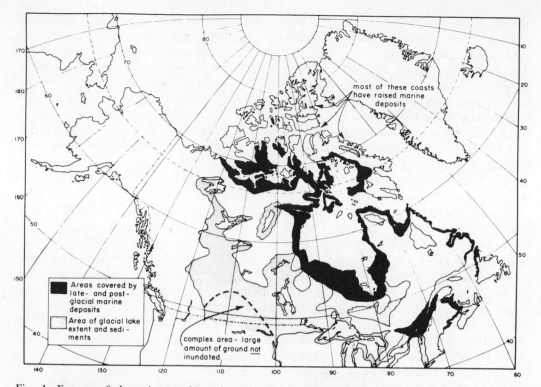

Fig. 1. Extent of deposits marking former glacial lakes and the marine overlap in North America (from Glacial Map of Canada, 1968, and other sources).

The extent of these glacial lakes in North America is a function of the broad topography of the area, with the geographic center of the Laurentide Ice Sheet occupying the interior shallow sea basins of Hudson Bay and Foxe Basin and the watershed (divide) for these seas that lay well to the south and east. Thus as the ice retreated from its maximum Wisconsin position, lakes were ponded between the ice margin and the watershed. Extensive glacial lakes were also present west of the Atlantic/Ungava Bay divide in Labrador. The great extent of terrain affected by marine or glacial lakes in North America is illustrated in Fig. 1, which shows quite clearly the importance of glacial isostasy and glacial geology in understanding various environmental factors in Canada and the northern United States. Glacial lakes have also been studied in Fennoscandia, notably in Sweden and Finland.

Glacial lakes frequently are marked very clearly by a limited number of former lake levels. These levels are delimited by wave-cut terraces (sometimes cut into bedrock), massive shingle storm ridges, and other deposits and shoreline forms. Because of the progressive nature of ice retreat, crustal rebound involved progressive tilting. The main control on any former lake level was the drainage col in use at the time; if ice retreat exposed a new col, rapid drop of the lake level resulted. Hence shorelines of any specific lake level are clearly distinct one from another.

In contrast to the lakes, the marine coasts consist, potentially, of an infinite number of traces of former shorelines as the sea fell intermittently due to glacio-isostatic recovery. It is no surprise, therefore, that the great contribution of studies on glacial lake shorelines (see, e.g., Goldthwait, 1908; Elson, 1967, this volume) has been the defining of the pattern of crustal warping in response to the glacial load. This knowledge has been put to good use by the geophysicists (see Broecker, 1966; Walcott, 1970; and Brotchie and Silvester, 1969, this volume) as tests of various crustal models or as a basis for determining such things as the thickness and rigidity of the crust (Walcott, 1970).

The presence of distinctly isochronous shorelines that can be mapped was an advantage to researchers working in the region of Glacial Lake Agassiz and in the area of the Glacial Great Lakes. Such mappable features are less obvious in the coastal record in North America within the region affected by glacio-isostatic recovery, although significant terrace levels occur in the Maritimes and lower St. Lawrence Valley. However, as previously noted, transgressive shorelines were recognized and mapped early in the last century in Europe and formed the basis for much of the early theoretical work.

In the Canadian Arctic and Subarctic, the only really mappable feature was the marine limit, which is *metachronous*; that is, it has different dates in different areas. Hence, it does not provide per se the same quality of information on the form and rate of crustal warping and relaxation that isochronous shorelines provide. Partly as a result of this, and partly as a result of the formidable difficulties and inconvenience of arctic research in the early part of the twentieth century, the study of glacial isostasy in this period was very strongly oriented toward the Glacial Great Lakes, Glacial Lake Agassiz, and the marine shorelines of Fennoscandia. The northern area of Britain was also suitable for study but advancement was severely curtailed by the stress on hard-rock geology and by the stereotyping of the shorelines as the "100-ft," "50-ft," and "25-ft strandlines," regardless of the probability that they were warped (see Sissons, 1962, this volume).

The various investigtors usually presented their shoreline data in three forms; (1) isobase maps; (2) shoreline (equidistant) diagrams; and (3) shoreline displacement curves. It is to be stressed that strandlines are *isochrons,* lines of equal date; in contrast, the word *isobase* refers to a line joining sites of equal uplift. Donner (1969) provides a useful review of various graphical methods of interpreting raised shorelines along with a good historical survey of their origins and uses.

By the early twentieth century, sufficient research had been undertaken in Scandinavia to show that: (1) the greatest amounts of uplift, within a specific period, had occurred within the area of the inferred center of each ice sheet or cap; (2) the crust was progressively warped during recovery, producing a peripheral up-bulge around the retreating ice margin that lead to a seaward tilt on the ocean side and a landward tilt toward the ice center; and (3) various historic and geodetic survey evidence indicated that the area of current most rapid uplift was in the Gulf of Bothnia, and that this probably coincided with the former area of the maximum ice sheet thickness.

The decades of the 1920s through the late 1940s saw tremendous strides being taken in the study of postglacial uplift in Fennoscandia, with such publications as Nansen's "The Strandflat and Isostasy" (1921) and the work of Von Post, Tanner, Sauramo, Grønlie, Florin, Hyppä, and others which began to outline in considerable detail the time-stratigraphic systems of strandlines in Norway, Sweden, and Finland. However, the Fennoscandian scientists also developed new ways of analyzing their data. The shoreline relation diagram was primarily developed by Von Post in 1929 and Tanner in 1930 and is based on the observation that there was a direct proportional relationship between the elevations on some low strandline and the elevation of points on a high strandline (see Nansen, 1921, this volume).

During this same period, workers in Fennoscandia (Nansen, 1928; Lidén, 1938; Sauramo, 1939) constructed graphs of uplift as a function of time (uplift curves); thus, for the first time, geophysicists were able to "view" the relaxation properties of the crust due to glacial unloading. This is a continuing interest. Table 1 (from Andrews, 1970) lists some of the categories of crustal movement that can be studied and the types of diagrams that are being used in studies of glacio-isostatic recovery.

Recent Studies

During the last two decades the literature on Quaternary studies of glacio-isostatic recovery have increased enormously. A large part of this increase has taken place in areas that were virtually scientifically unknown prior to the 1940s. Some of the most significant results of field studies of the response to glacial isostasy have come from the islands of Spitsbergen (see Schytt et al., 1967, this volume), Greenland, and Iceland, as well as Antarctica and the huge area of the Canadian Arctic. At the same time, research in Fennoscandia has produced fresh information and ideas and a wealth of very detailed reports. In Britain the field study of glacial isostasy, static for over a century, was revitalized by Sissons' paper (1962) and the work of Synge and others of the last decade, including those published in 1966 in a special volume of the Transactions of the Institute of British Geographers devoted to *The Vertical Displacement of Shorelines in Highland Britain.*

Of considerable importance to our understanding of postglacial history and glacio-isostatic recovery has been the reconnaissance mapping of much of Arctic Canada by the Geological Survey of Canada and the former Geographical Branch of the Canadian Government. Thus, by 1962 Farrand and Gajda were able to produce, for the first time, a preliminary map of the marine limit height for Canada and the northern United States. Although inaccurate in certain sectors (cf. Ives, 1963), the map showed the tremendous area covered in ten years or so of mapping. A recent version (*National Atlas of Canada*, 1972, p. 35–36, by this writer) is reproduced here as Fig. 2.

Although the marine limit is a fundamental geological marker, it is, as noted earlier, time transgressive. This may be readily seen by comparing the isolines, or contours, of Fig. 2 with maps by either Bryson et al. (1969) or Prest (1969) which show isochrones on the retreat of the margin of the North American ice caps. The problem of constructing strandline diagrams was succinctly stated by Bird (1954, p. 461):

Table 1. Methods of presenting data on studies of raised shorelines
(Elevations a.s.l. unless noted)

Diagram	x axis	y axis	z axis	Limitations/comments
Equidistant	Distance	Elevation	—	Sensitive to the direction of the projection plane. Should be normal to local isobases. Sensitive to isobase curvature if geographical dispersion of points is too great. Sensitive to selection of points and model of form assumed, rectilinear or curvilinear.
Relation diagram	Elevation of the reference level	Elevation above reference level of strandlines		Sensitive to errors in the dating of the strandlines and to correct elevation measurements.
Isobase (Okko, 1967)	Present rate of uplift	Elevation of strandlines		Sensitive to the additional factor of date of deglaciation. Thus, only useful in restricted geographical area.
Isolines on marine limit	Geographical space		Elevation of marine limit	Time-transgressive analysis of crustal deformation.
Isobase map	Geographical space		Elevation of a feature as a result of uplift/emergence in unit time	Considered to represent the pattern of glacial loading.
Up = (D, A) (Andrews, 1968a)	Distance (D)	Date of deglaciation (A)	Postglacial uplift (Up)	Shows importance of date of deglaciation on amount of postglacial uplift. Given Up and D, can predict A.

From Andrews (1970).

One of the most important achievements in Scandinavia and the Great Lakes region of North America has been to show the tilt of individual strand lines and hence the warping of the earth's crust. Nowhere in the Canadian Arctic has this yet proved possible. In the region under discussion (*Hudson Bay*) the only datum plane recognizable over any considerable distance is the upper limit of submergence. As it is difficult, if not impossible to show that this was formed contemporaneously within even short distances, tilt cannot be deduced from it.

There are very few good varve sequences in the New World. In Scandinavia

9

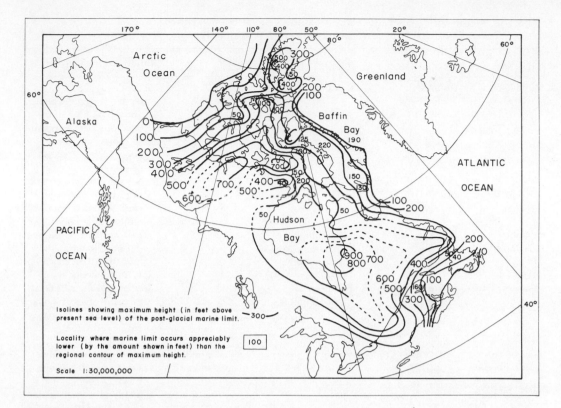

Fig. 2. Isolines on maximum marine limit elevations. Map constructed by filtering low values that mark late deglaciation (from National Atlas of Canada, p. 35–36 by J. T. Andrews).

the chronology had been supplied by varve counting (a technique developed by De Geer) combined with pollen analysis and other paleontological control. The field mapping in Spitsbergen, Greenland, Arctic Canada, and elsewhere coincided with the development of the radiocarbon method of dating organic materials. Accordingly, raised marine deposits that contain marine bivalves, wood, whalebone, detrital plant remains, and marine algae can now be dated in terms of radiocarbon years before present (BP) defined, by convention, as A.D. 1950. Infilled or shallow lakes ponded behind beach ridges and basal peat samples can also be dated. If the stratigraphic material is adequate, the advent of radiometric dating now means that any series of samples can be dated to provide an accurate picture of the emergence of the site through time (see Løken, 1965, this volume).

During the same period, however, numerous workers (e.g., Fairbridge, 1961; Shepard, 1963; Mörner, 1969; Walcott, 1972a) were studying the postglacial movement of worldwide sea level that is related to the mass transfer of water from the former ice caps into the oceans. The "eustatic" sea level curve must be known in order to correct crustal uplift against relative sea level changes as seen on a glacio-isostatically rising coast. The necessity for correction may be illustrated as follows: two areas have marine limits of 100 m but one was deglaciated 12,000 years ago

10

and the other only 5,000 years ago; in the last 12,000 years world sea level has risen by about 50 ± 10 m, whereas in the last 5,000 years this figure is 0 ± 3 m. The actual uplift at these two sites is thus c. 150 m and 100 m, respectively. With dating and corrections of this sort added to relative sea levels, Farrand (1962) and Washburn and Stuiver (1962) constructed postglacial uplift curves from North America and Greenland, respectively. Several workers have shown that rebound is approximately exponential and of the form $U = U_0 e^{-kt}$, where U_0 is the total post-glacial uplift (150 m or 100 m in the above examples), t is the time ($\times 10^3$ yr), k is the decay constant, and U is the uplift in time, t. At t_0, uplift is just beginning and $U = 0$. Many workers have commented on the similarity of the various curves and Andrews (1968) suggested that the decay constant for 21 curves from Arctic Canada was the same (within experimental error). Ten Brink (1971) noted that four curves from Greenland had very similar decay constants, although the value of the decay constant is very different from that characterizing the Laurentide postglacial rebound. In Greendland (Ten Brink, 1971), values for k are about 0.7×10^{-4} yr^{-1} compared to 0.44×10^{-4} yr^{-1} from Arctic Canada. The half life, $t_{1/2}$, of postglacial uplift curves is the time taken for 50% of U_0 to be recovered and can be calculated from $0.693/k$. Thus in Greenland $t_{1/2} = 0.9 \times 10^3$ yr (Ten Brink, 1971) and for Arctic Canada $t_{1/2} = 1.5 \times 10^3$ yr, clearly indicating that postglacial rebound in Greenland has been extremely rapid in comparison with the published results from Canada.

The suggestion that the decay constant for Arctic Canada is the same at all sites (Andrews, 1968) has important practical implications because it means that, given the elevation and the age of a marine limit, the postglacial recovery of a site can be predicted with an accuracy of 500–1000 years and to within a few meters. With this information available at hundreds of sites it is feasible to construct uplift curves at these sites; then by taking the estimated elevation of sea level, relative to the present level, for a series of specific dates (say every 1000 years), isobases on postglacial uplift can be drawn (Andrews, 1970). Furthermore, tentative maps of present and past rates of glacio-isostatic recovery can be constructed. The results of this wholly empirical approach provide a first approximation of the crustal

Table 2. Summary of Fennoscandia concepts of land and sea movements

Author	Land	Sea
De Geer	Up + down + up	Constant
Ramsay	Up	Variable
Von Post and Florin	Rapid changes of large amplitude	Rapid changes of large amplitude
Sauramo	Hinge lines	
Nilsson and Hyppa	Constant uplift and simple tilted shorelines over large areas	
Mörner	Low-amplitude changes caused by shifts in glacial and water loads, migration of center of uplift	Rapid low-amplitude fluctuations

Mörner (1972, personal communication).

warping and vertical movements within the area of the Laurentide Ice Sheet. They are, however, only approximations and need further field checking and further analysis of the validity of the assumptions (e.g., that k is constant. See Fillon, 1972, this volume.). For an excellent review of the Canadian data see Walcott (1972b). A review of the Fennoscandian concepts, provided by a Scandinavian colleague, is given in Table 2.

Geophysical Studies on Glacial Isostasy

Knowledge about the deep structure of the earth has recently come from the worldwide seismic station network. The shocks created by earthquakes give significant information on the short-term response of the earth to stresses. This type of information relates primarily to the elastic properties of the crust and mantle but tells very little about the response of the earth to long-continued stresses operating with a periodicity of thousands and tens of thousands of years. Stresses of this duration are adjusted to principally by slow viscous creep at depths of 100 km or more.

The concept of the earth as being composed of a thin elastic crust overlying a viscous fluid (see Brotchie and Silvester, 1969; Walcott, 1970, this volume) is a simple but useful approximation of a much more complex, layered earth (see McConnell, 1968, this volume). Earth responses to the Quaternary glacial loading/unloading sequence can be modeled by the effect of a weight on a spring (elastic crust) and a piston filled with oil, or "dashpot," (viscous interior) [see, e.g., Johnson (1970)]. The rheology of an elastic/viscous earth (Fig. 3) thus takes on aspects of two fundamentally different properties. An elastic substance is one that deforms

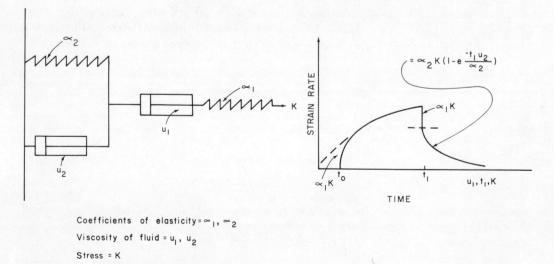

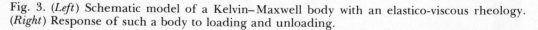

Fig. 3. (*Left*) Schematic model of a Kelvin–Maxwell body with an elastico-viscous rheology. (*Right*) Response of such a body to loading and unloading.

instantaneously due to shear stress, but on release of the stress all the strain, or deformation, is recoverable. A viscous substance flows as soon as stress is applied and the rate of flow is proportional to the applied stress. However, upon removal of the stress, a viscous material does not recover completely from the deformation. In earth models a linear viscous (Newtonian) model is assumed. This can be represented in one dimension (Johnson, 1970, p. 14) by $\tau s = R(d\Sigma s/dt)$, where τs is the shear stress, R is the coefficient of viscosity, Σs is strain and t is time. The Kelvin–Maxwell configuration of spring and dashpot (Fig. 3) may be applicable to a simple modeling of the earth and researchers have also considered the properties of Maxwell and Kelvin analogs by themselves. A complex, layered earth can, of course, be simulated by a series of springs of different elastic responses and dashpots filled with oils of varying viscosities. Although a linear viscous model is assumed, there is a possibility that nonlinear responses may apply.

In considering the loading/unloading of the ice caps and the related crustal responses it is permissible to ignore the viscous nature of the earth and consider the equilibrium deflection of a shell of particular thickness and elastic properties under the load. These last mentioned properties and crustal thickness govern the rigidity of the crust. If the ice caps had overlain a viscous earth, the deflection or downwarp at the edge of the ice load would be zero; in reality, however, the rigidity of the crust is such that at the margin of the former ice load, measured deflection varies between 70 and 200 m. Beyond the ice margin (Fig. 4), the depressed region rises to zero deflection within distances of 100 to 300 km and, theoretically, then rises above sea level in the vicinity of the forebulge (Newman et al., 1971, this volume). The rigidity of the crust as illustrated in Fig. 4 means that isostatic equilibrium is not achieved near the margin of an ice cap. On the other hand,

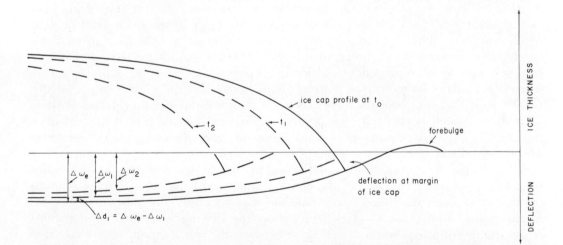

Fig. 4. Schematic diagram of the deflection of the crust beneath and beyond an ice sheet. At time t_0 the load has deflected the crust to its equilibrium position $\Delta\omega_0$, whereas during glacial retreat new equilibrium deflections are possible at t_1 and t_2; but because of the characteristic relaxation time related to the plastic inflow of material, the crust does not attain these equilibrium positions immediately, so uplift continues after removal of the ice load.

toward the center of the ice cap, the very gentle gradients of the ice cap profile allow isostatic equilibrium to be achieved if the load is applied for a long enough period of time. Thus the Antarctic ice sheet, in existence at least 2 million years, seems to be isostatically compensated (Crary, 1971, p. 344). One of the important questions yet unanswered is—how long does it take for isostatic equilibrium to be achieved either during the loading or unloading parts of the glacio-isostatic process?

Three processes are probably involved in the take-up and relaxation from the application of an ice cap load. First is the effect of elastic compression. The amount of elastic depression at the center of the ice cap is dependent on the height and density of the ice, the radius of the load, and the rigidity in the earth (Gutenberg, 1941, p. 752). For the Laurentide Ice Cap, the elastic depression might have amounted to 100 m. The second process is viscous creep under low stresses, and the third process is plastic flow at depths of 100 to 200 km and probably confined to the low velocity layer. The initial boundary condition for plastic flow is the strength of the layer. When stresses exceed the strength of the material, flow occurs (or if the material is brittle it will fracture). The problem of plastic flow is complicated by the suggestion that viscosity changes as a function of the size of the load; it also appears to be time dependent. Broecker (1962) has also estimated that up to 100 m of surface depression might be caused by a phase change between basalt and eclogite.

Studies of the elevation and age of raised beaches in areas once covered by the ice of the last deglaciation show only the effect of plastic inflow of material at depth because the elastic depression is recovered immediately as the load is decreased. The relaxation time for postglacial uplift, defined as the time taken to recover $1/e$ of the depression (about 64 percent), varies between about 1000 and 3500 years. However, this is only the apparent relaxation time for recovering isostatic equilibrium. On the basis of large negative free gravity anomalies still present in Hudson Bay (about -45 m gals) Walcott (1970) has suggested that full recovery would have to take tens of thousands of years. However, since the area of Hudson Bay has undergone repeated glacial loading episodes (possibly as many as eight) it may be that the large free air gravity anomalies reflect the plastic nature of the mantle such that complete recovery was never attained during interglacials (lasting 50,000 to 100,000 years each), and a large deficiency of mass has developed over the course of the Quaternary glaciations, because, by definition, a plastic deformation is not fully recoverable upon removal of the yield stress.

The shape of the equilibrium deflection profile under and beyond an ice cap (Fig. 4) is primarily a function of the elastic properties of the crust; a static solution of this sort has been developed by Brotchie and Silvester (1969, this volume). However, in the case of uplift due to glacial unloading, the driving force behind the uplift is the amount of unequilibrated deflection between the initial equilibrated deflection at t_0 and that at t_1 when the ice has retreated and thinned. At any point, the initial equilibrium deflection $\Delta\omega_e$ is replaced by $\Delta\omega_1$ (Fig. 4). Part of this deflection will be regained immediately by elastic rebound, but the remainder will recover according to $U = (\Delta\omega_e - \Delta\omega_1)e^{-kt}$, where $\Delta\omega_e - \Delta\omega_1 = \Delta d_1$ on Fig. 4. However, if the rate of ice retreat is faster than the mantle response to this deficiency of mass (which is probably

14

the case during the last period of deglaciation of the northern hemisphere ice caps), then after some further time t_2 the ice has retreated to a new position (Fig. 4) and a new equilibrated position is determined ($\Delta\omega_2$). In the interval between t_0 and t_1, however, only a proportion of the unequilibrated deflection was recovered; the remainder has still to be recovered when the ice retreats between t_1 and t_2. The rate of rebound at any site is proportional to Δd_1 (Fig. 4) and the magnitude of the decay constant, k, in the above equation. In the simple earth model of a viscous half-space (Heiskanen and Vening-Meinesz, 1958) the decay constant is derived as $k = (\rho mg/2\pi n)L$, where L is given by $L = XY/(X^2 + Y^2)^{0.5}$, with X and Y the dimensions of the load $\times 10^6$m, ρm = mantle density, $g = 980$ cm sec^{-2}, and n = dynamic viscosity. Thus relaxation time $(1/k)$ is inversely proportional to the size of the load and proportional to the dynamic viscosity. According to this model, recovery from large loads occurs more rapidly, all things being equal, than recovery from a smaller load. If a dynamic viscosity of 10^{22} poises is used in the equation above and $X = Y = 3.4$ for the Laurentide Ice Sheet, a relaxation time of about 8000 years is derived and $k = 0.128 \times 10^{-3}$ yr^{-1}. This is about ¼ of that ($K = 0.44 \times 10^{-3}$ yr^{-1}) shown by postglacial uplift curves (Andrews, 1968).

More recently Anderson and O'Connell (1967) have suggested that the reverse is true and they have proposed that smaller deformations are recovered fastest. However, part of modern geophysical research has been the investigation of variations in relaxation time for a single load. Glacio-isostatic deflection may be considered to be made up of the sum of several superimposed depressions having the general shape of sine waves. McConnell (1968, this volume) suggests that surface deformation will react to a less viscous layer at depth d (Fillon, 1972, this volume). Conversely, Broecker (1970) points out that if the crustal flexural parameter is nearly 200 km in length, as suggested by Walcott (1970), relaxation time will be sensitive to the rate of ice retreat. Here, then, we have two possible explanations for regional variations in relaxation time, the first being variations in viscosity at depth 1 to 6 times the deformational wavelength, and the second being related to the rate of ice retreat. The earth's mantle is relatively insensitive to loads less than 100 km wide as these are largely taken up by elastic processes.

Although there are good theoretical reasons why relaxation time should vary across a loaded area [the size of Fennoscandia or the Laurentide area and the data have been interpreted to suggest this is true (Fillon, 1972; McConnel, 1968)], it is not completely clear if this is really so. Considering the errors in construction of postglacial uplift curves from raised marine deposits (Andrews, 1970), it is difficult to know how to interpret variations in k. Ten Brink (1971) also has evidence that postglacial recovery in Greenland is characterized by proportionally similar rebound.

Geophysical studies to date on the problem of glacio-isostatic rebound have used a series of models where certain of the assumptions are patently incorrect. This procedure, however, is the standard scientific method of successive approximation. The earth has been assumed to be a viscous half-space, and elastic–viscous shell and mantle, or layered; the load has been assumed to be radially symmetrical or simply linear, and the earth has been assumed to be flat or curved. Future research will probably concentrate on numerical modeling of the actual field data and the iterative fitting of reasonable earth models to this real data array. This procedure will be expensive in computer time. The work of Cathles (1971) provides the first insights in this direction. Despite the sophistication of this approach, it is being increasingly evident that the limitations are not at the geophysical modeling end of the process but in the need for more and better field data.

The eventual solution to the geophysical problems offer an answer to the age old questions about the viscosity of the upper mantle, the depth of viscous flow, and the elastic strength of the lithosphere.

Organization of the Volume

The above sketch of the history of the field and geophysical interest in glacial-isostasy should give the reader sufficient background to read the selection of reprints that follow. Clearly, with a literature of the size and scope of that dealing with glacial-isostasy there are the twin problems of which papers to select and how the selected papers are to be organized into groups. The problem of selection is made more difficult by having to give adequate representation to very early work and yet reserve enough space to deal with current papers that represent frontiers in this particular research problem.

The papers included in this volume are divided into three main parts. The first includes the introductory papers by Jamieson (1882) and Shaler (1875); a synthesis of existing data on the marine limits in the eastern part of North America by De Geer (1892); a magnificent survey of glacial lake shorelines in the Lake Michigan basin by J. W. Goldthwait (1908); and extracts from Nansen's 1921 classic entitled "The strandflat and isostasy."

The second part opens with a review of postglacial uplift by Kupsch (1967) and includes papers published within the last 10 years on field problems, methods, and results. The selection was chosen to represent research from North America, Great Britain, and Fennoscandia. This section is subdivided into studies of crustal deformation and postglacial uplift. As examples of studies of the first problem there are the papers by Donner (1969) which illustrate the postglacial updoming of Fennoscandia along an east to west cross section and the paper by Schytt et al. (1967) who use the radiocarbon-dated samples to reconstruct the isobases in the region of the Barents Sea to suggest that a major ice cap was once situated over this shallow shelf. Field studies on postglacial uplift include the initial synthesis of North American data published by Farrand in 1962, and a detailed field study of one fiord head in Baffin Island, N.W.T., by Løken (1965). Some papers do not fit neatly into these two categories—an example is the paper by Mörner (1971), which uses glacio-isostatic data to reconstruct the eustatic sea level curve over the last 12,000 years or so.

The third, and final section introduces a number of papers that deal with the geophysical interpretation of postglacial crustal warping and uplift. McConnell (1968) uses strandline data from Fennoscandia to develop the relaxation spectra for different deformational wavelengths. Of considerable interest is a comparison of the papers by Broecker (1966), Brotchie and Silvester (1969), and Walcott (1970), all of which use the well established form of the Lake Algonquin shoreline (dated about 10,500 BP) to test different geophysical models. Walcott's model is essentially a static one

16

involving elastic flexing in the vicinity of the ice edge with a long flexural parameter that implies a thick crust. In contrast, Broecker takes a crust with a very short flexural parameter and associates the form of the lake shoreline to rates of glacial retreat. Brotchie and Silvester take a model with a crustal thickness of 32 km and include the effect of the rate of glacial retreat and the inflow of material.

All or part of each paper is included in this volume. Preceding each is a commentary which is aimed placing each one in perspective and providing commentary on what ideas and concepts emerged from the paper and how they have been taken into the modern literature and perception. Twenty-eight papers are included in this volume; however, each one has its own reference list, therefore, these papers and their attendant bibliographies include a major portion of the existing literature.

It should be realized, however, that the publications overemphasize the contribution of English-speaking scientists. At the present time a good deal of the Fennoscandian research is published in English and is thus accessible to most students. However, the classical Fennoscandian papers, such as that of Tanner in 1930, are published in Swedish, Norwegian, etc., and are unfortunately not available to many because of the language barrier. The student should be aware, therefore, of the great amount of research accomplished in Fennoscandia between 1920 and 1950. This legacy is, of course, incorporated into the English-papers emanating from that area that have been reprinted in this volume.

Any selection of papers is bound to invoke some criticism; these have been chosen on the basis of my own interest and research in glacial-isostasy. They are, by and large, papers that I feel represent significant contributions to knowledge or illustrate a specific problem or technique.

Acknowledgments

I should like to thank all the contributors to this volume for their assistance. R. W. Fairbridge ably edited the Introduction and Comments. Jane Bradley did much of the work in sending out requests, typing, and collating the final version.

Early Work

I

Editor's Comments on Paper 1

Shaler: *Preliminary Report on the Recent Changes of Level on the Coast of Maine*

This paper and that by Jamieson (Paper 2, this volume) form the initial scientific presentation on the problems of glacial isostasy. Although both Shaler and Jamieson report similar observations and speculate on the mechanisms of ice loading and unloading, each study was, in fact, conducted independently.

Shaler's paper first introduces the various items of physical evidence that support the concept of changes of sea level, including such evidence as marine shells raised well above present sea level, as well as foreset beds now below sea level. Shaler then discusses the different types of evidence that are used to delimit the elevation of former sea levels—stratified sands, silts, and clays and wave-washed bedrock, which are still used today as criteria for mapping raised marine deposits. He relates these deposits to the deposition of materials in front of and beneath a receding ice cap. Shaler also notes that glacial striations are remarkably fresh in the area once covered by the sea and he uses this to deduce that the recovery from the ice load was rapid. (Pages 324 to 334 have been omitted as they contain only additional detailed descriptions.)

Shaler thus firmly establishes the correlation between depression of the coast of Maine and New England and the glacial period. In his subsequent discussion, he attacks the theory of Adhémar, who had proposed that the observed evidence for sea level changes in northern Europe and elsewhere were caused by the gravitational attraction to the sea in an ice cap that occupied alternately the north and south poles. Shaler's conclusion is that it is the mass of the ice load that causes the depression of the continents. He uses an analogy of a weight placed on a sheet of lake ice—the ice is depressed around the weight and an elevated area occurs around the "sunken point" (p. 339). This may be the first suggestion of the existence of a forebulge! Shaler thus envisages the crust as being rigid but having an underlying region where loads are accommodated by the outward flow of material. This is basically the concept that is still in use today (e.g., Brotchie and Silvester, 1969).

20

Reprinted from *Mem. Boston Soc. Nat. Hist.*, **2**, 321–323, 335–340 (1874)

1

X. PRELIMINARY REPORT ON THE RECENT CHANGES OF LEVEL ON THE COAST OF MAINE: *with reference to their origin and relation to other similar changes.*[1] By N. S. SHALER.

INTRODUCTION.

FOR many years there have been frequent reports of changes in the level of rocks and shoals on the coast of Maine. Generally these accounts have been of an untrustworthy nature, inasmuch as they seemed framed in palliation of some blunder of seamanship. A vessel being cast away in some well-known waters, it was natural enough for her captain to claim that the rock was not there before; or that the depth of water had greatly diminished within recent times. Every sailor who had excused his own blunders in this fashion, was naturally inclined to foster the opinion that the rocks were growing nearer the surface year by year, and without any intended deception one can easily imagine that in time a decided conviction might thus be reached. Experience in other regions has shown, however, that a prevailing opinion of this kind is apt to have some foundation. For some centuries the popular belief of the inhabitants of the Scandinavian coast concerning the changes of level of the shore, received no scientific examination; when, however, these questions were approached in a determinate fashion, it soon was made evident that the popular opinion was singularly correct, both as regards the extent and character of the movement. The following considerations, and the facts upon which they are based, result from a tolerably careful study of the record of changes along the New England coast, and from several years of summer labor.

Before giving a detailed account of the facts observable on the coast of Maine, it will be well to consider the general nature of the changes of level which have been observed in neighboring countries. I shall therefore give a brief résumé of the facts concerning elevations and depressions of shore lines in the north Atlantic section of our own continent, and those of the opposite continent of Europe. From the time that geology began to exist as a science, it has been a well-accepted fact that the surface of the dry land has been continually changing its level with reference to the sea. It is, however, among the later investigations that we begin to find anything like a careful study of the last, and therefore the most easily determinable, changes which our continents have undergone. Although by no means complete, these investigations enable us to assert, with a precision which can rarely be obtained in the science, that a great movement of elevation has taken place throughout the whole northern section of the Atlantic coast line, in the most recent times. The quantity of the vertical movement varies greatly in different places; it seems, however, to be quite certain that the elevation is generally greater as we advance towards the north.

[1] Published by permission of the Superintendent of the U. S. Coast Survey.

On our own coast it varies from a few feet, in the neighborhood of New York City, to one thousand or more on the coast of Labrador, and two or three times this amount on the Greenland coast. On the European shore this movement has not yet been proved to increase with the same regularity as we pass to the northward. Something of the kind, however, is distinctly traceable, though the local character of the phenomena is much more decided than in America. For instance, there is some proof of a depression of over a thousand feet in Wales, though northern France shows only slight evidences of very recent submergence. It will be necessary to refer to this local character of the movement in Europe, when we come to consider the cause of the submergence which took place about the time of the glacial period; for our present purpose it is only necessary to urge that the whole coast line of the northern Atlantic, with local exceptions, (which we shall see are not difficult to explain), is marked by indications which show that it is just recovering from a period of great depression. At many points the evidence is pretty clear that this movement of elevation is still in progress. It is evident, therefore, that the general character of the changes which have taken place in the immediate past in the northern Atlantic, quite harmonizes with the supposition that there is still some change in progress along the coast of North America. At least it may be worth while to give the subject a careful study. In the following section will be found the results of a journey along the coast of Maine, made with the especial intention of observing the evidence of change which might be found there, while the closing section of this Memoir will be given to the consideration of the physical causes of the changes of level of shore lines, with special reference to the great changes which ushered in the present geological period.

ON THE PHENOMENA OF ELEVATION OF THE COAST OF MAINE.

Of the three possible conditions of any coast line, elevation, stability, and depression, the former gives by far the clearest evidence of its action. Depression can only be shown by level marks observed at long intervals, or by the existence of a contour of surface of determined aerial origin beneath the surface of the water. Stability can be shown only by the obscure indications made by the long continued action of water at its point of contact with the land; extensive submerged tables of rock lying just at the height where the cutting action of the water would have left them, afford the best evidence of this condition. Elevation, however, is shown, whenever the circumstances are favorable, in a very remarkable manner. The following natural indications of this movement may be taken as valuable in the order in which they are named, the most trustworthy being given first.

1. The remains of marine animals lying embedded in stratified deposits above the level of high tide mark.

2. The existence of extensive stratified deposits of sand, gravel or mud, at points where fresh water lakes could have had nothing to do with their formation.

3. The existence of a topography not explicable on any other supposition than marine action above the level of high tides.

The first of these proofs cannot be reasonably expected to occur on all shores which have been recently elevated. Even at the present time a good deal of the shore sand and gravels making along our coast are quite wanting in animal remains; and in the glacial period, and the changes which immediately followed it, during which times the great depression of our shores occurred, the physical influences must have been very much against the existence of animal life in any abundance. We shall therefore have to rely, in the main, upon the evidence which can be obtained from the non-fossiliferous gravels and sands of the coast.

The third sort of evidence of elevation which the shore line can give us, though not without its value in the problem before us, we shall find to hold but a subordinate place. We shall see that the duration of the subsidence on the coast of Maine was not great enough to admit of a great amount of marine work, or at least that part of the time of submergence during which the shore was bare of ice, and exposed to the wear of the sea, was not long enough to permit any great amount of erosion. How slight it was may be inferred by the fact that along nearly the whole shore the glacial scratches, which were certainly formed before the gravel beds were laid down in the stratified form which so clearly evinces the action of the seas, are still unworn, down to the water's edge.

This feature of the persistence of the glacial scratches, even to the margin of the sea, is not by any means peculiar to the coast of Maine; it is clearly seen at Newport, Rhode Island, where the scratches, which seem to have had no protection against the action of the weather, retain an admirable clearness, showing, beyond a doubt, that even in portions exposed to the full brunt of the storms, the erosion was not enough to take even a small fraction of an inch from the stone. Along the coast of Massachusetts we have the same feature distinctly shown at various points. On Cape Ann, one of the most exposed promontories of our stormy northern coast, the evidence is very clear.

Taking then the beds of stratified drift as the only acceptable and abundant evidence of depression, we must look at the question of the origin of these beds, and the possibility of their being formed by other agents than those which are at work in the sea. Some slight amount of stratification is certainly not inconsistent with the action of ice in the form of extensive sheets, especially when we consider that the plainest mark of the ice work would be that left by the water derived from the melting of the glacial mass, as the conditions changed to those which are now in action. But when the stratified drift is distributed in extensive sheets of approximately horizontal materials along the shore, all doubt of marine action may fairly be put aside.

[*Editor's note:* In the next 11 pages Shaler primarily describes the results and interpretation of his field work. Although of some interest, this material has been omitted for the sake of brevity.]

ON THE CAUSE OF DEPRESSION IN THE GLACIAL PERIOD.

That there is a necessary connection between the accumulation of great masses of ice, such as the glacial periods brought upon the surface of the circumpolar regions, and the depression of those regions, is a point most easily demonstrated. Over the whole of the shore of the North Atlantic the evidence is complete; at the southernmost point where the glacial action has been observed, we find slight evidences of subsidence. This increases steadily, though irregularly, as we go northward, until we come to the highest latitudes where civilized man has penetrated, where we have evidences of subsidence amounting to two thousand feet or more. The only hypothesis as yet advanced to account for this irregular movement, is that of which Adhémar was the originator.

The accumulation of a great weight of ice at either pole in succession, would result in the displacement of the centre of gravity of the earth, which would be drawn the nearer to the pole where the accumulation took place. The result of this change upon the altitude of the sea would be very great. The water being free to obey the changed conditions would rearrange itself with reference to the new centre of attraction. The result would be the deepening of the sea by the amount of the displacement of the centre of gravity about the pole where the ice accumulated. That such a result is in good degree a necessary consequence of the accumulation of masses of ice about either one or the other pole, is easily believed. But there is every reason to suspect that the conditions were not fulfilled during the last glacial period. The observations of Darwin, and more recently, and more especially, of Agassiz, have shown that the southern extremity of South America bears marks of as recent and extensive glaciation as the Northern Hemisphere. Until it is proven to the contrary by evidence which has not been seen, we must suppose that the glaciation in the two hemispheres was simultaneous. If this be the case, the idea of a displacement of the centre of gravity is no longer so available as an explanation; for although the tendency to accumulate the seas about the poles would exist in a very diminished way under this condition, its amount would not be sufficient to account for the phenomena of depression. Moreover, any result of this kind would probably be more than compensated by the substrata of material from the sea, and the consequent reduction of its depth. If we suppose both hemispheres laden with ice down to the parallel of 45°, then the thickness to be on the average one mile, we shall have subtracted from the ocean water enough to lower its depth by over half a mile of altitude. Something of this would be compensated by the attraction of the mass of ice about either pole ; but it is clear that inasmuch as the seas do not rise under the influence of the masses of the continents against which they rest, at least more than a trifling amount, so the water could not rise against the polar ice caps to any great height.   The double polar ice cap, even were it a mile thick, could not affect the gravitation of the sea more than the high lands of Western South America. Moreover, the fact that in high northern countries some time elapsed after the disappearance of the ice cap before the re-emergence of the land took place, while on the theory of disturbed gravitation it should have disappeared *pari passu* with the waning of the ice, is a strong argument against the sufficiency of this explanation.   Probably the most insuperable objection which can be made against this hypothesis of depression through a change of the centre of gravity arising from the magnitude of the ice sheet, is found in the fact that the depression does not increase with regularity throughout the whole of the Northern Hemisphere, which it should have done if this view be correct.   The increase from a few feet (not exceeding fifty) along Long Island Sound to three hundred on the coast of Maine, is excessive.   This rate, if kept up, as it must have been if this hypothesis be true, would have made the depression at the poles many miles of depth.

In view of these arguments I find myself compelled to abandon this view concerning the origin of the glacial depression.   In seeking another explanation of the phenomenon, I have endeavored to arrive at something which should be easily connected with the general facts of continental and other mass movements of the earth's crust.   In pursuing this object the following opinions have been forced upon me.

The constant movements of sea and land show clearly that the surface of the earth and the solid matter, for a considerable depth, are subject to movements which vary much in direction and intensity.   At first sight, however, it would seem as if this variety of movement was far more considerable than it is in fact.   The following considerations will serve to limit the phenomena in the range of its action in an important way.

As the sea is the region of constant deposition, and the land of constant erosion, there must be in the long run quite constant upward movement of the land areas and depression of the floors of earth beneath the seas.   For instance, the region of the Ohio valley was near the sea level in the silurian and carboniferous times, and is a few hundred feet above it at the present day, notwithstanding the constant erosion which has affected it nearly ever since.   We cannot reasonably reckon the time which has elapsed since the coal period at less than twenty million years.   Now the rate of wear on this region, as shown by the discharge through the Mississippi River, will carry away about one foot in seven thousand years. This may be reduced to one foot in ten thousand years, if we would keep within bounds. But even this slow rate will require a steady rise of at least twenty-five hundred feet since the close of the carboniferous period, relatively a very modern period.   Further to the northward, the Laurentian Hills, judged by their age, must have lost several times as much height, even supposing they waste only at the rate of the plain, which is far below the south.   This class of facts entitles us to suppose that the lands are, on the whole, constantly rising.   On the other hand, all the known facts, such as the continuous deposition of strata to the amount of thousands of feet in waters always shallow, the increasing evidence of ancient zoological barriers in the sea, under circumstances which require us to suppose that depth of water was the obstacle to the exchange of life ; and other facts which cannot be succinctly stated, give us evidence of the truth of the natural supposition that concurrent with the rise of the land there has been a constant increase in the depth of the sea.

If we try, however, to reduce all great changes of level to the two movements, sinking of the sea and rising of the land, we are met by the difficulty, which will suggest itself to all geologists, that the sea and land have in many regions changed places in alternation. It seems to me, however, that this difficulty is in a great measure overcome by the careful study of the necessary conditions of the movement. In diagrams Nos. 3, 4 and 5, we have represented the line from the centre of any sea to the interior of any continental area. Assuming that the constant tendency of the movements is to depress the seaward part of this section and lift the landward part, we have to notice that there must be a fulcrum point, or point of rotation of the movement. This point may occupy either of three relations in reference to the sea. It may be to the seaward of the shore; it may fall at the shore line, or it may lie to the landward of the shore. Supposing the sea floor to be constantly sinking, and the land constantly rising, it is evident that in the first of these suppositions the land would continually gain on the sea; in the second case, great changes might take place without any effect on the position of the shore; while in the third case the sea would seem to gain on the land. The fact that in most of the land areas where we knew the geology well enough to form an opinion on the matter, there are centres of ancient upheaval about which the oscillations of level have taken place,

In the diagrams 3 and 4, similar letters denote corresponding points.

In figures 3 and 4, the straight lines A B, A′ B′, are diagrammatic expressions for sections extending across the shore. For convenience of delineation, the action of the movement of the small segments of the crust represented is supposed to be like that of a rigid bar.

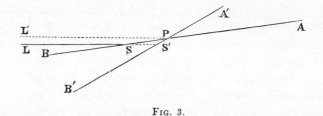

FIG. 3.

In Fig. 3, the pivot point, P, is to the landward of the shore, S, the line A B indicating the surface of the continent near the coast. Let the depression of the sea floor and the elevation of the land go on until the continental surface is in the position indicated by the lines A′ B′, and the shore will be removed to the point S′, and the sea gains. S L, indicates the sea level. If we suppose, however, that the dotted line P L′, denotes the sea level, then the pivot point will fall just at the shore line, and all the changes in the position of the line A B will not affect the position of the water lines.

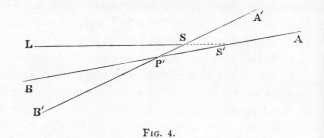

FIG. 4.

In Fig. 4, the pivot point is the seaward of the shore line A B, indicating the original position of the continental surface, and A′ B′, the position of the change. Inspection will show that in this diagram the change has caused the shore to move seaward, and the land gains.

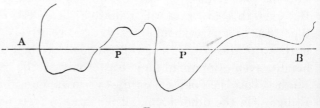

FIG. 5.

Fig. 5 represents a shore line with an axis of rotation, A B, cutting it in such a manner that the points P and P′ may be taken for the pivot points of the diagrams 3 and 4, respectively. All that portion of the shore line above the straight line would be sinking; all below, rising.

is much in favor of this theory of the changes of sea and land. For our present purpose it is not necessary to argue this question in all its phases, it is enough that it enables us to coördinate the changes of the sea and land, and so far to aid us in the conception of the instability of the equilibrium of the crust which is one of the most important facts brought to light by the study of the geological record. The constant changes in the position of the fulcrum point of the movement of sea and land, show that the position of this point, with reference to the whole line, is a matter liable to constant change. In the determination of the position of this fulcrum we may be reasonably certain that whatever the source of the force which brings about the upheaval and subsidence, the matter of weight must have a certain value. Physicists who have attentively considered the question of the origin of the continental folds and ocean furrows, are quite in accord in considering them as folds of the earth's crust, or, to keep clear of assumptions, let us say the outermost part of the earth, which has been compelled to adapt itself to a diminished interior. Without pretending to claim this conclusion to be so well founded as to put an end to controversy, we may use it to complete our conception of the relation of forces to the section from the interior of the continent to the centre of the sea. It will be evident that this section must have its present place determined by two main factors, the energy of the uplifting force and that of the restraining weight. If the amount of this weight at particular points can be made to vary considerably, the whole effect of the movement may be materially changed. A careful consideration of the accompanying diagram will show that a change in the position of the point of rotation or fulcrum of the section will completely alter the result of the movement. When the pivot point is just at the shore line, a great deal of sinking of the sea floor and elevation of the centre of the continent may take place without affecting the relations of sea and land. When the pivot point is to the land side of the shore, then the movement will make the sea appear to gain on the land; when the pivot is to the seaward of the shore, then the land will gain. Elsewhere I have followed these considerations in a more detailed fashion with special reference to the question of the origin of continents. The conclusion is, however, so evident that it is hardly necessary to trace the whole matter here.

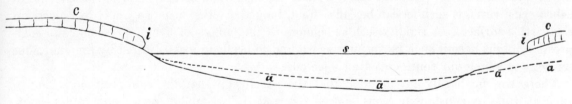

FIG. 6.

The dotted line, *a, a, a, a,* indicates the change of position of the surface after the ice caps have been formed. *c,* Continents. *i,* ice sheet. *s,* sea level.

It is very important to notice, however, that in the case of a rigid mass, such as we suppose the crust to be, supported on material having sufficient mobility to give way under strain, much as a fluid would do, then the imposition of any weight upon one extremity of a given section, such as is shown in Fig. 6, would necessarily produce a change in the position of the pivot point. Now in the ice accumulation of the glacial period we have just such a change of weight as would be likely to bring about considerable effects of this kind. A great mass of water is taken from the sea and heaped to the depth of a mile or more

27

upon the land. A mile in depth of ice weighs about as much as half a mile of ordinary rock, so that by covering the continent of North America with a deposit of this kind we more than double its altitude above the sea. Now if the weight of the mass uplifted be an element in determining the height to which the continents are raised, then we must allow this ice mass a decided influence in depressing the continental areas.

The reader may help himself to form a conception of the nature of these effects if he will float a rigid substance, such as a bar of wood, upon some semifluid, such as tar or treacle, weighting one end so that by its depression it may sink beneath the surface, thus elevating the other end. Now if he will add weight to the elevated end, subtracting it, if he please, from the other end, he will see the change in altitude which I conceive to have occurred by the accumulation of glacial matter. Another illustration, more satisfactory on the whole, is afforded by a little study of any of our frozen lakes. Let a weight be accumulated at any point on ice of moderate strength. The result is the depression of the ice at that point, and its elevation around the sunken point. In this condition the ridge and depression may well represent the state of the tensions which give its continent and sea basin. Now if we take a part of the weight from the depression and place it on the elevated part of the ice, we will thereby change the altitude of the elevation. If the hollow of the ice were filled with water this change would bring about an extension of the area covered with water, representing the invasion of the land by the sea during the glacial period. The fact that in a substance as rigid as ice these movements can take place without fracture, shows us that there is nothing in the solidity of the earth to prevent such movements.

It must not be thought that this view requires us to suppose the interior of the earth is a fluid mass, although it is quite reconcilable with that theory. I am inclined to cling to an opinion set forth by me some years ago that the central region of the earth is quite rigid, while between the outer fifty miles of strata and basement rocks, and the inner core of solid matter, there may remain a section which has not yet been completely solidified, and still admits of sufficient movement to give us the rise of the continental folds and the sinking of the sea floors. Nor am I satisfied that the condition of the mass on which the so-called crust rests, is such as can be called fluid, judged by the tests we apply to objects on the earth's surface. A rigidity such as belongs to the metals of average resistance to compression might permit such movements as we find to occur in masses urged by the enormous tensions to which our continents are subjected.

There can be hardly any question that the conclusion that the continents are kept in their attitude by constant tensions, is necessary and indisputable. As a factor in determining the position which any point occupies with reference to the centre of the earth, the weight of the mass supported must be of importance. Unfortunately the data for determining the value of this element are quite wanting; it seems evident enough, however, that we may more reasonably look to the weight of the ice accumulated on the continents during the glacial period for the depression of the land areas it occupied, than to any other cause. Only in this way can we account for the local character of the depression in many places, or for the coexistence of extensive subsidences in the southern as well as the northern hemisphere.

If this theory of the glacial depression be accepted we may obtain thereby a basis on which to compute the rigidity of the earth's crust as well as other important features connected with those parts of the earth, which though quite close to the surface, are beyond direct observation.[1]

[1] This report is meant to be preliminary to a more special and exhaustive study on the glacial and other dynamic phenomena of the coast of Maine.

Editor's Comments on Paper 2

Jamieson: *On the Cause of the Depression and Re-elevation of the Land During the Glacial Period*

Jamieson published a paper in 1865 in which he drew attention to the correspondence between submergence and the presence of glaciers. This view was challenged by Croll and in two short papers Jamieson answered Croll's challenge and provided some remarkable insights into the process of glacial-isostasy. In the first of these papers, reprinted here, Jamieson first outlines the field evidence for submergence, discusses the fact that the submergence is not equal and that the former beaches are in fact tilted, and provides estimates of the weight of the ice on the ground.

In the next section he proposes that the depression is caused by the weight of the ice and the subsequent crustal response. An important concept mentioned by Jamieson is that a body may be rigid if pressure is applied to it for a short period of time, but that it will yield if the force is applied over a long period of time. The position of the earth's surface is considered to be in a state of "delicate equilibrium" so that additions or subtractions of mass will cause readjustments. Jamieson even considers that the movement of material from beneath the ice caps may have caused volcanic eruptions. Thus Jamieson includes in his concept of glacial-isostasy both an elastic flexing of the crust and a deformation caused by straining in response to the ice load.

Reprinted from *Geol. Mag.*, **9**, 400–407 (1882)

2

III.—On the Cause of the Depression and Re-elevation of the Land during the Glacial Period.

By Thomas F. Jamieson, F.G.S.

IN a paper published in the Quarterly Journal of the Geological Society for January, 1865, p. 178, I drew attention to the remarkable fact that in various parts of the world the presence of glaciers had been attended by a submergence of the land, and I suggested that the enormous weight of ice laid upon the surface of the country might have caused a depression, while the melting of the ice would also account for the rising again of the land which seems to have everywhere followed some time after the ice disappeared.

The paragraph in question, if it did no other good, was of service in so far as it attracted the attention of the thoughtful mind of Dr. James Croll, who in a letter dated 22nd Aug. 1865, addressed to Professor Ramsay, and published in the pages of "The Reader" of the 2nd Sept. in the same year, commented on the suggestion I had made. But, while admitting the fact that submergence had generally accompanied glaciation, he doubted the sufficiency of the cause I had assigned, and maintained that the circumstance would be better explained by the influence of a great polar ice-cap whose mass would effect a slight shifting of the centre of gravity of the earth and thus draw the waters of the ocean northward or southward, according as either pole happened to be under ice at the time. This theory had previously been advanced by M. Adhemar and developed by him at considerable length, in a volume entitled "Revolutions de la Mer."

Dr. Croll's letter gave rise to a lengthened correspondence and controversy in the pages of the periodical called "The Reader," in which a part was taken by Mr. S. V. Wood, jun., Mr. O. Fisher and others; but so far as I remember, no further notice was taken by any of them of the cause I had suggested.

Dr. Croll's maturer views on the subject will be found set forth in his remarkable work called " *Climate and Time*," and also in a paper published in the Geological Magazine for July and August, 1874, " *On the Physical Cause of the Submergence and Emergence of the Land during the Glacial Epoch*." He there assumes that the Antarctic continent is at present covered by a cap of ice 2800 miles in diameter and six miles thick at the pole, with an average depth of about two miles all over; and he maintains that the submergence of the northern lands during the Glacial period was caused by the transference of this great mass of ice from the southern to the northern hemisphere.

It seems to me that we have no sufficient proof of the present existence of a south polar ice-cap of the dimensions Dr. Croll assumes; and the evidence, so far as I can judge, seems to be decidedly against the notion that there was any such ice-cap at the North Pole during the Glacial period; but without meaning to question the adequacy of a great polar ice-cap, supposing it did exist, to affect the centre of gravity of the earth to some degree, and thereby to cause a sensible change in the sea-level, I shall here mention a few objections that present themselves to the application of this theory made by Dr. Croll.

1st. The evidence of submergence we find in the shape of raised sea-beaches, and beds of clay and sand containing marine fossils, seems to show that during the time of their formation no such immense cap of ice could have existed in the North Polar area, nor in any of the glaciated regions of the northern hemisphere with which we are acquainted. These elevated marine beds have been traced by the Arctic Expeditions along the shores of Greenland and Grinnell's Land up to lat. 82°, within a few hundred miles of the Pole itself, proving that during the time they were laid down no great ice-cap could have existed there, and showing that even in Greenland much of the coast was covered with sea-water and not by ice. Again, in Norway and Sweden the marine glacial beds show that during the time of their formation, the land-ice in Scandinavia was much circumscribed, and had retreated far within the present line of shore. In our own country also the Arctic shell beds found in Wales, Lancashire,[1] and various parts of Scotland and Ireland, demonstrate that the ice had shrunk into comparatively small dimensions, and did not cover anything like the present area of the British Isles. The same holds good in America, where the marine clays and sands found along the banks of the Ottawa, the St. Lawrence, Lake Champlain and Labrador afford convincing proof that the great Canadian glacier had dwindled into comparative insignificance at the time they were deposited. I am therefore unable to see where the enormous mass of ice could have lodged that Dr. Croll calculates upon; for during the time of submergence to which these marine beds belong the land-ice of the northern hemisphere, instead of being at its maximum, was so far reduced as to be evidently inadequate for the purpose he requires. The submergence according to his theory must have diminished step by step with the diminishing ice; for the action of gravitation he invokes operates instantaneously, and the submergence could not have endured a day after the ice vanished.

2nd. If the submergence was caused by an ice-cap drawing the ocean northward, the sea-water should have risen to equal heights along the same parallels of latitude, but the facts hitherto reported do not agree with this at all, for they indicate a very unequal amount of submergence in places situated in the same latitude. The best-known high-lying marine beds of New England and Canada lie between lat. 44° and 52°, which corresponds to that of France and the South of England, just where evidence of submergence is conspicuous by its absence. Again, we find in Scandinavia evidence of submergence to the extent of 600 or 700 feet, but in the same latitude along the eastern side of the Baltic and in Finland no evidence of a like submergence has been found. In the neighbourhood of Dublin, Lancashire and North Wales, sea shells are found in the superficial beds of sand and gravel to heights of 1200

---

[1] Macclesfield, Cheshire, see GEOL. MAG. 1865, Vol. II. pp. 293-299.

DECADE II.—VOL. IX.—NO. IX. 26

and 1350 feet, but no evidence of submergence to anything like this extent has been detected on the eastern side of England or in the neighbouring parts of Europe. In Canada the marine shell beds reach up to 470 feet at Montreal, but, says Mr. Thos. Belt,[1] "going eastward from Montreal the elevation of the marine beds, marking the former submergence of the land, gradually decreases until in Nova Scotia it reaches zero." Dana, in the 2nd edition of his Manual of Geology, tells us that the altitude of the marine deposits on the southern shores of New England is 40 or 50 feet, at Lake Champlain (which is in the same lat. as Nova Scotia) they occur up to 393 feet. It would therefore appear that on the Eastern coast of North America the submergence, like the glaciation, extended farther South than it did in Europe. This is quite intelligible if there was a depression of land caused by the glacier, but is unaccountable by a rise of the ocean.

3rd. Again, according to Dr. Croll, the amount of submergence should have diminished very very gradually as we trace it from north to south. For example, he calculates that if the submergence was 485 feet at the North Pole, it would be 434 feet in the lat. of Edinburgh, or only 51 feet less. The evidence, however, does not agree with this, for in Holland, Belgium, France, and the South of England little or no submergence has been discovered, although a large amount occurred a few hundred miles to the north of these places. In Scandinavia the highest lying shell-beds are in the southern half of the peninsula, where they attain an altitude of 500 or 600 feet, but in going northward from Trondhjeim they seem to decrease in elevation, and in Finmark none have been discovered at nearly so great a height. At Hammerfest, according to M. Bravais, the highest of the old sea beaches is only 92 feet. This is inconsistent with Dr. Croll's theory, according to which the submergence should have increased to the north instead of growing less, but it is intelligible on the ice pressure hypothesis, as the amount of precipitation is greatest along the south half of the Norwegian coast and therefore the weight of ice would be heaviest in that quarter.

Moreover, Bravais found that the old sea beaches in Finmark are not horizontal. If this be correct, it is a good proof that the movement has been due to a rise of the land, for had it been caused by a fall of the ocean it is evident the old beaches should have been horizontal. Bravais found his uppermost beach line decline in level from 221 feet at its southern extremity in Altenfiord to 92 feet at its northern end at Hammerfest, thus lowering in level from south to north.[2]

These objections to the theory of Dr. Croll induce me to think that the cause assigned by him is not the true one, and that an explanation must be sought in some other direction. I may further remark that Dr. Croll's theory assumes that the glacial conditions were not simultaneous in the Arctic and Antarctic regions. But in the present state of our knowledge it seems to me that we are not entitled to

[1] Trans. of Halifax Institute, 8th May, 1866.
[2] Quart. Journ. of the Geol. Soc. vol. i. p. 541.

make this assumption, for I do not know of any trustworthy evidence to show whether the glacial conditions were or were not simultaneous in both hemispheres. At present we cannot tell.

Supposing, however, we grant an ice-cap at one pole of the size Dr. Croll requires, viz. six miles thick at the centre, or about 32,000 feet. This would give a weight of 12,000 lbs. on the square inch, which is equal to 800 atmospheres, and would amount to more than 21,000 million tons on the square mile. Could the bottom layers of ice sustain such a weight without liquefying? Supposing even this to be possible, would not such a weight depress the ground beneath it? Would not the centre of gravity of the earth be more likely to pull the cap down than the cap to shift the centre of gravity? The earth would require to be a very rigid body indeed to sustain such a weight for thousands of years without yielding.

I, however, prefer resting my objections on the geological evidence, which I think conclusively shows, from the unequal amount of submergence in adjoining areas, that the circumstance was not occasioned by a rise of the sea, but by a fall of the land. I shall afterwards point out that we have also a good deal of evidence of depression in regions where there was no submergence, and therefore where movements of the sea do not enter into the question at all.

HYPOTHESIS PROPOSED.

It seems to me that the facts agree better with the notion that the submergence was due to a sinking of the land, and my idea is that the ice-covered regions were depressed by reason of the great weight of ice placed upon them, and that when the ice disappeared they rose again with extreme slowness, and may have eventually attained nearly their former level; but in most cases, I believe, some amount of permanent depression probably occurred.

The amount of depression would depend on two things, viz. :—

1. The weight of ice.
2. The elasticity or yielding nature of the area on which it lay.

The amount of ice that accumulated upon the land during the period of maximum glaciation is now admitted on all hands to have been very great. Whether we ask the geologists of Scandinavia, or of Switzerland, or of Great Britain, or of Canada, or of the United States, all of them will tell us that the thickness of the ice in these regions must be reckoned by thousands of feet.

Assuming the specific gravity of the ice to have been 875 compared with water as 1000, or in other words to have been seven-eighths of the weight of water,[1] then the weight of a mass of ice 1000 feet thick, would be 378 pounds to the square inch, or equal to fully 25 atmospheres, and would amount to 678,675,690 tons on every square mile. If the ice was 3000 feet thick, it would at this rate amount to over 2000 million tons on the square mile. If 4000 feet thick, it would give a pressure of a hundred atmospheres, or 1500 lbs. on the square inch.

---

[1] A. Helland found the sp. grav. of the Greenland icebergs to be 886. See Quart. Journ. of the Geol. Soc. vol. xxxiii. p. 155, Feb. 1877.

Now we have every reason to believe that from 1000 to 3000 feet was quite a common thickness in many of the glaciated regions of Northern Europe. Some authorities indeed place it at a higher figure, while in America both Agassiz and Dana calculate that a far greater depth of ice existed. The latter thinks that on the Canadian watershed it must have been 11,000 feet thick (Manual of Geology, 2nd ed. p. 538), and on the northern border of New England he supposes it had a mean thickness of 6500 feet.

In some of the Fjords of Norway Mr: A. Helland puts it at 6000 feet.

But without insisting on these high figures, it is evident that a thickness of even 3000 feet of ice will give us a weight by no means despicable, a weight which would require a marvellous rigidity indeed in the earth beneath it to sustain such a load without yielding in some degree.

But no substance we are acquainted with is absolutely rigid; it always yields more or less to pressure, and when the pressure is removed it tends to resume its former position, and will do so more or less perfectly according to the elasticity of its nature. Some bodies are so elastic as to regain their original form completely, without suffering any permanent change; and physical authorities tell us that practically speaking almost every solid body may be considered perfectly elastic up to a certain point. That is to say, there is generally a limit of constraint from which it will recover when the strain is removed. But if the strain is carried beyond this limit, the body undergoes a permanent change in shape or size, and acquires what is technically called a *set* from which it does not recover.

Now, that the crust of the earth is flexible and elastic the phenomena of earthquakes sufficiently demonstrate. The surface heaves like the billows of the sea, sometimes causing trees to bend so as to touch the ground with their tops, or tossing up flagstones into the air so as to make them come down bottom upwards. Waves of elastic compression are transmitted many hundreds of miles from the seat of disturbance, as in the case of the great earthquake at Lisbon, when the water of lakes in Scotland rose and fell two feet in vertical height as the shock passed along. All this implies some amount of flexibility, which no doubt will vary a good deal according to the nature of the strata and materials of which the earth is composed. Geologists seem inclined to believe that whether the great internal mass of the earth be fluid or not, there are, at all events, great pools or lakes of melted matter here and there in the interior, at various depths below the surface; and if this be so, a region lying above an elastic cushion of this sort would be more likely to yield to pressure than where the substructure is more solid. A very small amount of yielding is all that is required, spread over so wide an extent as to be quite inappreciable to the eye when viewed in a section drawn on a true scale; and that the earth is so absolutely rigid as to sustain with indifference the imposition of a weight of ice thousands of feet thick over areas of almost continental dimensions seems to me very improbable indeed. It appears more likely that the position of the surface is in a state of delicate equilibrium, and that any considerable transference of pressure will cause a re-adjustment of levels. If upheavals and depressions of the land have not been caused by changes of pressure, it may be asked what is it they have been caused by? In volcanic eruptions even so slight a difference in pressure as the varying weight of the atmosphere has been thought to have some influence on the subterranean forces. The elasticity of the ground is so great indeed as to respond to very insignificant agencies. The rumbling of a waggon along the street, the galloping of a troop of horse, or even the measured tread of armed men, causes a vibration of the surface which is plainly felt.

If beneath that part of the surface which was affected by the heavy pressure of the ice there happened to be a quantity of lava in a fluid state, the result might be to cause an outburst of the lava to take place at some more distant point. This would relieve the tension and lead to a permanent depression of the ice-covered area. For example, in North America, the great fields of ice that lay on certain portions of that continent by their downward pressure may have occasioned some of those extensive eruptions which seem to have taken place in the region of California after the commencement of the Glacial Period.[1] The volcanic phenomena of Iceland in like manner may have been affected by similar causes. That there has been a considerable permanent depression of some of the most heavily glaciated regions since the commencement of the Glacial period I think there is much reason to believe. The features of the Fiord districts of Norway and the West Highlands of Scotland and of British Columbia, for example, seem to show this; for these coasts have all the appearance of depressed mountain lands, which have been cut and carved by streams and glaciers far beneath the present level of the sea. The glaciation of the United States also presents some singular features that seem to point to an elevation of the Canadian ridge between the lakes and Hudson's Bay much greater than now exists.

But in regard to the effect of pressure, time is an important element in the problem.

Bodies that seem absolutely rigid to pressure applied for a short space of time yield perceptibly to a force which is long continued. The effect may be so minute as to be quite imperceptible at first, yet when multiplied a sufficient number of times it will give us all that we require. And this is just the lesson that Geology has so often impressed upon us. Now in the case we are considering we can draw largely on the Bank of Time.

In regard to the pressure exerted by the ice, we must bear in mind that it probably continued for very many thousands of years. If we suppose that it caused a depression of only an inch in the year, this would give us a foot in 12 years, 100 feet in 1200 years and 500

[1] See J. Le Conte, On the Old River Beds of California, in the American Journal of Science, for March, 1880.

feet in 6000 years; and this might be much more than we require, for allowing longer time far less than this might be enough. Now an inch in the year is less than a hair's breadth in the day.

The earth's diameter is about 7900 miles, which is equal to rather more than 500 million inches. Suppose then we had a pile of books consisting of a million volumes, each containing 1000 pages, a compression of an inch would be represented by the thickness of only a single leaf or $\frac{1}{500,000,000}$ part of the whole. Now for a depression of an inch we may allow at least a whole year and perhaps even five years. The action we invoke therefore does not seem to be an unreasonable or extravagant one.

The recovery of level after the ice disappeared would depend upon the elasticity of the materials beneath the depressed region. The rise would probably be very slow and gradual, like the depression, and in many cases the recovery of level would probably be incomplete. The unequal strength of the earth's crust and the irregular way in which it yields to pressure are exemplified by what we know about *faults*.

Faults are often of great magnitude, extending sometimes to many hundred feet and occasionally, it has been supposed, to even thousands of feet; and they show that the subterranean structure may be such as to give way very unequally to pressure. It has been thought that subterranean cavities exist which might allow a permanent amount of subsidence to be effected with comparative ease.

The general paucity of animal remains in the marine beds of the Glacial period would seem to indicate that the actual occupation of the surface by the sea-waters had not been very prolonged, for had it been so, we should expect to find remains of whales, seals, and fishes, far more abundantly than we do, and also great beds of sea shells.

Animal life swarms in many parts of the Arctic Ocean, and we can hardly suppose the sea to have occupied our coasts for thousands of years without leaving remains of marine life in a degree of plenty far beyond what we find in any of the Glacial-Marine beds of this country. During the commencement and earlier stages of the depression the submerged tracts which are now above water would probably be occupied by glacier-ice, and it would only be after the ice broke up and floated off that the sea would take its place, by which time the land would be on the rise again. The subsidence of the land no doubt contributed much to the dispersion of the ice by enabling it to float away and get melted in warmer water.

If the glacial conditions were simultaneous in the north and south hemispheres during the age of ice, there can be little doubt that the abstraction of so great a mass of water from the ocean, together with the contraction in bulk of the water owing to the colder temperature, would have very sensibly lowered the sea-level, as Mr. Alfred Tylor pointed out.[1] But owing to the want of proper data

[1] GEOL. MAG. 1872, pp. 392 and 485. Mr. Alfred Tylor calculated that a deposit of snow and ice 1500 feet thick over an area of land one-tenth of that of the sea, would reduce the level of the ocean 150 feet. He supposes a subsidence of 600 feet altogether.

on which to calculate the total cubic contents of the sea, and of the supposed glaciers, it is difficult to form any good estimate of the change of level that would result from such a cause. If there was an ice-cap at both poles at the same time, the centre of the gravity of the earth would remain much as it is, and the cause of submergence invoked by Dr. Croll and M. Adhemar would not come into operation.

(To be concluded in our next Number.)

Editor's Comments on Paper 3

De Geer: *On Pleistocene Changes of Level in Eastern North America*

Sweden's greatest Pleistocene geologist, the Baron G. de Geer, visited North America in the latter part of the nineteenth century. During this visit he went on a number of field trips with the principal object of determining the local marine limit. This paper presents the first syntheses of observations on postglacial movements of eastern North America and compares them with data from northern Europe and Spitzbergen. Particularly useful is de Geer's explanation of methods that can be used to delimit the local marine limit. He also notes the coincidence of the uplifted areas in both Scandinavia and North America with the great shield areas and poses the question as to how the influence of uplift due to ice loading and that of long continued tectonic movements can be differentiated.

De Geer summarizes all available data on a map that attempts to present for the first time a large-scale regional synthesis of uplift from both the area of the Glacial Great Lakes and Glacial Lake Agassiz and the coastal area from New York north to Newfoundland and thence to James Bay and Hudson Bay. Marine limit elevations are shown rising from the St. Lawrence Valley to James Bay, where a contour of 800 ft is shown. (Note: The original map is in color and therefore easier to read than the black-and-white version reproduced here.)

Reprinted from *Proc. Boston Soc. Nat. Hist.*, **25**, 454–461, 470–477 (1892)

3

MAY 18, 1892.

President W. H. NILES in the chair.   Sixty-two persons present.
The following papers were read :—

ON PLEISTOCENE CHANGES OF LEVEL IN EASTERN NORTH AMERICA.

BY BARON GERARD DE GEER,

of the Geological Survey of Sweden.

(Walker Prize Essay, 1892.)

One of the most important principles upon which geology is
founded is the theory of continental changes of level.   The main
points of this theory seem, in several cases, to have been well established by American geologists.   Thus it has been pointed out,
that to account for sandstone several thousand feet in thickness,
and other deposits in shallow water, it is necessary to assume
that the sea-bottom was sinking at the same rate that the sediment

was accumulating. Again, as the continents in certain instances for long periods of time have not lost in height, and this notwithstanding their immense denudation, they must have been gradually rising. It is also very generally admitted, that many of the abundant alternations of strata deposited during different bathymetrical conditions, as well as the breaks between them, were caused by the oscillations of the earth's crust. In many instances, however, it can not be decided whether the change of level was really due to movements of the land, or whether it was only the surface of the sea that rose and fell. Since the eminent Austrian geologist, E. Suess, in his grand work "Antlitz der Erde" has in a very ingenious way tried to refer most of the oscillations to the latter cause, denying the rising of the continents altogether, and since his views have been adopted by many geologists, it seems particularly desirable to get more positive facts for the final settlement of the question.

For the present at least it is hard to get such facts concerning the older formations, as they are very often eroded away to a greater or less extent and concealed by younger deposits. It is thus in most cases impossible to determine the original extent of a certain layer and especially of the sea in which it was formed. Consequently we cannot determine with sufficient accuracy in what way the corresponding geoid-surface has been deformed.

In regard to the Pliocene and Pleistocene formations it is of course less difficult, but in many parts of the world it seems as if the old shore-lines, which once marked the limit of these formations, were not very well developed or easily recognized. It may sometimes be due to the fact that when the shores are low and consist of loose marine deposits, it is often very difficult to distinguish the new from the old formations, and also to ascertain the depth of the last submergence. The beach is also easily effaced, partly perhaps through wind-blown sand.

In the glaciated regions, however, the conditions are often different and more favorable for the formation of enduring shore-lines, as the land is there generally covered with till and angular, stony debris, forming an excellent material for recording the action of the waves. Most of the old shores are described as situated in the glaciated regions, though this may perhaps depend upon another and deeper cause.

Although a great many marine shore lines, shell deposits, and sediments of Pleistocene age have been leveled in Europe and in America, it is nevertheless very hard to find any method for determining their limits with great accuracy.

As far as I can see, that described below is the most suitable and perhaps the only possible method for this purpose. I have tested it in the northern part of Europe during the last ten years and by way of comparison in the eastern parts of North America during the autumn of 1891. In this paper I shall especially describe and discuss the results of the last named investigations, but as these point to a very close analogy with the corresponding phenomena in northern Europe, it seems appropriate to give first a general view of the latter.

INVESTIGATIONS IN EUROPE.

During the summer of 1882 I spent three months in Spitzbergen studying the glacial deposits and the raised beaches. Though these were very instructive for the study of the origin of shore-lines in general, the conditions were not favorable for an accurate determination of the uppermost marine limit, the land being rather mountainous, precipitous, and destitute of till.

Since that time I have used every opportunity to discover and determine the marine limit in Sweden. The method I have used is the following : On every occasion I start with the most recent fossils in marine deposits ; with the aid of the topographic map I select a drift-covered hill of sufficient height with moderate slope and with a situation as open as possible to the ancient water-body and as near as possible to a point previously leveled.

Above the water-laid clay and sand there is in most cases a belt of gravel, and still higher, where sediment is almost wanting, there are more or less conspicuous marks of erosion and water-wash up to a definite horizontal limit. In favorable localities this can be determined with an accuracy of from a few feet to less than one foot. Later on I will give a more detailed account of the methods I have employed for these determinations. I will only emphasize here that while geologists have too often measured the highest conspicuous shore-lines which happened to be developed in a certain locality, I have used such figures only when nothing else was available. On the other hand I have always tried to choose hill-slopes so uniform that evidently the till above the measured limit, had it also been submerged, must necessarily have shown traces of water-action in addition to the assorted, washed, and rolled material below.

Up to the present time I have thus leveled the marine limit at about seventy different points in the southern and central parts of Sweden and in a few places in southern Norway. For northern Sweden I have three or four approximate but important determinations by Högbom, Svenonius, and Munthe. For the other parts of the Scandinavian region of uplift the uppermost marine limit is not yet determined, but there are in geological literature a great number of measurements of the height of raised beaches and marine sediments, and from these I have tried to ascertain the highest available minimum-figures for different tracts in the region. While they are only preliminary, they nevertheless point very clearly to the same laws for the upheaval of land that I found to prevail in Sweden and it seems allowable to use them for the present, of course with due reservation, as the principal conclusions drawn from them will probably not be essentially changed by future more accurate determinations.

To get a general view of the warping of land since the formation of the marine limit I have used the graphic method of Mr. G. K. Gilbert (see his admirable work on Lake Bonneville) and have connected with lines of equal deformation, or as I have called them *isobases*, such points of the limit as were uplifted to the same height.

Among the results of the investigation the following may be mentioned as being of especial interest for comparison with the conditions in North America.

All the observations evidently relate to one single system of upheaval, with the maximum uplift in the central part of the Scandinavian peninsula, along a line east of the watershed, or nearly where the ice-sheet of the last glaciation reached its greatest thickness. Here the land must have been upheaved somewhat more than a thousand feet (more than 300 meters), and around this center the isobases are grouped in concentric circles, showing a tolerably regular decrease in height in every direction toward the peripheral parts of the region, until the line for zero is reached, outside of which no sign whatever of upheaval is to be found.

The considerable height at which the uppermost marine marks are found, and the places where they occur, in the central parts of the land, show at once that no local attraction of the land ice could have been sufficient to raise the water to any such amount, had the ice been many times as thick and extensive as it probably was even at its maximum. Such an explanation seems less possible, as there could be very little room for any attracting land ice when the sea covered the parts of the land mentioned above.

But as no local changes of the sea level can account for the phenomenon, so it is also impossible to explain it either by the general oscillations of the sea, perhaps from the one hemisphere to the other, produced by changes in the situation of the center of gravity of the earth — according to the assumption of Adhémar and Croll — or by oscillations to and from the equator caused by changes in the rotation of the earth, as has been supposed by Swedenborg and Suess. If this theory were true, all the shorelines would slope in a single direction, but as they in fact slope as well to the south as to the west, north, and east, it is evident that the phenomenon must be explained by a real rising of the land.

Moreover the region of upheaval is practically about the same as that of the last glaciation; especially is it worthy of notice, that the maximum of both seems to have occupied about the same place.

Still more remarkable is the coincidence of the uplifted area with the Scandinavian azoic region, or what Suess has called "the Baltic shield." This comprises Sweden, Norway, Finland, and the Kola peninsula, or a well defined tract where the old rocks are laid bare by erosion and the surrounding lands thickly covered with younger sediment. The limit of the Baltic shield, where it has been directly observed, and perhaps everywhere, is marked by great faults. Now the isobase for zero, or the boundary for the uplifted area, seems all the way a little outside of the above named limit and follows very conspicuously its convexities and concavities. Likewise all the other isobases point to a close connection between the upheaval and the geological and to a certain extent the topographical structure of the land. Thus it is commonly found that higher tracts have been raised more than lower; and the basins of the great Swedish lakes, Wener and probably also Wetter, have been less uplifted than their surroundings, which might indicate that they were originally more depressed and very probably formed by unequal subsidence.

The coincidence between the areas of erosion, glaciation, and upheaval may be thus explained : as in continental areas in general, this old tract of erosion has probably in the main been one of upheaval, while the contrary was the case with the surrounding regions, where the sediment was accumulating to a very considerable extent.   During the Ice age, among other high lands, the Baltic shield received an ice-sheet equal in weight to more than a thousand feet in thickness of the rocks which had been eroded away during previous periods.   As Jamieson long since suggested, it is very probable that the crust of the earth must yield and subsidence of land take place beneath this added load.   Therefore it is reasonable that the movements in the crust should be chiefly dependent upon its geological structure.

When the ice-load disappeared, the land partly re-emerged, until a balance was reached, which seems to have happened before the original height was attained, a part of the change having become permanent.

If the ice-load was the essential cause of the submergence, a still larger subsidence must be supposed to have followed after the earlier and greater glaciation.   It is true that very few traces of unquestionable interglacial marine deposits have been found, and that these are all along the boundary of the late glacial region of upheaval, or in southern Denmark and along the Baltic coast of Germany : but this is just what would be expected.

Then, as Dana first pointed out with reference to the fiords as submerged river-valleys, the land had probably in the beginning of glacial time a much greater elevation than at present.   Thus it is quite possible, that during the great glaciation a considerable subsidence from the highest elevation occurred, followed during interglacial time by a partial re-elevation of the land ; while the early marine deposits during the late glacial subsidence might have been a second time so deeply depressed below the sea-level that they have not since been uplifted sufficiently to appear above it.   According to this explanation, it is easy to conceive why the interglacial marine deposits are accessible just in the tracts which were least affected by the late glacial subsidence.

I take this opportunity to remark that in my opinion the marine sediments which Murchison, Verneuil, and Keyserling[1] found at Dwina and Petschora in northern Russia, and which have been

---

[1] Murchison, Verneuil, und Keyserling, Geologie des europäischen Russlands; bearbeitet von G. Leonhard. Stuttgart, 1848, pp. 348–351.

lately traced over large areas by Tschernyschew[1], are probably of interglacial age, though they are not covered by till, as are those occurring at the border of the last glaciation. But as their fauna contains such boreal and southern species as *Cyprina islandica* and *Cardium edule*, it is not probable that they could be contemporary with the arctic fauna of the late glacial subsidence in Scandinavia. On the other hand, it is difficult to believe that the considerable oscillation of land in northern Russia could have taken place so lately as in postglacial time. Hence there is some reason to infer that the deposits in question belong to the interglacial period, and it is my opinion that, like the undoubtedly interglacial deposits before-mentioned, they are still accessible above the sea-level only outside of the region which was affected by the late glacial submergence.

Before leaving the changes of level in Scandinavia I must add a few words about the latest oscillation, though this is not yet quite so well known, and can only to a certain extent be compared with the conditions in America.

After the late glacial upheaval in Scandinavia had proceeded so far as to isolate the Baltic basin from the sea, thus forming a lake with a true fresh-water fauna, characterized by *Ancylus fluviatilis* Linné, and after this lake, following the general unequal movement of the region, had been partly emptied, then, as I have succeeded in showing, a new subsidence of land occurred, by which the outlets of the *Ancylus* lake were changed to sounds, and a marine though scanty fauna migrated into the Baltic. The deposits formed along the Baltic, as well as along the western coasts of the land during this last subsidence, are now partly uplifted, less in the peripheral and more in the central part of the region, but nowhere more than 200-300 feet above the sea level. They contain a true post-glacial fauna, with many southern forms which are never found in the late glacial beds. Between these two marine deposits, peat bogs, river channels, and other traces of erosion are observed in many places in southern Sweden, showing conclusively that at least this part of the land was uplifted between the two subsidences. Several of these peat bogs and river channels continue below the level of the sea, and such signs of

Th. Tschernyschew, Travaux exécutés au Timane en 1890; Petersburg, 1891, pp. 27, 52.

submergence are also found at the southern shore of the Baltic and around the North Sea.

It is not yet possible to say anything with certainty about the nature of this last oscillation; but while there seem to be some difficulties in such an explanation, it may be possible that we have to deal here only with oscillations in the situation of the pivot point of the crust-movement or the isobase for zero.   Professor N. S. Shaler has suggested,[1] that while the continents are as a rule rising and the sea-floors sinking, yet it may happen that the pivot point, when it lies somewhat at the inside of the shore, will take part in the sinking of the sea-bottom, and then it will seem as if the continent were sinking, though in fact it may very well be rising in the interior.   If it should turn out that this ingenious explanation could be applied to the Scandinavian oscillations of land, then the whole phenomenon would be more easily understood; according to this theory, in the center of the region after the removal of the ice-load a constant rising of the land occurred, and at the same time probably a sinking of the surrounding sea-bottom, in which latter movement some portions of the land for a time took part during the post-glacial subsidence. After that, the portions mentioned began again to participate in the great continental upheaval, which seems to be still going on though probably at a much reduced rate.

INVESTIGATIONS IN NORTH AMERICA.

Observations.

The very interesting and valuable investigations of Gilbert, Upham, and Spencer have shown that the shore-lines along the great lakes in the interior of eastern North America have been unequally uplifted more toward the north than toward the south, and this seems to be quite in accordance with the generally adopted opinion in regard to the marine deposits along the Atlantic coast.   This opinion seems to have been founded principally upon the different heights to which marine shells could be traced in different tracts, this kind of evidence being the most indisputable, though on the other hand affording only minimum figures.

[1] Recent changes of level on the coast of Maine. Mem. Bost. soc. nat. hist., v. 2, p. 337, 1874.

Plate XIII.

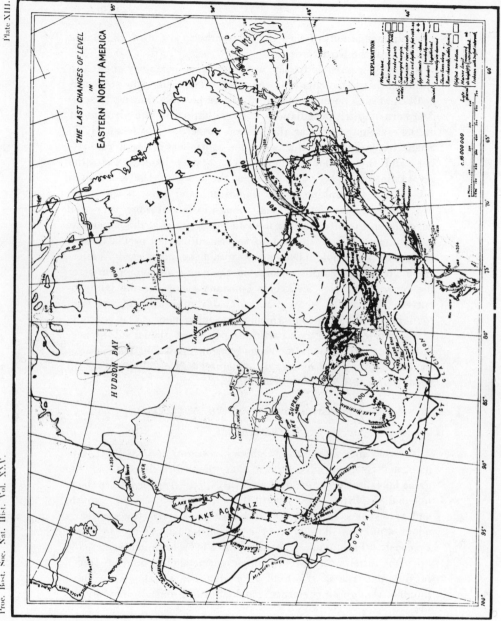

THE LAST CHANGES OF LEVEL
IN
EASTERN NORTH AMERICA

48

[*Editor's note:* In the next 8 pages de Geer tables previous estimates of marine limit elevations in New England, St. Lawrence Lowlands, and Labrador, and describes his own results from field work conducted during his visit to North America. These data are then used to construct the map reproduced in this selection.]

CONCLUSIONS.

All the observations on the height of the marine limit have been put down on the accompanying general map, and with aid of interpolation isobases have been drawn through equally uplifted points with an interval of 200 feet (60 m.).

Concerning the extension of the isobases into the interior of the continent, where the marine limit could not be directly determined, I have tried to use interpolation in the following manner. As has been stated, it is probable that the geoid-surface, which in the submerged regions is marked by the marine limit, is situated in the tract northeast of Lake Ontario at a height somewhat less than 75% of the older high-water level recorded by the Iroquois beach. From the figures given by Professor Spencer[1] we find that this beach is situated at about 36% of the Ridgeway beach at the three localities where both occur near each other.

[1] High level shores in the region of the Great Lakes and their deformation. Amer. journ. of sci., March, 1891.

Consequently the geoid of the marine limit should be found at 27% of the latter beach, if the deformation of both had been proportionate.

To see to what degree this has been the case for the different beaches, I have also reckoned in percentages the proportion between Forest, Ridgeway, and Maumee beaches, and from the figures thus obtained as far as we can judge from the material at present available the differential uplift of the highest or the Maumee beach was somewhat greater than that of the Ridgeway beach, the same being the case with the Ridgeway beach in comparison with the Forest beach, but the lowest one, or the Iroquois beach, seems to have a proportionately steeper slope than the Forest beach and to be in this respect more proportionate to the Ridgeway beach.

As this has been explored for the greatest distance and seems to be the easiest of identification, I have thought it advisable to use it for this preliminary interpolation, without attempting to make any correction for the divergence from the proportion of 27% which may occur in the southern part of the region.

Thus of the figures on the map indicating the interpolated height of the marine limit all those along the Iroquois beach represent 75% of its height, and those along the Ridgeway beach 27 % of its height.[1]

Concerning the westernmost part of the glaciated area we owe accurate information about the gradient of the warped beaches to Mr. Warren Upham's excellent investigation of Lake Agassiz. However, until the deserted beaches around Lake Michigan and Lake Superior are more fully explored and the damming ice-border is continuously traced between the different basins, it is difficult to form any opinion about the absolute amount of the upheaval of the land since the formation of the marine limit.

In the mean time we must content ourselves with the following facts. As Prof. J. E. Todd and Mr. Upham have stated, the deserted shores of Lake Dakota, situated close to the southwest of Lake Agassiz, show no or only a slight unequal deformation. As

---

[1] This method of interpolation can of course be accurate only when the change of level has been successive and regular, as may perhaps to some extent have been the case with the sea, but probably much less with the ice-dammed lakes. Still the present state of our knowledge does not seem to allow any more satisfactory method, and this may be sufficient for the present purpose.

the longer axis of this lake trends in nearly the same direction as the greatest warping of Lake Agassiz, it seems probable that the limit for this warping and at the same time for the upheaved area lies just between Lake Agassiz and Lake Dakota or through Lake Traverse. It is by no means certain that the limit for the uplifted region or the isobase for zero remained at the same place when the marine limit in the St. Lawrence valley was formed; but we may assume it for this part of the continent, since we cannot, at present, expect to get more than a general idea of the direction of the isobases and their maximum gradient. To judge from the probable thickness and direction of the receding ice-border, it appears that the formation of the highest or Herman beach of Lake Agassiz was probably antecedent to the geoid surface which is traced in this paper. Moreover it is quite possible that the ice had not receded from the St. Lawrence valley before Lake Agassiz received the northward outflow. Yet to be quite sure of maximum figures for the gradient, I have used the measurements of the highest or Herman beach, though they may be too high.

As to the probable position of the marine limit in the other northern portions of the area very little can be added. Some explorers, believing that every kind of drift is deposited in the sea, have not paid due attention to the determination of the limit for the real marine deposits; others seem to have estimated only the height of beaches accidentally discovered and their most reliable observations are made with a barometer often at a long distance from any known level or base-barometer.

From Murray's paper on the glaciation of Newfoundland[1] it seems that marine deposits are found on that island at a height of about 200 feet. According to R. Bell[2] distinct beaches are seen at Nachvak in Labrador at an estimated height of 1,500 feet; but I have not found more precise measurements for these tracts. Even if this measurement should be over-estimated, these beaches may be among the highest in all the uplifted area. But the low levels at which marine deposits are found in the relatively well explored southern and western parts of Greenland, make it improbable that the extraordinary high levels, reported as occurring

[1] Proc. and trans. Roy. soc. Can., 1883, I, pp. 55-76.

[2] Rept. geol. surv. Can., 1885, p. 8 DD; and Bull. geol. soc. Amer., 1889, I, p. 308.

along Smith Sound, should belong to the same system of upheaval as that of the Canadian region.

East of the middle of Hudson Bay between the coast and
Clearwater Lake, A. P. Low has found sediments and terraces
probably of marine origin up to about 675 feet above sea level.[1]

Southwest of James' Bay, on the Kenogami River, a tributary
of the Albany, Bell[2] has found marine fossils about 450 feet, and
west of Hudson Bay at Churchill River about 350 feet above the
sea level.

As is easily seen from the above statements, the observations
at present available do not allow the drawing of even approximate
isobases over a large portion of the area; but from the part sufficiently studied, it seems possible to form a general idea concerning the nature of the changes of level; these point to a remarkable analogy to the conditions in Scandinavia. Thus the
greatest subsidence has taken place in Labrador—probably near
the watershed—where the ice accumulation had its center. But
as the ice in the northern part of this land, according to Bell, had
a northward movement, it will probably be found that the
amount of subsidence also decreases to that side, about as it did
in all other directions in which the ice covering thinned out.

The conformity between ice-load and subsidence seems to have
been still greater here than in Scandinavia, and in this respect it
will be very interesting to see what will result from a continued
investigation of the warped beaches in the lake basin with its
marked ice lobes.   It can already be seen that the isobases in
the peninsula southeast of the St. Lawrence River, which we
will here for brevity's sake call the Atlantic peninsula, follow very
closely the extension of the last glaciation.   Especially is it note
worthy that the amount of subsidence was small along the
Gulf of St. Lawrence in connection with the fact, stated by
Chalmers, that the ice thinned out in that direction.

Nova Scotia, which probably only in its western portion and to
a small amount participated in the subsidence of the mainland,
seems from this fact not to have been wholly ice-covered during
the last glaciation, and the local glaciers might not have been
thick enough to produce any noticeable changes of level.

[1] Rep. on expl. in James' Bay and country east of Hudson Bay. Geol. and nat.
hist. surv. Can., Ann. rept., 1887, III, p. 59 J.

[2] Geol. and nat. hist. surv. Can., Ann. rept., 1886, II, pp. 34, 38 G.

The gradient of the deformed geoid surface was evidently steepest on the Atlantic side of the continent, where the slope of the ice-sheet must also have been greatest; it is here generally 1 : 1,400, with the exception of the St. Lawrence valley, where its direction is oblique to its general trend against the Atlantic and its amount is not larger than about 1 : 4,900.

The steep gradient will probably be found also at the coast of Labrador, which in many respects is analogous with the high, fiord-cut coast of Norway in Scandinavia. In the interior of the American continent, where the ice spread out over a large area, the isobases are far more distant and show a smaller gradient just as in Scandinavia. Thus the mean gradient from Georgian Bay toward the southwest to the limit of the area seems about 1 : 3,400, being much steeper at the border of the azoic region and smaller at the outside.

The connection between the subsidence and the geological structure of the earth's crust is perhaps not quite so striking as in Scandinavia. Still it seems probable that the Canadian azoic or Archaean region has changed its level more than the surrounding tracts, though this is not yet sufficiently proved in regard to Hudson Bay. The general conformity between the ice covering and the old azoic plateau makes it difficult in the present state of our knowledge in many cases to discern between the influence of these two circumstances. Thus it may be remarked that the above mentioned convexity of the isobases around the Atlantic peninsula may also have some connection with the Atlantic mountain ranges, and that the most uplifted part lies near the Adirondacks, consequently at quite a distance west of the iceshed at Quebec. The fact that Newfoundland, which at least during the last extension of the glaciers may have been only locally glaciated, also shared in the submergence may in some degree be accounted for by its geological structure.

All the above statements concerning the late glacial upheaval are based upon the height to which the marine deposits are uplifted, but as we generally cannot tell whether this rising of the land has been continuous or partly counteracted by subsidence, it would be more correct to speak of it as the final result of the changes of level since the maximum of the late glacial submergence.

Along certain parts of the Atlantic coast many facts were ob-

served long since which show that these tracts in very modern
times have been and perhaps still are sinking, and it is of interest that these signs of subsidence are found along the Atlantic
coast plain outside of the glacial region of uplifting as well as
somewhat within its boundary, just as has been the case in
Scandinavia.   Thus submarine peat-bogs are known in New
Jersey and Nantucket Island as well as at the northeastern end
of the Bay of Fundy and at the mouth of Bay Chaleur. These
last localities show that if the rising of land is still going on in the
interior, the isobase for zero, or, to use Shaler's expression, the
pivot point between the continental upheaval and the oceanic
subsidence, has moved at least more than fifty miles toward the
land side. The amount of this subsidence is not yet known,
but at the Bay of Fundy it must have been at least 40 feet and at
Nantucket probably 10 feet. Even the numerous small buried glacial river valleys at the southern shores of Cape Cod, Nantucket,
Martha's Vineyard, and Long Island afford evidence of a slight
submergence. The same is the case, as Merrill has pointed out,
with the Hudson River estuary, which must have subsided
somewhat since the channel was cut out of the glacial clays in
the valley.

Another question is whether the deep submarine river valley
southeast of New York harbor, so well described by Professor
Dana, belongs to so late a period. The fact that its upper end
down to a depth of about 100 feet has been entirely filled up at
the outside of Sandy Hook seems to indicate that the Hudson
River leveled the adjacent part of the pre-existing channel
during the maximum of the post-glacial elevation, having its
mouth here for a considerable time. The other analogous submarine channel described by Dana from the north side of Long
Island may perhaps afford a possibility of determining their
age. Having crossed Long Island Sound in an oblique direction,
it becomes during the last 10 miles more and more shallow, ending abruptly at Long Island against the terminal moraine. Here
it may be possible to ascertain with a few borings, whether the
channel, as it appears, has been overridden by the moraines of
the last glaciation, and perhaps also whether it is younger than
those of the first glaciation.

Though the abrupt ending of this last channel is very likely
due to the terminal moraine, which, to judge from Dana's obser-

vations, has not quite filled it up, yet there appears to be no continuation of it on the other side of Long Island, even beyond the glacial deltas.

This curious fiord-like shoaling of the submarine channel, before it reaches the edge of the continental plateau, is repeated by the submerged river channels described by A. Lindenkohl from the Delaware and Chesapeake Bays.

This phenomenon might perhaps be explained according to T. F. Jamieson's suggestion for certain fiords, as a consequence of the unequal and intense subsidence of an iceloaded continent.

But concerning these channels, as well as the one described by Chalmers in the St. John River estuary and the large channels reported by Spencer from St. Lawrence Bay and several other places, we must allow that at present we know very little indeed of their history and precise age, with perhaps the general exception that they may point to a high elevation in early glacial time.

In this connection it is of interest that in America as well as in Europe the interglacial marine deposits at present accessible above the sea level are found only near the margin and at the outside of the region which during the last glaciation has been exposed to subsidence. I refer here to the interesting fossiliferous deposits described by Shaler, Upham, and others from Nantucket, Martha's Vineyard, Long Island, and Boston. The question whether the Columbia formation belongs to the same subsidence cannot safely be discussed, before its marine origin is conclusively shown by fossils, boundary shore-lines, or other indisputable evidence. The same is true of the lower beds of sand and clay about 50 feet thick which Lyell in the report of his first voyage to North America describes from Beauport near Quebec. The clay with boulders, which he observed resting upon these beds and covered by fossiliferous, late glacial deposits, is as I could myself ascertain a true till, probably belonging to the last glaciation.

Though the situation of these possibly interglacial deposits is open to the St. Lawrence estuary, their marine origin is very questionable, since no fossils have been found.

But if it is difficult to get any idea of the interglacial geoid deformations from the marine deposits, it is still more so with respect to the scanty remnants of lake sediments. As compared with these the buried river channels seem to be easier to trace, though of course affording less accurate information.

In this connection and as possibly pertaining to the general interglacial hydrography of the Great Lake basin, I may perhaps mention the common occurrence of waterworn pebbles in the drumlins west of Syracuse as these` may very likely be derived from buried shore-lines belonging to the same interglacial lake as the interesting deposits east of Toronto.

Finally I will emphasize, that the purpose of this paper is much less to give an ultimate solution of the different complex problems connected with the continental changes of level, than to show a way by which, I think, such a solution can be reached with as little loss of time as possible.

From the details already determined in North America as well as in Europe, it is evident that the changes of level are closely connected with the local structure of the earth's crust and with the local extension of the glaciations; and thus it is conclusively shown that no changes whatever in the level of the sea can account for the phenomenon.

Notwithstanding all doubts as to the possibility of vertical uplifts of the great continental portions of the earth's crust, we may already be fully justified to use about this with a new meaning the well known words of Galilei: " Yet it does move."

Editor's Comments on Paper 4

Goldthwait: *A Reconstruction of Water Planes of the Extinct Glacial Lakes in the Lake Michigan Basin*

In 1907 J. W. Goldthwait published his classic study entitled "The abandoned shorelines of eastern Wisconsin." In a paper published in 1908, this survey is expanded to include the area between Lake Michigan and Lake Huron. The paper is a classic in terms of the detailed field observations, the presentation of these data, and the interpretation of the shoreline diagrams. Goldthwait's paper marks, in many ways, the culmination of work on the glacial lake shorelines of the Glacial Great Lakes that commenced in the late nineteenth century. Although later work added details and certain interpretations have been challenged, the reconstructions of these shorelines has not changed dramatically over the last 60 years. Of particular note is the concise procedure outlined for drawing strandline or equidistant diagrams (p. 468). Note that de Geer (Paper 3) uses the word "isobase" to refer to metachronous marine limits, whereas Goldthwait's use suggests a restriction to synchronous features.

The paper begins by describing the features on which the evidence is based, mainly raised beach ridges and wave-cut terraces, and their relationship to former water planes. The heights of the beaches above datum were measured with a spirit level. Goldthwait defines and discusses the lake shoreline complexes related to three stages of Lake Chicago and the Lake Algonquin beaches. A major argument that he advances is the need for a "hinge line," or zone of structural weakness, because of the manner in which several beaches converge to the same geographical area. Implicit in Goldthwait's discussion is the suggestion that the shorelines are formed of individually tilted segments, each with a specific gradient. Later work, as well as the argument advanced by Robinson in 1908 in the *Journal of Geology*, indicates, however, that the shorelines of these ancient lakes have been warped, with the gradient continuously increasing toward the former ice margin. This fact is well shown in his Fig. 3, where the spacing on the reconstructed Lake Algonquin water plane increases markedly toward N 5° to 15°E. The Lake Algonquin beach as defined by Goldthwait (plus modifications) is used by Broecker, Walcott, and Brotchie and Silvester (see their papers in this volume) to model crustal response to ice loading and unloading.

_____

[1]*Wisconsin Geological and Natural History Survey, Bull.* **17**, Sci. Ser. No. 5, 134 pp.

Reprinted from *J. Geol.*, **16**, 459–476 (1908)

4

A RECONSTRUCTION OF WATER PLANES OF THE EXTINCT GLACIAL LAKES IN THE LAKE MICHIGAN BASIN[1]

JAMES WALTER GOLDTHWAIT
Dartmouth College

CONTENTS

The Raised Beaches.
Method of Investigation.
The Shore Lines of Lake Chicago.
The Algonquin Beach.
Construction of a Profile of Water Planes.
Significance of the Fan-like Profile.

THE RAISED BEACHES

Around the borders of Lake Michigan are many fragments of abandoned shore lines, which stand at different heights above the present lake. They mark a series of stages of extinct lakes of late glacial times, known as Lake Chicago, Lake Algonquin, and the Nipissing Great Lakes. Near the south end of the lake, where erosion has been slight and the accumulation of shore drift has been going on since the earliest times, there is a record of nearly all the stages through which the lake has passed. On both the east and west sides of the lake, however, in Michigan and Wisconsin, where the cutting back of cliffs at the present level of Lake Michigan has been vigorous, the old shore lines have been partly or wholly destroyed for stretches of five to twenty-five miles. Even where the present lake has not cut away the record, it is usually incomplete, because the higher beaches have been destroyed by cliff recession during the lower of the extinct stages. It follows that the old shore lines preserved at one locality do not necessarily correspond with those at a neighboring locality, either in number or in order. The matter of correlation is not a very simple one; the highest beach at a given locality may not correspond with the highest at a neighboring locality, even though it may be less than a mile away. Moreover, while the

[1] With the permission of the Director of the U. S. Geological Survey.

beaches around the south end of Lake Michigan are still horizontal, having been undisturbed by earth movements since they were formed, those in the more northerly portions have been affected by repeated differential uplifts. Each beach rises northward at a rate different from those above and below it, and at a rate which increases repeatedly in a northward direction. At one or more points, the planes marked by the inclined beaches split, vertically, so that a single stage in the southern part of the lake represents fifteen or twenty stages in the northern part. This is the result of the repeated tiltings of the northern district. The problem of proper correlation of the fragments, then, and of the complete reconstruction of the old water planes is a very difficult one. Not only must as many of these fragments as possible be discovered, but at each locality every beach and terrace of the series must be noted, its strength and peculiar characters recorded, and its altitude measured with all possible precision.

The raised beaches about the south end of Lake Michigan have been described in detail by Leverett,[1] Alden,[2] and others. The beaches along the west side of the lake, in eastern Wisconsin, were first studied in detail by the present writer,[3] in 1905. The planes which were recognized there have since been traced farther north in the upper peninsula of Michigan, by Hobbs.[4] On the east side of the lake, Taylor and Leverett have for several years been accumulating detailed information concerning the beaches. It was with the purpose of supplementing this work by a series of more detailed and precise measurements, and thus establishing more definitely the identity of certain beaches, and their relations, that the writer, under Mr. Taylor's direction, undertook a six weeks' survey of the shore lines along the east side of Lake Michigan in July and August, 1907,

[1] Frank Leverett, "The Illinois Glacial Lobe," *U. S. Geol. Surv.* (Monog. XXXVIII), 1899; also earlier papers (see *op. cit.*, p. 419).

[2] W. C. Aiden, "Chicago Folio," Geologic Atlas of U. S., *U. S. Geol. Surv.*, Folio 81, 1902; "The Delavan Lobe of the Lake Michigan Glacier of the Wisconsin Stage of Glaciation, and Associated Phenomena," *U. S. Geol. Surv.* (Prof. Paper 34), 1904; "Milwaukee Special Folio," Geologic Atlas of U. S., *U. S. Geol. Surv.*, Folio 140. 1906.

[3] J. W. Goldthwait, "Correlation of the Raised Beaches on the West Side of Lake Michigan," *Jour. Geol.*, Vol. XIV, pp. 421–24, 1906. "Abandoned Shore-Lines of Eastern Wisconsin," *Wis. Geol. & Nat. Hist. Surv.* (Bull. xvii), 1907.

[4] For the Mich. Geol. Surv. Results not yet published.

for the U. S. Geological Survey. The method of measurement and of assembling the data secured in this study is the same which had been used in eastern Wisconsin in 1905. It was unnecessary in this case, however, to spend much time in exploration; for Mr. Taylor had selected a large number of localities where measurements could be made most advantageously. Considering the shortness of the season, therefore, the field covered was a large one, and the results obtained were unusually complete.

The old water planes, or imaginary surfaces of the extinct lakes, are marked by a variety of shore features. One type which is common on both the past and present shores is the cut bluff and bench. As developed along the present shore of Lake Michigan, the steeply sloping bluff or cliff rises from the water's edge, while the bench or terrace at its base reaches out under water. The point at the base of the bluff or the top of the bench is approximately the highest point at which erosion by storm waves is effective. It is usually a little above the normal lake level. Where bedrock cliffs instead of clay bluffs form the coast, however, the bench is perhaps likely to be a little lower than lake level. There is a constructional variation here of a few feet, but of only a few feet, in the case of Lake Michigan. In making measurements on such a bench, to determine the altitude of the old water plane, the base of the bluff was always taken. It is the only determinable point that one can take as a standard. Care was taken, of course, in making these measurements, to avoid places where the bench had been built up by landslides, or by alluvial wash down the face of the bluff, or where it had been gullied by streams. A range of error of five feet would probably be quite enough in this region to allow for discordances of benches due to original constructional variation in height.

Quite a different feature of topography is the beach or beach ridge —a line of shore drift banked up by the waves at or close to the water's edge, and rising only as high as storm waves can fling material. In exposed places on the shore of Lake Michigan, beaches have been observed whose crests stand fully six feet above calm water level. As a rule, however, they stand only three or four feet above it. In rare cases, where local conditions favor a heavy surf, the beaches probably attain a height of eight or even ten feet. The crest of a beach, however, is the only point which one can take as a criterion for measuring a water plane; so in the study of these raised beaches it is the crest that has been measured each time. Care was taken to make sure of the presence of gravel on the surface of these beaches, in order to eliminate the effects of wind-blown sand, which often raises a beach, locally. If we allow a range of five feet for original variation in height of beach crests, we have probably satisfied all discordances except those which can be recognized as due to peculiar local conditions.

Other varieties of the beach need scarcely be mentioned, such as the bar or bàrrier, built between headlands, in comparatively deep water. Its height is liable to be more extreme on that account. The object in the foregoing remarks is to show that the points selected for measurement (the base of a bluff, or the crest of a beach or barrier) were chosen for convenience, not to say from necessity, and that an original constructional variation in height of about five feet is fully recognized.

METHOD OF INVESTIGATION

In the measurements the spirit- or Y-level was used almost exclusively, and to decided advantage. While much has been accomplished with the hand level and aneroid, in skilful hands, neither of these instruments has the accuracy or reliability of the Y-level. The influence of weather conditions, especially lake breezes, on the aneroid, and the personal equation in the hand level are liable to cause mistakes in such work as this. Where there is a whole series of shore lines to be measured in one locality and these beaches and benches follow one another in short vertical intervals, all possible accuracy in measurement is needed to correlate them, individually with members of a similar series at a neighboring locality. The Y-level, then, is almost indispensable for work in the central and southern portions of the Great Lake region, and desirable in all parts of it. Upham and Tyrrell used the Y-level[1] in measurements of altitude of the raised beaches of Lake Agassiz before 1890, Spencer[2] used it in

[1] Warren Upham, "The Glacial Lake Agassiz," *U. S. Geol. Surv.* (Monog., XXV), p. 9, 1896.

[2] J. W. Spencer, "Deformation of the Algonquin Beach and Birth of Lake Huron," *Am. Journ. Sci.* Vol. cxli, pp. 12–21, 1891, and other papers.

Ontario at about the same time. Lawson[1] used it along the north coast of Lake Superior in 1891; but between that time and 1905, the hand level and aneroid were very generally used, instead; and accordingly the correlation and identity of the beaches of the Lake Michigan basin were imperfectly known.

Absolute accuracy is of course impossible even with the Y-level. On the one hand are original variations in height of the beaches and benches, due to local conditions under which they were constructed or cut; for which five feet has been allowed. On the other hand a certain amount of error is involved in the process of leveling; first, through the slight inaccuracy in the use of the instruments, and second, through the use of Lake Michigan as the datum or starting-point. It frequently happened, on days when levels had to be run, that a strong on-shore wind was blowing and the waves were running high, so that one could not tell within half a foot what the normal level of the lake would be at that place. To be quite fair, then, we must expect in these measurements occasional discordances due to a combination of these errors of six feet or so.

THE SHORE LINES OF LAKE CHICAGO

The old shore lines fall into two distinct groups. There is an earlier group, well registered around the south part of Lake Michigan, but unknown in the northern part. These belong to the so-called Lake Chicago, a lake which was confined to the Michigan basin, with its outlet at the extreme southwest corner, into the Desplaines valley at Chicago. A later group, represented by but a single shore line in the southern part of the basin, below the beaches of Lake Chicago, but rising to a considerable height northward, and splitting into a large vertical series, records the complex history of Lake Algonquin and the Nipissing Great Lakes.

The shore lines of Lake Chicago, as distinguished at the south end of the basin, and described by Leverett and Alden, mark three distinct stages, to which the names Glenwood (or 60-foot), Calumet (or 40-foot) and Toleston (or 20-foot) stages have been given. To be more exact, the average altitudes of these three shore lines are 55,

[1] A. C. Lawson, "Sketch of the Coastal Topography of the North Shore of Lake Superior, with Special Reference to the Abandoned Strands of Lake Warren," *Minn. Geol. Surv.*, 20th Ann. Rept. p. 231.

38, and 23 feet above Lake Michigan, or approximately 636, 619, and 604 feet above sea level.   In the following table, the altitudes of these beaches is given for six selected localities, where spirit-level measurements have been made by the writer.

Locality	Miles North of Chicago	Glenwood	Calumet	Toleston
Evanston and Niles Center, Ill...........	15	636′	619′	605′
Zion City, Ill.........................	43	634′	621′	x
State Line, Ill., and Wis...............	46	634′	616′	x
Line between Racine and Kenosha counties, Wis.........................	57	637′	621′	...
Holland, Mich........................	65	638′	621′	605′
Spring Lake and Eastmanville, Mich.....	87	633′	(613′)	602′

It will be noticed that the first four localities are on the west side of Lake Michigan.   The last two are on the east side.   Spring Lake and Eastmanville are near Grand Haven.   The symbol "x" in the table indicates that the shore line is missing because of destructive cliff cutting at a lower stage.   Other measurements might be given, less accurate than these, but confirming them, almost without exception.   They indicate that as far north as a line through Grand Haven, Mich., and Milwaukee, Wis., the three beaches of Lake Chicago are horizontal.   The terrace at Eastmanville, given at 613′ in the table above, is so obscure a feature that its exact altitude cannot be estimated within several feet.   Disregarding this one measurement, then, the data show a remarkable accordance in the altitude of beaches, the variation being not more than five feet for each stage.   That much ariation, as has already been remarked, must be expected in the original construction of the beaches.   There is no indication of system in the slight departures from uniformity of height from north to south, hence no reason to suppose that any of these water planes are inclined at all south of Grand Haven and Milwaukee.   Apparently, then, these beaches, representing the earliest stages of the lakes of late glacial times have been unaffected by any of the earth movements which are known to have deformed the central and northern portions of the Great Lake region since the ice withdrew.

On the accompanying map (Fig. 1) the northern limit of horizontality for these beaches of Lake Chicago is indicated by a line;

and this line is extended east and southeast through Lake St. Clair and Ashtabula, Ohio, so as to mark the corresponding limit for the beaches of Lakes Maumee and Whittlesey in the Erie basin, as determined by Leverett and Taylor. South of this line, then, there seems to have been no subsequent deformation in either the Michigan or the Erie basins.

Beyond Grand Haven, on the east side of the lake, the data recently collected at several localities, as far north as Ludington,

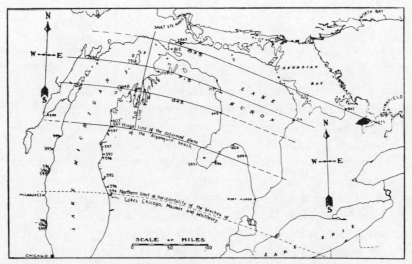

Fig. 1.—Map of Lakes Michigan and Huron, showing northern limit of horizontality of the beaches of Lakes Chicago, Maumee and Whittlesey, altitudes of the Algonquin beach at selected localities, and isobases of deformation and line of maximum inclination of the warped Algonquin water plane.

indicates that the beaches rise northward and increase in number; but the precise correlation has not been possible for several reasons. The measurements are as follows:

Locality	Distinct Beaches at
Muskegon	604', 609', 613'–616', 628'
Montague.	632', 634', 656', 658.
Bass Lake	628'–630', 639', 642', 649'–650'.
Ludington and Amber	636'–640', 673'–675'.

The fragments north of Grand Haven are scarce and generally obscure, becoming sandy and irregular as they go north. It appears as if they were approaching the ice border for those stages, and fading away among the moraines. Those beaches which locally show strong development cannot be grouped into a single system of diverging planes like the Algonquin beaches presently to be described. The rate of inclination from south to north is hardly one foot per mile. It is possible, then, that the southward slant of these beaches is due wholly to ice attraction, according to the calculations of R. S. Woodward.[1] Assuming an ice sheet of reasonable extent and thickness, Woodward found that its attraction might be sufficient to raise the surface of the lake nearby into a curve, concave upward, with an apparent inclination of several inches to the mile for the first fifty miles or so from the ice border. If this be the right explanation for the Lake Chicago beaches north of Grand Haven, it accounts for the apparent lack of harmony between the measurements; for the relation of successive water planes to each other would be similar to the branchings of a feather (see Fig. 2) and the few points where

Fig. 2.—Diagram showing how a series of ice-attracted water planes might look, in profile.

measurements have been made might happen to lie on several different divergent surfaces. If, on the other hand, the northward ascent of the beaches is due to a series of differential uplifts of the region, these uplifts must have occurred before the formation of the next lower beach, the "Algonquin" beach, for that beach is horizontal over the whole southern half of the Michigan basin.

North of Ludington little is known of these beaches, and no measurements have been made. While the Glenwood and Calumet shore lines might expectedly terminate at any place, against a moraine, the Toleston or lowest shore line of Lake Chicago ought to extend northward to the place where the ice border first uncovered a low pass leading across Michigan into the adjoining Huron basin, probably east of Little Traverse Bay. No traces of such a beach are known. It is probable that the record is too weak and obscure to be recognized.

[1] R. S. Woodward, "On the Form and Position of the Sea Level," *U. S. Geol. Surv.* (Bull. 48. p. 88), 1888.

THE ALGONQUIN BEACH

Below the beaches of Lake Chicago, encircling the south end of Lake Michigan, is the Algonquin beach. It rises northward, in the central and northern part of the basin, as shown by the map, Fig. 1. The identity of this beach, the highest shore line on Mackinac Island, as the "Algonquin" beach[1] of Spencer was long ago recognized by Taylor. On the map, Fig. 1, can be seen the altitude of this Algonquin beach at about thirty-five selected localities in the Huron and Michigan basins. The warped attitude of the water plane which passes through these points is indicated by the system of isobases and the line of maximum inclination which runs perpendicular to them. The data have been taken from several sources. The measurement on the Garden peninsula (725') is one of many made by Hobbs in 1907. The twenty or more remaining measurements around Lake Michigan and the Straits of Mackinac were made by the present writer, in company with F. B. Taylor, in 1907. On the west side of Lake Huron the measurements were made by Frank Leverett, A. C. Lane, W. M. Gregory, W. F. Cooper, and C. A. Davis. East of Lake Huron the measurements are all Spencer's except the one at Beaverton, which was recently made by Taylor.

Recent investigations by Taylor in Ontario, supplementing earlier studies, indicate that this beach marks a period of activity of *two* outlets, one at Port Huron and one east of Kirkfield, Ontario, where there was an overflow into the valley of the Trent River. This Algonquin beach, of the "two-outlet" stage, seems to be the highest beach common to the Huron and Michigan basins. This gives reason to conclude that when the ice border south of the Straits of Mackinac withdrew so as to let the waters of the Michigan basin merge with those of the Huron basin, the "Trent outlet" was already running. Had the lakes merged before the Trent pass was uncovered and while the Port Huron outlet alone was active, then the plane of that beach, adjusted to the Port Huron outlet, would have been temporarily abandoned when the Trent pass was uncovered, and the waters fell to a low level; and the subsequent uplifts which raised the

[1] J. W. Spencer, "Notes on the Origin and History of the Great Lakes of North America" (abstract), *Am. Assoc. Adv. Sci., Proc.*, Vol. XXXVII, pp. 197-99, 1889, and later papers.

Trent pass up to the level of the one at Port Huron would have raised this old plane (within this region of deformation) well above any subsequent level of the waters. No such plane above the Algonquin beach of the "two-outlet stage" is known, unless it be the Toleston beach. As already noted, this beach is lost, north of Ludington, and no beach is known in the Huron basin which connects with this in the northern part of the lower peninsula of Michigan.

The altitudes of the Algonquin beach at about fifty localities in the northern part of the Michigan basin is shown in Fig. 3. This is a miniature copy of a large scale chart (composed of sheets of the U. S. Lake Survey) on which the data collected in 1905 and 1907 were plotted for final inspection. Through and among these points, a series of isobases was drawn with a vertical interval of ten feet. These were found to be parallel to each other, and to be in harmony with the similar lines across the Huron basin, as shown in Fig. 1. Across the isobases was then drawn a line in the direction of maximum inclination—a gentle curve that changes gradually from N. 15° E. near the Straits of Mackinac to N. 5° E. south of Grand Traverse Bay. With this map as a basis for locating points, a profile of the Algonquin beach and the complex series of shore lines below it was then drawn, as follows: Upon the line of maximum inclination was plotted the position of each station on the east side of the lake where measurements had been made. Each one was then transferred directly to a sheet of co-ordinate paper, on which distances from left to right represented distances from south to north. A horizontal line at the base was taken to represent the present level of Lake Michigan (581.5 feet A. T. in July and August, 1907). With a vertical scale of 20 feet to the inch (500 times as large as the horizontal) the altitude of every beach and bench was recorded by an ordinate. These ordinates served to reconstruct the water plane in profile. A reduced copy of this profile constitutes Plate I. In it, between Hessel and Onekama, a distance of 125 miles, over 25 different lines of levels at as many localities along the east side of Lake Michigan are represented. In other words, the stations are on the average 5 miles apart. About 190 measurements are on well-formed beaches or benches, and

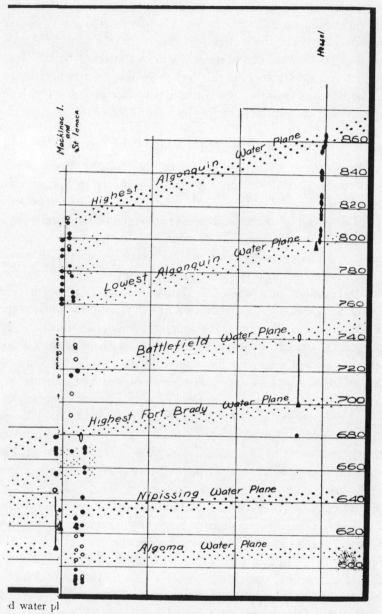

about 50 on beaches or benches that are less distinct, though not to be neglected in a fair consideration of evidence. A distinctive symbol is used for the two contrasted types of topography, benches and beaches; and the symbol for a well-formed beach or bench is different from the one for a beach or bench that is faint. The size of the spot is intended to cover the probable range of error in measurement. The height of a cut bluff is shown by a wriggling vertical line.

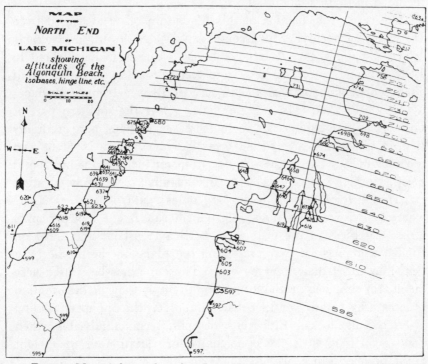

FIG. 3.—Map of the north end of Lake Michigan, showing the warped attitude of the Algonquin water plane.

Through the highest beach thus recorded a line or band was drawn with a thickness (according to the scale) of 6 feet, to represent the range of variation in height to be expected in the original construction of the beach or bench. The line or band, when drawn as a gentle curve, so as to pass most directly through the highest ordinates, includes 21 out of 24 of them. Of the 3 which are either too high or too low, to fall within it, one at Carp Lake, fully 8 feet too low, is easily accounted for. It is a beach ridge on a height of land, where no opportunity exists for the record of a higher stage. The other two discordant points, one at Petoskey and one at Leland, are five feet below and 4 feet above the center of the band, respectively, instead of being within 3 feet of it. This discordance of one or two feet is not a serious one; for it might be due to an error in leveling which augments an original variation in height, instead of being included in that variation.

The rate of inclination of this Algonquin plane measured from Hessel (at the ancient "Munuscong Islands") southward to Mackinac Island (15 miles) is 3.73 feet per mile. From Mackinac Island to Beaver Island (24 miles) it is 3.30 feet per mile. A rather rapid change of inclination near the isobase of Beaver Island introduces a tilt rate of about 2.00 feet per mile. Over the southern part of Grand Traverse Bay the rate again decreases, perhaps more gradually than in the former case, so that near Traverse City it is about 1.00 foot per mile. The further change from an incline to a horizontal position, which is accomplished near Onekama, seems to be a rather rapid one; for Onekama is only 25 miles south of Traverse City, where, as we have just remarked, the tilt rate is one foot per mile. The abruptness of the changes near Beaver Island and Onekam could be emphasized by representing the plane on the profile by a bent line rather than a curved one. The curve has been used here merely for simplicity, without meaning to imply that the deformation is necessarily a warping rather than an uplift by the tipping or jostling of large fault blocks. Either sort of uplift seems admissable, when due weight is given to the opportunity for variation in the height of the beaches or benches.

Below the highest shore line are a number of others. Some of them are equal to the Algonquin in strength; others are comparatively faint. They can best be recognized in the northern part of the region, where the vertical space between them is greater. On Mackinac Island, for instance, strong benches and beaches record ten lower stages of considerable importance. The Algonquin beach is the highest of a closely spaced group of ridges which are well displayed on the short target range back of Fort Mackinac. This "Algonquin group" of beaches occupies an interval at that place of about 50 feet.

The lowest of them and two intermediate ones are especially prominent. Forty feet below the Algonquins, and separated from them by an interval in which there are no plainly developed beaches or benches, is a beach of remarkable strength, called by Taylor the "Battlefield beach." It was so named because of its conspicuous development as a great ridge of cobblestones and gravel on the old battlefield, on the north slope of the island. Below the Battlefield beach another interval of 40 feet, unoccupied by any very persistent beaches, leads down to the group known as the "Fort Brady beaches." There are several of these ridges. Four of them fill a vertical interval of 25 feet; below them are at least two others of which we have a distinct record. A little below the Fort Brady beaches is the "Nipissing shore line," one of very remarkable strength. It is probably the most conspicuous of all the shore lines, and it is peculiar in consisting usually of a bench and bluff rather than a beach ridge. On Mackinac Island it stands 53 feet above Lake Huron, being represented there, however, by a great deep-water barrier which runs southwest from the Episcopal church. Below the Nipissing is one stage of importance, marked by the "Algoma shore line." At the Straits of Mackinac this stands about 25 feet above the lake.

All these shore lines and groups of shore lines—the Algonquin group, Battlefield beach, Fort Brady beaches, Nipissing shore line, and Algoma shore line—can be traced southward from the Straits of Mackinac, on this profile (Pl. I), with certainty for at least twenty miles, to Beaver Island, where they are all represented. But as they are followed further southward the record of them is found to become more and more imperfect and incomplete. The planes gradually converge until the discrimination between them becomes difficult; and what is more troublesome, in the development of cliffs at the lower stages (especially the Nipissing and the present stage) many of the higher beaches have been cut away and no record of them is left. The exact position of the planes between the Algonquin and the Nipissing, therefore, cannot be absolutely demonstrated, though a reasonable amount of confidence is placed in the reconstruction here given. In the case of the Nipissing shore line, however, there is no uncertainty. Its exceptional strength and peculiar character make it possible to follow this plane southward, down its gentle inclination

of 0.75 feet per mile, diminishing to less than 0.50 feet per mile near
Beaver Island, to Onekama, where it becomes horizontal and seems
to unite with the Algonquin plane to form the single 596-foot plane
already mentioned. While the data for the intermediate planes do not
permit an unqualified statement, they seem to indicate that the lowest
of the Algonquins, the Battlefield, and the Fort Brady beaches all converge to the same point, Onekama, instead of being overlapped,
one after another, by the Nipissing. The Algoma plane seems to
constitute another member of this split series; but its exact position
south of Petoskey is somewhat in doubt.

It is perhaps possible that the highest Algonquin beach becomes
horizontal at about 24 feet above Lake Michigan, near Herring Lake.
This is suggested by the work of the present writer in eastern Wisconsin, and by the data thus far collected by those who have worked
in the Huron basin. If so, the 14-foot "Nipissing shore line" seems
to have very generally destroyed the 24-foot "Algonquin shore line"
along the east side of Lake Michigan, south of Herring Lake. The
evidence here seems rather to indicate that the Algonquin and
Nipissing shore lines coincide to form the single 14-foot shore line.
Figures 1, 3, and 5, and Plate I, embody this idea.

SIGNIFICANCE OF THE FAN-LIKE PROFILE

The steepness of inclination of these water planes, and their convergence to a single point, affords a basis for choosing between differential uplifts and ice attraction, to explain their present condition.
At the Straits of Mackinac the calculated rate of inclination for the
highest Algonquin is 3.73 feet per mile; for the lowest of the Algonquin group, 3.00 feet; for the Battlefield, 2.10 feet; for the highest
Fort Brady, 1.29 feet; for the Nipissing, 0.75 feet; and for the Algoma,
0.33 feet. All those above the Nipissing slant too steeply to be
accounted for by ice attraction. They must be explained by a series
of earth movements which repeatedly raised the shore line of this
region out of water. The Nipissing and the Algoma shore lines are
inclined no more steeply than the surface of the lake close to the ice
border might be inclined by ice attraction; but they are probably
inclined more steeply than they could be so far from the ice border
for the stages they represent. The ice front must have been at least

200 miles to the northeast of this region; for it had withdrawn from the Mattawa valley, east of North Bay (see Fig. 1), and at this distance its attraction on the waters would hardly have raised the lake surface more than an inch or two to the mile, according to Woodward's computations. Furthermore, the water planes come together at a single point. They should converge in turn to a series of points, feather fashion, if they marked the attraction of the lake to successive positions of the retreating ice front (see Fig. 2). Evidently, then, the inclined position of the planes, and their fan-like relation, must be attributed to earth movements.

The line from which the planes diverge (the "hinge line" on the map, Fig. 1), or as seen on the profile (Plate I) the point at which they split, might be determined, it seems, in either of two ways. It might mark the southern limit of a series of deformations, acting thus as a hinge on which the tiltings took place. If so, it is evident that here were no less than ten or fifteen distinct tilting or warping movements, all of which hinged on the same line. On the other hand, the point of splitting might be located at an outlet (or along the line of equal deformation through an outlet), which together with the region north of it had been raised by tiltings; for as the outlet rose, the horizontal surface of the lake would rise, south of it, drowning the old shore lines there, while to the north the former shore lines would be raised out of water each time and would come to form a fan-like series. This process is illustrated diagrammatically by Fig. 4.

The choice between these two explanations for the case at hand may be quickly made, if one considers the information shown on the map (Fig. 1). There are two outlets, only, which could possibly be associated with the splitting, viz., the Kirkfield and the Port Huron outlets. The isobase through the former (if it were drawn an 875-foot line, parallel to the 835-foot isobase on the map) would pass nowhere near Onekama, but rather through the upper peninsula of Michigan. That is a district as yet not critically examined. The Port Huron pass lies south of the "no tilt" line (or "hinge line" as it is called in Fig. 1); that is, it lies within the district which has been unaffected by uplifts; consequently no fan-like splitting can occur at it or in line with it. It seems necessary to conclude that Onekama lies on the hinge line of the tiltings which raised all the planes into their inclined positions. That so many tiltings, separated no doubt by considerable intervals of time, should have had the same hinge suggests that this line, for reasons unknown, is one of structural weakness.

In order to show the significance of the fan-like profile of water planes, in Plate I the following series of diagrams (Fig. 5) is introduced. These present in a very conventional and much simplified way the relation which the successive planes of Lake Algonquin and the Nipissing Great Lakes should bear to each other according to the present generally accepted interpretation of the history of the great Lakes, worked out by Mr. Taylor.   Since that history is

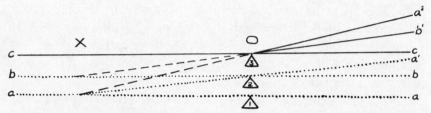

FIG. 4.—Diagram showing in profile how a fan-like group of water planes might be produced by a number of differential uplifts which tilted a lake basin and its outlet. Outlet at *O*. Uplifts affect region to the right of *X*. Three stages are shown. First stage, outlet at *1;* horizontal water plane of the lake at *aa*. Second stage, tilting on right side of *X* has raised outlet to *2;* water plane *aa* has been inclined to *aa¹;* the lake has risen to plane *bb*, drowning that part of it which lies on left side of outlet. Third stage, another uplift has raised outlet to *3;* has tilted *bb* to *bb¹*, and increased tilt of *aa¹* to *aa²;* lake has risen to *cc;* on left side of outlet planes *a* and *b* have been drowned; on right side, they rise, splitting at outlet, fan-fashion.

recognized as subject to revision, through further study, the diagrams should be taken to represent simply the conditions which seem most probable, in the light of the evidence already at hand. Although in the actual case the planes have been warped to curved profiles, as shown in Plate I, they are represented in the diagrams as simply tilted. This is wholly for the sake of simplicity, and must not be thought to imply that the crustal movements were actually tiltings instead of warpings.

The final diagram (6, in Fig. 5) shows the present position of these planes. That portion which lies south of the Trent outlet is based on recent detailed work in the Michigan and Huron basins, including that described in this paper. The large profile, Plate I, corresponds with that part of Diagram 6. The remaining part, north of the Trent outlet, covers a large field in which studies have been less detailed and the reconstruction on that account cannot be regarded

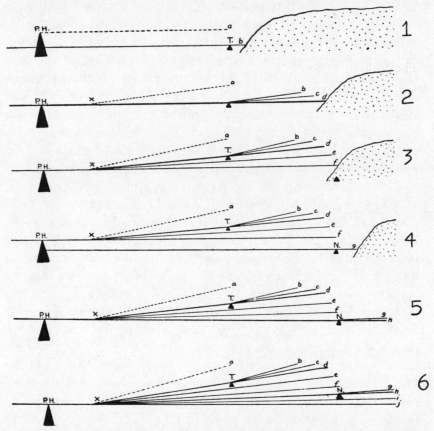

FIG. 5.—Diagrams showing how the water planes of Lake Algonquin and the Nipissing Great Lakes should be related, as seen in profile, if the generally accepted history of the lakes is correct. 1, 2, 3, 4, 5, 6, six selected stages in the lake history during the retreat of the ice border and the differential uplifts. X, the point on which the differential uplifts hinged, Onekama. The triangles represent outlets, the controlling point being the apex. *P. H.*, Port Huron outlet. *T*, Trent outlet. *N*, Nipissing outlet. *a, b, c, d, e*, etc., successive water planes of the lakes, in the order of their age. *a*, early Algonquin beach, unknown in the Michigan basin, but theoretically necessary in the Huron basin, *b, c*, planes of temporary low water stages through the Trent outlet; *d*, plane of the "Algonquin beach;" *e*, plane of the Battlefield beach *f*, plane of a Fort Brady beach; *g*, plane of a temporary low-water stage through the Nipissing pass, possibly marine submergence; *h*, plane of the Nipissing shore line; *i*, plane of the Algoma shore line; *j*, present plane of the Lake Michigan.

as actually established.   From this series of diagrams it may be seen
that (1) the "Algonquin" beach of the Michigan basin marks the
"two outlet" stage when the discharge was shared between the pass
at Port Huron and that at Kirkfield.   Before it there had been a
"low-water" stage, adjusted to the Trent (i. e., Kirkfield) outlet
in its original low position; but earth movements had lifted the outlet,
and consequently the lake level south of it to the level of the Port
Huron pass.   The records of this low water stage (southern continu-
ations of planes *b* and *c*) have been drowned in the Michigan and
Huron basins.   The lower members of the group of Lake Algonquin
beaches, the Battlefield and Fort Brady beaches, represent successive
stages after the restoration of the Port Huron outlet (planes *e* and *f*).
Each seems to record a differential uplift in which the northern part
of the region was raised out of water, from a hinge near Onekama.
The Nipissing plane (*h*) represents, like the Algonquin, a "two-outlet"
stage, with the discharge divided between Port Huron and North
Bay.   Previous to it the low Nipissing pass east of North Bay had
been opened and a second low water stage had occurred.   The data
thus far collected seem to admit a possibility that this was a stage of
temporary marine submergence; but further careful studies and
more accurate measurements in that most critical field will be needed
before the truth is known.   With repeated uplifts, the waters of the
Huron and Michigan basins rose from this low level, the "low water"
beaches were drowned (except north of the Nipissing pass), and the
Port Huron outlet was re-established.   Since then the Nipissing
plane has been raised far out of water.   The Algoma beach (*i*) seems
to mark the most important pause in this period of comparatively
recent uplift.

Editor's Comments on Paper 5

Wright: *The Quaternary Ice Age*

Wright's book is dedicated to Jamieson, "originator of the isostatic theory of the Quaternary oscillations of sea level" (frontispiece). In many ways *The Quaternary Ice Age* is a remarkable book and it presents a first-class study of the prevailing ideas concerning many aspects of the Quaternary, particularly for studies of postglacial sea level changes. Wright very clearly recognized the dual nature of the late- and postglacial sea level changes, affected as they were by both the isostatic recovery of the land and the rise of world-wide sea level caused by deglaciation. These ideas are set forth in the chapter entitled "The isostatic theory of the Quaternary oscillations of sea-level" (Chapter XVIII), where he develops the isokinetic theory for sea level changes.

I have selected three different segments of Wright's book, namely, those dealing with results from Scandinavia, from North America, and his summary chapter (Chapter XVIII) on the isostatic theory.

Scandinavia

The first selection is a remarkable testimony to the vigor of Scandinavian research on Quaternary marine shorelines. The map of the Littorina Sea shown by Wright should be compared with recent determinations (see map in Flint, 1971, p. 351, or paper by Donner, this volume); it will be found that the agreement is remarkably good. The map of the Yoldia Sea shorelines is quite different; current ideas indicate that a significant part of the ice cap covered the central region of the Baltic.

North America

Wright's analysis of the North American data indicates a considerable difference between that continent and northern Europe in terms of scientific direction. In North

America very little work had been carried out along the coastal areas, but great attention had been paid to the shorelines and the history of the great interior lakes. In contrast, the northern European evidence was almost completely biased toward the raised marine deposits. A comparison between Section 1 and this one are most instructive.

The Isostatic Theory

This selection first discusses and dismisses the old bugaboo of the attraction of the water plane by the ice mass itself. It then proceeds with a general defense of the theory and concludes by showing in graphic form (Figs. 148, 149, and 150) Wright's ideas about the relative movement of land and sea required to produce the features and deposits of northwestern Europe.

Reprinted from *The Quaternary Ice Age*, 326–333, 387–390, 416–420 (1914)
By permission of Macmillan (London and Basingstoke)

5

CHAPTER XV

THE LATE QUATERNARY OSCILLATIONS OF
LEVEL IN FENNOSCANDIA

The Yoldia Sea—Rise of the sea in the wake of the retreating glaciers—
The deposits of the Christiania Valley—The Yoldia maximum of
submergence—The isobases of the late-glacial depression—The late-
glacial recovery—Wave-like progress of the uplift—The post-glacial
shell-banks of the Christiania region—The Tapes submergence—Its
northern limit in the Christiania region—De Geer's estimate of late-
glacial and post-glacial time—History of the Baltic—The Baltic Ice
Lakes and the Zanichellia Sea—The distribution of Yoldia in the
Yoldia Sea—*Idothea entomon*—The marginal terraces of Vaberg—
The Ancylus Lake—The Ancylus transgression in Gotland—The
climate of Ancylus time—The late Ancylus emergence—The Littorina
submergence—Significance of the submergence of the late Ancylus
outlet—Munthe's researches on the salinity of the Littorina Sea—
Isobases of the Littorina Sea—Identity of the Littorina and Tapes
stages—The kitchen-middens and the Campignien immigration—
Identity of Campignien time with the Littorina-Tapes maximum—
Brögger's correlation of the stages of retreat with the sub-stages of
the Neolithic Period.

THE honours of post-glacial geology rest undoubtedly
with the Scandinavians. Their elaborate and careful
investigations into the recent geographical developments
of their country have made them justly famous. Living
in the midst of problems of an interest perhaps unequalled
in any other region, they have attacked them with a
scientific spirit worthy of the highest admiration, and
evolved from them a story which is one of the romances
of modern geology. The researches of Munthe, De Geer,
and Holst into the history of the Baltic, and Brögger's
investigations in the Christiania region have a world-
wide fame, and no less widely known is the work of

326

Nathorst and Gunnar Andersson on the succession of floras in the post-glacial beds of Sweden.

When the great European Ice-sheet withdrew finally towards its centre of distribution east of the Scandinavian mountains, it left the north-west of Europe under very different conditions from those which prevail at present. Denmark, the southern low-lying parts of Sweden and Norway, and most of the coastal lands of *Late glacial submergence of N.W. Europe.*

FIG. 111.—*Portlandia (Yoldia) arctica* and *Arca glacialis* from the Late Glacial Marine Clays of Norway.   (After Brögger.)

the Baltic, were submerged beneath the sea. A great open arm of the Atlantic stretched at that time into the basin of the Baltic, and extended over the Russian plain, joining in all probability with the Arctic Ocean by way of the White Sea. Raised beaches and stratified clays with marine shells are found up to considerable heights in many parts of the area. The shells are those of molluscs quite different from such as are now found *The Yoldia Sea.*

FIG. 112.—*Yoldia hyperborea* from the Late Glacial Marine Clays of Norway.   (After Brögger.)

living round the coasts of Sweden and Norway, and more closely allied to forms at present confined to the colder parts of the Arctic Ocean. One of them, *Yoldia arctica*, Gray, now only lives in sea-water at a temperature below 0° centigrade, and thrives in the muddy waters discharged at the mouths of glacier streams. Its remains are abundant in the deposits of the sea which occupied the basin of the Baltic in late-glacial times.

This has in consequence come to be known as the Yoldia Sea. A little crustacean, *Idothea entomon*, L., found along with Yoldia in these deposits, is also a characteristic form of the Arctic Seas. Unlike the molluscs which accompanied it, it has not become extinct in the Baltic basin, but persists in a stunted form in that sea itself and in some inland lakes formerly connected with it.

Glacial character of the Yoldia Sea.

The deposits of the Yoldia Sea show every trace of the presence of floating ice. The beds are often contorted and disturbed as if by its grounding. That the disturbances were contemporaneously produced is proved by the fact that the disturbed beds are often overlain by others absolutely undisturbed. Boulders are often found embedded in the sediments as if dropped in from icebergs or floe-ice.[1]

Relation of the sea to the retreating glaciers.

The exact relations of the sea-level to the retreating glaciers have been worked out in detail by Professor Brögger [2] in the Christiania region. He has shown that during the first stages of the retreat, when the outer moraines were deposited, the land became submerged beneath the sea as the glaciers retired. This submergence continued for some time, and then a reverse movement set in, and from that on the land emerged continually. We will best exhibit the relations by tracing in succession the stages of retreat of the ice-margin, and the position of the sea-level in regard to them.

The moraines of the Christiania Valley, taken in order from without inwards, are :

The outer ra-moraine.
The inner ra-moraine.
The moraine between the inner ra and the epiglacial moraine.
The epiglacial moraine.
The moraines between the epiglacial moraine and the uppermost moraine.
The uppermost moraine.

---

[1] See Gerard de Geer, " Om Skandinaviens Geografiska Utveckling efter Istiden," *Sveriges Geologiska Undersökning*, Ser. C. No. 161a.

[2] W. C. Brögger, " Om de Senglaciale og Post-glaciale Nivåforandringar i Kristianiafeltet," *Norges Geologiska Undersgelse*, No. 31 (1901).

The retreat of the ice from the epiglacial moraine signalised the commencement of the emergence of the land which continued into post-glacial times, and brought to an end the so-called Christiania Period which was characterised throughout by progressive submergence.

The initial period of progressive submergence.

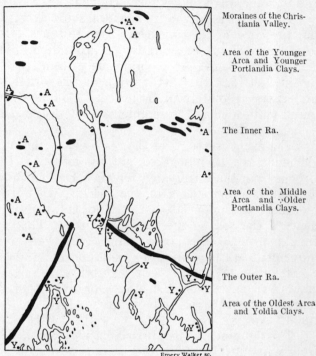

Moraines of the Christiania Valley.

Area of the Younger Arca and Younger Portlandia Clays.

The Inner Ra.

Area of the Middle Arca and Older Portlandia Clays.

The Outer Ra.

Area of the Oldest Arca and Yoldia Clays.

Emery Walker sc.

FIG. 113.—Map of the Christiania Fjord, showing the successive moraines and the intervals on which were deposited the various marine clays. Localities of the Yoldia Clay are marked Y, and of the Arca or Portlandia Clays, A. (After Brögger.)

That the ras of Christiania are not terminal moraines in the sense of marking the extreme limit of advance in a glacial period, is considered by Brögger to be proved by the fact that the country outside and inside them does not differ, the striae on both sides being equally fresh and in the same direction. They merely mark a halt in the retreat of the glaciers of the great Ice Age.

The Older Yoldia Clay.

Resting on the till and glaciated surfaces outside the outer ra-moraine are stratified marine clays, from which have been obtained twenty-six species of mollusca, all high arctic and nearly all living in the Kara Sea. This is the Older Yoldia Clay. It is deposited against the outer slope of the ra-moraine, and is never found inside it. Its altitude in the neighbourhood of the ra is 40 to 50 metres, but it also occurs right down to sea-level. Now the typical and common species obtained from it live in abundance in the Arctic Seas only at depths of from 10 to 30 metres. Deeper water species have never been found in it. It is clear, therefore, that the vertical range of the fossil shells in the Christiania region is greater than their possible vertical range on the assumption of a constant sea-level. From the occurrence of the clay at an altitude of 40 to 50 metres, it may be concluded that the sea stood at 50 to 80 metres above its present level, whereas from its occurrence at sea-level the figures 10 to 30 metres are arrived at. The only way of explaining these apparently contradictory results is to assume that at the beginning of Older Yoldia times the sea stood at the lower level (say 20 metres), and that it gradually rose until it attained the higher level (say 65 metres).

Change of sea-level during (1) the Older Yoldia Clay ;

But we have evidence of still further submergence before the glaciers abandoned the outer ra, for resting upon the Older Yoldia Clay outside the moraine, but never inside, are found two other clay deposits, the Younger Yoldia Clay and the Oldest Arca Clay. These occur in direct superposition thus :

> Oldest Arca Clay,
> Younger Yoldia Clay,
> Older Yoldia Clay,

and they pass imperceptibly one into the other.

(2) the Younger Yoldia Clay ;

In the Younger Yoldia Clay, which is only 1 to 2 metres thick, many of the arctic species are of reduced size, indicating an amelioration of climate. They also indicate a greater depth of water, namely, 40 to 60 metres.

The sea was therefore still rising, but the characteristic deep-water forms of the succeeding Arca Clay were not yet able to exist in it.

The Oldest Arca Clay contains a fauna which is almost quite distinct from that of the Yoldia Clay, only a few forms being common to the two. It is essentially a deep-water fauna, indicating some 80 to 100 metres of water just outside the moraine. It shows that the sea-level must at this time have attained an altitude of about 135 metres above its present level. (3) the Oldest Arca Clay;

The glaciers now abandoned the outer ra and receded some 25 kilometres to the inner ra, leaving a clean space between for the reception of subsequent marine deposits.

On the fresh striated surfaces of this space were deposited the Middle Arca and Older Portlandia Clays. Their fauna indicates an amelioration of climate. *Arca glacialis*, for example, is of smaller size than in the Lower Arca Clay. The Middle Arca Clay is never found inside the inner ra. (4) the Middle Arca and Older Portlandia Clays;

A further recession of the glaciers now took place. They abandoned the inner ra and retreated to the moraines of the Christiania Valley, leaving a stretch of country as before for the reception of marine sediments.

On this were deposited the Younger Arca and the Younger Portlandia Clays. The former is a deep-sea clay occupying the lower parts of the valley up to about 100 to 130 metres above the sea, and the latter a shallow-water clay in the higher parts at some 100 to 175 metres above sea-level. The fauna of the Younger Arca Clay is considerably less arctic than those of the preceding clays. (5) the Younger Arca and Younger Portlandia Clays.

On the area left bare by the further recession of the glaciers to the epiglacial moraines south of the large lakes of Central Norway masses of clay and sand were deposited, but only in rare instances do they contain marine mollusca. Inside the epiglacial moraines no marine strata occur. The sea, in fact, at this stage ceased to rise in the wake of the retreating glaciers, and there set in, in the Christiania region, an almost con- The epiglacial moraines the limit of marine deposits.

tinuous emergence of the land which has gone on up to the present day.

Let us consider now what general ideas we have arrived at. We have seen that throughout a great part of Scandinavia and Russia in late-glacial times the sea stood at a considerably higher level relatively to the land than at present, and Brögger has established in the Christiania region the all-important fact that it rose as the glaciers retreated. That is to say, that the land, on the departure of the ice, did not find itself in the maximum state of submergence and rise therefrom in consequence of the removal of the weight of the glaciers. The relation is by no means so simple as this. On the contrary, the submergence increased at first as the glaciers diminished, and it was not until a fairly advanced stage of the retreat that the elevation, which appears to have followed upon the disappearance of the ice, first made itself felt. We shall return to this point when discussing the more theoretical aspects of the question of oscillations of sea-level during Quaternary times.

Extent of the submergence. Throughout the whole of Scandinavia and the basin of the Baltic this late-glacial submergence has left its marks. All along the north-west coast of Norway are found terraces and strand-lines up to considerable heights on the land.[1] In some places several terraces can be seen one above the other, marking different levels of the sea, and it is a matter of some difficulty to correlate these from place to place. It is not even safe to conclude that the highest terraces in different districts are quite contemporaneous. Nevertheless, by drawing a series of isobases for the highest terraces a fairly accurate idea of the nature of the late-glacial and post-glacial warping is obtained. These isobases are lines of equal depression of the land, with regard to the present sea-level, in the

[1] See Gerard de Geer, "Om Skandinaviens Geografiska Utveckling efter Istiden," *Sveriges Geol. Unders.*, Ser. C. No. 161a (1896). Rekstad and Vogt, "Sondre Helgelands Kvartär Geologi," *Norges Geol. Unders.*, No. 29, p. 62 (1900). Helland, "Strandliniernas Fald," *ibid.* No. 28, part 2 (1900). V. Tanner, "Studier öfver kvartärsystemet i Fennoskandias nordliga delar," *Fennia*, vol. xxiii. No. 3 (1906), and vol. xxvi. No. 1 (1907).

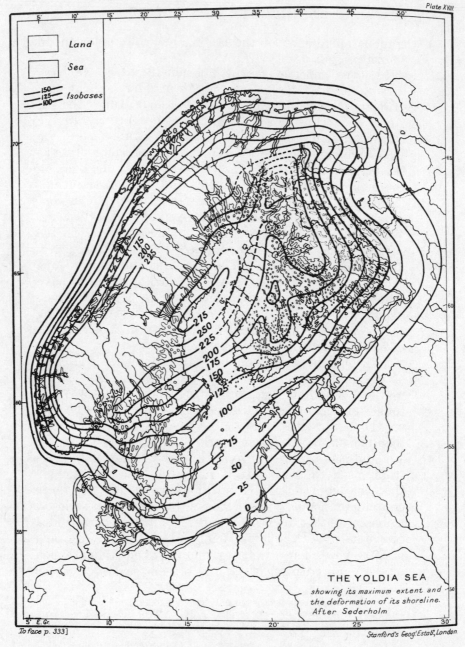

Plate XVIII

Land

Sea

150
125 Isobases
100

175
200
225

275
250
225
200
175
150
125
100

75

50

25

0

THE YOLDIA SEA
showing its maximum extent and
the deformation of its shoreline.
After Sederholm

5′ E. Gr.

To face p. 333]

Stanford's Geog¹ Estab¹, London

London : Macmillan & Co., Ltd.

86

late-glacial period of submergence.   They are constructed
by measuring the altitudes above sea-level of the highest
terraces in various places and joining up the localities
where the altitude is the same.   When drawn they are
found to form sweeping curves round about the central
portion of Scandinavia, as shown in Plate XVIII.   It is
at once apparent from the figure that the depression was Greater
least in the outlying districts and increased towards the depression
centre of the affected area.   That is to say, for example, areas.
that if we were to visit some of the outlying islands off
the Norwegian coast we would find the highest terraces
on them no very great height above present sea-level,
but if we then proceeded to the coast of the mainland
they would there lie at a somewhat greater height, and
this height would increase as we passed up the fjords
into the interior of the country.   The general result
may be stated thus.   In late-glacial times an immense
oval area with its centre in the Baltic basin, east of the
Scandinavian mountains, was depressed to a considerably lower level than that at which it lies at the present
day.   It can hardly be a mere coincidence that the Coincid-
central area of this depression was the great centre of depression
dispersion of the European Ice-sheet.   This has given with
rise to the idea that the depression was caused by the dispersal.
weight of the ice, and that on the removal of the latter
the land has tended to rise once more to its former level.
This theory does not, however, explain the period of
progressive submergence which has been proved by
Brögger in the Christiania region to have preceded the
elevation.   We must ask the reader to assume for the
present that this was due to a general rise of the sea-
level, the absolute necessity of which, as a result of the
returning to the ocean of the immense quantities of
water bound up in the ice-sheets, we shall demonstrate
later.

* * * * * * *

CHAPTER XVII

THE LATE GLACIAL CHANGES OF LEVEL IN
NORTH AMERICA

The Champlain Epoch—Its late-glacial age—The ice-dammed lakes of
the Great Lakes region—Tilting of their shore-lines—The Erie-Ontario
basin—Lakes Maumee, Whittlesey, Warren, and Iroquois—Northern
limit of horizontality of the shore-lines of Lakes Maumee and
Whittlesey — The Michigan-Superior-Huron basin — Lake Chicago
—Northern limit of horizontality—The Algonquin shore-line—Lake
Nipissing—The Onekama hinge-line—The effect of outlets on the
northward splitting and tilting of the shore-lines—Lake Agassiz—
Gilbert's measurement of the recent tilting in the Great Lakes.

In the two preceding chapters some account has been
given of the remarkable oscillations of level which
affected Scandinavia, the British Isles, and the northwest of Europe generally during late-glacial and postglacial times.   Canada and the United States underwent
similar changes, the records of which are preserved to us
in the elevated marine sediments along the coast, and
the ancient warped and tilted shore-lines around the
Great Lakes.   Unfortunately, in the study of their postglacial marine sediments the Americans are far behind,
but they have carried out a series of researches on the
shore-lines around their lakes, which rival in interest the
magnificent results attained by the Scandinavians.

From New York northwards along the American The
coast there are preserved here and there very extensive Champlain
deposits of marine clay and old sea beaches at consider- deposits.
able heights above present tide level.   These remnants
of submergence vary greatly in altitude, but the upper

387

limit rises, as far as is known, fairly steadily towards the north. The period of formation of these deposits is generally known as the Champlain Epoch, from Lake Champlain, around which they are typically developed. An arm of the sea at this stage extended up the St. Lawrence Valley to Lake Ontario. It filled the basin of Lake Champlain, and probably connected southward with the sea at New York by a narrow strait extending along the Hudson Valley. Marine shells and bones of the whale and other marine animals have been obtained from these deposits. They are found around Lake Champlain at altitudes varying from 400 feet or less about the south end of the lake to 500 feet at the north end, and also at an altitude of 600 feet near the east end of Lake Ontario. Deposits of this stage are known in Connecticut, but they have been best studied in Maine, where they have a well-defined upper limit of about 230 feet above the sea, and have yielded many marine fossils, including species of Mya, Astarte, Leda, and Yoldia. G. H. Stone has given an excellent account of the Maine deposits, and demonstrated very clearly their intimate connexion with the glaciation. He points out that the kames and eskers on coming down to the 230-foot level spread out into great gravel deltas, the material of which is found to become finer and finer at greater distances from the old shore-line, passing first into sand and ultimately into marine clay. There appears to be little doubt that the sea stood at the 230-foot level while the ice-sheet still covered the State of Maine, and that the subglacial streams in many cases discharged directly into the sea.[1]

Even less is known about the Champlain deposits of Canada than about those of the United States. Bell[2] has described post-glacial deposits with marine fossils which occur up to elevations of about 600 feet above

[1] G. H. Stone, "Classification of the Glacial Sediments of Maine," *American Journal of Science*, 3rd Ser. vol. xl. p. 122 (1890) ; and " The Glacial Gravels of Maine," *Monograph xxxiv.* U.S. Geol. Survey (1899).

[2] R. Bell, " Proofs of the Rising of the Land around Hudson Bay," *American Journal of Science*, 4th Ser. vol. i. p. 219 (1896).

St. James' Bay, and traces of post-glacial submergence are reported at still greater heights in Labrador.[1]

We will now turn to the much more satisfactory question of the late-glacial and post-glacial warping of the Great Lakes region. The principal contributions towards its solution have been made by Spencer, Gilbert, Warren Upham, Leverett, Taylor, and Goldthwait. The main upshot of these studies has been to prove that immediately after the retreat of the ice the whole region was subjected to a slow and probably intermittent differential uplift, which was greatest in the north and died away gradually southwards. This is shown by the study of the tilting of the shore-lines of the glacially dammed lakes. As the ice withdrew northward, a series of immense lakes were held up between its margin and the high ground to the south, draining over the lowest passes. On the further retreat of the ice-margin these coalesced with one another, and lower outlets becoming available the surface of the water was reduced at intervals to lower levels. A whole series of shore-lines was thus produced, each of which shows a different degree of tilting, the oldest invariably rising quickest when traced northwards. This uplift, which was going on more or less continuously during the whole complex history of the lakes, has the appearance, however, of having had a somewhat intermittent character, for individual shore-lines corresponding to a definite outlet are often found when traced northward to split up into two, three, four, or more subsidiary shore-lines, all distinct from one another, and evidently corresponding to pauses in the uplift. In order to give a better conception of the evidence on which this splendid generalisation is based, and to bring out certain refinements which recent researches have made possible, an attempt will now be made to trace in detail the history of

The Great Lakes region.

Tilted shore-lines of the ice-dammed lakes.

[1] For general account of Champlain deposits, and further references, see Chamberlin and Salisbury, *Geology*, vol. iii. p. 399 (1906). See also Gerard De Geer, "On Pleistocene Changes of Level in Eastern North America," *Proc. Boston Soc. Nat. Hist.* vol. xxv. p. 454 (1892) ; and Warren Upham, "The Glacial Lake Agassiz," p. 504, *Mon. xxv.* U.S.G.S. (1895).

the different basins in which the successive lakes were formed.

The Erie-Ontario Basin

The earliest of the glacial lakes held up in the Erie

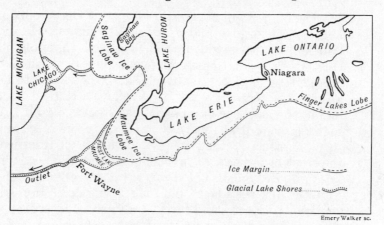

FIG. 134.—Map of First Lake Maumee and contemporary glaciation. (After Leverett and Taylor.)

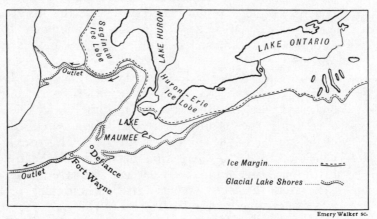

FIG. 135.—Map of Lake Maumee at its greatest extent when it possessed two outlets. (After Leverett and Taylor.)

Lake
Maumee. Basin was Lake Maumee.

* * * * * * *

The reader will now no doubt have a clear conception both of the nature and approximate magnitude of the two factors with which it is proposed to explain the Quaternary oscillations of sea-level. One of these factors, the depression of the sea-level, has been shown to be an absolutely necessary result of glaciation. The other factor, the isostatic depression, has attached to it a very high degree of probability, since, in all of the cases known to us in which a temporary load has been imposed on the crust of the earth, the phenomenon of warped shore-lines is observable. The theory is not merely an inference from the observed facts, but an *a priori* deduction which is shown to meet these facts at almost every point. In this respect it has a distinct advantage over any attempt to explain the facts by deformations of the earth's crust of the nature of folding and faulting. The various regions will now be considered in detail, and an attempt made to bring out clearly the points for which the theory affords an explanation as well as those which are still obscure.

Cause of submerged forest oscillation.
Now, in order to understand the complicated relations of the land and sea in Northern Europe during the Quaternary, it is important in the first place to realise thoroughly the nature of the oscillation which allowed of the growth of the submerged forests. This oscillation, which occurred in the period between the Yoldia and Littorina Seas, and also between the late-glacial and early Neolithic submergences in Scotland (see p. 377), was responsible for the cutting off of the Baltic from the ocean, whereby it was converted into the Ancylus Lake. In the fact that the submerged forests have been proved over a wide area, almost the whole of North-western Europe, they differ essentially from the raised beaches

which are grouped round the glacial centres. The wide-spread character of the oscillation, which appears also to have affected North America (see p. 425), is good ground for assuming that it was caused by a general lowering of sea-level. This, according to our theory, must have been brought about by a recrudescence of some one of the ice-sheets, possibly that of the Antarctic Continent. When this ice-sheet melted again the sea rose once more, and appears to have been rising slowly since. Now in thinking of the effects of this oscillation we must conceive of it as referred to an ideal sphere of unalterable radius concentric with the earth and not to

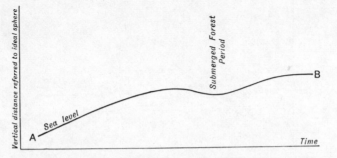

FIG. 148.—Curve showing the supposed oscillation of sea-level in north-western Europe since the retreat of the ice-sheet.

the land, which, within the depressed areas, was moving itself. Referred to this ideal sphere the motion of the sea-level in glacial and post-glacial time passed through the following phases—

(1) Low sea-level during Glacial Period.
(2) Rise owing to melting of ice-sheets.
(3) Fall owing to recrudescence of some ice-sheet (submerged forests).
(4) Rise owing to remelting of same ice-sheet.

This motion of the sea is expressed as a curve in Fig. 148, in which the ordinates represent vertical heights, and the horizontal distances time. This curve represents correctly the motion of the sea with respect

2 E

to the land outside the areas affected by the isostatic depression, but not within them. Within these areas we have to consider the effect of the rise of the land, which was at first slow, then more rapid, and finally slower again (see p. 412). In Figs. 149 and 150 this motion is represented by the line CD, the rise being, of course, greater and the curve steeper in the central areas, the history of which is that of the northern Christiania district north of the isobase through the Mjösensee (see p. 338). The line AB in these figures represents the motion of the sea transferred from Fig. 148, both

Combined motions of land and sea.

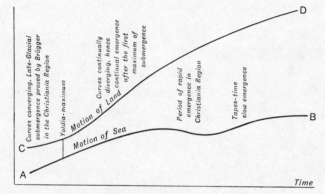

FIG. 149.—Curves showing the absolute and relative motions of the land and sea in the central isostatically affected areas since the retreat of the ice.

motions being, of course, referred to the ideal sphere just mentioned. Now a moment's thought will make it clear that the change in the vertical distance between the curves AB and CD represents the relative motion of the land and sea. Where the curves are diverging the land is emerging; where they are converging the land is becoming submerged; where they are parallel the relative level of land and sea is undergoing no change, and there is in consequence a tendency towards the formation of well-marked shore-lines.

Central Scandinavian district.

Let us consider first the central area which is represented in Fig. 149. On the left of the figure the isostatic recovery has not got properly started, while the rise of the sea-level is in full swing. The curves are approaching one another. This represents the submergence proved by Brögger to have followed, in the Christiania region, immediately upon the retreat of the ice-margin (p. 330). Soon, however, the curves become parallel. This is the Yoldia maximum. They then diverge for a while slowly, representing slow emergence. This is the period of deposition of the late-glacial shell-banks, representing 0-40 per cent of the uplift (p. 334). There follows then for a time a rapid divergence, representing

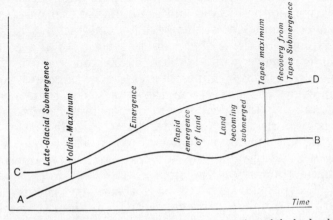

Fig. 150.—Curves showing the absolute and relative motions of the land and sea in the marginal isostatically affected areas since the retreat of the ice.

rapid emergence of the land. This is the period which in the Christiania region is represented by very few shell-banks, *i.e.* 40-66 per cent of the uplift (see p. 336). Next there comes an interval of slow divergence of the curves, representing slow emergence. This is represented by the upper Tapes banks, corresponding to 60-85 per cent of the uplift (p. 334). Finally, the emergence quickens up a little in the time of the lower Tapes banks, 85 to 100 per cent of the uplift.

Now let us turn to Fig. 150, representing the history of the peripheral district, south of the Mjösensee zero line in the Christiania region. The story is the same until the maximum of the submerged-forest depression of sea- Peripheral district.

level is passed.   There then comes a difference, however, for the curves are converging.   This represents the Littorina-Tapes submergence of the peripheral regions. The curves soon cease to converge and become parallel; this is the Tapes maximum.   They then diverge slightly, representing the recovery from the Tapes submergence.

The two districts compared.

Now the only difference between this figure and the last is in the gentler slope of the upper curve, representing the necessarily smaller and slower elevation of the land in the peripheral region of the isostatically depressed area, yet this introduces an important difference in the relative motion of land and sea.   In the first case there is a continuous, though irregular, emergence after the maximum of the Yoldia submergence.   In the second this emergence is interrupted by a submergence.   The so-called Littorina-Tapes depression is therefore not a depression of the crust in the true sense, that is with reference to the ideal sphere of constant radius.   The crust was rising all the time, and the shore-line of the Littorina-Tapes maximum was cut at the period when the sea was rising at the same rate and so became for a while stationary on the land.

The peculiar merit of this theory is that it starts with two simple *a priori* conceptions, both in themselves inherently probable and supported by evidence in other areas, and by combining them explains an extremely complex series of changes in the relative level of land and sea.

Editor's Comments on Paper 6

Nansen: *The Strandflat and Isostasy*

This great monograph should be glanced at by all interested in glacial-isostasy. For this volume I have selected a number of pages dealing with various aspects of the problem, based both on the theoretical viewpoint and on field data.

Field Data

The first extract presents a discussion and table of the gradient of the late glacial shorelines. Such figures provide an initial basis for regional comparisons of the degree of glacio-isostatic deformation. Steep gradients suggest either a relatively thin lithosphere or rapid deglaciation. Rates of tilting in Norway are quite similar to those in a recent table presented by Andrews and Barnett (*Geogr. Ann.* 1972) for tilts from marine and glacial lake shorelines from around the former Laurentide Ice Sheet. Nansen also shows (p. 259–262, not reproduced) the evidence that supported the development of the relations diagram.

Isostasy

Nansen gives, in this second excerpt, an unusually clear exposition of the principal of isostasy as it was understood at the time. He discusses the degree to which the earth can attain perfect isostatic compensation and notes that Gilbert and others believed that the rigidity of the crust pemits only partial compensation over some critical regions. Although Nansen is aware of questions concerning the plasticity of the lithosphere and vertical density gradients, he states: "A more natural way may be to examine as closely as possible what has actually happened at this surface (lithosphere), and from this we may draw some conclusions as to the state of the deeper strata." Although we have a better grasp of these problems today than in 1921, nevertheless, Nansen's statement is still applicable; more and better field obsrvations are required for adequate geophysical modeling based on glacio-isostatic recovery data.

The Norwegian strandflat, an extensive area of planed bedrock at 10–17 m above sea level and considered to be of marine origin, is used by Nansen to demonstrate the crust returns to its former level after glacial loading and unloading. On his page

301 Nansen illustrates his concept of glacial-isostasy. The ice cap rests on a rigid lithosphere overlying a viscous fluid. Note, in particular, his concept of a rising wave around the ice cap which will sink upon deglaciation as material moves into the central area to compensate for the deficiency of mass. This is a clear reference to what we now call the "forebulge." Nansen notes that in coastal Norway, equilibrium has been achieved in approximately 11,000 years.

Reprinted from Fridtjof Nanses, "The Strandflat and Isostasy," *Det Norske Videnskaps-Akademi i Oslo, Series: Mathematics and Natural Science*, **11**, 251–253, 297–304 (1921)

6

Gradient of the Lateglacial and Postglaciel Upheaval of the West and North Coast of Fenno-Scandia.

A general impression of the postglacial upheaval of the land may be obtained by comparing the gradients of this upheaval in the different regions of the coast of Norway and of the Kola Peninsula as given in the following table. The values of the gradients are computed from the heights of the highest terraces and raised beaches which are supposed to indicate the upper limit of lateglacial submergence. The heights were measured by the authors given in the last column. The heights of the lower level of the strandflat have not been deducted from the figures used for the computation of these values.

In several regions it is difficult to decide which raised terraces and beaches actually indicate the upper limit of lateglacial submergence. In the outer coastal regions of Finmark Tanner [1906, 1907] has found raised shore-lines and terraces situated considerably higher than those which have been used for the computation of the gradients in our table. Similar higher shore-lines have also been found at several levels by Grønlie [1914] in the Tromsø distrikt and by W. Ramsay [1898] on the Kola Peninsula. These shore-lines, however, are less distinct than the lower shore-lines and have a much older appearance. They do not seem to correspond to the shore-lines which are supposed to mark the upper marine limit (the upper limit of lateglacial submergence) further south in Norway, and it is difficult to understand that they can have been formed during the last glacial period. They may more probably be survivals from a previous period of submergence, as is assumed by Ramsay and Grønlie.

For my computations I have therefore used the heights of the shore-lines and terraces belonging to the level which Tanner calls $I\varepsilon$, as this level seems to correspond to the upper level of the two conspicuous raised shore-lines of the Tromsø—Hammerfest region and Vesterålen, and to the generally accepted upper limit of the lateglacial submergence farther south in Norway.

These gradients given in the table differ so much locally that one might doubt their correctness; but on the whole there seems to be a certain system in their variations. If they be introduced in a map of the Norwegian coast where the submerged continental shelf is also outlined [cf.

99

Locality	Gradient of Elevation in *per mille*	Observer
Northern Österdal	0.7 — 1.6	G. Holmsen [1917]
Christiania Fjord	ca. 0.7	Øyen
Coast between Larvik and Christiansand	0.68	Danielsen [1912]
Hardanger Fjord, outer part	ca. 0.78	Rekstad [1906]
— — central part.................	0.68	— —
— — Halsenøi to Rosendal	0.94	— —
Sogne Fjord, between Vik and Vadheim	1.0	— [1910]
Sønd Fjord, outer part	1.0	— [1906]
— inner part	0.9	— —
Nord Fjord, coastal zone 25 km..............	0.0	Kaldhol [1912]
— middle 36 kilometres	1.2	— —
— inner 32 —	2.0 (?)	— —
Sondmør, Hardeidland	1.2	Rekstad [1905]
Coastal region northeast of Ålesund	1.1	— [1906]
Nordmør, Reinsvik (Christiansund) to Tingvoll ..	1.48	Kaldhol [1913]
— Reinsvik—Vågbø (Halsa Fjord)	2.00	— —
— Reinsvik—Bruset (Todal)	1.32	— —
— Tingvoll—Bruset	1.23	— —
— Henden (Arisvik Fjord)—Valsøibotten..	1.87	— —
Southern Helgeland, average	0.73	Rekstad [Vogt 1900]
— — region of Dønna	0.67	— [1904]
Dunderland Valley to Træna	0.91	— [Vogt 1907]
Mainland to Lofoten	0.76	J. H. L. Vogt [1907]
Ofoten to Kvæ Fjord	0.71	Vogt [Rekstad 1905]
Region between Andøi and Senjen	0.90	Helland [1900] [1]
Region of Senjen	1.22	— —
— between Senjen and Ringvasøi	0.85	— —
— — Ringvasøi and Skjervøi	0.90	— —
Fjord south of Hammerfest	0.67	R. Chambers [1850]
Porsanger Fjord	0.60	Tanner [1907]
Lakse Fjord	0.48	— —
Tana Fjord	0.58	— [1906]
Varanger Peninsula	0.51	— —
Region south of Varanger Fjord	0.60	— — [2]
Region of Ribachi (Fisker) Peninsula	0.53	Ramsay [1898]
— » Kildin	0.65	— —
— » Woronye River, Kola Penins.	0.72	— —
Southeastern coast of Kola Penins.	0.32	— —

[1] O. T. Grønlie gives in his text [1914, p. 224] values (between 1.35 and 1.77 per mille) of these gradients for the region between Andøi and Skjervøi, which are much higher than those computed from Helland's observations; but as far as I can see Grønlie must have made some strange mistakes in his distances. His map of the isobases gives much the same gradients as Helland's observations.

[2] As will be mentioned later (p. 263) it seems to me that Tanner has probably mistaken his level *Iε* in this region, and I have used his highest terraces for the computation of the gradient.

Nansen, 1904, Pl. XI], it is seen that, as a rule, the steepest gradients occur in those regions where the edge of the continental shelf is nearest to the outer coast of the land, *e. g.* in the region of Senjen and along the coast from Nordmør to Nord Fjord (cf. Fig. 166). In Sønd Fjord it is also steep (1.0 per mille) although this region is farther away from the edge of the continental shelf. But here the deep outer part of the submeged Norwegian channel is very near the outer coast, while this channel is shallower outside the region of Hardanger Fjord where the gradient is less steep (0.7—0.8 per mille).

The lowest gradients occur along the south-east coast of Norway (0.68 per mille), in Helgeland (0.67—0.73 per mille), in the region from Hammerfest to the Varanger Peninsula (0.48—0.60 per mille), and on the Kola Peninsula. These regions are farthest away from the outer edge of the submerged continental shelf.

It is also noteworthy that the gradient seems to increase somewhat towards regions where there are deep submerged fjords on the continental shelf outside the coast, *e. g.* outside Træna [cf. Nansen, 1904, Pl. XI], where the gradient was found to be 0.9 per mille, while it was 0.67 and 0.73 per mille[1] in the region of Dønna and Vega to the south, where the shelf outside is broader and less dissected.

As A. G. Høgbom has pointed out to me in a letter, the investigations of several geologists obviously show that there may be appreciable local variations in the upheaval of the land.

By careful levelling of a very distinct raised beach (from the Tapes period), formed of boulders, along the coast of Lake Venern in Sweden, R. Sandegren [1916] has found that along a distance of about 11 kilometres, from the region of Kleven (height 65.2 metres above the sea) to the region north of Otterbäcken (70 metres above the sea), the gradient of upheaval is 0.44 per mille, and along the next 19.5 kilometres to the NNE, between the region north of Otterbäcken to the region east of Vall in Visnum (height 92.1 metres above the sea), the gradient is 1.13 per mille. Hence, in the latter region, the land has been elevated about 13.5 metres more than it would have been, if the gradient of upheaval had been uniformly 0.44 per mille along the whole distance of 30.5 kilometres.

By his measurements of the levels of the ancient beaches left by the ice-dammed lakes in Northern Österdal, Gunnar Holmsen [1915, 1916, 1917] has found that the gabbro mountains may have had an influence upon the postglacial upheaval of the land which has taken place in this region. The isobases are deflected by, and go round, the intrusive masses of gabbro, and it looks as if the latter have risen somewhat more than the surrounding regions, and the gradient in these regions may thus locally be increased from 0.70 per mille to between 0.85 and 1.63 per mille.

[1] J. H. L. Vogt [1907, p. 30], however, remarks that according to later investigations by A. Hoel, these values are probable somewhat too low.

How are the Isostatic Vertical Movements of the Lithosphere effected?

Pentti Eskola has recently [1920] put forward some ideas regarding transformations in the chemical composition of minerals and rocks effected by changes in pressure, which, if correct, may prove to be of great importance when the above question comes to be answered.

He draws attention to the fact that pressure influences the nature of minerals "by moving the equilibria towards the associations and modifications which have the smallest volume", and he points out that certain rocks, belonging to what he calls the "eclogite facies", which have probably been suddenly pressed up from very great depths, are all of them very heavy with a specific gravity of about 3.3 to 3.5, and contain heavy minerals like garnet and diamond. These rocks are unstable at the pressure at which they are now found near the surface, and have, as it were, been "quenched" by the sudden reduction of the pressure. If the pressure had been reduced slowly, as would have been the case if these deep-seated rocks had been brought up towards the earth's surface by gradual denudation, they would gradually have been transformed into rocks composed of more voluminous and lighter minerals, conditioned by the existing pressure.

Eskola points out that "eclogites occupy a volume about 15 per cent smaller than that of corresponding gabbros, and the magma probably has a still larger volume. The volume of jadeitite (sp. gravity = 3.33) is as much as 22 per cent smaller than that of the corresponding molecular mixture of albite and nephelite (sp. gravity = 2.61)".

He also maintains that as the melting point of eclogites is raised by pressure, it seems "likely that there exists a zone in the deepest parts of the earth's crust where gabbroid material exists stable in the form of eclogite, at temperatures under which a gabbro would melt if the pressure were reduced".

Eskola mentions that this conclusion was formerly drawn by L. L. Fermor, who states that eclogite must be the high pressure form of gabbro, and that "we must, therefore, assume that in the infra-plutonic zone the basic rocks are present as eclogites and the more acid rocks as garnetiferous granites". Fermor thinks that this infra-plutonic zone may form "a continuous shell round the earth, the whole of which shell is a potential magma. This shell, being composed of rocks of the consistency of a plastic solid, may afford a cushion upon which the isostatic operations of the earth, believed in by some geologists, have their foundation".

It is in my opinion obvious that if changes in volume, of the order mentioned above, can be effected by changes of pressure, appreciable vertical movements of the surface level of the lithosphere may be thus produced.

The magnitude of these movements would depend on the thickness of the layer of those deep-seated rocks, the pressure of which is changed beyond the critical point. If we suppose, for instance, that, by the denudation of 100 metres of the continental surface (sp. g. = 2.6), the pressure in a layer of deep-seated eclogite (sp. g. = 3.45), 75 metres thick, be reduced below the critical point, and this layer of rock gradually be transformed, it may cause an upheaval of the continental surface of about 13 metres.

This vertical movement would be effected without any "flow" of the plastic substratum underlying the rigid crust. If the transformed rocks form parts of the lithosphere the rise of the crustal surface caused by their expansion would be added to the isostatic upheaval of the crust caused by the denudation.

If it is rocks of the plastic substratum underlying the lighter, rigid crust, which are transformed, they may by their expansion become parts of the lighter crust, and will then increase the upheaval of the latter caused by the denudation. In that case the thickness of the lighter crust would not be much reduced by denudation.

If the transformed rocks remain parts of the plastic substratum, their expansion may reduce the isostatic upheaval of the latter, because the expansion makes the layers of the substratum lighter, and the crust may sink deeper into it, which would, however, necessitate a "flow" in the substratum.

If instead of a reduction of the weight of the crust (e. g. by denudation) we suppose an increase of its weight by deposition of sediment, or by the formation of an ice-cap, the deep-seated rocks may be inversely transformed, and a gradual increased sinking of the crustal surface would be caused.

As, however, we still know much too little about the actual conditions at these depths below the earth's surface and about the possible transformations that may take place there, we must leave the crustal movements, thus caused, out of our present considerations.

If we try to form an idea of the isostatic movements of the earth's crust caused by a load, e. g. by an ice-cap, the simplest supposition is that the crust is floating on a semifluid or viscous, molten magma, in a manner similar to that in which an ice-sheet floats on water. In that case the conditions may be considered as being practically hydrostatic below a certain depth. I. e. there is practically a uniform pressure in all directions, and when the crust is depressed by an additional load, it will gradually sink down into the magma, in a manner similar to that in which a loaded ice-sheet will bend under a load and sink deeper into the water, until it attains its new level of equilibrium. The plastic magma under the sinking crust will be gradually pressed out towards the sides, and an "undertow" will arise. This will continue till the crust has reached its new level of equilibrium. The volume of the magma displaced sideways by the "undertow" will be equal to the total volume of the depression of the crust below its initial level, the possible changes in the volume of the magma caused by pressure not being taken into consideration.

We do not know what the state of matter may be at the depth of the zone of compensation, perhaps at 120 kilometres, or more, below the earth's surface — whether it is solid, viscous, fluid, or some other state unknown to us. The experimental researches by F. D. Adams [1912, 1917] and also those made at the Carnegie Geophysical Laboratory indicate that the problem is a complicated one [cf. Bailey Willis, 1920]. As the interior friction of solid rock, and thus its absolute strength or rigidity, is increased by pressure to a certain limit, where the rock is potentially crushed, the "zone of flowage", where the rock may be considered to be plastic, lies much deeper than was generally estimated.

We may, however, assume that this zone begins a great deal higher than the zone of compensation. As high temperature is essential to the mobility of the rock, and as the temperature of the lithosphere increases rapidly with depth, we may assume that at a depth of 50 to 60 kilometres the rock is in what may be called a plastic state, and that its plasticity increases with the depth.

We shall not here try to discuss whether at a certain depth there is a continuous substratum of molten magma or not. The chief point for our consideration is that at some depth under the rigid surface of the lithosphere there is a zone of flowage, where the rock material, in whatever state it may be, is plastic and mobile. That it must be so, and that this plastic substratum behaves to a certain extent like a viscous fluid, seems to be fully corroberated by our investigations of the strandflat and the crustal movements of Norway after the last glacial epoch.

The coefficient of viscosity of this plastic substratum is probably very high, but even if it be as high as estimated by Schweydar [1921], i. e. ten thousand times that of sealing-wax at normal room temperature, we must keep in mind that the pressure is also extremely high, and the substratum, therefore, is responsive to changes of pressure, and possesses a certain degree of mobility, so that in the course of time, as is proved by our observations, it gradually adapts itself to the conditions of equilibrium.

It is, however, obvious that this must require a very long time. On the one hand because the rigidity of the crust will offer great resistance to deformation, and it will give way only very slowly, probably by shearing. On the other hand because, as was just mentioned, the internal friction of the plastic substratum is so very great.

The flow in the plastic substratum will, therefore, be extremely slow, and besides it will meet with great resistance, because, for instance, the crust, in the zone surrounding a depressed area, will have to be lifted in order to make room for the displaced plastic matter.

Let us try to imagine what will happen, if the crust be gradually pressed down by an increasing ice-cap, provided that the conditions are fairly hydrostatic at a certain depth under the earth's surface.

Supposing the ice-cap begins to be formed in the central area of an extensive region like Fenno-Scandia, the load of the ice-cap will press the crust down in this central area, and in a zone surrounding it the crust will be pressed up and will there form a kind of concentric wave, as in-

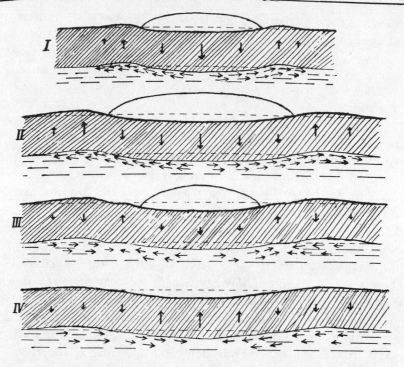

Fig. 170. Diagram showing, with much exaggeration, the depression of the crust under an ice sheet, and the upheaval on the sides. The vertical arrows indicate the vertical movement of the crust, the horizontal arrows the "undertow" in the plastic substratum.

dicated in Fig. 170, I. As, however, this wave will not represent a state of equilibrium, it will gradually extend outwards, and will be flattened down as it becomes wider and wider.

If now the ice-cap increases in thickness and in extent, the crust will continue to be pressed down, and the surrounding upheaval wave will increase somewhat in height, while it will be moved outwards by the advance of the ice-cap (Fig. 170, II).

The depression of the crust will continue a long time after the ice-cap has ceased to increase, and will only very slowly and asymptotically approach its level of isostatic equilibrium. The wave of upheaval surrounding the depressed area, will continue to widen, and the real level of isostatic equilibrium will not be fully reached, before this surrounding wave is entirely flattened out, and the depression of the ice-covered area is fully compensated for by the upheaval of the floor of the Ocean, as was mentioned on p. 233, but this is a state which is never reached.

When the ice-cap begins to decrease towards the end of the glacial period, the underlying crust will probably not yet have been fully depressed to the level of equilibrium conditioned by the weight of the ice. The subsidence will, therefore, continue inside the area which is still covered by the retreating ice-cap, and it will not stop as long as the weight of the ice masses is in excess of the load corresponding to the amount of depression. After that time an upheaval of the crust will gradually begin.

Meanwhile an upheaval of the land will start in the outer zone of the previously ice-covered region, soon after it has been left free by the retreating ice. This upheaval may be facilitated by a double "undertow" of matter coming from the still subsiding area inside, under the ice-cap, as well as from the upheaved peripheral zone (the upheaved wave) outside, as is indicated in Fig. 170, III. This may possibly have been what happened during the period before the Tapes-Littorina period in the regions of Skagen and Kalmar, &c. (cf. above p. 284).

When the upheaval of the more central area began, an undertow of matter towards this rising area from the surrounding previously upheaved zone would arise, and this would cause a sinking of the land in that zone (Fig. 170, IV), corresponding to the previously mentioned sinking of the land in the region of Skagen and Kalmar, &c. (cf. p. 284), before or at the beginning of the Tapes-Littorina period.

As, however, the upheaval of the depressed area advanced, the undertow of matter from the peripheral, formerly upheaved regions (outside the ice-cap at its widest extent) would be increased, and a general upheaval of the whole depressed region would be developed, and this would now continue, till the upheaval was completed. At the same time, the land in the peripheral, formerly upheaved zone, surrounding the ice-covered region, would gradually sink.

Where the retreating ice-cap was to some extent replaced by a transgressing sea, the upheaval of the crust would be retarded, as was previously pointed out (cf. p. 280).

It seems to me, that a development as here indicated agrees well with what we now know about the late-glacial and postglacial crustal movements which probably have taken place in Fenno-Scandia and the surrounding regions.

Along the coast of Norway, where the retreat of the margin of the ice-cap was very much slower e. g. than in Southern Sweden and in Denmark, and where the ice remained near the coast till late-glacial times, the crustal movements have probably been less complicated, and the upheaval may have been fairly continuous from its beginning.

In regions where the ice-cap left great quantities of moraine material, as for instance in Denmark and in Northern Germany, the crust was naturally depressed by the weight of these deposits, and this fact would also cause a sinking of the land which may have continued long after the ice retreated. The quantity of moraine material, however, carried by the last ice-cap, may probably not have been very great, and the sinking of the land thus caused after the last glacial period may, therefore, have been less considerable than after the previous glacial periods.

Of much importance are the two questions of the time required for the isostatic readjustment of the crust, and of the size of the area within which it can take place.

How long a time does the earth's crust require to reach its new isostatic level after a disturbance of its equilibrium?

As the internal friction in the plastic or mobile substratum, under the rigid crust, is, at any rate, extremely great, whatever the state of this matter may be, we may expect that when the substratum is exposed to stress it will take a long time before this friction is gradually overcome, and motion is started. In addition to this there is naturally also an enormous resistance to overcome in the rigid crust itself, before it can be depressed or upheaved. Hence the isostatic movements of the crust will always show a great deal of lag.

For this reason many geologists have assumed that the establishment of isostasy at the earth's surface will require extremely long geological periods. I think, however, that our study of the strandflat and the raised shore-lines of Norway may prove that the time required for the attainment of approximate equilibrium is very much shorter than is generally believed.

Although it may be very difficult to estimate the length of time elapsed since the ice-cap of the last glacial epoch actually began to decrease, there are now many careful researches by Gerard de Geer and his school in Sweden as well as by others, especially in North America, which will help to estimate the length of the late-glacial and postglacial periods. According to the results of De Geer's investigations, it is not more than 13,000 years since the margin of the retreating ice-sheet stood in Southern Scania. Even if we take other estimates we cannot possibly assume the time since then to have been more than 16,000 or 20,000 years. During this period nearly the whole of the late-glacial and postglacial upheaval of Fenno-Scandia has been accomplished. A very considerable part of the upheaval has even been accomplished in about half that time, since the ice-margin had retreated to Northern Sweden, 8,000 years ago, *e. g.* the land near Bottenviken has been upheaved about 270 metres during that space of time.

We know that along the coast of Norway the upheaval of the land was practically completed at least before the beginning of the Christian era. Hence we may assume that along the Norwegian coast the upheaval of the land and the re-establishment of equilibrium was completed during a period of probably 11,000 years, and at any rate of not more than 18,000 years. This is a remarkably short period, and seems to indicate that the plastic substratum of the earth's crust is more mobile, than many geologists are prepared to allow.

What is the extent of the smallest area within which isostatic movements
may occur?

This is a very difficult question, which we cannot answer at present.
It is obvious that the question of time is here of much importance. The
smaller the area is within which the equilibrium is disturbed, the longer
it will take before it can be re-established. It may be possible that in the
course of a very long time equilibrium may be more or less attained
within quite small areas, although the process is so extremely slow that
it is not yet the case in many localities examined.

We have seen (pp. 253 f.) that in Norway and Sweden there may be
quite considerable differences in the upheaval of the land within small
distances of no more than a few kilometres. As, however, we do not know
the causes of these differences, they can hardly be used as proofs of a
great local adaptability of the crust to isostasy.

It may, however, be pointed out, that the comparatively small ice-cap
of Scotland, of the last glacial period, has caused a considerable depression
and subsequent upheaval of the land, within an area with a diameter of
less than 500 kilometres. The last glacial ice-cap of Iceland also caused
a depression and subsequent upheaval of that island.

Lake Bonneville in the region of the present Great Salt Lake, in Utah,
with a diameter across of probably about 230 kilometres, caused a depression and subsequent upheaval of the flooded land, although the depth
of the water above the present level of Great Salt Lake may have been
no more than 320 metres [cf. Gilbert, 1890]. The depression caused by
the load of water may possibly have been about 45 metres in the central
area of the lake.

These facts indicate that considerable isostatic movements may take
place within areas no more than a few hundred kilometres wide, and
probably even much smaller.

Editor's Comments on Paper 7

Daly: *The Changing World of the Ice Age*

This extract from a general textbook on the Quaternary by Daly concludes the selection of papers by early workers. R. A. Daly was the Sturgis Hooper Professor of Geology at Harvard University and he had worked and published on sea level problems for many years. The combined effects of economic depression and the Second World War produced a significant decline in the number of publications on glacial isostasy between the late 1930s and early 1950s. Daly epitomizes one aspect of the early work—that is, an interest in both the field and the theoretical issues. All the papers reproduced so far in this volume contain documentation of field data and discussion of the mechanisms of glacial isostasy. By and large, publications after 1950 are more clearly differentiated into two groups: those concerned with field evidence, and those using field material from geophysical interpretation and modeling.

Daly's book contains an excellent synthesis of field observations from northern Europe and North America on the extent and location of the major ice caps, the effects on the crust of the ice caps, and the mechanisms of uplift. The second part of the book is concerned mainly with the questions of changing eustatic levels. Recent workers (e.g., Walcott, 1972; Newman, 1971) have begun to reaffirm some of Daly's concepts, specifically those related to the development and migration of the forebulge (see his Fig. 69 in this paper). His conclusions are summarized in Chapter IV, which is reproduced here in part.

Daly recognizes that the effect of ice loads on the earth's crust are twofold; an elastic, or immediate, response and a slower plastic one. Figures 40 and 41 from Daly's book illustrate the relationship between ice unloading and uplift and the relative contribution of elastic and plastic uplift as it affected central Fennoscandinavia. Two primary models of upwarping of glacial tracts are proposed by Daly; the first (Fig. 69) is the "viscous-bulge" model and the second is the "punching" one. Objections to the first model are pointed out by Daly, who apparently favored the "punching" model of Fig. 70. Although recent work has answered some of Daly's criticisms, it must be noted that his third criticism (that the migration of the forebulge should be observable in glacial lake shorelines) has not been satisfactorily answered.

Reprinted from *The Changing World of the Ice Age,* 70–71, 119–126 (1934)

7

The Changing World of the Ice Age

R. A. DALY

Nansen later estimated the central uplift, millennium by millennium, with the result illustrated by Table XV and by Figure 40. At the top is a horizontal line indicating the passage of time from 16,000 B.C. to 2000 A.D. The vertical line at the extreme left is scaled to show amount of uplift, in meters. Ignore the other lines and con-

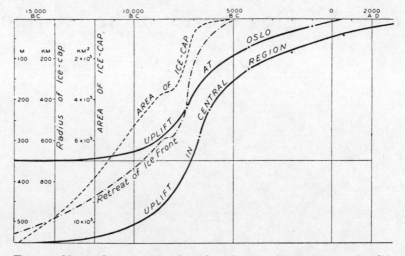

FIG. 40. *Nansen's curves showing his estimates of the progress of uplift of the earth's crust at Oslo and in the central part of the Fennoscandian tract (meters), during the last 18,000 years; also changes in the radius (kilometers) and area of the ice-cap, and the progress of retreat of the ice front.*

centrate attention on the lower of the two curved, S-shaped lines. This gives Nansen's preliminary (minimum) estimates for the rates of uplift at the center of the glaciated tract. Here the land was 530 meters lower than now in the year 16,000 B.C. For several thousands of years before that, there was no important uplift, and the movement had been but slight even at epoch 10,000 B.C.[8]

The long lag in the plastic recoil of the earth's crust is illustrated by another diagram (Fig. 41), which, based on one of Nansen's, indicates, approximately, the variation in the rate of uplift at the center of the Fennoscandian tract. The horizontal dimension of the drawing represents the time from 25,000 B.C. to 2000 A.D. The rate

[8] F. Nansen, *Avhand. Norske Videnskaps-Akad.*, Oslo, Kl. I, No. 12, 1928, p. 63.

of uplift, in meters per century, is shown in the vertical dimension. The more closely shaded area expresses Nansen's conclusion as to the approximate rate of plastic uplift. Accordingly the indicated rate was practically zero before the year 17,000 B.C. It slowly increased to the year 10,000 B.C., and then grew rapidly, until at 7000 B.C. it reached the maximum of 13 meters per century. Then it fell about as rapidly, to become only 1 meter per century at the present time.

The purely elastic part of the uplift—of rate indicated on the diagram by the broken line—began with the melting of the ice, about 25,000 B.C. With the acceleration of melting, the rate of elastic uplift increased, reaching a maximum about 16,000 B.C., and

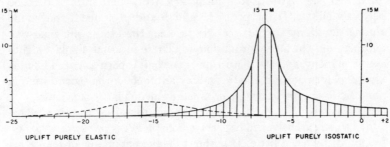

UPLIFT PURELY ELASTIC   UPLIFT PURELY ISOSTATIC

FIG. 41. *Rates of elastic and non-elastic uplift at the center of the glaciated tract of Fennoscandia (meters per century).*

becoming zero when melting of the ice was practically completed, namely, about 7000 B.C.

The elastic uplift was never very rapid, but its effects were cumulative for seventeen thousand years or more, and their total would be about one fourth of Nansen's estimate of 530 meters for the central uplift.

Ramsay assumed a central thickness of 4000 to 5000 meters for the ice-cap, and also assumed the earth's material that was forced out of the glaciated sector when the plastic basining took place, to have been located immediately below the earth's crust. Taking 3.0 as the density of this displaced material, he found 1350 to 1500 meters for the non-elastic depression of the floor of the ice-cap at its center. Sauramo states its amount to have been "probably two or three times greater than what is indicated by the highest shores," and probably reaching 900 meters. Tanner, reasoning as Ramsay did in this matter, arrived at a result that was much the same. In

*   *   *   *   *   *   *

MODES IN WHICH THE EARTH'S CRUST YIELDS TO A LOAD

THE rocky floor under any of the Pleistocene ice-caps sank for two different reasons: because of the earth's elasticity, and also because of the earth's plasticity. Its elasticity may be crudely illustrated with a toy balloon or a soccer football. If you press your finger on either, a dimple appears. You remove the finger, and the dimpled surface flies back to its original form. A closer analogy is the dimpling of a steel ball by localized pressure on its surface. Such a ball has great but finite rigidity, and yields a little, without breaking and without internal flow of the steel. When the pressure is removed, the ball instantly resumes its original shape almost to perfection. If the restitution of form is perfect, the ball is said to be perfectly elastic. If the pressure is high and prolonged, the bonds among the atoms of the steel are broken, the atoms slip past one another, and the steel actually flows. These internal displacements are permanent, and the distortion of the ball is permanent. The first kind of deformation is elastic and instantaneous; the second kind is non-elastic or plastic, and is not instantaneous but takes time.

So it is with the earth. Under the weight of each meter of ice added to an extensive ice-cap, the whole planet is *immediately*, elastically, distorted a little. If without delay that meter of ice be removed, the earth *immediately* takes on its original shape. Both reactions are those of an almost ideally elastic body. But against extensive loads of *prolonged* application the earth's materials flow and the crust is basined, *plastically*, under the load.

Suppose, on the other hand, that an extensive ice-cap completes its basining of the earth's crust and then melts away. The uncovered region rises, at first by an elastic response of the earth, and later by a plastic response. The rise is greatest in the central region where the ice was thickest, so that the ultimate result is an updoming of the glaciated tract. Such updoming is indicated not only in Fennoscandia and North America, but also in many other regions with separate ice-caps, including the Alps, Great Britain, Iceland, Siberia, Spitsbergen, Novaya Zemlya, Franz Josef Land, Patagonia, and New Zealand. The large number of instances assures us that the earth is everywhere sensitive to loads of wide span. Further, there is no reason to doubt that one set of mechanical principles controlled not only all the updomings, but also the previous basinings. The mechanics will be discussed particularly in relation to the case of updoming. The choice is made because we can measure actual effects of the last Pleistocene meltings and unloadings, and can use the measurements for checking theoretical conclusions about the mode of recoil.

PLASTIC RECOIL

WE shall first consider the non-elastic or plastic distortion of the crust; then the elastic.

Discussion of the plastic distortion will be easier if two existing theories about it be reviewed.

One, which may be called the *bulge hypothesis*, will be considered first, because it is accepted by many geologists, though without proper warrant from the facts of Nature. The theory may be illustrated by a series of sections.

Let the stippled band of Section No. 1, Figure 69, represent the earth's flexible crust. Below it is a weak substratum, part of a complete earth-shell. Section No. 2 shows the crust plastically basined under a full-grown sheet of ice. According to the *bulge hypothesis*, the basining is accompanied by outward, horizontal flow in the substratum and just beneath the crust. The place of this flow is indicated by the two horizontal arrows. Because the substratum is highly viscous, its material moves but a relatively short distance away from the ice-covered region. Hence a peripheral bulge, B', is formed. In the drawing there is great vertical exaggeration of both ice-cap and bulge.

After much of the ice has melted away, we have, by the hypothesis, a situation like that portrayed in Section No. 3. Return flow in the substratum, just below the crust, is supposed to have reduced the bulge to the condition represented by B''. The crest of the bulge is assumed to have moved inwards, as if carried on a wave of the substratum material, the wave advancing through such positions as B'' and B''' (Section 4) toward the center of the ice-cap.

Continued melting and continued progress of the substratum wave are assumed to lead on to the still later stages indicated by

Sections No. 4 and No. 5. The last stage is reached when the earth's crust has attained its pre-Glacial position of equilibrium.

This *bulge* or wave hypothesis, which might also be called the hypothesis of horizontal flow at *small* depth, is faced with three objections.

1. In the first place, the bulge, once formed, should have been

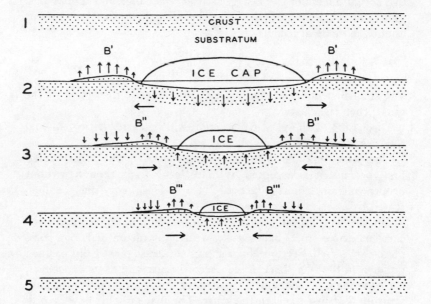

FIG. 69. *Sections illustrating the "wave" or "viscous-bulge" hypothesis to explain the plastic upwarping of a deglaciated tract.*

preserved to this day. The bending strength of the earth's crust is known to be so great that the comparatively narrow bulge could not be flattened out by its own weight; there is no other way in which it could have been annulled. But neither in Europe nor in America is there any known trace of such a stereotyped, peripheral uplift. Drainage patterns do not declare it. Other field evidence shows the earth's crust bordering the glaciated sectors to lack discernible relics of the assumed narrow bulge.

2. The second objection will be more fully developed when we study the alternative hypothesis. Meantime the main point may be briefly stated. Since the great central part of a fully grown ice-cap

varies relatively little in thickness, its weight would there produce almost no stress-difference of a kind that would cause outward flow near the earth's surface. The steel-like rigidity of the material just below the crust would be long preserved and prevent horizontal flow of that high-level material from beneath the load. On the other hand, vertical shearing stress along the edge of the cap is incomparably greater than in the central region. Hence, at the edge, the decay of the initial rigidity of the earth is rapid. There the crust yields by *down-punching*, the compensating horizontal flow being concentrated in the *deeper*, hotter, and more plastic shells of the earth.

3. Field observation supplies the third and specially cogent objection to the bulge hypothesis. The progression of the assumed subcrustal wave would have notably and permanently tilted the lake beaches formed on the glaciated tract during the recoil. But the older Late-Glacial beaches are almost perfectly dead-level over wide belts.

The second and better explanation may be called the *punching hypothesis*, or hypothesis of horizontal flow at *great* depth. Again, a series of diagrammatic sections will illustrate the argument (Fig. 70).

Section No. 1 shows with much vertical exaggeration the thickness of a circular ice-cap (shaded), that has basined its floor to the

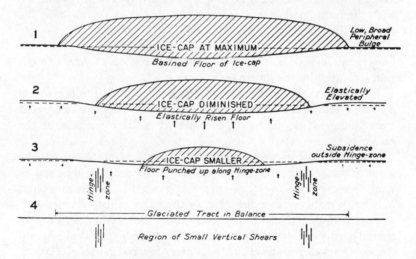

Fig. 70. *Sections illustrating the "punching" hypothesis to explain the plastic upwarping of a deglaciated tract.*

limit set by the weight of the ice. The position of the pre-Glacial floor, assumed to have been level, is marked by the broken line; the position of the floor after the basining, by the lower continuous line. The plastic basining is supposed to have been accompanied by an extremely slight uplift of the continent surrounding the glaciated tract. This uplift is of a much smaller order of magnitude than that assumed on the bulge theory. To be able to show it at all on the scale of the drawing, the corresponding displacement of the crust is there specially exaggerated.

Section No. 2 shows what happens when the ice-cap loses some area and thickness. There is an immediate, elastic rise of the whole country, ice-cap plus a broad marginal belt, the rise being greatest at the center of the ice-cap and least at the ends of the section, say 500 kilometers from the ice. This elastic effect will be considered later.

Section No. 3 represents the stage when the crust, still further unweighted by the melting of ice, has finally yielded to the stress and rises plastically. The floor of the ice-cap has been up-punched along a circumferential zone of vertical fractures. This belt, named "hinge zone," is assumed to run all the way round the ice-cap and, of course, appears twice in the cross section. Because of the actual breaking of the crust at the hinge zone, the superelevated belt outside the glaciated tract is no longer held up by a firm connection with the basin, and is snapped down by gravity. Arrows show the opposed directions of movement of the earth's crust.

Complete melting of the ice and continued slippings within the hinge zone finally lead to equilibrium for the earth's crust, a condition indicated by the fourth cross section.

The up-punching could have taken place only if the horizontal flow of material into the basined sector of the earth was concentrated at great depth—a depth equal to a large fraction of the radius of the ice-cap.

The sections of Figure 70 were drawn on the assumption that the ice-cap was symmetrical. However, the Pleistocene ice-caps of Europe and North America were not symmetrical, either in ground plan or in transverse profiles. Hence the vertical displacements at and near the earth's surface should have been unequal, and we should hardly expect definite, clean-cut hinge zones to have been continuously developed all around the center of each region of glaciation.

This second, preferred explanation of the Late-Glacial and Post-Glacial deleveling of the earth's crust is supported by the discovery of actual hinge zones in both North America and Europe. The peripheral stresses in the crust, due to a circular load of material with uniform density and thickness, increase with the radius of the burdening mass. Hence the broader an ice-cap of given thickness, the more numerous should be the hinge zones within its circumference. This appears to be one of the reasons why these zones were first found in North America.

Figure 55, page 99, shows parts of two hinge lines in America, to which the names Whittlesey and Algonquin have been applied. They were discovered when the present altitudes of the old, abandoned beaches of the Great Lakes were determined by leveling.

The older beaches south of the Whittlesey hinge line are now horizontal. North of that line the same beaches are tilted toward the south. Somewhat younger beaches behave in exactly the same way on the two sides of the Algonquin hinge line.

Those relations are explicable on the punching theory. With the removal of burdening ice, not only outside its retreating edge but also by some contemporary melting all over the residual part of the ice-cap, the shearing stresses at the Whittlesey line increased. Suppose, for example, that by the time the ice front had receded about to the Whittlesey line, the glaciated tract as a whole had lost, by melting and evaporation, ice to the average thickness of 500 meters. That relief of load would produce vertical shearing stress along the Whittlesey line, equal to at least 400 kilograms per square centimeter, or probably twice what the crust can permanently support. Hence the earth sector capped by the residual ice would be punched up along a sheath of circumferential fractures grouped along the Whittlesey line. Relatively little movement would occur along any one fracture, but the vertical displacement in the belt of fractures might well total 75 meters or somewhat more. By so much would the crust inside the Whittlesey line be moved upward, relatively to the part of the crust that was "snapped down" on the other side of the line. The country south of the line, having lost elastic connection with the central region of the glaciated tract, was henceforth unmoved, because in equilibrium with the earth's crust in general. Lake beaches thereafter built south of the Whittlesey line remained level. North of that line the same beaches were permanently tilted to the south, as the ice continued to lose volume and weight and plastic uplift continued in the central region.

Again the peripheral stresses around the still diminishing ice-cap increased to the breaking-point, and, about 150 kilometers north of the Whittlesey line, the second, Algonquin hinge line was fixed by up-punching on the north and down-snapping on the south. Hence younger beaches are horizontal south of the second hinge line and tilted north of it.

The punching hypothesis implies a more or less sudden and interrupted recoil of the earth's crust from its former depressed attitude. Taylor's evidence that the movements were actually spasmodic is therefore significant.[2]

The Whittlesey hinge line has been traced nearly to the Pennsylvania boundary (Fig. 55, page 99). Farther to the eastward it has not yet been mapped. However, Salisbury has furnished a relevant observation. He found the strandline of the Late-Glacial Lake Passaic in northern New Jersey to be tilted southward, with a maximum deleveling of about 20 meters in a profile 40 kilometers long.[3] Hence, if the Whittlesey hinge line extends to the Atlantic, it must pass south of New York City. Still farther east its trace would be on the continental shelf, running perhaps 150 kilometers to seaward of the point where the coast, near Boston, meets the zero isobase or line of zero height above present sealevel for the oldest marine beach of Late-Glacial origin.

It is a question whether a hinge line north of the Algonquin shall ever be found. After the displacements at the Whittlesey and Algonquin lines, the potential for vertical fracturing of the earth's crust had been largely destroyed. Later uplift is more likely to have taken place by bending, rather than fracturing, of the crust.

Fennoscandia seems to have its hinge zone, represented by a set of closely spaced breaks in the tilts of beaches along the Arctic coast. Figure 71 illustrates the relations of the strand profiles. The tilts of the older and higher strandlines are seen to be sharply accentuated at certain points, as if the glaciated area to the south, on the right, had been sheared up along fractures. If so, the region to the north of this zone of rapid change in tilt, on the left of the section, was snapped down. Thereafter this northern region was in permanent equilibrium and its strandlines have escaped tilting.

[2] F. B. Taylor, *Monograph* 53, U. S. Geol. Survey, 1915, pp. 466, 468.
[3] R. D. Salisbury, *Final Report State Geol. Survey*, New Jersey, vol. 5, 1902, p. 224. Recently C. A. Reeds (Guidebook 9, Internat. Geol. Congress, 1933, p. 61) has described similar warping of the strandline of the Glacial lake Hackensack, east of the site of Lake Passaic.

An analogous condition is found in the region of Nord Fiord, which appears on the map of Figure 72. The Atlantic water covers the stippled areas. Numbers show, in meters, the heights of the marine limit. This strandline was recorded by the ocean, because Norway was slow in recovering from its Glacial basining. Each of the broken lines are isobases, joining points of equal uplift since the recovery began. From the 17-meter isobase to and beyond the 70-

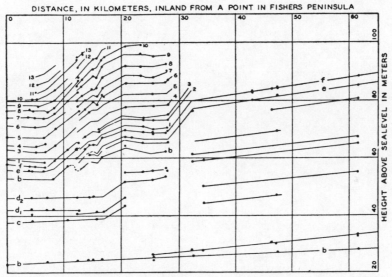

FIG. 71. *Deformation of Late-Glacial beaches along the Arctic coast of Scandinavia.*

meter isobase, the land has been tilted seaward, away from the center of glaciation. A little to the westward of the 17-meter iso-base, the marine limit is at 16 meters, and it remains at almost exactly that level all the way to the farthest sea-cliffs on the off-shore islands—a distance of about 20 kilometers. Throughout this belt, the old strandline is not tilted but is horizontal. The contrast of the two belts of land, one tilted and the other unmoved, means a relation like that on the Arctic coast to the northeast, and also like that at the Whittlesey and Algonquin hinge lines of North America.[4]

[4] For details regarding the tilted beaches along the Arctic coast see V. Tanner, Bull. '88, Comm. Géol. Finlande, 1930. Concerning the case at the Nord Fiord see F. Nansen, *The Strandflat and Isostasy*, Christiania, 1922, p. 248; H. Kaldhol, *Bergens Museums Aarbok*, 1912.

Recent Field Research in Quaternary Aspects of Glacial–Isostasy

II

Editor's Comments on Paper 8

Kupsch: *Postglacial Uplift—A Review*

The next two sections concern field evidence on glacio-isostatic recovery gathered since the early 1950s. These data can be subdivided into studies that are concerned with spatial aspects of warping and deformation (through an analysis of the deformation of former lake shorelines, for example) and those that concentrate on the changing rates of rebound through time.

The first paper is a useful and critical review of the evidence for postglacial uplift. Kupsch states in the introduction that the review is primarily a collection of questions asked by a geologist who has worked extensively in the field of Quaternary geology but not in the area of postglacial uplift. The absence of any specific field experience is not judged a handicap because it gives the author freedom to discuss and scrutinize field evidence and concepts that might be accepted by a worker engaged in the specific problem. Kupsch does, indeed, raise some major criticims of both the field data and the interpretation of these data. Of interest is his use of the Greenland and Antarctic ice caps as analogues of the Laurentide and Fennoscandian ice sheets, especially in view of the evidence that Antarctica is essentially in isostatic equilibrium with its ice load. In the conclusion, Kupsch raises the questions of the (existence?) of hinge lines and the ". . . presence and size of the forebulge." Neither of these questions has been answered with any certainty.

Walter O. Kupsch was born in the Netherlands in 1919 and is now a Canadian citizen. He started his university education in Amsterdam and received his Ph.D. from the University of Michigan in 1948. He has a wide variety of experience in industry and university affairs and has served on a number of Government Committees in Canada. Currently Director of the Institute of Northern Studies, University of Saskatchewan he has a great personal interest in Arctic Canada and polar areas, in general.

Reprinted from *Life, Land and Water*, W. J. Mayer-Oakes, ed., 155–160, 169–186 (1967)

8

POSTGLACIAL UPLIFT -- A REVIEW

Walter O. Kupsch

Introduction

Postglacial uplift is a broad subject dealt with in introductory geology classes as prime evidence for the reality of isostasy and isostatic compensation, "An equilibrium condition in which elevated masses such as continents and mountains are compensated by a mass deficiency in the crust beneath them. The compensation for depressed areas is by a mass excess." (Howell, *et al* 1960:156)

It should be stated that I have never done any research on either the concept of isostasy or glacial uplift and that the following review is based on secondary sources only. The literature on the topic is voluminous and within the short time available to me it has been possible only to scan some of the more comprehensive works. The present paper is therefore not to be regarded as a review of any substance but only as a collection of some questions which arose when reading general accounts, mainly on the textbook level, dealing with postglacial uplift. I am raising these questions at the peril of exposing my own ignorance. Persons with a better knowledge of the subject may very well be able to answer most of them or to show that some are based on a misunderstanding on my part of the basic concepts and working methods involved.

General Concept

Because I am using Arthur Holmes' (1965) textbook for classes in general geology, it appears appropriate to start the review with the following quotation:

> *During the recession of the continental ice-sheets of Europe and North America conditions were highly favourable to the development of widespread marginal lakes [such as Lake Agassiz] (Fig. 515) [here reproduced as Fig.41]. At the time of maximum extension of the ice the less mountainous parts of the underlying floor were depressed into a shallow bowl by the isostatic effect of the load of ice. The thickness of the ice reached 8,000 feet or more, tapering off towards the margins. The corresponding subsidence of the crust, where the load was greatest, would therefore be over 2,000 feet; sufficient, that is, to depress vast areas of the rock surface well below sea-level. Such is the condition of Antarctica today. During the retreat of the ice the crust was gradually unloaded*

155

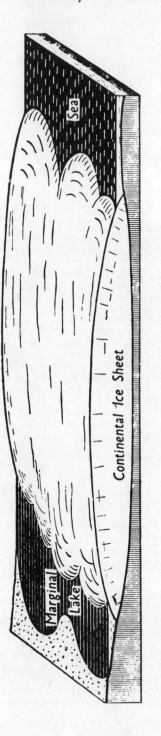

Figure 41. Crustal depression. The diagram illustrates the isostatic depression of a land surface loaded by a continental ice sheet and the consequent development of marginal lakes, such as Lake Agassiz, during the recession of the ice. Vertical scale and slopes are greatly exaggerated. The diagram fails to show the earth's curvature which on this scale (the North American ice sheet having covered as much as 40° of arc in its longest dimension) should have been taken into consideration. After Holmes (1965:684).

*and isostatic recovery worked in from the margins, though
with a considerable lag. Consequently, for thousands of
years there were large tracts, abandoned by the ice, that
sloped towards and beneath the receding ice-front. Many
of these became giant lakes while others were invaded by
the sea. The isostatic recovery already achieved since
the disappearance of the ice is clearly demonstrated by
the emergence of beaches now locally preserved at various
heights above sea-level, and by the tilted attitude of
many lake terraces. Moreover, the fact that the shores
of Hudson Bay and the Gulf of Boothia are steadily continuing to rise shows that the process of restoring
isostatic equilibrium is still going on (Holmes 1965:
683-684).*

The general concept of the earth's crust depressed under the
weight of a continental glacier and readjusting itself after the
disappearance of the ice follows a discussion on isostasy and
isostatic readjustment which in part reads as follows:

*It may happen that certain processes disturb the pre-
existing isostatic balance much more rapidly than it
can be restored by deep-seated rock-flowage in the
mantle. For example, when the last of the thick
European and North American ice-sheets melted away
between, say, 11,000 and 8,000 years ago, these regions
were quickly relieved of an immense load of ice. The
resulting uplifts which then began are still actively
in progress. Far above the shores of Finland and
Scandinavia there are raised beaches which show that
a maximum uplift of nearly 900 feet has already occurred
(Fig. 40)* [this is also evident from Fig. 42 this paper]*,
and every twenty-eight years another foot is added to
the total all around the northern end of the Gulf of
Boothia. The region is still out of isostatic balance,
and it can be estimated that it has still to rise another
700 feet or so before equilibrium can be reached.*

*Similarly around the northern shores of Hudson Bay new
rocky islands have appeared within the memory of the
older Eskimos, and the land is known to have risen at
least 30 feet since the Thule Eskimos first established
themselves there, indicating an average uprise of not
less than three feet per century (Holmes 1965:59).*

In the following sections we will consider the varied evidence
on which the concept of isostatic readjustment following continental
glaciation is based.

Hudson Bay

The "emergence of beaches" (Holmes 1965:684) was observed in
Hudson Bay as early as 1631 by Captain Luke Foxe (Warkentin 1964:
13-14). His explanation that the abandoned strandlines represent
extraordinarily high-tide storm beaches is no longer acceptable
now that it is known that some strandlines are as high as 900 feet
(Innes and Weston 1966) above the average present level of Hudson
Bay. Many textbooks in geology (see quotations from Holmes
presented above) leave the impression that the abandoned beaches
of Hudson Bay are above present sea-level solely because of
isostatic movements following deglaciation, that they would be
even higher above the water of the bay were it not for the eustatic
rise in sea-level which resulted from the melting of the continental
glacier, and that the out-pacing in vertical rise of the land over
the sea is still going on today. A direct analogy with the much
more intensely studied Fennoscandian region is generally suggested.

However, when we consider isostatic readjustment as the sole
cause of the high-level beaches around Hudson Bay, no account is
taken of the preglacial geological history of this sedimentary
basin. King (1965: 837) stated: "Beneath Hudson Bay the Precambrian
rocks of the Canadian Shield are extensively covered by Paleozoic
strata, downwarped into a shallow basin . . . . The bay owes its
submergence largely to effects of the last glaciation, but it has
been a long-persistent negative area as well." A review of the
geology of the Hudson Bay Basin (Nelson and Johnson 1966) suggests
that the basin, which was a strongly negative element in Paleozoic
time, became land sometime in the late Jurassic or early Cretaceous.
It is interesting to note that some Russian geologists maintain that
Fennoscandia was being uplifted before glaciation (Coulomb and
Jobert 1963:122).

Although, therefore most of the Hudson Bay region was probably
no longer below sea level just prior to the first Pleistocene
glaciation, it was lower in elevation than western Canada. It is
generally held that the preglacial drainage from that part of the
North American continent was to the northeast toward the lowland
now covered by the sea water of Hudson Bay (Barton *et al* 1965:195,
197; Flint 1957:170). It follows from this that a depression of
the topographic surface of the earth's crust already existed between
western Canada and the Hudson Bay region before the continental
glacier developed. The weight of the ice emphasized this bowl-shape
but did not create it. To show the position of the crust before
loading by glacier ice as a level line is therefore misleading,
at least in the case of a cross section between western Canada
and Hudson Bay (Fig. 52).

The "fact that the shores of . . . the Gulf of Boothia are
steadily continuing to rise shows that the process of restoring
equilibrium is still going on." (Holmes 1965:684) This is based
on observations of level changes as recorded by tide-gauges at
several places where long term records are available. Gutenberg
(1941: 733-739), who constructed a map showing the present rate

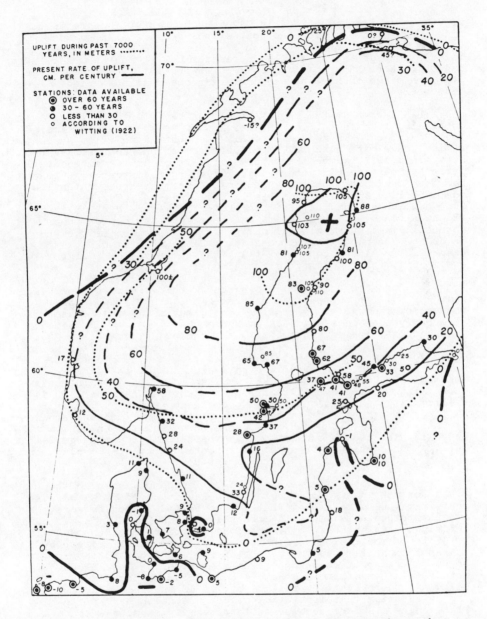

Figure 42.   Fennoscandian postglacial uplift. The map shows the
postglacial uplift which has taken place during the
past 7000 years in meters as well as the presently
continuing uplift expressed in centimeters per century.
After Gutenberg (1941:738).

of uplift in Fennoscandia (Fig. 42), used observations from 98
tide-gauge stations with records extending over time periods
varying in length from less than 30 years to as much as 126 years.
Calculations of the presently continuing uplift in the Great Lakes
region of North America are based on fewer (19) lake-level gauge
stations of which nine exceed a 60-year observation period
(Gutenberg 1941:742).

The "fact that the shores of Hudson Bay . . . are steadily
continuing to rise . . . " (Holmes 1965:684) is based on limited
local evidence. In Hudson Bay only one control point is available,
a tide gauge, at Churchill, Manitoba, for which records have been
kept since 1928, suggesting a present rate of uplift exceeding 1
meter per century (Gutenberg 1941:717). Other historical evidence
has been used both to refute and to substantiate the contention
that the land around Hudson Bay is still rising. Tyrrell (1896:
205) concluded from the position above the water-level of inscrip-
tions on rocks made by sailors on the eighteenth century near
Churchill that "post-glacial uplift of this portion of the shore
of Hudson Bay has virtually ceased." He thus refuted Bell's
(1880:21C) earlier contention that "the relative level of the sea
and land in this vicinity is changing at the rate of about seven
feet in a century." The problem was again investigated in the
field by Johnston (1939:97) who agreed with Tyrrell in concluding
that ". . . the prime cause of post-glacial uplift was the
removal of the weight of the ice sheet; but . . . this cause has
apparently long ceased to act at Hudson Bay . . . ." He went on
to suggest by analogy that ". . . the recent uplift of the Great
Lakes basins is due to some recent and local cause, and is not a
direct continuation of post-glacial tilting due to glacial
melting." (Johnston 1939:98) A re-evaluation of the field-evidence
was made by Gutenberg (1941:749) who regarded the information as
uncertain, supported Bell's earlier contention that the land is
still rising, and added that this conclusion was ". . . confirmed
by the very recent tide-gauge data", referring to the one station
at Churchill. It is worth mentioning here that the Churchill tide
gauge is "located on a wharf within the harbor" (Gutenberg 1941:
747), which does not appear to be a particularly stable place on
which to locate such a reference point.

From the above it follows that the evidence on which the
"fact" of continuing uplift of Hudson Bay is based is slender,
that it may be unreliable, that it has been disputed by some,
and that statements implying presently continuing uplift are based
largely on analogy with other regions, particularly Fennoscandia.
The rate of the implied present uplift is arrived at either by
extrapolation of rates in the better-studied Great Lakes region
or by regarding it as being of the same order as the values
obtained in Fennoscandia. Some geophysicists attempted to
calculate the prevailing rate of uplift of the land around Hudson
Bay by assuming that isostatic balance prevailed before the
Pleistocene glaciers developed, that depression of the crust
resulted entirely from the weight of the ice, and that uplift
will halt on re-establishing crustal equilibrium. The precepts
in any such calculations appear to be tenuous, however, as they
ignore the possibility of earth movements having no connection
with glaciation.

[*Editor's note:* In pages 161 to 168 Kupsch reviews some of the literature on the form of strandline deformation and discusses the question of whether "hingelines" occur, problems of the accuracy of isobase determinations, and quotes Flint on some of the unresolved problems of the Fennoscandia marine shoreline history.]

Geophysics

If, as some contend, changes in the rate of uplift are abrupt and hinge-lines are the surface traces of faults, an expression of this fracturing may be noticeable in the field and movement along the hinges may have caused historically recorded earthquakes. Moreover, if the hinges resulted from the periodic retreat and intervals of halting or readvance of the glacier margin (Fig. 46) the fracture pattern in the lacustrine and marine areas should show some correlation with the glacial deposits recording that history on the land areas. That the history of postglacial uplift as determined from water-covered areas is intimately intertwined with the late glacial and postglacial history of land areas and that attempts at correlation between the two are imperative is suggested by King (1965:834) follows from the following:

> *The perfection of some of the shorelines suggests that*
> *they were formed during lengthy interludes between times*
> *of tilting, as though regional uplift alternated with*
> *stillstand, perhaps because unloading was interrupted*
> *by pauses in ice wasting or by renewed ice accumulation.*

In the Lake Agassiz region where Johnston (1946:13) recognized at least seven hinge lines of deformation of the beaches there are no apparent topographical or geological features which indicate that the location of the hinge line was influenced by a line of weakness. "The absence of any such feature suggests that the earth crustal movements were very deep seated in origin." (Johnston 1946:13) However, a future more detailed study of the topography and surficial geology using airphotos may reveal some subtle features which escaped the attention of previous workers.

As far as seismicity is concerned, the area of maximum postglacial uplift in the interior of the Canadian Shield appears to be almost completely aseismic (Gutenberg and Richter 1949:91) and the thinly-covered Shield, to which the Lake Agassiz region belongs, shows only a very low to low frequency of occurrence of earthquakes which, moreover, have no direct discernible relationship to any hinge lines.

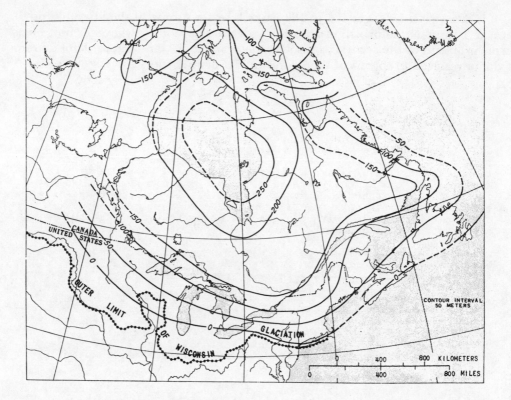

Figure 47.   North American postglacial uplift. The map of north-
              western North America illustrates postglacial uplift
              by means of contours (isobases) on the highest observed
              marine and lacustrine strandlines indicating the minimum
              amount of uplift which has taken place since deglaciation.
              The contour interval is 50 meters. After King (1965:836,
              Fig. 4A).

As shown in Fig. 47, the zero isobase or first hinge line of
North American postglacial uplift closely corresponds to or roughly
approximates the position of the outer limit of Wisconsin glaciation
(King 1965:836). That there is no closer correspondence between
the limit of glaciation and the zero isobase is the result of
various factors such as the impossibility to determine uplift, or
lack of it, where there are no indications of former levels preserved in beaches or other suitable features. It should also be
kept in mind that in the Great Lakes Region, for instance, the
Maumee zero isobase represents only the outer limit of measurable
crustal warping in that region (Fig. 43 and Fig. 47). This is so
because, as Flint (1957:244) points out

> . . . as soon as the ice began to thin, the crust began
> to rise. But actual displacement of the glacier margin
> had to take place, permitting lake or sea water to occupy
> part of the region formerly covered with ice, before
> shoreline-making could start. Hence it is inevitable
> that some recovery of the crust by upwarping must have
> occurred before even the earliest strandlines were
> fashioned. As there are no means of measuring the
> amount of this early recovery, the value of total measured
> recovery . . . is a minimum and total actual recovery was
> greater than this by an unknown amount.

The discovery of negative isostatic anomalies in Fennoscandia,
suggesting that the land there still has to rise about 200 metres
before equilibrium has been restored, led to gravity investigations
in Canada designed to determine if analogous conditions exist in
the Hudson Bay region (Innes and Weston 1966:171). The separation
of the gravitational effects due to deep-seated sources from those
arising from near-surface mass distributions is a difficult problem
and usually an intractable one. Nevertheless it is clear that
gravity variations due to density changes within the upper parts
of the crust are superimposed upon strong regional trends in which
the anomalies are systematically more negative in the vicinity of
Hudson and James Bay and reach minimum values in an area believed
to have been the locus of maximum glacial loading (Innes and Weston
1966). Although gravity studies support the contention that postglacial and presently continuing uplift resulted from the removal
of an ice load, Innes and Weston (1966:175) recognized ". . . that
both the Canadian and Scandinavian Shields were rising prior to
the onset of glaciation, as the result of some fundamental but
unknown process. It may be concluded, therefore, that their recent
uplift is the combined result of short term glacio-isostatic effects
of large amplitude, superimposed upon tectonic events having a much
longer time-scale." The same thought is also expressed by King
(1965:837): "All the Quaternary isostatic movements were superposed
on a much greater, long-term trend of epeirogenic movements in the
Canadian Shield and its surroundings."

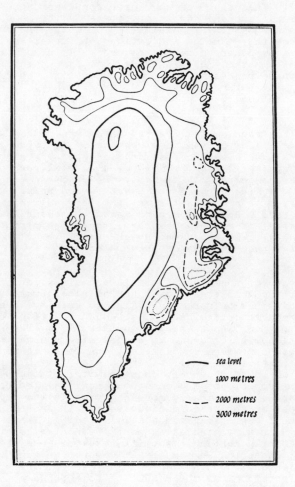

Figure 48.   Greenland, bedrock topography. The map is a contour map
             in meters of the ice-rock interface showing a large
             central part below sea level. After Hamilton (1958:121).

The failure to recognize processes other than glacio-isostatic uplift to account for raised and tilted strandlines may lead to misinterpretations of the postglacial history, particularly in places where anomalously high uplift conditions exist within deglaciated and risen regions. Sim (1961) in describing high-level marine shells from Ellesmere Island, mentioned the possibility that there, marine submergence may extend well above the elevations usually suggested. He did not discuss the possibility that local or regional earth movements other than glacio-isostatic uplift could account for this anomaly and other abnormally high levels in Greenland and Ellesmere Island.

Analogies

A contour map of the ice-rock interface presented by Hamilton (1958:121), here shown as Fig. 48, shows that a large part of Greenland lies below sea level. If a cross section through this part is presented (Fig. 49) it clearly shows the present bowl-shape. Similarly, parts of central Antarctica are below sea-level (Fig. 50) in places as much as 750 meters, comparable in value to local depression under the thickest ice in Greenland. The cross-section used to illustrate this does not show such a pronounced regional down-bending as the Greenland one. It has to be kept in mind, however, that a direct comparison between the two cross-sections can not be made because of differing scales. The original sections have about the same vertical scale; the Greenland section has a vertical exaggeration of 40 times. The Antarctic section having a horizontal scale almost half that of the Greenland section has therefore a vertical exaggeration of about 80 times.

The ice-bedrock interface in both Greenland and Antarctica was determined during traverses across the ice caps, determining their topography, and by seismic reflection shooting their thicknesses. Because only a limited number of shot points can be established under the difficult conditions of travel, details of the ice thickness profiles are filled in with measurements of the gravity field which is particularly sensitive to changes in ice thickness because of the large density contrast between ice and rock (Bentley *et al* 1964:1). The chief source of error in the seismic method, because of the absence of velocity surveys in drill holes penetrating the ice from surface to bottom, is an uncertain variation in seismic wave velocity with depth in the ice sheet. Where the ice near the base may have a large load of debris the seismic velocity may approach that of the rock underneath. If in such places the first change in velocity is regarded as indicating solid rock the ice thickness may actually be greater than calculated. Such difficulties described by Hamilton (1958: 115-116) from south Greenland where " a layer of frozen moraine underlies the ice and that the reflections normally obtained are from the interface between the ice and the moraine" are mentioned here only to demonstrate that a certain amount of interpretation (rather than direct observation as is the case with the topography

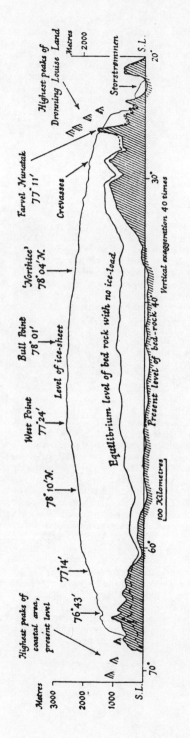
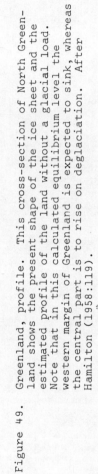

Figure 49.   Greenland, profile. This cross-section of North Greenland shows the present shape of the ice sheet and the
estimated profile of the land without a glacial load.
Note that in this calculated equilibrium level the
western margin of Greenland is expected to sink, whereas
the central part is to rise on deglaciation. After
Hamilton (1958:119).

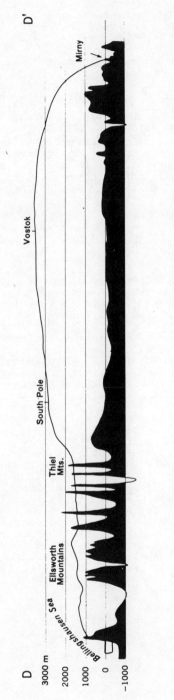

Figure 50. Antarctica, profile. This cross-section from the
Bellinghausen Sea across the South Pole to Mirny shows
the present topography of the ice sheet and the ice-rock
interface. After Bentley *et al*., (1964: plate 2,
section D-D').

of the ice) goes into the reconstruction of the ice-rock
interface. It should not be assumed, however, that such errors
are large enough to invalidate the regional reconstruction of
that interface.

Even if the existence of low topographic basins.below the
ice caps of Greenland and Antarctica is accepted on the basis of
the seismic and gravity evidence it does not necessarily follow
that it is soley the result of a loading phenomenon caused by the
ice. The possibility of independent earth movements should be
considered as well as the possible effects of deep erosion as
shown in Fig. 51, if the model presented by Dapples (1959:407)
is correct. It is generally held, however, that ice-loading
has been the main cause of the depression of Antarctica and that
any erosional effect, if present, has been only a minor contributing factor because the continent is now essentially in isostatic
equilibrium as determined from seismic and gravity data (Innes and
Weston 1966:175).

                            Mechanics

If the cross-section of Greenland (Fig. 49) is studied it
can be seen that whereas the central part would rise in the event
that the ice-load were removed it is held, in this particular
reconstruction, that the margins of Greenland would sink. In this
view the margins are regarded as having been bowed up too high at
present and thus the concept of a forebulge underneath the ice-cap
margin is introduced. It is mainly in this respect that an analogy
between Greenland and North American Pleistocene glaciation does
not hold. As already mentioned above, in the Great Lakes region
the postglacial movements are either up or non-discernible (Fig.43),
the presently continuing movement is positive. No negative movements have so far been found distributed in a systematic manner
surrounding the region of postglacial uplift. In Europe superelevation of southern Denmark, northern Germany, The Netherlands,
and the floor of the North Sea (regions which by now have "snapped
down" nearly to their original preglacial positions but are still
sinking at present as suggested by tide-gauge measurements) has
been assumed by some authors (Daly 1926:200; Gutenberg 1941:738).
Recent studies in The Netherlands cast doubts on the interpretation
that downsinking of the Dutch coast is to be regarded as solely a
negative glacio-isostatic movement. A large part, if not all, of
the downward movement may be the result of compaction ("inklingking"
in Dutch) of the unconsolidated sediments characteristic of the
Danish, German, and Dutch coasts and of the North Sea Basin. Non-
glacial epeirogenic movements may also be involved in this sedimentary basin. For instance, it has been suggested by some that downward movements have resulted from solution of the Permian Zechstein
salts.

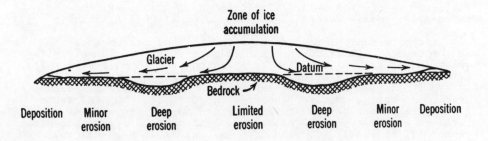

Figure 51. Ice sheet model. This conceptual illustration of an
            ice cap shows zones of accumulation, deep erosion, minor
            erosion and deposition. After Dapples (1959:407).

The absent or tenuous evidence for any presently negative
earth movements in the marginal area of former continental glaciers
should be kept in mind when the mechanics of crustal response to
glacial loading are considered. Although it is generally stated
that crustal uplift in formerly glaciated tracts is due to horizontal flow of subcrustal material tending to restore equilibrium
following the melting of the ice (Innes and Weston 1966:175) it
should also be pointed out that "the question of what happened to
the displaced substance remains unanswered" and that although
"displacement of rock material at depth is implied, evidence of
compensating uplift during glaciation is negative." (Flint 1957:257).

It can be demonstrated for the Great Lakes region that the
rate of crustal uplift during glaciation changes sensitively with
variations in mass of the glacier. In this region an inward shift
of the outer limit of deformation with time can be noticed. If
Fig. 43 is studied it can be seen that the outer limit of warping
for each of the successive beaches lies north of its predecessor.
If time determinations of the beaches are taken into account a
northward migration of 160 miles during about 8,000 years becomes
evident (Flint 1957:254). In North America it is believed that
very likely the same sort of migration took place all around the
center of glaciation, although not exactly contemporaneously. In
Fennoscandia such a migration is, however, not clearly demonstrable:

> *There the zero isobases of successive water bodies seem
> to occupy nearly the same position. The difference
> between North America and Fennoscandia may be a result
> of the fact that in Fennoscandia the water bodies lay
> nearer the center of the ice sheet than did the North
> America lakes. Therefore they did not form until a
> greater proportion of the maximum volume of the ice had
> wasted away, and hence a greater proportion of the total
> crustal adjustment had taken place, than had occurred in
> North America before the great glacial lakes appeared.
> (Flint 1957:254)*

The mechanics of the postglacial uplift as the result of the
removal of the continental glaciers in Europe and North America
are discussed at length by Daly (1934:119).

> *The rocky floor under any of the Pleistocene ice-caps
> sank for two different reasons: because of the earth's
> elasticity, and also because of the earth's plasticity
> . . . . Under the weight of each meter of ice added to
> an extensive ice-cap, the whole planet is immediately
> elastically, distorted a little. If without delay that
> meter of ice be removed, the earth immediately takes on
> its original shape. Both reactions are those of an almost
> ideally elastic body. But against extensive loads of
> prolonged application the earth's materials flow and the
> crust is basined, plastically, under the load.*

*Suppose, on the other hand, that an extensive ice-cap
completes its basining of the earth's crust and then
melts away. The uncovered region rises, at first by
elastic response of the earth, and later by a plastic
response. The rise is the greatest in the central region
where the ice was thickest, so that the ultimate result
is an updoming of the glaciated tract.*

Two theories about the mode of plastic recoil are reviewed by
Daly (1934:120-126). They can be summarized as follows·

Bulge or wave hypothesis

Basining is believed to have been accompanied by outward,
horizontal flow in the substratum and just beneath the crust
(Fig. 52).A comparatively narrow bulge is formed and horizontal
flow takes place at small depth, not far from the 100 kilometer
level. The bulge hypothesis demands pure bending of the crust,
but not localized zones of vertical fracturing.

Punching hypothesis

Outward horizontal flow is believed to have taken place at
great depth, below the 1000 km. level. The plastic basining is
supposed to have been accompanied by an extremely slight uplift
of the continent surrounding the glacial tract. The low, broad
peripheral bulge is of a much smaller order of magnitude than
that assumed in the bulge theory (Fig. 53). The crust, on unweight-
ing by the melting of ice, yields to the stress and rises plastically. The floor of the ice-cap is punched up along a circumferential
zone of vertical fractures, the "hinge-zone"

Daly (1934:123-124) prefers the second hypothesis of the
deleveling of the earth's crust because it is "supported by the
discovery of actual hinge zones in both North America and Europe."
He believed that in the hinge zones "the individual displacements
in the vertical sense should be small and not represented by
conspicuous fractures and 'faulting' at the surface of the earth."
(Daly 1934:147)

Whichever theory is preferred, "the disposition of the displaced
rock material during glacial maxima remains a mystery". (Flint 1957:
241) Both theories require some uplift outside the glaciated tract
during times of glaciation. They differ only quantitatively in this
respect. But the evidence for any peripheral elevation is "wholly
negative, even where shore features and stream terraces should
reflect such movement and afford a basis for measuring it."
(Flint 1957:241)

To substantiate his contention that the punching hypothesis,
which implies sudden displacements in the outer shell of the earth,
is to be preferred over the wave hypothesis, Daly (1934:127-128)
calls attention to the position of earthquake centers in Denmark
and also states that" . . . occasional earthquakes centering in the
Province of Quebec may possibly be connected with . . . postglacial

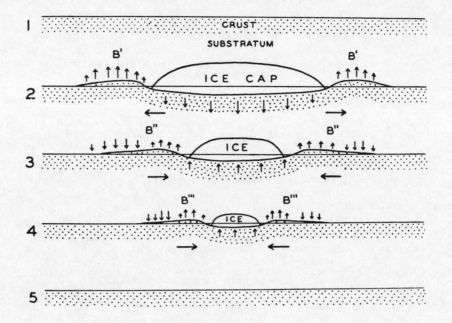

Figure 52. Bulge Hypothesis. The sections illustrate the bulge
            or wave hypothesis to explain the plastic upwarping of
            a deglaciated tract. After Daly (1934 [1963]: 121).

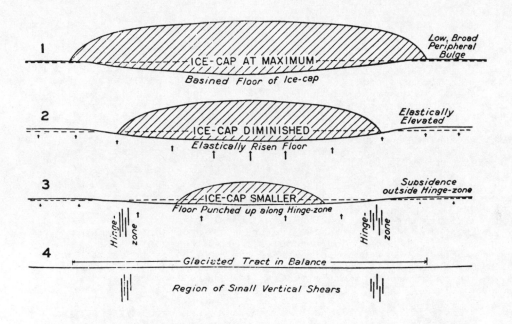

Figure 53. Punching hypothesis. The sections illustrate the
            punching hypothesis to explain the plastic upwarping
            of a deglaciated tract. After Daly (1934 [1963]: 122).

recoil." As far as seismicity in North America is concerned it
should be reiterated that most of the glaciated tract has few, if
any, earthquakes and that those in the St. Lawrence valley are
generally regarded as of tectonic origin unconnected with glacio-
isostatic uplift. (Kumarapeli and Saull 1966:641)

                            Summary

     The idea that postglacial uplift of formerly glaciated tracts
in Fennoscandia and North America is mainly the result of the
removal of the superimposed weight of ice on the crust is not
criticized in principle but a review of the observations on which
this idea is based shows that in places detailed information is
still lacking, and that other possible causes are generally not
considered. The various observations and conclusions are summarized in the following series of statements.

1. Field evidence of postglacial uplift is provided by abandoned
   strandlines substantially above present ocean or lake levels.

2. Altitude determinations on these abandoned strandlines show
   that they are higher above the present level surface where,
   according to other independent evidence, the continental
   glaciers were thickest than near the margins of those glaciers.
   Regionally, contour lines drawn on the elevated strandlines
   are therefore concentric to the area where the ice was the
   thickest (Fig. 42; Fig. 47).

3. In both Fennoscandia and North America the zero contour of
   uplift lies generally north of and approximately parallel to
   the limit of the latest glaciation.

4. The elevated strandlines are believed to have been originally
   level surfaces and any gravitational attraction of the water
   by the ice mass and any possible surface gradient of flowing
   lake water is regarded as being quantitatively negligible.

5. The height of the abandoned strandlines above the present level
   surface is believed to provide only a minimum measure of the
   actual uplift which has taken place for various reasons: (a)
   when the ice thinned some uplift took place already but the
   covering of the land by water and the attendant creation of
   strandlines had to await the total disappearance of the ice,
   and was possible only because of a time lag between water
   invasion and full recovery; (b) postglacial uplift and deleveling
   was accompanied by an eustatic rise in sea-level.

6. Postglacial uplift data are most complete for the Fennoscandian
   region. Fewer data are available for North America where the
   most detailed work has been done in the Great Lakes area. The
   well-developed raised beaches of Hudson Bay and the Canadian
   Arctic Archipelago await further study and even in the Lake

Agassiz region some necessary detail is still lacking. "Of the strandlines of Lakes Regina, Souris and Agassiz . . . we can say that they are warped up toward the north and east, the highest ones the most steeply in the case of Agassiz. But the unraveling of the discontinuities awaits detailed measurements that have not been made." (Flint 1957:255)

7. In the Lake Agassiz region and elsewhere detailed studies are beset by uncertainties in tracing wave-cut cliffs and beaches, and by relating such shore line features to the former water level. Introduction of reasonable margins of error in height determinations of former water level may eliminate or weaken the evidence for some hinge-lines. The question ". . . whether warping on a large scale or only tilting of blocks of the earth's crust occurred" (Johnston 1946:1) has not yet been answered. It is possible that only in some localities actual failure of the crust along hinges occurred whereas in others a slight bending or warping took place.

8. Surface expressions of hinge lines have not been recognized and the record of earthquakes in North America does not suggest any movement along such hinges in historical times. The Canadian Shield is the region of lowest seismicity on the continent and earthquakes in other more mobile regions, such as the St. Lawrence valley, are generally attributed to earth movements unconnected with deleveling following removal of the weight of an ice sheet. An area of higher than average seismicity in Denmark has been related to the position of a hinge line by Daly (1934:128).

9. Detailed study of the region surrounding the Glacial Lake Agassiz basin is required to establish if the postulated hinge lines in the basin can be related to ice marginal positions in the land area. A correlation of the postglacial history of the lake basin and that of the surrounding land area is needed to determine if the hinges resulted from the periodic retreat and intervals of halting or readvance of the glacier margin.

10. Although the weight of the continental ice sheet depressed the crustal surface in North America, that surface was not perfectly level before the glacier developed. The preglacial drainage pattern indicates a gradient sloping from western Canada down to what is now Hudson Bay.

11. The deep erosion by the continental ice sheet peripheral to its areas of accumulation probably contributed only insignificantly to the bowl shaped surface of the earth underneath the ice.

12. Field evidence of continuing uplift of formerly glaciated regions is provided mainly by level gauges along the present shores of the ocean or lakes.

13. Abundant data of continuing upward movement are available
    for Fennoscandia; fewer data are available for the Great
    Lakes area; minimal data, including, besides one gauge station,
    some historical records, are available for Hudson Bay.

14. Data suggesting that the land surrounding Hudson Bay is at
    present still rising are slender, may be unreliable, and have
    been disputed by some investigators.

15. A rate of present uplift of about 1 meter in 100 years has been
    calculated for Fennoscandia (Fig. 42) and the average rate of
    warping of the Great Lakes area in North America is given by
    Flint (1957:249) as "a little less than 1 mm/100 km/yr, not
    quite as much as the rate in the Baltic region." In the
    absence of reliable data for Hudson Bay the present rate of
    uplift of that region can be given only by analogy with the
    better studied regions of Fennoscandia and the Great Lakes
    region.

16. The contention that the Hudson Bay region is not yet in isos-
    tatic balance, and therefore still subject to rising until
    that balance is restored, is supported by gravity measurements.
    However, the observed negative anomaly which increases toward
    the locus of greatest former ice thickness may be interpreted
    as the combined result of short term glacio-isostatic effects
    of large amplitude, superimposed on tectonic events having a
    much longer time-scale.

17. The failure to recognize the influence of earth movements other
    than glacio-isostatic uplift may lead to misinterpretations,
    particularly where anomalous uplift conditions exist within
    formerly glaciated regions.

18. Cross-sections used to illustrate that interior parts of
    Greenland and Antarctica are below sea level and that the
    land underneath the ice is presently bowl-shaped are based
    on seismic and gravity measurements. Possible errors in the
    ice thickness determinations are not believed to be of such
    an order of magnitude that they invalidate the concept of a
    presently downwarped surface underneath the glaciers.

19. The effects of glacial erosion on the present position of the
    bedrock below sea level are believed to be minor in the
    interior regions of Greenland and Antarctica but probably had
    a profound influence on the depths of fiords occupied by outlet
    glaciers in the marginal areas (Fig. 50).

20. Ice-loading is believed to be the main cause of the depression
    of Antarctica because the continent is now essentially in
    isostatic equilibrium.

21. The results of future unloading of Greenland can be anticpated
     as shown in Fig. 49, which presents a prediction introducing
     the concept of a fore-bulge, where on deglaciation negative
     earth movements would take place. In some geology textbooks
     (Zumberge 1958:73) post-glacial uplift of Greenland, accom-
     panied by an eustatic rise of sea level, is shown without
     any change in shape from the presently prevailing one under-
     neath the ice (Fig. 54).

22. The evidence for presently continuing downward movements out-
     side the glaciated tract in Western Europe is weak. The
     interpretation that subsidence resulted as a compensating
     movement to glacial unloading is controverted by some
     investigators.

23. Evidence for negative movements in North America outside the
     formerly glaciated area is absent and the question as to what
     happened to the supposedly displaced crustal material has not
     yet been answered.

24. Theoretical considerations of the mechanics of glacial load-
     ing and postglacial uplift are only in part supported by
     acceptable field evidence. The main points still being
     disputed are, (a) the importance of hinge lines, and (b) the
     presence and size of a fore-bulge.

Acknowledgements

Attendance of the Lake Agassiz Conference and the consequent
preparation of the paper were made possible from grants by the
National Research Council and the Saskatchewan Research Council.

References

BARTON, R. H., *et al.*,
   1965   Quaternary. In "Geological History of Western Canada,"
          edited by R. G. McCrossan and R. P. Glaister. *Atlas*
          Alberta Society of Petroleum Geologists, pp. 195–200. Calgary

BELL, ROBERT
   1879   Report on exploratins on the Churchill and Nelson Rivers
          and around God's and Island Lakes, 1879. *Geological Survey*
          of Canada, Report on Progress for 1878–79, part C, 72 pp.
          Ottawa.

BENTLEY, C. R. *et al*,
   1964   Physical Characteristics of the Antarctic Ice Sheet.
          American Geographical Society Antarctic Map Folio
          Series, Folio 2, 10 p., 10 maps. Washington.

COULOMB, JEAN and GEORGES JOBERT
   1963   *The Physical Constitution of the Earth.* Hafner Publishing Co.

DALY, R.A.
   1926   *Our Mobile Earth.* Scribner. New York.
   1934   [Reprinted 1963], *The Changing World of the Ice Age.*
          Hafner Publishing Co., New York.

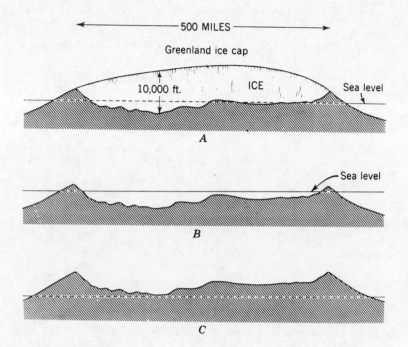

Figure 54.   Eustatic and isostatic rise. (A) The land beneath
              Greenland is depressed below sea level because of the
              weight of the overlying ice. (B) If the ice melted,
              the crust would begin to rise, but would lag behind
              the sea-level rise caused by return of glacier ice to
              the hydrologic cycle. (C) Final isostatic balance,
              reached long after the ice melted, would raise the
              entire land surface above sea level. In these diagrams
              only the effects of an eustatic rise of sea level and
              an isostatic rise of Greenland without a change in shape
              from a bowl to that of a dome are considered. After
              Zumberge (1958:73).

DAPPLES, E. C.
    1959 *Basic Geology for Science and Engineering.* Wiley, New York.
FLINT, R. F.
    1957 *Glacial and Pleistocene Geology,* Wiley and Sons, New York.
GUTENBERG, BENO
    1941 Changes in Sea Level, Postglacial Uplift, and Mobility
          of the Earth's Interior. *Geological Society of America,
          Bulletin.* Vol. 52, Part 1, pp. 721–772. New York.
GUTENBERG, B. and C. F. RICHTER
    1949 *Seismicity of the Earth and Associated Phenomena.*
          Princeton University Press, Princeton.

HAMILTON, R. A.
    1958   *Venture to the Arctic.* Penguin Books Inc., Baltimore.
HOLMES, ARTHUR
    1965   *Principles of Physical Geology.* Thomas Nelson, London.
HOWELL, J. V. *et al.*
    1960   *Glossary of Geology and Related Sciences.* AMerican
            Geological Institute, Washington. (Second Edition).
INNES, J. S. and A. A. Weston
    1966   Crustal Uplift of the Canadian Shield and its Relation
            to the Gravity Field. *Annales Academiae Scientiarum Fennico*
            Vol. No. A. 111 90, pp. 169–176.
JOHNSTON, W. A.
    1915   Rainy River District, Ontario. Surficial Geology and
            Soils. *Geological Survey of Canada. Memoir* No. 82.
            Ottawa.
    1939   Recent changes of level of the land relative to sea
            level. *Amer. Jour. Sci.* v. 237, No. 2, p. 94–98.
    1946   Glacial Lake Agassiz, with Special Reference to the Mode
            of Deformation of the Beaches. *Geological Survey of
            Canada, Bulletin* 7. Ottawa.
King, P. B.
    1965   Tectonics of Quaternary Time in Middle North America.
            In "The Quaternary of the United States," edited by H. E.
            Wright Jr. and D. G. Frey, pp. 831–870. Princeton.
KUMARAPELI, P. S. and V. A. SAULL
    1966   The St. Lawrence Valley System. A North American
            equivalent of the East Africn Rift Valley System.
            Canadian Journal of Earth Sciences, Vol. 3 No.
            pp. 639–658. Ottawa.
NELSON, S. J. and R. D. JOHNSON
    1966   Geology of Hudson Bay Basin. *Canadian Petroleum Geologists,
            Bulletin.* Vol. 14, No.   pp. 520–578. Calgary.
SIM, V. M.
    1961   A Note on High-level Marine Shells on Fosheim Peninsula,
            Ellesmere Island, N.W.T. *Geographical Bulletin,* No. 16,
            pp. 120–122. Ottawa.
TYRREL, J. B.
    1889   Notes to Accompany a Preliminary Map of the Duck and
            Riding Mountains in North-western Manitoba.
            *Annual Report of the Geological and Natural History
            Survey of Canada for 1887–88,* No. III, Part I,
            pp. 1–16. Ottawa.
UPHAM, WARREN
    1880   Preliminary Report on the Geology of Central and Western
            Minnesota. *Eighth Annual Report of the Year 1879 of
            the Geological and Natural History Survey of Minnesota,*
            Vol.   pp. 70–125. St. Peter, Minnesota.
    1896   The Glacial Lake Agassiz. *United States Geological Survey*
            Monograph No. 25. Washington.
WARKENTIN, JOHN
    1964   *The Western Interior of Canada.* McClelland and Stewart
            Limited. Toronto.
ZUMBERGE, J. H.
    1952   The Lakes of Minnesota—Their Origin and Classification.
            Minnesota Geological Survey Bulletin 35. Minneapolis.
    1958   *Elements of Geology.* Wiley, New York.

Editor's Comments on Paper 9

Elson: *Geology of Glacial Lake Agassiz*

This paper by Elson provides the most authoritative statement to date on the sediments, beach, and shoreforms, as well as the sequence of strandlines, ice recession, and advance phases, of Glacial Lake Agassiz. For its sheer size above, Glacial Lake Agassiz must be included in any discussion of the field evidence for glacial-isostasy. Elson has built on, and expanded, earlier work on Glacial Lake Agassiz, most notably by Johnston. The importance of radiocarbon dating (developed in 1959) on glacio-isostatic field interpretations is apparent in this, and subsequent, papers and provides a major break between the earlier papers and the research of the 1960s and 1970s.

Figure 5 of Elson's paper illustrates the way a strandline is deformed along a line normal to the regional isobases. Most strandlines are significantly inflected and only the lower ones appear to be tilted rather than warped. The highest and oldest shorelines are broadly deformed into nonlinear forms and no marked break, or hingeline, is shown. However, in southern Manitoba between the Campbell and Stonewall waer planes (10,500 to ca. 8300 BP) a marked flexing of the waterplanes occurs near the vicinity of the International Boundary. The younger strandlines here are only tilted. This complex of shoreline deformational patterns is intriguing. Walcott (1970, *J. Geophys. Res.*) has used Elson's data to estimte the length of the crustal flexural parameter and arrives at a figure of about 180 km.

The second excerpt of the paper selected for reprinting begins with a table of available ^{14}C dates and then proceeds to use the dates to establish a picture of the age and sequence of the various water planes and ice-front positions. These are schematically shown as Figs. 6 and 7 through 13 (not reprinted). In its later stages, the history of Glacial Lake Agassiz becomes part of the story of the deglaciation of Hudson Bay. The Glacial Lake Agassiz record is unique in terms of its geographical coverage from former margin to ice center and in the length of time that it existed. The Campbell readvance of about 10,000 BP may be correlative with the European Younger Dryas cold period.

John A. Elson is Professor of the Geological Sciences at McGill University, Montreal, Canada. He has worked extensively in the field of Quaternary geology with special attention to the problems of Glacial Lake Agassiz and the Champlain Sea of the St. Lawrence Valley.

9

GEOLOGY OF GLACIAL LAKE AGASSIZ*

John A. Elson

Introduction

Evidence of Glacial Lake Agassiz occurs in an area of roughly 200,000 square miles in the provinces of Ontario, Manitoba, and Saskatchewan, and the states of Minnesota, North Dakota, and a small portion of South Dakota. The history of our knowledge of this Lake is summarized by Morgan Tamplin (this volume). It should be emphasized that the recent work of V.K. Prest (1963) and S.C. Zoltai (1965a, 1965b) in northwestern Ontario has added much information essential to new interpretations of the Lake's history. Also, the Surveys and Mapping Branch of the Department of Energy, Mines, and Technical Surveys has nearly completed (late 1966) a series of maps at the scale 1:250,000 with 100 foot contours covering the Lake basin in Canada. Similar maps for the part within the United States have been available for about ten years. These maps make possible the delineation of water planes and the search for related ice margins and outlets. The completion of aerial photographic coverage and construction of new highways extending into parts of the basin hitherto reached only by aircraft and canoe, also have aided greatly.

Although Lake Agassiz sediments occur within an area of 200,000 square miles, that area was not all submerged at any one time. As the ice margin retreated, the southern outlet eroded deeper causing the lake to become shallower and hence to contract in the south while it expanded in the north. Subsequently, new, still lower outlets opened into other drainage basins, mainly to the east. The surface areas of most phases of Lake Agassiz probably did not exceed about 80,000 square miles, at any one time.

General Features of the Glacial Lake Agassiz Basin

Boundaries

The accuracy of the boundaries of Glacial Lake Agassiz (Fig.2) varies considerably. In western Minnesota and along the escarpment extending from South Dakota into northern Saskatchewan the strandlines are well developed beach ridges and wave cut scarps and

---

*This preliminary report is based on the presentation made in Winnipeg in November, 1966 with substantial revisions. Because of rapid evolution of thought while new data were being incorporated, several minor inconsistencies in the historical part of the text have appeared.

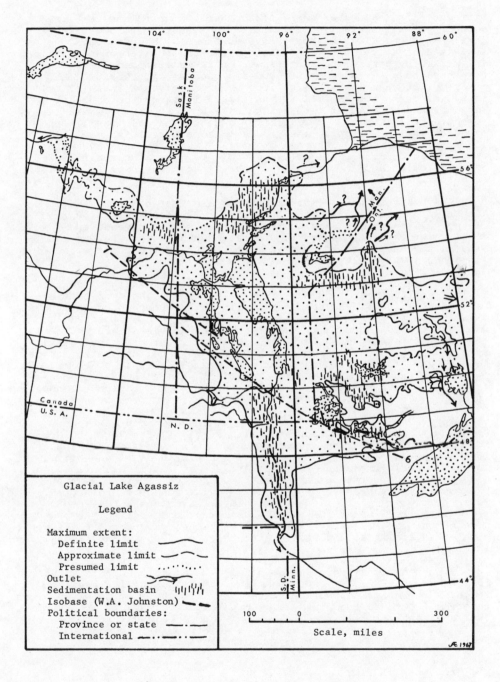

Figure 2. Glacial Lake Agassiz.

terraces. In the western part of Northern Ontario, however, the
boundary is particularly difficult to map because the coast was
mainly an archipelago of drift-covered bedrock islands that
inhibited wave action. The positions of former water planes there
are marked by strandline features (such as wave formed scarps and
terraces) only on thick drift deposits such as end moraines and
eskers. Elsewhere they are defined by the limits of areas of well-
sorted sand resting on till (which itself is sandy) or on bedrock.
Both Prest and Zoltai interpret the Lake boundary to be where this
washed deposit has an upper limit at a consistent altitude. Other
evidence is less direct. The crests of some eskers and moraines
are flat (wave washed) whereas others are hummocky and have kettles
(not washed). From these it is possible to determine upper limits
of wave action and to extrapolate the presumed water plane to its
intersection with the topography.

The northern Lake limit in Ontario and Manitoba is based on
scattered geological reports containing references to: (1) varved
clays, (2) the presence of washed eskers, and (3) several end
moraines which have wave-formed terraces on one side whereas the
other sides have knob and kettle topography and therefore were not
subjected to wave action.

A basin area north of latitude 56° in the vicinity of longitude
100° in northern Manitoba has previously been included as part of
Lake Agassiz but is excluded from it in this report in concurrence
with Antevs (1931:46). This area is excluded because the basin,
which is in the Churchill River drainage basin, is separated from
the rest of Lake Agassiz (about 50 to 100 miles to the southeast)
by high ground and an end moraine that is unwashed on the west side
but has wave-formed features on the east side. The lake formerly
present in the Churchill drainage does not seem to have been confluent with Lake Agassiz according to present information, although
it may have been contemporary.

Between longitudes 100° and 104°, intensely folded metamorphosed
Precambrian rocks form topography with a relief of several hundred
feet and the lake basins enclosed by the rock ridges do not appear
from air photograph interpretation to contain lacustrine clays.
The geological literature for that area is of little assistance in
determining the northern Lake Agassiz boundary. Farther west the
rock basins do appear to contain clays. The boundary shown (Fig.2)
is in part based on the intersection of a water plane (defined by
the well-marked strandline on the escarpment that formed the south
side of the lake -- Wapawekka Hills) with the topography on the north
side of the basin at and west of Lac LaRonge, Saskatchewan (longitude
105°).

A previously unreported outlet of Lake Agassiz near longitude
109° is shown (Fig. 2). The strandline on the south side has been
traced a few miles east of longitude 106°. If the sloping water plane
inferred from the strandline is extrapolated westward it intersects
the topography at Flatstone Lake (longitude 108°, just north of
latitude 56°). There the lake discharged through four channels west
from Flatsone Lake and possibly also through the Aubichon arm of

Lac Ile a la Crosse, into the Churchill Lake basin. The Churchill
Lake basin is comparatively small and formerly contained a lake
that discharged northwestward through anomalously large channels
into the Clearwater River system. Clearwater River flows westward
to join Athabasca River where it turns north toward Lake Athabasca.
Thus there may have been a waterway extending from the Arctic Ocean
by way of the MacKenzie River, through Lake Agassiz and south by
the Mississipi River system to the Gulf of Mexico for a short time
(possibly several decades) roughly 10,000 years ago. The possibility
of nearby Methy Portage serving as an outlet was considered by Upham
(1895:231-232) and rejected because of its high altitude. The
present interpretation is based on new topographic data.

Topography

     Local relief in the Lake Agassiz basin ranges from about a
foot or two per square mile in parts of the southern (Red River)
clay belt to several hundred feet per square mile in areas of Precambrian rock in the eastern and northwestern parts of the basin.
A generalized topographic map (Fig. 3) showing the general configuration of the basin was compiled from the 1:250,000 maps mentioned
above.

     The east-facing escarpment that defines the west side of the
Lake is known as the Coteau des Prairies in the south, and as the
Manitoba escarpment farther north. It increases in height above
the lake floor from about 400 feet in the south to more than 1100
feet at Riding and Duck mountains, decreases to about 800 feet
south of Lac LaRonge, and fades out from there toward the northwest.
Seven major reentrants in the escarpment were formed by preglacial
river systems and divide it into a series of cuestas each from
20 to 100 miles long. From south to north these are: Pembina
Mountain at the International Boundary; Riding Mountain; Duck
Mountain; Porcupine Hills; Pasquia Hills; Cub Hills; and Wapawekka
Hills, southeast of Lac LaRonge. The escarpment is formed by
Mesozoic rocks, primarily the Cretaceous Riding Mountain formation
which is a silicious shale (equivalent to the softer Bear Paw
formation and the Pierre shales elsewhere).

     The reentrants are now drained by river systems, from south
to north respectively, Assiniboine River (between Pembina and
Riding Mountain), Valley River, Swan River, Red Deer River,
Saskatchewan River, Bear River and Montreal River. Of these,
the Saskatchewan is largest and the Assiniboine is next largest.
Both of these last systems are parts of extensive preglacial
drainage systems that extended across the Great Plains from the
Rocky Mountains and discharged northeastwards into the Hudson Bay
region. They formed major drainage lines during deglaciation.

     The low-lying, low relief terrain that forms the part of the
Lake Agassiz basin occupied by the present-day lakes Winnipeg,
Manitoba, Winnipegosis and Cedar Lake, extends west to Lac LaRonge
and is underlain by Paleozoic dolomite and limestone formations
that dip west or southwest at a few feet per mile. In the south

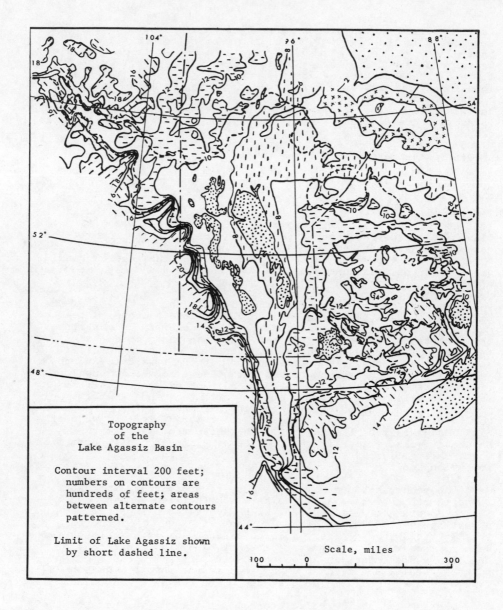

Topography
of the
Lake Agassiz Basin

Contour interval 200 feet;
numbers on contours are
hundreds of feet; areas
between alternate contours
patterned.

Limit of Lake Agassiz shown
by short dashed line.

Scale, miles

100   0   300

Figure 3. Topography of the Lake Agassiz Basin.

and west most of this rock plain is concealed under glacial and lacustrine sediments. But in the north, extending almost as far west as the Wapawekka Hills from Lake Winnipeg and Lake Winnipegosis, are many low cuestas with the scarps facing north and northeast.

On the north and east sides of this elongate lowland, the Precambrian rocks have a diversified topography. Along latitude 54°, west of longitude 100°, intricately folded belts of Precambrian metamorphic rock have a local relief of roughly 200 feet. From this area southeast to the angle in the boundary between Manitoba and Ontario, the cover of drift is relatively thick and preglacial relief of the Precambrian rocks must have been low. Much of it is an exhumed erosion surface from which Paleozoic rocks have been stripped. The relief in most of this region is usually on the order of a few tens of feet with some outcrops projecting higher. The low relief Precambrian area extends south along the east side of Lake Winnipeg in a belt extending east from the lake to longitude 96°. A straight portion of the 800 foot contour just east of Lake Winnipeg (Fig. 3) trends south southeast and joins a similar straight portion extending east from the north end of that lake. A similar straight section forms part of the 600 foot contour trending southeast just east of longitude 96°, latitude 54°. These straight contours probably represent the pre-Paleozoic erosion surface. Flat-lying Paleozoic rocks underlie the Hudson Bay lowland for roughly 100 miles or more southwest of Hudson Bay.

The topography on the Precambrian rocks in western Ontario is too diversified for summary description. In the south are belts of folded metamorphic rocks with relief as great as 500 feet. In the area north of Lake Superior and west of Lake Nipigon, through which extend several of the outlets of Lake Agassiz, are tablelands comprising Keweenawan sills and flows resting on nearly horizontal interbedded red shales and sandstones that lie unconformably on older Precambrian intrusive and metamorphic rocks. Here local relief is as great as 500 feet, although 200 to 300 feet is more general. In the region south of latitude 52°, centered around longitude 92°, local relief is generally less than 100 feet. In this area, more or less homogeneous intrusive rocks tend to form the high ground, and minor folded metamorphic rock series form lower ground. Some local relief here is due to moraines and eskers, but most of it represents bedrock.

End moraines and interlobate moraines form most of the highest ground in the area north of latitude 53° along longitude 92° and in a belt extending 100 miles west and 200 miles southeast from there (Fig. 4). A moraine forms the peninsula on the west side in northern Lake Winnipeg and extends west to form the divide between Cedar Lake and Lake Winnipegosis (Fig. 4).

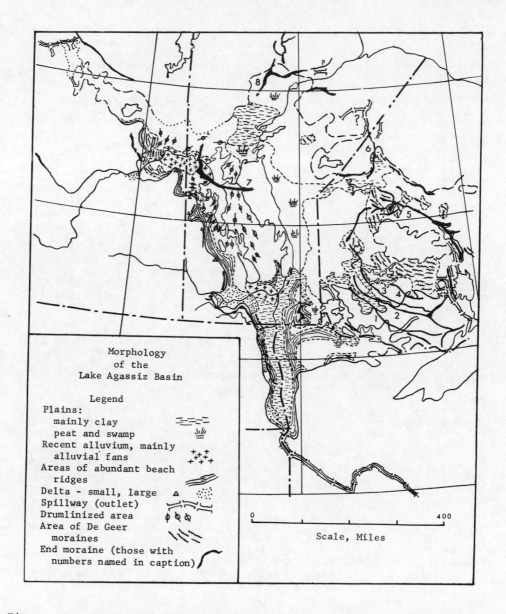

Morphology
of the
Lake Agassiz Basin

Legend

Plains:
    mainly clay
    peat and swamp
Recent alluvium, mainly
    alluvial fans
Areas of abundant beach
    ridges
Delta - small, large
Spillway (outlet)
Drumlinized area
Area of De Geer
    moraines
End moraine (those with
    numbers named in caption)

Scale, Miles

0   400

Figure 4.   Morphology and surface deposits of the Glacial Lake
             Agassiz basin.
             Numbered moraines are: 1, Eagle-Finlayson; 2, Hartman;
             3, Lac Seul; 4, Sioux Lookout; 5, Agutua; 6, Sachigo;
             7, The Pas; 8, Burntwood - Etawney.

[*Editor's note:* In the omitted pages that separate the two excerpts from Elson's paper, the author discusses the deposits within several of the major basins and describes some of the near-shore features, such as deltas. The final 16 pages are not reprinted, although they provide an extremely useful synthesis of the interpreted history of Glacial Lake Agassiz by means of maps showing the ice margin, lake extent, and lake outlets for the major glacial lake stages that Elson recognizes.]

TABLE 6

RADIOCARBON DATES[a]

Sample Number	C14 Age Years Ago	Location Lat.	Location Long.	Notes
		I. Dates directly related to Glacial Lake Agassiz.		
Y-165	12,400±420	49°47'	98°35'	Rossendale, Man., peat in alluvial fill. Science 122, 457.
Y-1327	11,740	-	-	Below Herman beach in NW Minn. Wright & Frey, 1965, Quaternary of U.S.A., p. 39.
Y-166	[11,230±480]	49°44'	98°34'	Rossendale, Man., clam shells in alluvial fill. Science 122, 457.
W-723	10,960±300	47°56'	97°22'	Grand Forks, N.D., wood in sand overlying till. Radiocarbon Supplement 2, 152.
GSC-383	10,600±150	49°46'	98°45'	Lavenham, Man., marl from valley fill. Publication pending; Radiocarbon 9.
Y-411	10,550±200	49°46'	98°45'	Lavenham, Man., wood from valley fill, locality of GSC-383. Science 126, p. 912.
GX-498	10,310±260	48°05'	93°30'	Koochiching Co., Minn. Peat from base of raised bog. Radiocarbon 8, p. 144.
W-900	10,080±280	47°50'	97°20'	Grand Forks, N.D., wood in sand. Radiocarbon Supplement 3, 88.
W-1005	10,050±300	47°46'	97°07'	Thompson, N.D., wood in gravel. Radiocarbon 6, 47.
L-563c	10,000±1000	48°49'	91°39'	Steeprock Lake, Ont. disseminated carbonate in varved clay. Radiocarbon Supplement 3, p. 145.
GSC-391	9,990±160	49°00'	95°14'	Buffalo Point, Man., wood in gravel. Publication pending; Radiocarbon 9.
W-388 C-497	9,930±280 [11,283±700]	- -	- -	Moorehead, Minn., wood in clay. Science 127, 1478. Libby, 1955, Radiocarbon Dating, p. 121.
W-993	9,900±400	46°55'	96°45'	Fargo, N.D., wood below 28 ft. clay and silt. Radiocarbon 6, 45.
W-1361	9,820±300	47°37'	97°10'	Blanchard beach, N.D., wood. Radiocarbon 7, p. 378.
W-1360	9,810±300	47°38'	97°05'	Hillsboro beach, N.D., wood. Radiocarbon 7, 378.

TABLE 6-- Continued

Sample Number	C14 Age Years Ago	Location Lat.	Location Long.	Notes
GSC-384	9,580±220	48°33'	93°29'	Roddick Tp., Ont., carbonaceous matter in marl under beach gravel. Publication pending.
W-1057	9,200±600	48°53'	95°03'	Lake of the Woods, Minn., wood from beach. Radiocarbon 6, 44.
Y-415	9,110±110	49°40'	99°33'	Treesbank, Man., wood. Science 126, 913.
GSC-9	8,860±250	51°26'	93°43'	Nungesser Lake, Ont., gyttja below alt. 1335'. Radiocarbon 4, 18.
Y-416	8,020±100	49°37'	99°26'	Stockton, Man., wood and peat. Science 126, 913.
SM-696-2	7,861±423	49°02'	94°18'	Morson, Ont., organic carbon from antler in wave-formed terrace. Can. Jour. Earth Sciences 2, 238.

II.   Minimum age for the drainage of Glacial Lake Agassiz.

GSC-92	7,270±120	58°11'	95°03'	Churchill, Man., marine shells from emerged beach, Radiocarbon 6, 170.

III.   Dates in alluvium younger than Glacial Lake Agassiz.

W-860	6,200±320	49°48'	97°12'	Winnipeg, Man., wood from silt above till. Radiocarbon Supplement 2, 175.
W-862	6,750±320	"	"	
GSC-215	3,650±140	49°45'	97°08'	Winnipeg, Man., shells and wood respectively beneath 25' silt. Radiocarbon 7, 30.
GSC-216	3,660±130	"	"	
C-723	2,684±200	48°35'	98°10'	Robbin, Minn., charcoal, Libby, 1955, Radiocarbon dating, p. 126.
C-722	2,150±400	48°35'	98°10'	Robbin, Minn., charcoal, Libby, 1955, Radiocarbon dating, p. 125.
W-1185	2,540±300	46°32'	97°14'	Sheyenne R., Richland Co., N.D., wood. Radiocarbon 7, 378.
Y-11	2,830±130	49°30'	99°04'	Cypress River, Man., wood, Science 122, 457.
Y-64	2,560±200	49°30'	99°04'	Cypress River, Man., wood, Science 122, 457.

TABLE 6--Continued

Sample Number	C14 Age Years Ago	Location Lat.	Long.	Notes
S-94	3,200±70	49°42'	98°50'	Holland, Man., wood from landslide. Radiocarbon 4, 71.
GSC-346	1,670±130	51°02'	100°31'	Grandview, Man., charcoal, Radiocarbon 8, 107.
S-178	770±50	53°34'	102°07'	Bainbridge Creek, Sask., wood, Radiocarbon 7, 231.

IV. Dates from peat deposits in or near the Glacial Lake Agassiz basin.

Sample Number	C14 Age Years Ago	Location Lat.	Long.	Notes
W-562	4,360±160	48°27'	94°00'	Lindford, Minn., Radiocarbon Supplement 2, 148.
GX-429	3,160±75	48°05'	93°30'	Koochiching Co., Minn., Radiocarbon 8, 144.
S-129	9,570±130	50°43'	99°38'	Riding Mtn. Nat. Park, Man., Radiocarbon 4, 75.
Y-418	1,400±80	51°10'	100°15'	Ashville, Man., Science 126, 913.
GSC-10	4,670±130	52°53'	99°08'	Grand Rapids, Man., Radiocarbon 8, 107.
WIS-1	2,380±90	54°07'	101°17'	Root Lake, Man., Radiocarbon 7, 405.
S-122	5,050±80	55°36'	105°17'	La Ronge, Sask., Radiocarbon 4, 73.
WIS-72	6,530±130	56°50'	101°03'	Lynn Lake, Man., Radiocarbon 8, 531.

[a]Square brackets indicate dates made by the solid carbon method.

from 1150 to 1165 feet. The cut is about 500 feet long and is
mostly slumped. The following composite section is based on
observations made in 1955 and repeated in 1964. The upper five
to twenty feet is medium grained sand of dunes which have masked
the original surface morphology of the locality, but are now
stabilized. Below this is a relict soil profile -- developed
mainly on horizontally bedded sand -- a dark grey humified "A"
horizon about two feet thick overlies a "B" horizon composed of
an iron oxide-cemented zone 1.5 feet thick at the top containing
reddish brown sand with caliche tubes formed around root openings.
This grades downward into brown sand mottled with red and grey
about one foot thick overlying horizontally bedded brown sand.
Fossil terrestrial snail shells are abundant in the cemented part
of the "B" horizon. Beneath the relict soil are lenticular silty
and sandy beds containing abundant fossil wood fragments, chiefly
transported logs, some of which have insect borings. Locally,
peat is present in a thin layer. Bones, most probably *Bison*, were
also found. The organic material is most abundant in a layer
about two feet thick in the upper part of these beds. No complete
section could be observed because of the slumping, but these beds
appear to be from three feet thick at the east end to twelve feet
at the west end of the exposure.. Wood from near the base of this
layer gave a radiocarbon age of 9,110 years (Table 6, Y-415).
Underlying the fossiliferous sand and silt are up to five feet of
sand and gravel resting on till. The contact is erosional and the
surface of the till has a relief of fifteen feet.

A tentative interpretation of this sequence can now be presented.
After an episode of downcutting and erosion of the till, the river
aggraded and shifted its course so that the fossiliferous silt and
sand was laid down in relatively quiet water. Deposition was followed
by some downcutting and lowering of the water table, in order for the
soil profile to develop. The soil, formed in a humid to slightly
subhumid climate, was followed by an episode of eolian action,
possibly resulting from a drier climate. This interpretation is
compatible with climatic events that caused the transition from
pollen zone 1 of Ritchie (1964) to pollen zone 2 in the southern
part of the Riding Mountain area that occurred sometime after 9,570
years ago (Table 6, S-129). However, the geological data are
insufficient to discriminate among three possible hypotheses of
aggradation. The cause may have been (1) a change of base level
resulting from a fluctuation of Lake Agassiz, (2) downstream extension of the river by delta growth, or (3) a result of climatic events.
This deposit is younger than the Treesbank Ferry deposits and does not
appear to correlate with the valley fill farther downstream which is
radiocarbon-dated at 1500 years older. Contemporaneity, assuming an
aberrant date, would require an unreasonable and unaccountable river
gradient of 4.5 feet per mile between this point and Steels Ferry.
Perhaps this enigmatic aggradation episode may be explained by
temporary blocking or diversion of the river as a result of a land-
slide or sand dune activity.

Rossendale gully

Some gullies on the eastern front of the older delta, such as one about three miles south-southeast of Rossendale at latitude 49°47', longitude 98°35', also contain alluvial fills. In this area silty sand at least thirteen feet thick overlies peat exposed in a dugout (reservoir) now filled with water. The silt and sand are interbedded, poorly sorted and contain wood fragments below about seven feet (as observed in an augur hole). The buried peat is at an altitude of about 1055 feet and gave a radiocarbon date of 12,400 years (Table 6, Y-165). This single date is considerably older than was expected and interpretation has been held in abeyance. However, recently J.C. Ritchie (personal communication) has obtained a date of a similar order from a lake south of the delta which adds weight to the possible validity of date Y-165.

Strandlines, Including Beaches

The term "beach" is often used loosely by geologists to designate any indication of a former shore. In the strict sense it means only the sand and gravel ridge deposited by waves at the shore. The word "strandline" includes all forms that can be used as indicators of the boundary between land and water. The principal waterplanes of Lake Agassiz represented by well developed strandlines are listed in succession from oldest (top) to youngest (bottom) in Fig. 6.

Beaches

Beach ridges are the most common strandline forms in Lake Agassiz and are abundant on both sides of the Red River basin and along the Manitoba Escarpment north to latitude 55° (Fig. 4). Their usual form is a ridge two to fifteen feet high, but locally as high as about thirty feet where spits have extended across embayments. Width ranges from about 150 to 500 feet but commonly several are grouped together into complexes a half-mile or more wide. Their lengths between gaps (such as embayments and deltas), are mostly tens to scores of miles. Where coastal slopes change, beaches may change into terraces or wave-cut cliffs. Many beaches are partly buried under alluvial fans near the foot of the Manitoba Escarpment. Some have been destroyed by erosion of younger, lower shores. Beach ridges are easily traced on air photos because they have a vegetation that contrasts with growth on adjacent deposits. For beach formation, waves must attack a material (generally till) that will yield sand and gravel-size particles. Locally, longshore transport or migration has extended sand and gravel beaches beyond their parent sources into areas of silt and clay.

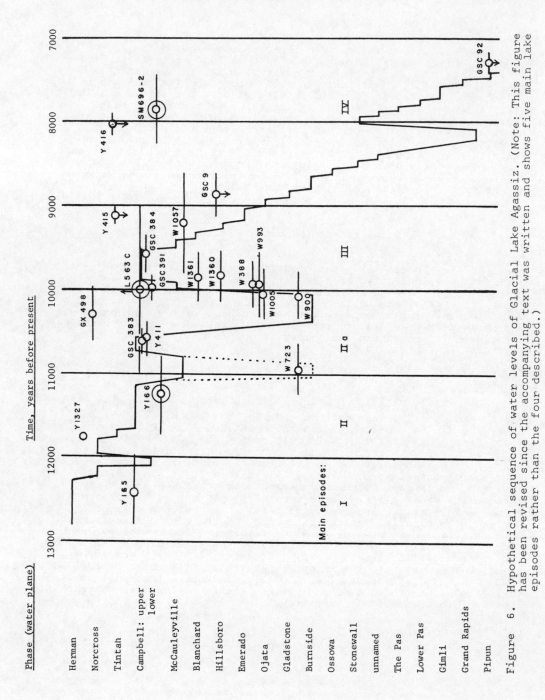

Figure 6.   Hypothetical sequence of water levels of Glacial Lake Agassiz. (Note: This figure has been revised since the accompanying text was written and shows five main lake episodes rather than the four described.)

Strandline Morphology

In addition to beaches, strandline forms include low scarps five to fifteen feet high, where waves attack silt and clay or gently sloping coasts. Sand may be moved in, laterally, by beach drift. Some low scarps that have sand at the base also have it at the top, thrown up by violent wave action. At least one such scarp (at Elm Creek, Manitoba) has been misinterpreted by Johnston (1946) as representing two water levels. He interpreted both sand bodies to be beaches. The low scarps are common in the Swan River and Carrot-Saskatchewan River reentrants.

Higher scarps, from 25 to more than 100 feet high are common only along the Campbell strandline, apparently because insufficient time was available for their formation at other levels. The scarp of the Campbell water plane has a sand beach at the base and stands highest in the silts of the Assiniboine delta. Lower scarps (25 to 40 feet) occur in till. In some places, the Campbell strandline follows bedrock scarps but these are exhumed preglacial, lithologically controlled features.

Wave-cut terraces armoured with a lag concentrate of boulders are common on the southern part of the Cretaceous escarpment.

Although ice-pushed ridges resulting from the pressure of floating ice driven by the wind are common around the modern lakes of the region, they have not been certainly identified among the Lake Agassiz strandlines except on the crest of the Eagle-Finlayson moraine in Ontario. There, boulders averaging 2.5 feet in diameter form ridges from 3 to 10 feet high and as wide as 75 feet and several miles long.

Factors influencing the form of strandlines cannot be discussed in detail here but they include:

(1) the slope of the coast -- this determines whether eroded detritus will remain in the zone of wave action to form a beach or will be deposited in deep water out of reach;

(2) the depth of water as determined by the slope governs the width of the area across which wave energy is dissipated;

(3) the fetch, and hence the size, of the waves depends on the wind direction as well as the size and shape of the lake;

(4) the type and erodability of material attacked by the waves, e.g., till, clay, or sand;

(5) the direction of wave attack which depends on both the effective storm wind directions and the configuration of the coast;

(6) the duration of the water level:

(7) the frequency of storms.

The identification of a water level represented by relict
strandlines is basic to tracing them across gaps, as well as basic
to theorizing about crustal uplift. Usually the points measured
by levelling are the crests of beach ridges and the toes of scarps,
but Upham (1890:90E) used the base of the lakeward slope and
recognized that the crests were five to fifteen feet above the
lake level. According to Bagnold (Bascom 1964:198-199) the crest
of a beach ridge is generally 1.3 time the deep-water height of the
wave that built it above the water level. However, the toe of a
wave-cut scarp is less consistent and may be above or below water
level, depending on wave height and also on the geological structure
of the coast. Hence, tracing strandlines across gaps is not a
simple matter of extrapolating the crest of a beach, especially if
the beaches form changes.

Duration of water levels

Radiocarbon dates bracketing Lake Agassiz indicate a duration
of roughly 4,500 years. About 55 water planes are represented by
strandlines (not including five or more water planes that existed
during low-water phases). Hence the maximum average duration of
the waterplanes was less than 80 years, not allowing for the time
taken for the water levels to change. Wide departures from this
hypothetical average are certain.

The duration of the Campbell water planes (the lowest phase of
Lake Agassiz to discharge south through the Minnesota River valley)
was obviously longer than any others because the massive beach
(and the escarpment behind it), which together can be followed for
about 1500 miles, required more energy for their formation. The
properties of the sediments in beaches is indicative of the wave
energy expended on them, hence their duration. At any given locality,
the sands and gravels of the Campbell beach are well rounded and well
sorted compared with other beaches in the vicinity. The degree of
rounding of pebbles may be indicative of duration. Sorting involves,
as well, the violence (energy) of individual storms and may be more
related to wave height, hence fetch.

A preliminary study of roundness of three samples of beach
gravel from 4 to 32 mm. in size gave the following results:

(1) Herman beach, Larimore, N.D. -- roundness 0.56, fetch
    150 miles;

(2) Campbell beach, Dauphin, Man. -- roundness 0.62 fetch
    250-300 miles;

(3) Burnside beach, Amaranth, Man. -- roundness 0.58 150-200
    miles.

The scale of roundness ranges from 0.1 for a sharply angular particle
to 1.0 for a sphere.

Effect on beaches of lake size, shape and orientation

Some relationships between wind velocity, duration, fetch,
waves and beaches are shown in Table 7. An index of wave energy
and the minimum water depths for which these relationships hold
are also given. In speculating on the form of coastal features it
is assumed that the effective storm wind directions were mainly
from the northeast and northwest as at present.

The variation in the effectiveness of waves on various segments
of a given strandline is illustrated by combining the information
in Table 7 and Fig. 9 (Campbell phase). Waves attacking the Pembina,
Riding and Duck mountains had a fetch of 200 to 300 miles. The
energy index of these ranges from 90 to 180 and beaches 12.5 to 17
feet above water level were formed. In this region the Campbell
scarp is well developed and the beach is massive. On the west side
of the Red River basin south of latitude 48°, the fetch for the same
winds (from the northeast) was 25 to 100 miles, the energy index
only from 4 to 30, and corresponding beach heights are from 2.5 to
7 feet. In marked contrast, waves generated by northwest winds with
a fetch of 400 to more than 500 miles attacked the east side of the
Red River basin north of latitude 48° in Minnesota. These were no
doubt somewhat impeded by shallow water offshore and interference of
an island in southeastern Manitoba, but for part of the coast the
energy index was greater than 275 and beaches 21 feet high could
have been formed. In this area the Campbell beach is a spectacular
ridge system 90 miles long.

The effects of the lowering of lake level are apparent if Table
7 is examined in conjunction with Figs. 7 to 10. For example, consider the west shore at latitude 48° which was attacked mainly by
waves from the northeast. The fetch changed from about 150 miles
during Herman time to 200 miles or more during Norcross and Campbell
times, although then partly impeded by an island in the east; it was
then reduced to 50 miles in McCauleyville time. The energy index
was initially 60, increased to 100 and then decreased to 10. Corresponding beach heights are 10, 13, and 4 feet respectively. This
locality is obviously not one in which to compare beach characteristics for purposes of estimating duration of various waterplanes,
except for the highest ones. A suitable locality for comparing
beaches is along Riding Mountain where the basin was deep and the
fetch was more than 200 miles during most of the lake's existence.
Unfortunately, alluvium covers much of the beaches there, but Duck
Mountain beaches are not covered so extensively and are in almost
as strategic a position. A study of them is in progress.

TABLE 7

CONDITIONS OF WAVE AND BEACH FORMATION

1 Fetch Miles	2 Wave height feet	3 Beach height feet	4 Energy index	5 Wind Velocity mph	6 Generation time hours	7 Deep water limit feet
25	2	2.5	4	14	4.5	14
50	3	4	10	18	7	22
100	5.5	7.5	33	24	11	40
150	7.5	10	59	27	14	54
200	9.5	12.5	94	30	17	68
250	11.5	15	132	32	19.5	80
300	13.5	17.5	177	34	22	93
350	15	19.5	225	35	24.5	102
400	16.5	21.5	275	36.5	26.5	116

Notes:

This table was prepared by converting data given by Bascom
(1964:53, Table III) from nautical to statute miles and plotting
them as graphs. The figures above are interpolated from the graphs
and rounded off to the nearest half or whole number. They are, thus,
orders of magnitude, not precise data.

The column headings refer to the following facts.

(1) Fetch is the distance the wind blows across water in
     creating waves.
(2) Wave height is vertical distance between the trough
     and the adjacent wave crest.
(3) Beach height is measured from the mean water level to
     the crest of the ridge.
(4) The energy index is dimensionless, equal to the square
     of wave height, and is proportional to the amount of
     kinetic and potential energy stored in a wave.
(5) The velocity given in miles per hour is that which if
     acting on the corresponding fetch would produce waves
     of the height given. No greater waves could be
     produced by this wind.
(6) The time in hours required for the given wind to
     generate its maximum wave height. (5 and 6 together
     are relevant to frequency and duration of storms
     effective in beach construction.)
(7) Minimum depths of water necessary for development of
     these waves. In shallower water waves are impeded by
     drag on the bottom.

Appendix to Paper 9

The symposium volume *Life, Land and Water,* edited by W. J. Mayer-Oakes, was published by the University of Manitoba Press in 1967.

Revisions and Errata

Page 71: "Dates W-723 and W-900 are from wood buried in the "Ojata beach," 14 miles west of Grand Forks (Lee Clayton, personal communication). A corresponding adjustment is necessary in Fig. 6, p. 76.

Page 73, Fig. 9: The dark shaded areas west of longitude 104° and east of 92° (see Fig. 2) should be white; the area between the letters "L" and "C" south of latitude 52° at longitude 93° should be shaded light.

Bibliography

The editor worked the references of all papers in the Symposium Volume into one list at the end of the book to avoid excessive duplication. The following list was submitted with the manuscript of this paper:

A comprehensive list of 113 titles on Lake Agassiz prior to 1960 appears in Elson (1961). Only papers referred to in this paper are listed below.

Antevs, E., 1931, *Late-glacial correlations and ice recession in Manitoba;* Geol. Survey Canada, Mem. 168.
———, 1951, Glacial clays in Steep Rock Lake, Ontario, Canada; *Geol. Soc. Amer., Bull.,* **62**, 1223–1262.
Baker, C. H., Jr., 1966, *The Milnor channel, an ice-marginal course of the Sheyenne River, North Dakota;* U.S. Geol. Survey, Prof. Paper 550B, B77–79.
Bascom, W., 1964, *Waves and beaches;* Doubleday & Co., Inc., Garden City, N.Y.
Broecker, W. S., 1966, Glacial rebound and the deformation of the shorelines of proglacial lakes; *J. Geophys. Res.,* **71**, 4777–4783.
Christiansen, E. A., 1965, *Ice frontal positions in Saskatchewan;* Sask. Research Council, Geology Div., map 2.
Clayton, L., Laird, W. M., Klassen, R. W., and Kupsch, W. D., 1965, Intersecting minor lineations on Lake Agassiz plain. *J. Geology,* **73**, no. 4, 652–656.

Dawson, G. M., 1875, *Report on the geology and resources of the region in the vicinity of the Forty-ninth Parallel, from the Lake of the Woods to the Rocky Mountains.* Montreal, 1875, p. 248.

Dennis, P. E., Akin, P. D., and Worts, G. F., Jr., 1949, Geology and groundwater resources of parts of Cass and Clay counties, North Dakota and Minnesota. U.S. Geol. Survey and N.D. Geol. Survey, N. D. ground-water studies no. 10, p. 17–29.

Derry, D. R., and MacKenzie, G. S., 1931, Geology of the Ontario-Manitoba Boundary (12th Base Line to Latitude 54). *Ont. Dept. Mines, Ann. Rept.,* **XL**, 1–20.

Elson, J. A., 1957, Lake Agassiz and the Mankato-Valders problem. *Science,* **126**, 999–1002.

———, 1960, Surficial geology, Brandon, west of principal meridian, Manitoba; Geol. Survey Canada, Map 1067A in Memoir 300 by E. C. Halstead.

———, 1961, Soils of the Lake Agassiz region, in Legget, R. F., ed., *Soils in Canada,* Roy. Soc. Canada, Spec. Pub. 3, p. 51–79.

Goldthwait, J. W., 1910, *An instrumental survey of the shorelines of the extinct Lakes Algonquin and Nipissing in southwestern Ontario,* Canada, Dept. of Mines, Geol. Survey Branch, Mem. 10.

Horberg, Leland, 1951, Intersecting minor ridges and periglacial features in the Lake Agassiz basin, North Dakota; *J. Geol.,* **59**, 1–18.

Hurst, M. E., 1930, *Geology of the area between Favourable Lake and Sandy Lake, District of Kenora (Patricia Portion).* Ont. Dept. Mines, **38**, part II, 1929, p. 67–68.

———, 1933, *Geology of the Sioux Lookout area.* Ont. Dept. Mines, **41**, part VI, p. 16–18.

Johnston, W. A., 1915, *Rainy River district, Ontario. Surficial geology and soils.* Geol. Surv. Canada, Mem. 82.

———, 1916, The genesis of Lake Agassiz: a confirmation. *J. Geol.,* **24**, 625–638.

———, 1934, *Surface deposits and ground-water supply of Winnipeg map-area, Manitoba.* Geol. Surv. Canada, Mem. 174.

———, 1946, Glacial Lake Agassiz, with special reference to the mode of deformation of the beaches. Geol. Surv. Canada, Bull. 7.

Leverett, F., 1932, *Quaternary geology of Minnesota and parts of adjacent states.* U.S. Geol. Survey, Prof. Paper 161, p. 119–140.

Nikiforoff, C. C., et al., 1939, *Soil survey (reconnaissance) The Red River valley area Minnesota,* U.S. Dept. Agriculture, Bur. Chemistry and Soils, Ser. 1933, No. 25.

Norman, G. W. H., 1938, Last Pleistocene ice-front in Chibougamau district, Quebec. *Roy. Soc. Canada, Trans.* Ser. 3, **32**, sec. 4, 69–86.

Prest, V. K., 1963, *Red Lake-Lansdowne House area, northwestern Ontario. Surficial Geology.* Geol. Surv. of Canada, Paper 63-6, 23 pp.

Ritchie, J. C., 1964, Contributions to the Holocene paleoecology of west central Canada. 1. The Riding Mountain area. *Can. J. Bot.,* **42**, 181–197.

Rittenhouse, G., 1934, A laboratory study of an unusual series of varved clays from northern Ontario. *Amer. J. Sci.,* **228**, 110–120.

Rominger, J. F., and Rutledge, P. C., 1952, Use of soil mechanics data in correlation and interpretation of Lake Agassiz sediments, *J. Geol.,* **60**, 160–180.

Satterly, J., 1937, Glacial lakes Ponask and Sachigo, District of Kenora (Patricia Portion) Ontario. *J. Geol.,* **45**, 790–796.

Schwartz, M. L., 1967, The Bruun theory of sea-level rise as a cause of shore erosion. *J. Geology,* **75**, 76–92.

Tuthill, S. J., 1963, *Molluscan fossils from upper glacial Lake Agassiz sediments in Red Lake County, Minnesota.* North Dakota Geol. Surv., Misc. Ser. n. 20 (*N. Dak. Acad. Sci., Proc.,* **17**, 96–101).

Tyrrell, J. B., 1893, North-western Manitoba with portions of the districts of Assiniboia and Saskatchewan. *Geol. and Nat. Hist. Surv. Canada, Rept. Progress 1890–91,* **5**, Part E.

Upham, W., 1890, Glacial Lake Agassiz in Manitoba. *Geol. Surv. Canada, Ann. Rept.,* **IV**, Part E, p. 156.

———, 1895. *The Glacial Lake Agassiz.* U.S. Geol. Survey, Mon. 25, 685 pp (1896).

Wright, H. E., Jr., 1965, Glacial history of western Minnesota and adjacent South Dakota, p. 32–38 *in* Schultz, C. B., and Smith, H. T. U., eds. *Guidebook for Field Conference C. Upper Mississippi Valley, Internat. Assoc. for Quaternary Research,* Aug. 13–29, 1965, Nebraska Acad. Sci., Lincoln, Neb.

Zoltai, S. C., 1961, Glacial history of part of northwestern Ontario, *Geol. Ass. Canada, Proc.,* **13**, 61–83.

———, 1963, Glacial features of the Canadian Lakehead area. *Canadian Geographer,* **7**, n. 3, 101–115.

———, 1965a, *Kenora-Rainy River, surficial geology.* Ont. Dept. Lands and Forests, Map S165.

———, 1965b, *Thunder Bay, surficial geology.* Ont. Dept. Lands and Forests, Map S265.

Editor's Comments on Paper 10

Andrews: *Differential Crustal Recovery and Glacial Chronology (6,700 to 0 BP), West Baffin Island, N.W.T., Canada*

Field research in Arctic Canada on the age and form of raised marine strandlines has a very short history. The first major paper was published by Løken in 1962 (*Geogr. Bull.*, Ottawa) and subsequent work has been largely confined to Baffin Island and Labrador, with the notable exception of Blake's work in the Queen Elizabeth Islands (*Can. J. Earth Sci.*, 1970).

My own paper reprinted here illustrates some of the advantages and problems connected with arctic research. The strandlines are well preserved at individual sites, but there is, however, the problem of tracing laterally these water planes. They are not as continuous nor as well developed as the shorelines of Glacial Lake Agassiz or those in Northern Norway (see next paper). The association of the major strandlines with ice-front positions is well illustrated in the Flint Lake area and has been commented upon by many field workers. Whether the association is related to a relative sea level stillstand (of short duration) or to increased sedimentation and shore processes is not well known.

In the field area, the strandlines appear to be tilting downward to the northeast, indicating an uplift center in the vicinity of Southampton Island. The strandlines are not warped and they contrast with those of similar age immediately east (along the eastern Baffin Island coast) that are clearly curveliner (Andrews et al., 1970, *Bull. Geol. Soc. Am.*). The paper also illustrates the use of a number of empirical methods that enable a researcher to gain an insight into age/uplift and age/strandline gradient relationships and allow the geologist to expand his dated field sites to a larger area.

John T. Andrews was born in Millom, Cumberland, England, and was educated at the University of Nottingham (B.A. and Ph.D.) and McGill University (M.Sc.). He worked for the Canadian Government from 1961 to 1968 and is now Professor of Geological Sciences and Associate Director, Institute of Arctic and Alpine Research, University of Colorado, Boulder. He has worked in the Canadian Arctic and Subarctic since 1959.

Reprinted from *Arct. Alp. Res.*, **2**, 115–134 (1970)

10

DIFFERENTIAL CRUSTAL RECOVERY AND GLACIAL CHRONOLOGY (6,700 to 0 BP), WEST BAFFIN ISLAND, N.W.T., CANADA

J. T. ANDREWS

Institute of Arctic and Alpine Research and
Department of Geological Sciences
University of Colorado
Boulder, Colorado 80302

ABSTRACT

Eight marine strandlines are delimited on morphological evidence. They dip toward 050° at gradients that decrease with age to the present. The maximum ice load was therefore located southwest of Baffin Island, possibly between Baker Lake and Southampton Island. Five strandlines are associated with stillstands of the western margin of the residual Baffin Island ice cap. Each strandline is dated by reference to a radiocarbon-controlled emergence curve. The validity of the strandlines is tested by four methods; the strandlines appear as reasonable approximations of crustal deformation through time. Correlative geologic-climatic phases along the western margin of the late-glacial Barnes Ice Cap are indicated by the moraine evidence. The *Isortoq* Phase is dated about 6,700 BP. It was succeeded by a period of retreat with limited halts. Another major glacial phase, the *Flint*, occurred about 5,000 BP, and is correlative with the growth of the Ellesmere ice shelf and glacier readvance in other parts of the world. Younger prominent moraines are > 1,700, 700, and 250 years old.

STATEMENT AND OBJECTIVES

Studies of crustal deformation in response to glacial unloading have been most successful in areas where adequate time-stratigraphic (marine transgressions, pumice levels, etc.) or morphostratigraphic (continuous raised shorelines) criteria are present. It is no accident that the Glacial Great Lakes, Glacial Lake Agassiz, and the sequence of late- and postglacial Baltic seas and lakes were among the first areas to be studied and which resulted in coherent and reasonable data on crustal recovery. Difficulties in interpretation are, however, encountered in glaciated coastal areas which lack continuous raised littoral forms, and where the late- and postglacial rock units do not include alternating marine and terrestrial beds, or marine and freshwater beds. Studies have been pursued in such areas and the attendant difficulties have been commented on, as for example by Donner (1964) and Stephens and Synge (1966). There have been few if any new analytical methods developed since the 1920s, and few papers have attempted to test their strandline diagrams. In northern Europe and the British Isles certain

J. T. ANDREWS / 115

strandlines reflect marine transgressions and provide, by reasons of stratigraphy, radiocarbon dates, palynology, and cultural artifacts, a sequence of controls on both lower and higher strandlines. As yet, no equivalent degree of correlation can be achieved for any specific water plane in the Canadian Arctic.

The trace of a former water plane on the land surface is here called a strandline. Lines on a map joining sites of equal relative sea levels of similar age are called isobases. If a normal is drawn to a local isobase system and elevations are projected onto this plane the graph is called an equidistant diagram. Strandlines from an area that is rising isostatically represent a relatively short instance of time; they can be recognized on morphostratigraphic and straight stratigraphic grounds, and thus offer considerable scope as chronological markers throughout a region.

This paper is concerned with devising methods to test a series of strandlines delimited on the basis of morphostratigraphic criteria where individual strandline sections were short. Given good dating control, via a number of radiocarbon-dated samples, such a task is easy. However, in many parts of Arctic Canada such control is at best scarce. In such situations can the pattern of crustal deformation be adequately deciphered?

The close association between the late-glacial marine invasion of the field area and glacial chronology is critical for this paper; for 80 km the sea was in immediate contact with the retreating ice front and thus the study of sea-level changes is intimately related to the late-glacial chronology of the residual Baffin Island ice cap.

At the outset, problems connected with strandline reconstruction are considered; criticisms of existing methods do not invalidate their usefulness. An analogy can be made to the extensive use of ^{14}C dates, despite a very large number of complicating factors such as contamination, correct half-life, relation of calendar years to ^{14}C years, plus interpretative problems associated with the stratigraphic relations of dated samples. A threefold approach is adopted in this paper: (1) discussion of methods, (2) reconstruction of strandlines, and (3) tests to judge the validity of these time and elevation lines. The end product is thus an assessment of the difficulties involved in deciphering postglacial crustal movements in an area that lacks continuous morphostratigraphic features and has limited ^{14}C dating control. These two constraints appear to be common throughout the Canadian Arctic. The methods developed in this paper should have widespread applicability.

INTRODUCTION TO FIELD AREA

The west coast of Baffin Island (Figure 1) trends southeast-northwest and parallels the western margin of the Barnes Ice Cap, 100 km inland (Figure 2). Radiocarbon dates on marine shells indicate that Foxe Basin became ice-free between 6,700 and 6,900 BP (Sim, 1964; Craig, 1965; Blake, 1966). Studies around the Barnes Ice Cap (Figure 1) indicate that it is a relic of the Laurentide Ice Sheet (Løken and Andrews, 1966). Compressed, therefore, into the 100 km between the coast and the ice cap's margin, are nearly 7,000 years of late-glacial deposits.

This paper is based on 1965 field work along an 80-km traverse from the Tweedsmuir Islands (Figure 2) inland to Flint Lake. Sim (1964) prepared preliminary maps of glacial features and surficial deposits of the area at a scale of 1:500,000. Farther north, Ives (1964) discussed elevations of marine limits and the late-glacial history of the Windless Lake-Rowley River area (Figure 2). Andrews (1966 and 1968a) investigated the region between Isortoq Fiord and the northwest margin of the Barnes

FIGURE 1. Location of the field area. Center of postglacial uplift was determined from intersection of normals to isobases from two sites on west Baffin Island.

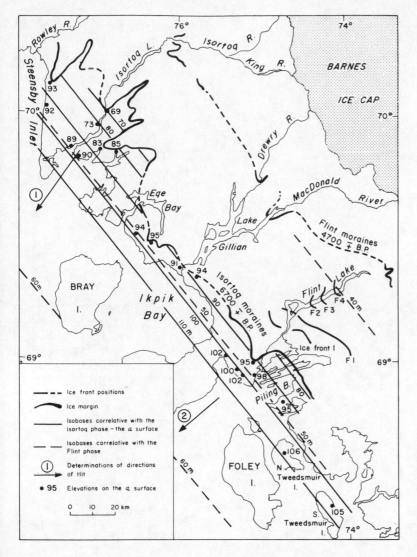

FIGURE 2. West Baffin Island showing generalized trend of the Isortoq and Flint moraines and their associated strandlines. Ice front positions along Flint Lake are also included. Tilt arrows are directed toward the uplift center.

Ice Cap. King and Buckley (1967) considered late-glacial events between Eqe Bay and Lake Gillian (Figure 2). They related elevations of glacio-marine deltas to the emergence curve for the west coast and were able to suggest a chronology for ice-front positions between 6,700 and 4,600 BP. The present field area (Figure 3) constitutes the southernmost in the above sequence.

TOPOGRAPHY AND CLIMATE

Elevations are moderate and relief is subdued, apart from over-deepened major valleys (Figure 4). An escarpment, overlooking coastal lowlands underlain by Paleozoic limestone, trends northwest from Longstaff Bluff with hilltops 160 m a.s.l. The upland surface rises inland from 250 m above the middle section of Flint Lake to 430 m at the head of the lake. A second escarpment parallels the coast and is breached between Flint Lake and Piling Lake (Figure 3). Between the two escarpments, elevations are low and bedrock is partly buried under marine silt. The Flint Lake trough has steep bedrock walls and structural control implicit in the sharp changes in trough orientation. Flint Lake lies 9.4 m a.s.l. and soundings show depths >110 m. Thus the trough is incised 360 to 540 m into the plateau surface. In contrast, the offshore zone has depths ≤50m.

J. T. ANDREWS / 117

175

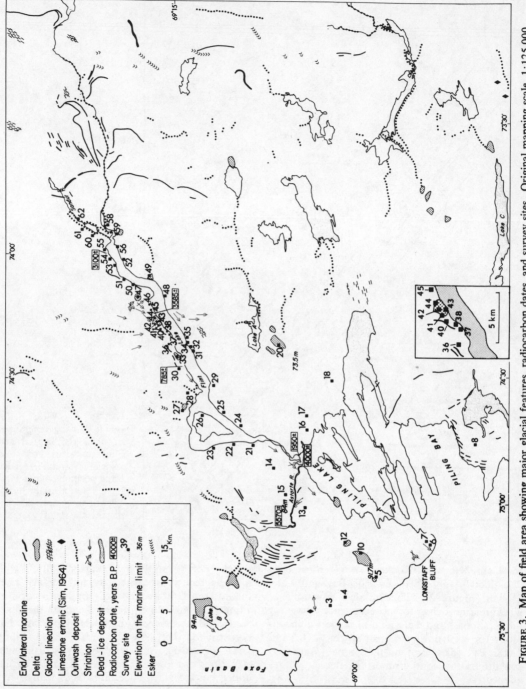

FIGURE 3. Map of field area showing major glacial features, radiocarbon dates, and survey sites. Original mapping scale, 1:125,000.

176

FIGURE 4. Air photograph looking northeast up Flint Lake (No. T221-R26). Delta of site 27 is in lower left of the photograph. Photograph shows the low relief of the upland plateau. Compare with Figure 3 for details.

Flat-lying Paleozoic limestone fringes the west coast of Baffin Island as in Baird Peninsula, and Bray and Foley Islands (Figure 2). Along the Longstaff Bluff-Flint Lake axis is a metasedimentary sequence tightly folded for the most part, although horizontally bedded, metamorphosed limestones were found. North of the metasedimentary rocks the bedrock is granite-gneiss.

The mean January temperature over the interior is −29°C, the mean July temperature is 4°C. Only 17 to 18 cm of precipitation are received. Accumulation on the Barnes Ice Cap is between 30 and 50 cm w.e. (Løken and Andrews, 1966; Sagar, 1966). On the interior plateau about 75% of the precipitation falls as snow. Radiative energy surplus is the dominant ablation process, accounting for 75 to 100% of total melt on the Barnes Ice Cap (Sagar, 1966).

ICE MOVEMENT

Paleozoic limestone erratics occurring east of known outcrops, the form of glacial lineations, and crag-and-tail features west and north of the Barnes Ice Cap (Ives and Andrews, 1963; Sim, 1964) indicate that during the maximum of the last glaciation, ice moved east and northeast across Baffin Island. Folding of interglacial organic detrital beds along the Isortoq

River suggests that the earliest movement was, however, toward the west (Terasmae et al., 1966). Deglaciation of Foxe Basin was influenced by the entry of sea water through Hudson Strait. Striations on some of the islands in Foxe Basin (Mathiassen, 1945; Burns, 1952) are oriented north-south and I interpret these as representing the flow of ice toward a calving bay across Foxe Basin.

Evidence of glacial movement in the field area is shown on Figure 3. Striations on hilltops above Flint Lake indicate a movement bearing 200°, whereas near the lake flow was toward 230°. King and Buckley (1967) also noted striations bearing 213° south of Lake Gillian. The evidence of south-moving ice might reflect a short-lived response to the development of a calving bay across Foxe Basin. Between Flint Lake and the Tweedsmuir Islands (Figures 2 and 3) the predominant movement was southwest. Striations on Longstaff Bluff show an older set directed 230° crossed by others trending 245°. North of Piling Lake (Figure 3) the ice moved northwest up the Astarte Valley. On the shore of the lake one exposure showed two sets of striations, a series of short fine striations were directed north-south, while the larger forms were directed toward 230°.

J. T. ANDREWS / 119

Clayey silt containing marine molluscs (*see* Appendix) outcrops for 80 km between Piling Bay and the head of Flint Lake. In places the clayey silt is overlain by coarse deltaic sediments. The sea progressively invaded the area as the ice retreated toward the northwest.

Figure 3 shows the glacial features of the area: large moraines are located near the coast and the head of Flint Lake. Till cover is thin and no more than one till unit is found. Valleys above the marine limit generally contain outwash. Dead-ice deposits and small eskers suggest that the plateau margin of the ice cap was locally stagnant.

SURVEY ERRORS AND MORPHOLOGY

Elevations of raised shore features refer to a datum at Longstaff Bluff and are given in meters above high tide (m a.h.t.). Present tidal range is about 3.5 m and the high levels of Flint and Piling lakes are 9.4 and 1.9 m, respectively. All raised marine deposits were surveyed to one of these water bodies using a Zeiss Ni2 level, Wild T1 or T12 theodolite. Air photographs at 1:60,000 and maps at 1:125,000, contour interval 60 m (200 ft), were used for mapping.

The raised shore forms vary in origin, development, and preservation. Marine limits are often clearly defined as the lower limit of unwashed ground moraine or as the upper level of shingle. Numerous deltas were built into the late-glacial sea at the ice/sea contact (Figures 4 and 5A). Another type of delta deposit occurs at the mouths of the streams entering Piling and Flint lakes; they were not associated with ice contact deposition. Delta sediments are coarse, have poor stratification, and are nonfossiliferous and thus contrast with the fossiliferous deltas of east Baffin Island (Løken, 1965). Delta lips are 3 to 4 m lower than associated marine limits, and from modern analogues, I conclude that the deltas were graded to low tide. At many sites, well-developed terraces are incised into older, raised deposits (Figure 5B); in fewer cases, wide terraces may fringe steep bedrock walls.

This paper is based on 63 surveyed stations where elevations were determined on the marine limit and on lower deltas and beaches. The variability of elevations on littoral deposits of a water body is related to differences in fetch and origins (*see* Elson, 1967; Kupsch, 1967; Institute of British Geographers, 1967). Within the field area, a maximum fetch of 25 km is possible, although at most sites it is only

FIGURE 5A. Low-level oblique air photograph of site 37 with site 38 in background. The well-developed terrace shown as Figure 5B is visible on the delta front on the left (west) of the photograph. Photo by J. D. Ives.

FIGURE 5B. Detailed ground photograph of terrace cut into the delta face at site 37. Terrace is part of the F3 strandline.

1 to 5 km. Even on the open coast, the offshore islands provide a barrier to long-traveled waves. Raised cobble beaches are common on the open coast (King and Buckley, 1967) but are absent from protected inner sites. The elevation of terraces or beach notches was measured at the break of slope at the back of each feature. The relationship of this point to the former water plane is not consistent and can be either above or below the water plane (Elson, 1967). In this paper, delta levels and terrace elevations are compared by adding 3.5 m to the elevation of delta lips and by assuming that the beach notch reflects the maximum elevation of the

water plane. Because of the limited fetch, both past and present, and because of the short open-water season (sea ice never entirely leaves Foxe Basin, even in summer), great variations in elevation of the same strandline are not likely. Each surveyed form is therefore thought to be correct to within $\pm$ 3.0 m.

The standard graphical analysis of strandlines is undertaken on an equidistant diagram where the y axis is elevation and the x axis is distance. Sites are projected onto the plane of the diagram and the lines are "fitted" through clusters of points. Three questions arise in the interpretation of such a diagram:

(1) The suggested strandline is usually tilted but *not* warped. Andersen (1960), however, reconstructed marine strandlines in southern Norway which were warped, while Broecker (1966) showed that a glacial lake shoreline had an exponential warping.

(2) Despite the apparent objectivity of fitting a line of best fit to data points by regression analysis, there is an underlying circularity as input for each strandline tends to be limited to those sites that are considered to form part of the strandline.

(3) Sites are scattered geographically, and construction of an equidistant diagram requires the selection of a suitable projection plane. In the simple linear equation, $Z = A_0 + A_1X$, where A_0 and A_1 are regression coefficients, X is distance from a selected origin, and Z is elevation. The gradient of a strandline (give by A_1) is dependent upon the projection plane.

DELIMITATION OF STRANDLINES IN THE FIELD AREA

It is often difficult to learn from a published paper on marine strandlines why a particular set of points is considered to represent the trace of a former marine plane. In determining the strandlines on Figure 6, two stages of processing occurred, the first in the field and the second in the office. At each survey site, the origin of the raised marine features was noted and their degree of development was ranked as well, moderately, or poorly developed. Elevations were determined and this information was plotted in the field on a base map and on an equidistant diagram in the plane 035° to 215°, which is the direction of tilt in Isortoq Fiord (Andrews, 1966). As field work progressed, a correspondence was noted between lateral moraines and/or ice-contact deltas and well to moderately developed shore forms. In two

cases (sites where this could be done were 12, 11, 10, and 42 to 38) it was possible to walk, with limited breaks, distances of 2 to 4 km and still remain on the same shoreline. Leveling showed that elevations at the western ends of the traverses were higher than at the ice/sea contact and, furthermore, the strandline would be incised into the face of a higher and older delta farther west (Figure 5). In Norway, Andersen (1965) noted a similar association and stated: "All the late glacial shorelines terminate at ice-front positions."

On the basis of these observations, the following methodology was adopted (it is presented in a more rational manner than in which it occurred but it can provide a framework for others):

(1) Locate a sea/ice contact on the field map

J. T. ANDREWS / 121

and note the elevation of the marine limit.

(2) Examine the nearest site to the west in this area to see if there is a comparable elevation on a well to moderately developed feature. (From previous work (Andrews, 1966), a rule of thumb was adopted that limited elevation changes of < 1.0 m/km are permissible, with the realization that, because of genesis, fetch, etc., elevations on the same strandline could vary by ± 3.0 m, this allowed some latitude in the inclusion of individual points.) Repeat this search for the entire field area on a site-to-site basis.

(3) Project these sites (considered to lie along a former water plane) onto a number of planes and consider the geometry of the deformation.

(4) Compute least squares equation for the data set using both a linear and exponential function.

(5) Accept the plane with the minimum variance as the correct solution.

(6) Finally, locate another sea/ice contact interbase as shown by the juxtaposition of lateral moraines and glacio-marine deltas, and repeat procedures.

Steps 1 and 2 were started in the field. Step 3 initially consisted of plotting all marine limits on a series of rotated projection planes, which showed that the best fit line dipped toward 050° at 0.88 m/km (Table 1). (Similar values were obtained by King and Buckley (1967) on metachronous delta levels from the coast and inland along Lake Gillian.) This plane was then used for the inspection of the field data (return to steps 1 and 2). The marine limit ranges from 106 to 40 m and is composed of stepped linear segments; the breaks in gradient occur in the vicinity of lateral/end moraines or other ice contact deposits.

At least eight strandlines are present (Figure 6); for each, rotation of projection planes was carried out in increments of 10° between bearings 020° to 090°. The best solutions lie in the range 040° to 060° with 6/8 dipping toward 050° (Table 1). The effect of rotation on the correlation coefficient, r, and standard error, Syx, is shown in Figure 7. No conclusive evidence for warping could be detected using the method sketched above. The strandline nomenclature (α, F1, etc.) is discussed later.

Testing the Strandline Diagram

A diagram such as Figure 6 is difficult to interpret. The eight strandlines are drawn on the basis of their appropriate least squares equations and, although correlation coefficients are high, standard errors are ~ ± 3 m (Table 1). Each strandline is based on between 6 and 18 points whereas Figure 6 plots every measured feature which reduces the visual viability of the suggested strandlines (especially in view of a × 160 vertical exaggeration). I considered it necessary to plot all heights on Figure 6 because a variety of causes could result in a shoreline having features ranked in all three classes (well, moderately, and poorly developed). For example, strandline F4 is based on 6 points (Table 1) although about 13 points occur on or

Table 1

Best-fit direction of dips and resulting values from rotation of strandline data and simple linear regression analysis[a]

Strandline	Direction of Dip (°)	r	Syx (m) ±	N	A_1 (m/km)
Marine limit	050	0.99	2.56	47	0.88
α line	050	0.90	1.93	10	0.65
F 1	050	0.92	2.73	13	0.39
F 2	050	0.97	1.85	10	0.40
F 3	050	0.93	2.01	18	0.30
F 4	060	0.94	1.18	6	0.26
F 5	050	0.96	1.94	15	0.33
F 6	050	0.92	1.26	10	0.18
β line	040	0.95	0.71	10	0.10

[a]N = number of points; r = correlation coefficient.

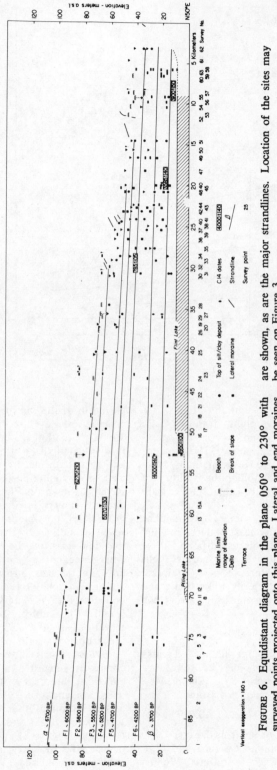
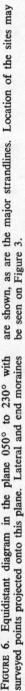

FIGURE 6. Equidistant diagram in the plane 050° to 230° with surveyed points projected onto this plane. Lateral and end moraines are shown, as are the major strandlines. Location of the sites may be seen on Figure 3.

181

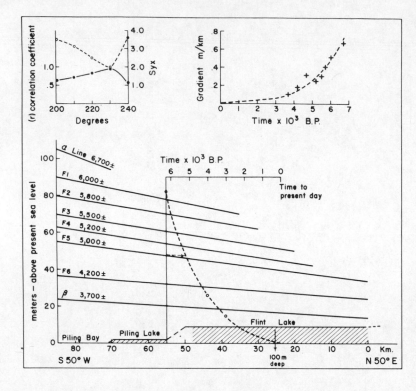

FIGURE 7. *Upper left:* Effect of rotation on r and *Syx* values for the α line, dated about 6,700 BP. *Upper right:* Gradients on the eight strandlines plotted against their age. *Bottom:* An emergence curve is shown in correct position on the projection plane. Seven strandlines pass through the curve and are dated.

about the line (Figure 6), but the other 7 are poorly developed. The eight strandlines are based on field evidence and not on the operational procedures outlined above. The evidence is not, in itself, conclusive. Four methods, however, can be devised enabling the validity of the strandlines to be tested (Table 2). They are outlined below.

Method (1) of Table 2 refers to strandline elevations determined by least squares method (Figure 6), and dated as discussed below. Five radiocarbon dates (Table 3) on marine shells collected along the Astarte River (Andrews, 1967), and one from Butterfly Lake (King and Buckley, 1967) define an uplift curve that can be approximated by:

$$U'_p = a + r \ln t \qquad (1)$$

where U'_p is postglacial uplift in time t (years) and a and r are constants derived from least squares procedure (Figure 8). Alternatively, postglacial uplift can be predicted from:

$$U'_p = A(1-0.677^t)/(1-0.677) \qquad (2)$$

where A is the amount of uplift in the first 1,000 years of postglacial rebound, t is time $\times 10^3$ yr, and i is a constant ~ 0.677 (Andrews, 1968b). Postglacial uplift is defined as the sum of the

elevation of the marine limit and the necessary eustatic sea-level correction (e.g., Shepard, 1963). Equation (2) can be used to determine postglacial uplift in time t, given the elevation and age of the marine limit (Andrews, 1968b). The predicted uplift curve for the Astarte River is shown on Figure 8 and the emergence curve is shown on Figure 7. On this latter diagram it is possible to determine the ages of seven strandlines by the points where they cross the time axis of the graph. Any number of emergence curves can now be drawn, for at any site there is information of the elevation and age of relative sea levels. Such curves will be referred to as "suggested emergence curves" (Figure 9). Table 2 lists suggested strandline elevations at 80, 53, 30 to 20, and 4 km from the arbitrary origin of Figure 6.

Method (2) is illustrated in Figure 9 which compares suggested and predicted emergence curves at three sites. The predicted curves are based on equation (2) and, therefore, the age and elevation of the three marine limits need to be known. At site 80 km the marine limit is 102 m (Figure 6) and date of deglaciation is about 6,700 BP. At site 53 km (Astarte River) values are not listed (Table 2) as here methods 1 and 2 are not independent checks. Heights of

TABLE 2

Comparison of strandline elevations, in meters, at four sites, based on three methods[a]

			Strandline			
Method	F2	F3	F4	F5	F6	β
Site 80 km						
1	78	68	62	53	36	24
2	69	62	54	44	37	24
3	80	69	61	52	37	24
Min. difference	2	1	1	8	0	0
Max. difference	11	7	8	9	1	0
Site 53 km						
1	69	59	54	47	32	21
2	—	—	—	—	—	—
3	70	58	54	44	32	22
Min. difference	—	—	—	—	—	—
Max. difference	1	1	0	3	0	1
Site 30-20 km						
1	—	—	47	42	25	18
2	—	—	47	39	30	24
3	—	54	48	34	26	20
Min. difference	—	—	0	5	1	2
Max. difference	—	—	1	8	5	6
Site 4 km						
1	—	—	—	35	24	15
2	—	—	40	35	28	21
3	—	—	40	32	23	17
Min. difference	—	—	0	0	1	2
Max. difference	—	—	0	3	5	6

[a]Sites at 80, 53, 30 to 20, and 4 km from 0 km of Figure 6. The three methods are outlined in the text.

the marine limits at 30 to 20 km and 4 km are known (Figure 6), whereas the date of deglaciation is estimated from Figure 7. These last two estimates of date of deglaciation are dependent on Figure 7 but estimates on lower levels are independent of method 1.

Method (3) for evaluating the strandlines is completely independent. Equation (2) can be developed into a shoreline relation (SR) diagram (*see* Donner, 1965), so that postglacial emergence can be predicted given the age and elevation of any point below the local marine limit (Andrews, 1969). At sites 20 km and 4 km two elevations are dated (Figure 6) below the marine limit. They can be used in the SR diagram to estimate the elevation of strandlines at these two sites (Table 2, method 3). Straight lines were then drawn between elevations of the

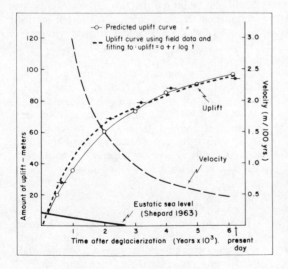

FIGURE 8. Postglacial uplift curve for Piling Lake. Comparison is made between a predicted curve based on equation (2) (*see* text) and a curve using all the points and fitting to $U'_p = a + r \ln t$ where a and r are derived. Velocity is based on r/t. Note x axis in in years after deglaciation. A eustatic sea-level curve is also shown.

J. T. ANDREWS / 125

183

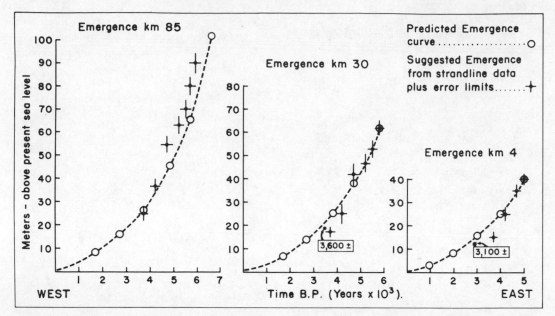

FIGURE 9. A comparison of three predicted and observed emergence curves at various distances from datum of Figure 6. Date on middle graph is rounded (i.e., Table 3, I-2830).

same age (i.e., the same strandline) and were projected westward through sites 53 km and 80 km. Elevations at these two sites were then determined (Table 2, method 3).

This paper is one of the few that considers the justification of a strandline sequence by a comparison with the form of postglacial rebound. Table 2 lists estimated strandline elevations and minimum and maximum differences for the three comparisons. At site 80 km, the maximum difference in elevation is between methods 2 and 3, whereas at sites 30 to 20 km and 4 km the maximum departures are not so simply related. At site 53 km methods 1 and 3 are almost identical. In 7 of 20 cases there is no difference between any two methods while 5 comparisons are >6 m. Maximum differences between methods must consider the size of the standard errors (Syx) listed on Table 1. On a 2σ criterion, these values lie between ± 5 to ± 1.4 m, which helps explain some of the variability of Table 2.

Another way to look at the importance of the elevation differences for any strandline is to judge them in terms of rate of postglacial rebound. The velocity curve on Figure 8 is instructive in this context. Consider the maximum error of Table 2, which is 11 m; it is related to the F2 strandline dated at about 5,800 BP when the velocity was about 3 m/100 yr, i.e., the 11-m elevation error represents only 360 years of uplift. Conversely, an error of 6 m

at 3,700 BP (site 4 km) represents about 600 years. These are maximum time errors and, compared to the 95% range of the standard error for ^{14}C dates (Table 3, i.e., 1,000 to 280 years), they are not unreasonable.

Method (4) of testing the strandline diagram is to examine the strandline gradient/time function (Table 1, Figure 7). In a study of strandlines within the Lake Vättern basin, Norrman (1964) found that strandline gradients decreased exponentially with age. The correlation between strandline gradient and time for the strandlines delimited in this present paper is acceptable with the coefficient of determination, r^2, equal to 0.877. The least squares fit is:

$$y = 64e^{-0.54t}; \; Syx \, (\log) = \pm 0.081 \quad (3)$$

where y is gradient in cm/km, $t = 0$ is the date of the first strandline (6,700 BP), and $t = \times 10^3$ yr. It is unlikely that such a good statistical relation could arise by chance. A final comment: the isobases trend 140° to 320° which is normal to the direction of ice movement during the maximum of the last glaciation, and therefore is not an unlikely orientation.

CONCLUSIONS

Two observations should be emphasized: strandlines dip 050° at gradients that decrease smoothly with age to the present; and the eight strandlines are reasonable estimates of both elevation and age. The word reasonable is used intentionally; there are differences in the origin and exposure of synchronous shore forms, and the term "synchronous" has to be used within the framework of the precision of the particular dating method. Thus strandlines represent time and elevation lines with precisions of about ± 250 years and ± 3 m; smaller estimates are unrealistic.

A water plane by definition represents an isochronous surface, but, as with other stratigraphic markers, Weller's (1960, p. 565) statement applies: "The determination of actual and exact age equivalence, or chronotaxis, is generally an unattainable ideal." Thus the strandlines on Figure 6 and 7 are more aptly thought of as zones, with widths controlled by time and elevation errors.

Although the estimates of the time and elevations involved are not small, they are such that a sea-level/moraine association can be dated with about the precision of one based solely on ^{14}C dates.

Table 4 lists correlative glacial phases (i.e., the geologic-climatic unit below rank of stade) and marine strandlines for west Baffin Island; the last column itemizes sea levels for the strandlines at a single reference site, Longstaff Bluff, whereas the preceding column gives the maximum range of sea levels for each strandline. The correlation of deltas, terraces, and moraines is based on the evidence discussed previously. The age of a glacial phase and associated strandline is based on Figure 7 and "synchroneity" is defined in the context of the precision of available dating methods.

Piling Lake Moraines and the α Strandline (ca. 6,700 BP)

End moraines near Piling Lake are divisible into two units (Figure 3) about 7 km apart. They head northward and merge into a belt of closely spaced moraines southeast of Lake B. In Piling Lake they were deposited subaqueously. They are composed of rounded cobbles and pebbles and are 50 m high in places. Elsewhere, particularly to the south, they are scarcely visible through a mantle of younger marine silt. Correlative massive ice-contact deltas are found north of Piling Lake at sites 10 and 12 and east of Lake B.

Elevations on the deltas at sites 10 and 12 and east of Lake B are between 93.0 and 94.0 m. King and Buckley (1967) recorded similar delta elevations at Eqe Bay and west of Lake Gillian (Figure 2). West from site 10, the marine limit rises to ~ 100 m and finally

reaches 106 m a.h.t. on South Tweedsmuir Island. The Piling Lake moraines are probably correlative with the Isortoq Moraine further north (Andrews, 1966). Consequently, elevations on ice-contact deltas and marine limits along the 200-km stretch of coastline (Figure 2) on the distal side of the Isortoq and Piling Lake moraines were analyzed by trend surface methods. The linear surface (Figure 2) accounts for 87% of the variability ($N = 20$). The maximum residual is 9.8 m at Windless Lake (northernmost site, Figure 2), but otherwise they are ≤ ± 3 m. Points lying on the distal side of the ice-front (Figure 2) are thus tilted toward 050° at 0.6 m/km which is the same gradient calculated by King and Buckley (1967, Figure 5). In Isortoq Fiord, a higher gradient of 1.0 m/km was suggested (Andrews, 1966). A date of 6,725 ± 250 years (I-406, Table 3) on a high shell deposit provides an estimate for both the Isortoq Phase and the associated strandline which was called the "α strandline" (Andrews, 1966).

In Isortoq Fiord, the equivalent strandline dips toward 035° ± 8° (Andrews, 1966), whereas King and Buckley (1967) derived an estimate of 049° ± 8° for the area between Lake Gillian and the Tweedsmuir Islands. Projection of these two orthogonals (Figure 2) suggests a center of uplift north of Southampton Island (Figure 1). Bird (1967) determined that southwest Baffin Island had been differentially uplifted with a consequent dip toward the northeast.

Ice Front 1 (ca. 6,300 BP)

Retreat from the Isortoq Moraine was marked both by tightly spaced moraines and by stagnation deposits (Figure 3). An important halt occurred at the Astarte Valley and is marked by a moraine parallel to the valley and east of the river. Marine limits at sites 14, 15, and 16 are 85 to 84 m.

Farther north, and parallel to the isobase, King and Buckley (1967) reported a date of 6,270 ± 210 BP from marine shells related to a delta at 84 m (Table 3). Relative sea level had fallen by 10 m in 400 years, and the rate of ice retreat in Piling Lake was ~ 37 m/yr.

In Eqe Bay, King and Buckley (1967) delimit a terminal position, Ice Front 2, associated with a relative sea level 80 m a.h.t. This front is probably synchronous with that at the head of Piling Lake (Table 4). The name Eqe Phase will be used for this glacial event.

J. T. ANDREWS / 127

185

TABLE 3

Radiocarbon dates from west Baffin Island

Location	Lab. No.	Lat. and Long. Correction	Relative Sea Level	Material	Date BP	Reference
Ikpik Bay	I-406	69°10′N, 75°28′W	89 m	Shell	6,725±250	Sim, 1964
Butterfly Lake	I-2410	69°21′N, 75°49′W	84 m	Shell	6,270±210	King & Buckley, 1967
Ikpik Bay	I-405	69°02′N, 75°02′W	74 m[a]	Shell	6,050±250	Sim, 1964
Lake Gillian	I-1833	69°28′N, 75°31′W	67 m	Shell	5,270±140	King & Buckley, 1967
Isortoq Fiord	I-1247	69°56′N, 77°02′W	18 m	Veg.	3,550±200	Andrews, 1966
Piling Lake	I-489	69°02′N, 75°02′W	9 m	Shell	2,050±170	Sim, 1964
Astarte River	I-1831	69°07′N, 75°02′W	64 m	Shell	5,570±130	Andrews, 1967
Piling Lake	GSC-557	69°06′N, 74°48′W	26 m	Shell	4,000±140	Andrews, 1967
Flint Lake (53% leach)	I-2830	69°17′N, 74°15′W	19 m	Shell	3,585±140	
Flint Lake	GSC-564	69°22′N, 73°54′W	13 m	Shell	3,100±150	Andrews, 1967
Piling Lake	I-1830	69°06′N, 74°47′W	5-6 m	Shell	1,950±100	Andrews, 1967
Flint Lake	I-1834	69°16′N, 74°22′W	41 m	Peat	785±105	Andrews, 1967

[a]Not 77 m as shown on Andrews 1966, Figure 2.

F1 Strandline (ca. 6,000 BP)

The western margin of the late-glacial Barnes Ice Cap retreated 15 km between 6,300 and 6,000 BP, and then halted. This stillstand is called the Gillian Phase (Table 4). It is not delimited by a moraine but by massive deltas graded to a sea level about 70 m a.h.t. East of Lake C (Figure 3) is a large delta at 69 m and several lower, distinct surfaces. To the northwest (Lake A) is a delta, several square kilometers in area; it has an elevation of 70 m (site 20) and lower, prominent surfaces are present (Figure 6). Within the Flint Lake basin the delta at 25 is considered correlative. It is graded to a low tide elevation of 66 m and the marine limit is 4 m higher. Deltas north of Piling Lake (sites 10 and 12) are cut by prominent terraces at 82 and 85 m a.s.l. and lie on the F1 strandline.

Ice Front CC (King and Buckley, 1967) in Lake Gillian is associated with a marine delta at 74 m a.h.t. and is probably synchronous with the F1 strandline.

F2 Strandline (ca. 5,800 BP)

A retreat of about 50 m/yr occurred between the distal ends of F1 and F2. The delta at 30 is 62 m a.h.t. On the proximal side are two lateral moraines. On the south side of the lake there is a delta at 58 m (site 31) and a broad terrace at 61 m a.h.t. (site 32). Immediately to the east of 32 and 34 is a glacial lateral terrace at 66 m, while below this the highest marine terrace is 53 m a.h.t.

There is a notable difference between elevations on the marine limits on either side of the lake, that is, between sites 32 to 34 and sites 30 and 33 to 36. On the south side, the break is abrupt, but the deltas at 33 and 36 have elevations of 60 m and 58 m and it is only between sites 37 and 42 that the marine limit falls from 58 m to 52 m. These findings indicate that in this area the ice margin was relatively stationary for a while along the south side of the lake, but that retreat along the north side allowed the sea to penetrate as far as 37. Retreat was not rapid, and lateral moraines, lateral terraces, and stripped bedrock mark successive positions of the northern margin. The difference in history is explained by the topographic differences in the two lake sides, as the northern shore is gentler and more embayed. The F2 strandline slopes 0.3 m/km and its intersection with the emergence curve dates it about 5,800 BP. The strandline is correlated with pronounced terraces cut into the deltas faces at sites 10, 11, and 12 at 75, 74, and 77 m, respectively.

F3 Strandline (ca. 5,500 BP)

The F3 strandline is terminated by an ice front deposit lying east of sites 46, 47, and 48 (Figure 6). Elevations on marine limits range from 49 to 50 m. Downvalley at 45, an in-

TABLE 4

Correlation of glacial phases and strandlines, west Baffin Island

| Andrews (1965, 1968a) | | King and Buckley 1967 | | | This Paper | | |
Moraine	Strandline	Ice front	Moraine	Strandline	Proposed phase and date (BP)	Range (m) of rel. sea level above present (Fig. 6)	Sea level(m) at reference site
Isortoq	α	Ice Front 1 and AA	Piling	α	6,700 Isortoq	94-107	98
		Ice Front 2	Ice Front 1		6,300 Eqe	84-97	92
		Ice Front CC		F1	6,000 Gillian	68-91	86
				F2	5,800 —	62-81	77
M2		Ice Front DD		F3	5,500 MacDonald	48-71	67
M3		Ice Front FF		F4	5,200 —	42-64	60
		Drewry Readvance	Flint ?	{ F5 F6	5,000 Flint 4,200	34-56 24-37	51
	β			β	3,700 —	14-25	23

distinct marine limit was surveyed at 50 to 53 m a.h.t., whereas on the south side the marine limit is clearly marked at 55 m a.h.t. Moderately rapid retreat occurred between site 41 and sites 46 to 47, but retreat was slow between sites 37 and 41. A lateral moraine slopes downvalley at 1:100 and grades to a glacio-marine delta at 58 m a.s.l. (site 37) (Figure 5A). Further retreat is evidenced by small lateral moraines that grade into progressively lower beach deposits between 58 and 52 m. The massive delta at sites 41 to 44 was deposited by a stream debouching into the late-glacial sea. The major part of the upper surface is related to a lip altitude of 44 m and is younger than F3. However, there are surface remnants up to 53 m a.h.t.

F3 appears to have been a significant water plane (Figure 6): deltas, terraces, and beaches of this plane now tilt 0.4 m/km toward 050°. A date of 5,500 ± BP is suggested for this phase. It correlates in time and general position with Ice Front DD of King and Buckley (1967) in Lake Gillian and with Ice Front M2 of Andrews (1968a).

F4 Strandline (ca. 5,200 BP)

There is strong evidence for this strandline, although the ice front is difficult to locate precisely. At 51 a delta at 45 m a.h.t. was observed on the distal side of a large lateral moraine. On the opposite side of the lake, a distinct wave-cut marine limit lies at 47 m a.h.t. On the proximal side of this suggested ice front, at 53, a lateral moraine trends southwest and grades into a subaqueously deposited end moraine.

The stillstand, tentatively correlated with the moraine at 51, was responsible for the cutting of beach notches on deltas 10, 11, and 12 and for the formation of constructional forms along Flint Lake. F4 may be associated with Ice Front FF of King and Buckley (dated 5,300 BP) and the M3 phase in Isortoq Lake (Andrews, 1968a).

F5 Strandline (ca. 5,000 BP) Flint Phase

As the ice retreated toward the head of Flint Lake, small marginal lakes were ponded in tributary valleys and it was often difficult to determine which beach denoted the marine incursion. Lateral moraines and terraces are common on the valley sides. Shells found at the head of the lake at 13 m a.h.t. are dated 3,100 ± 150 BP and testify to the reality of the marine invasion. The majority of the marine limit terminations range between 41 and 47 m and decline in elevation toward the east. Within Sandur Valley (Figure 3) sites 60, 61, and 62 have elevations of 44 m (distinct upper limit of water-washed bedrock), 47 m (lip of a coarse boulder fan) and 43 m (delta lip). Sandur Lake has an elevation of 17 m a.h.t., and its outlet cuts through major terrace deposits that rise to 35 m a.h.t. that are thus distinct in time from the cutting of the marine limit.

The hiatus between the marine limit and upper terrace is repeated at the first bend of Flint River; the marine limit was measured at 41 m but the upper outwash terrace rises to only 33 m a.h.t. The difference in elevations of the two features is interpreted as follows: retreat of the glacial lobe was initially rapid; a period of climatic deterioration led to the readvance of the late-glacial Barnes Ice Cap as witnessed by a number of massive moraines immediately upvalley and on the plateau surface (Figure 3). South of the valley, the plateau moraines trend north-south, but to the north they bear northwest or west-northwest. Note that Figure 3 indicates a large moraine south of the Flint Valley. It is younger than the F4 terminal position. It is possible that this moraine dates from 5,000 ± BP while the younger, more extensive system dates from 4,200 ± BP and relates to a 29-m level. The 33-m terrace cannot be followed upvalley to the termini of these lobes because of subsequent erosion. Furthermore, the tightly spaced moraines suggest that no single episode was involved, but rather that this period was one of limited frontal oscillations.

Correlative moraines of this phase, called here the *Flint Phase*, occur at the heads of Lake Gillian and Isortoq Lake. In the former, the Drewry Readvance was tentatively dated 4,600 ± BP by King and Buckley (1967), whereas at the head of Isortoq Lake, the M4 phase (Andrews, 1968a) was estimated to be 4,000 years old. These three moraine systems lie at the heads of their respective lakes and parallel the margin of the Barnes Ice Cap (Figure 2). Moraines between the Piling Bay Moraine and the moraines of the Flint Phase are not comparable in size or extent with either of these two systems. The readvance that produced the Flint Moraine represents a major climatic-glaciological break for the western margin of the ice cap.

F6 Strandline (ca. 4,200 BP)

There is a marked concentration of deltas, terraces, and beach deposits below the F5 line. A possible strandline (F6) extends southwest from a prominent river terrace that grades to 22 m a.h.t. The terraces cannot be correlated

with any specific moraine, but the relative displacement of sea level by 12 m suggests that a date of 4,200 ± BP is reasonable. The terraces might correlate with a sequence of large moraines 30 km east of the lake head.

β Strandline (ca. 3,700 BP)

Radiocarbon dates on marine shells indicate that the suggested level is > 3,100 ± BP and can be dated 3,700 ± BP (Figures 6 and 7). Elevations and suggested age are similar to the β line of Andrews (1966). Løken (1965) commented on the occurrence of detrital plant beds in the deltas of Inugsuin Fiord, east Baffin Island, and noted that Matthews had observed a similar phenomenon of equivalent age from northwest Labrador-Ungava (Matthews, 1967a). Correlative deltas have been observed by the writer from Home Bay, east Baffin Island, and from the Ottawa Islands, Hudson Bay, N.W.T. The β strandline has an estimated gradient of 0.1 m/km along Flint Lake compared to 0.3 m/km in Isortoq Fiord.

Recent Events

A brief summary of the history of the western margin of the ice cap is given for chronological completeness. Major readvance moraines, representing the King Phase, were traced along the western margin of the Barnes Ice Cap (Andrews and Webber, 1964). Lichenometrical studies suggested a date of at least 1,700 ± BP for ice withdrawal. The moraine trends obliquely to the present margin and appears to pass beneath it. Retreat and realignment of the ice cap's margin were followed by a major readvance of the southern dome about 700 BP (Løken and Andrews, 1966). Further retreat and changes in ice cap morphology preceded and were terminated by a readvance, the moraine of which closely fringes the present margin. Detailed lichen measurements indicate that this event dates from about 250 BP (AD 1700). Other moraines were dated at AD 1740, 1840, 1890, 1905 and 1920. Ives (1962) demonstrated that northcentral Baffin Island had suffered a recent period of severe nival conditions, with the formation of semipermanent and permanent snowbanks. Falconer (1966) dated moss collected from beneath the margin of a residual ice patch at 330 ± 75 (I-1204).

EXTRA-REGIONAL CORRELATIONS

Correlation of moraines along the western margin of the late-glacial Barnes Ice Cap (Table

4) implies periods of similar glacier response. The ice cap was not influenced by local topo-

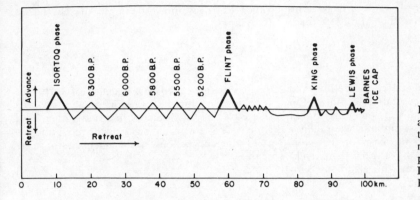

FIGURE 10. Schematic diagram showing the positions, dates and relative magnitude of glacial phases on west Baffin Island from about 6,700 BP to the present.

graphic factors and marginal fluctuations should reflect broad-scale climatic change. The chronology presented in this paper is based on radiocarbon-dated marine shells, on the association of strandlines and moraines, and on the predicted form of postglacial rebound. Up until about 2,000 BP our scale of measure cannot distinguish events less than the wavelength of the last worldwide glacial phase, that is, about ± 250 years.

Ten periods can be distinguished (Figure 10): (1) rapid disintegration of ice in Foxe Basin between 7,000 and 6,700 BP; (2) the Isortoq Phase (readvance or halt) about 6,700 years ago; (3) general retreat with halts and/or readvances; (4) a major readvance, the Flint Phase, about 5,000 years ago; (5) slow retreat until about 3,600 years ago; (6) fairly rapid retreat; (7) readvance at the beginning of the Sub-Atlantic period, the King Phase; (8) retreat, but a major readvance of the south dome of the Barnes Ice Cap about 700 years ago; (9) climatic deterioration leading to a readvance about 250 years ago; and (10) retreat with some stillstands. Investigations have indicated the Barnes Ice Cap is approximately in equilibrium (Løken and Andrews, 1966; Sagar, 1966). Although there is some "noise" in the sequence of Figure 10, it represents a fairly complete record of glacier response over 6,700 radiocarbon years.

Isortoq Phase (6,700 BP)

There is growing evidence for a glacial advance of this age (*see* Andrews, 1966, p. 182). In northwest Labrador-Ungava, Matthews (1967b) noted interfingering of marine and glacial deposits, and a radiocarbon date suggested a readvance occurred 6,700 ± 140 BP (NPL-82). Furthermore, moraines and outwash at the head of Frobisher Bay were probably

deposited at this time. Blake (1966) collected marine shells from sandy talus about 16 m a.s.l. which were dated at 6,750 ± 170 (GSC-464) and may date the marine limit. According to Matthews (1967c), marine temperatures were cold at this time, although this was followed by a warming trend between 6,000 and 6,500 years ago.

Flint Phase (5,000 BP)

The massive moraines at the head of Flint, Gillian and Isortoq lakes clearly represent the end product of an important climatic fluctuation. It is probable that they mark the end of the Hypsithermal in Arctic Canada. Crary (1960) suggested that the Ellesmere ice shelves developed about this time. On the basis of palynology from deposits in Keewatin, Nichols (1967a and b) suggested that at about 5,000 BP there was a short cooling trend during which time the treeline migrated south. Recently, Mercer (1967) reviewed the widespread evidence for glacial advances at the Atlantic/sub-Boreal boundary. The cooling phase evidently began about 5,000 BP and most glaciers had begun to retreat by 4,600 BP. The evidence from west Baffin Island supports the importance and worldwide nature of this cool phase.

β Strandline (3,700 BP)

Possible correlatives in Arctic Canada have been presented above. Nichols (1967a and b) noted a period of maximum wetness at 3,650 ± BP. On the other hand, Matthews (1967a) felt that this period represented the marine optimum in northwest Labrador-Ungava and is represented by his Lower Aporrhais (Gemma) Beach.

Correlations with the King Phase (Sub-Atlantic) and with the Lewis Phase (17th century) have been discussed by Andrews and Webber

J. T. ANDREWS / 131

(1964). Of special interest, because there is no widespread correlative, is the AD 1250 read-vance of the southwest margin of the Barnes Ice Cap.

SUMMARY

The concurrent events of glacial retreat and sea-level changes have been described along an 80 km west-to-east transect on west Baffin Island, N.W.T. Marine strandlines were identified in the field and on the basis of the equidistant diagram. They have been tentatively dated by their intersection with a radiocarbon-dated emergence curve. Because of errors in elevation and time, a strandline cannot be considered as an isochronous line, but rather as a zone. The form of postglacial rebound enables the conclusions to be checked to some degree.

ACKNOWLEDGMENTS

Field work was undertaken in 1965 with the former Geographical Branch, Department of Energy, Mines and Resources. Office studies were conducted in the former Geographical Branch and later in the Geological Survey of Canada, Ottawa. Vincent Coulombe, Art Froeze, Peter Lewis, and George Moroz provided valuable field assistance and Mrs. Lyn Arsenault ably assisted in data reduction. Finally, I would like to thank Drs. W. R. Farrand, J. D. Ives, O. H. Løken, and B. McDonald for reading and commenting on the paper and contributing to its improvement. The Geological Survey of Canada kindly dated two shell samples. The paper was read before the Canadian Association of Geologists at the 1967 meeting in Kingston, Ontario.

APPENDIX

Marine shells collected during the field work, 1965

Date × 10³ BP	5.5	4.3	4.0	3.6	3.5	3.0	2.0
Sample no.	27	29	30	28	20	21	26
Elevation of rel. s.l. (m)	65	30	26	20	19	15	7
Mya truncata Linne	X			X	X	X	X
Mya truncata var. *uddevallensis*			X				
Hiatella arctica	X		X				
Macoma calcarea	X						
Musculus sp.	X						
Portlandia arctica							X
Astarte borealis				X			X
Astarte montagui forma *typica*				X			X
Astarte montagui var *striata*			X	X	X		
Clinocardium ciliatum						X	X
Serripes groenlandicum				X	X		X
Balanus balanus			X				
Balanus crenatus			X				
Balanus sp.	X				X		
Hemithiris psittacea			X				
Neptunea despecta			X				
? *Trichotropis* sp.			X				

REFERENCES

Andersen, B. G.
  1960 : Sorlandet i sen-og postglacial tid. *Norg. Geol. Unders.* (English summary), 210: 142 pp.
  1965 : The Quaternary of Norway. *In* Rankama, K. (ed.), *The Quaternary*, I: 91-138.

Andrews, J. T.
  1966 : Pattern of coastal uplift and deglacierization, west Baffin Island, N.W.T. *Geog. Bull.*, 8: 174-193.
  1967 : Radiocarbon dates obtained through Geographical Branch field observations. *Geog. Bull.*, 9: 115-162.
  1968a: Late-Pleistocene history of the Isortoq Valley, northcentral Baffin Island, Canada. *In* Sporck, José A. (ed.), *Mélanges de Géographie offerts à M. Omer Tulippe*, Éditions J. Duclot, S. A., Gembloux, I: 118-133.
  1968b: Postglacial rebound in Arctic Canada: similarity and prediction of uplift curves. *Can. J. Earth Sci.*, 5: 39-47.
  1969 : The shoreline relation diagram: physical basis and use for predicting age of relative sea level (evidence from Arctic Canada). *Arctic and Alpine Res.*, 1(1): 67-78.

Andrews, J. T. and Webber, P. J.
  1964 : A lichenometrical study of the northwestern margin of the Barnes Ice Cap: a geomorphological technique. *Geog. Bull.*, No. 22: 80-104.

Bird, J. B.
  1967 : *The Physiography of Arctic Canada.* The Johns Hopkins Press, Baltimore, 336 pp.

Blake, W., Jr.
  1966 : End moraines and deglaciation chronology in northern Canada with special reference to southern Baffin Island. Geol. Surv. Can. Paper, 66-26, 31 pp.

Broecker, W. S.
  1966 : Glacial rebound and the deformation of the shorelines of proglacial lakes. *J. Geophys. Res.*, 71: 4777-4783.

Burns, C. A.
  1952 : Geological notes on localities in James Bay, Hudson Bay and Foxe Basin, visited during an exploratory cruise, 1949. Geol. Surv. Can. Paper, 52-25, 16 pp.

Craig, B. G.
  1965 : Notes on moraines and radiocarbon dates in northwest Baffin Island, Melville Peninsula, and northeast District of Keewatin. Geol. Surv. Can. Paper, 65-20, 7 pp.

Crary, A. P.
  1960 : Arctic ice islands and ice shelf studies, Part II. *Arctic*, 13: 32-50.

Donner, J. J.
  1964 : The late-glacial and post-glacial emergence of southwestern Finland. *Soc. Scient. Fennica Comment. Physico-Math.*, 30: 5-47.
  1965 : Shoreline diagrams in Finnish Quaternary Research. *Baltica*, 2: 11-20.

Elson, J. A.
  1967 : Geology of Glacial Lake Agassiz. *In* Mayer-Oakes, W. J. (ed.), *Life, Land and Water*, Univ. Manitoba Press, Winnipeg, 27-96.

Falconer, G.
  1966 : Preservation of vegetation and patterned ground under a thin ice body in northern Baffin Island, N.W.T. *Geog. Bull.*, 8: 194-200.

Institute of British Geographers
  1967 : Report of discussion on "The vertical displacement of shorelines in Highland Britain." *Inst. Brit. Geog. Trans.*, 42: 178-181.

Ives, J. D.
  1962 : Indications of recent extensive glacierization in north central Baffin Island, N.W.T. *J. Glaciol.*, 4: 197-206.
  1964 : Deglaciation and land emergence in northeastern Foxe Basin, N.W.T. *Geog. Bull.*, No. 21: 54-65.

Ives, J. D. and Andrews, J. T.
  1963 : Studies in the physical geography of north central Baffin Island. *Geog. Bull.*, No. 19: 5-48.

King, C. A. M. and Buckley, J. T.
  1967 : The chronology of deglaciation around Eqe Bay and Lake Gillian, Baffin Island, N.W.T. *Geog. Bull.*, 9: 20-32.

Kupsch, W. O.
  1967 : Postglacial uplift: a review. *In* Mayer-Oakes, W. J. (ed.), *Life, Land and Water*, Univ. of Manitoba Press, Winnipeg, 155-186.

Løken, O. H.
  1965 : Postglacial emergence at the south end of Inugsuin Fiord, Baffin Island, N.W.T. *Geog. Bull.*, 7: 243-258.

Løken, O. H. and Andrews, J. T.
  1966 : Glaciology and chronology of fluctuations of the ice margin at the south end of the Barnes Ice Cap, Baffin Island, N.W.T. *Geog. Bull.*, 8: 341-359.

Mathiassen, T.
  1945 : Report on the expedition. *Rept. of the Fifth Thule Expedition, 1921-1924*, 1(1), 134 pp.

Matthews, B.
  1967a: Late Quaternary land emergence in northern Ungava. *Arctic*, 20: 176-202.
  1967b: Late Quaternary events in northern

Ungava, Quebec: the glaciation of Deception Bay, Lac Watts and Sugluk areas. McGill Sub-Arctic Research Paper, No. 23: 42-62.

1967c: Late Quaternary marine fossils from Frobisher Bay (Baffin Island, N.W.T., Canada). *Palaeogeog., Palaeoclimat. and Palaeoecol.,* 3: 243-263.

Mercer, J. H.
1967 : Glacier resurgence at the Atlantic/sub-Boreal transition. *Quart. J. Roy. Met. Soc.,* 93: 528-534.

Nichols, H.
1967a: Central Canadian palynology and its relevance to northwestern Europe in the late Quaternary period. *Rev. Palaeobot. and Palynol.,* 2: 231-243.

1967b: Pollen diagrams from sub-Arctic Canada. *Science,* 155: 1665-1668.

Norrman, J. O.
1964 : Vätterbäckenets senkvartära strandlinjer (English summary), *Geol. Fören. Stockholm Förh.;* 85: 391-413.

Sagar, R. B.
1966 : Glaciological and climatological studies in the Barnes Ice Cap. *Geog. Bull.,* 8: 3-47.

Shepard, F. P.
1963 : Thirty-five thousand years of sea level. *In* Clements, T. (ed.), *Essays in Marine Geology in honor of K. O. Emery,* Univ. Southern Calif. Press, Los Angeles, 1-10.

Sim, V. W.
1964 : Terrain analysis of west-central Baffin Island, N.W.T. *Geog. Bull.,* No. 21: 66-92.

Stephens, N. and Synge, F. M.
1966 : Pleistocene shorelines. *In* Dury, G. M. (ed.), *Essays in Geomorphology,* Heinemann, London, 1-51.

Terasmae, J., Webber, P. J., and Andrews, J. T.
1966 : A study of late-Quaternary plant bearing beds in north central Baffin Island, Canada. *Arctic,* 19: 296-318.

Weller, J. M.
1960 : *Stratigraphic Principles and Practice.* Harper and Brothers, New York, 725 pp.

Wright, W. B.
1937 : *The Quaternary Ice Age.* 2nd edition, Macmillan, London, 478 pp.

Ms submitted December 1969

Editor's Comments on Paper 11

Marthinussen: *General View of Shore Levels in the Coast and Fjord Area of Finnmark*

The coast of Northern Norway is one of the classical areas for shoreline studies. From the literature and personal accounts, it seems as if the shore forms are amazingly well developed in this area. Marthinussen has been working in the area for many years, amassing an impressive amount of field data as is readily apparent on the plate included in the reprint. Marthinussen delimits 41 individual water planes on this diagram and subdivides them into three major age groups. Late-glacial shorelines (S) commence at a local marine limit and tilt steeply seaward. They are associated with ice-frontal positions. Below them are 12 postglacial shorelines (P) that gradually become less tilted. Both the S and P strandlines are cross-cut by the Tapes shorelines (N). The oldest of these, Tapes I, is ^{14}C dated at 7860 BP and hence is correlative with the Litornia transgression of Sweden. The start of this major European marine transgression is also radiometrically correlative with the final, catastrophis disintegration of the Laurentide Ice Sheet. Notice that Tapes I coincides with the upper limit of wave-rafted pumice; similar pumice has now been found in Greenland and the eastern Canadian Arctic.

Marthinussen's strandlines are only tilted and no evidence of warping of hinge lines has been noted. This, of course, contrasts with the warping of the Glacial Great Lake shorelines and those along the eastern Baffin Island coast. Why should the form of deformation be different in these ice marginal areas? Figure 144 shows the isobases for the P12 water plane. The isobases parallel the coast and are thus broadly curved in plan view; the parallelism between the coast and isobases suggests a crustal flexural parameter of 50–60 km.

Marthinussen is over 70 years old. He has worked in Northern Norway for 40 years as a schoolmaster and his scientific research is conducted primarily as an avocation. His knowledge of northern Norway's shorelines is incomparable.

Reprinted from *Geology of Norway*, No. 208, 416–432 (1960)

11

General View of Shore Levels in the Coast and Fjord Area of Finnmark

By M. Marthinussen

Introduction

The extreme northern part of Norway, bordering on the Arctic seas — and largely made up of less resistant rocks than the rest of the country — is classic ground in the history of shoreline studies. Here, in western Finnmark, Bravais in 1838 made his epoch-making observations on inclined shorelines, and here, in east Finnmark, Tanner, through his detailed shoreline investigations in the first decade of this century, could draw important conclusions concerning the deglaciation. And it was largely by following up this early work, and supplemented with more recent material from other northern districts, that he was able to build up his well-known shoreline system of 1930. Also a number of other geologists have published data on the shorelines of Finnmark and adjacent districts to the southwest, in later decades especially O. T. Grønlie and I. Undås.

During a great many years the author of the present has investigated the shorelines and other features of the Quaternary geology of Finnmark. Some results were published in 1945, and others have appeared in "Norges Geologi" (pp. 621, 641, 716—721, 745—747). Based on earlier as well as on more recent studies a short survey of some main features of especially the shoreline geology of Finnmark will be given in the following.

The discussion will to great extent be bound up with the equidistant diagram seen in Pl. 16. This is one of a series of diagrams which cover the fjord districts of Finnmark as well as some of more southwestern areas. Because the shoreline system of the author in some important respects does not agree with that of Tanner, it has been necessary to introduce special designations: S — Late-glacial (senglacial), P — Post-glacial, N — Neo-glacial.

The series of levels shown, especially from the middle fjord and coastal districts, fall into three groups separated by two particularly well marked shore level zones: "the main line", P_{12} (or S_0)[1] and the

---

[1] In future publications the term S_0 will be used.

Fig. 143. Sea-washed material at marine limit, 50 m a.s.l. Lille Omgang, Nord-
kyn Peninsula. (Phot. M. Marthinussen, N.G.)

uppermost Tapes level, which is a complex line made up of parts of four different lines (Tapes I—IV), described by the author in 1945.

Late-glacial.

The upper group of shore-lines, the S-lines, of which the upmost ones represent some of the oldest known shorelines of Norway, shows, as seen from the diagram, *falls* in the height of the marine limit in-wards (centrally) corresponding to successive stages in the ice-recession. Similar conditions have been found in the other diagrams, and thus we get a regional picture of the recession.

Moraines which can be correlated with the S_9—S_7 fall, occur inter al. in the outer Porsangerfjord district, in the middle part of Lakse-fjord and at Kongsfjord in the northwestern part of the Varanger Peninsula. The S_9—S_7 moraine stages in Finnmark probably cor-respond to the East-Andøy stage (Kirke-raet) of Grønlie (1940) in Andøya in Vesterålen. In this connection it might be mentioned that the relatively highest (and oldest) Late-glacial marine limit in northern

27

Norway is probably that found in the northern part of Andøy, at Stor-vatnet, 51,4 m (Bergstrøm 1950); but the highest and oldest marine shore-level in Finnmark, the S_{18} line, seems to come very close to it. A marine limit of 47,4 m in an end moraine n Æråsen, inside Store-vatn (identified by Bergstrøm, pers. com.), corresponds to the S_{17-16} levels. This glacial substage may possibly be of the same age as the supposed West-Andøy stage of Grønlie.

As to the S_4—S_2 fall, there are moraine deposits inter al. at Porsa in the Altafjord, at Kistrand—Smørfjord (Porsangerfjord), and in the inner part of Laksefjord.

"The main line", P_{12}, is the "upper line" of Bravais. Characteristic of P_{12}, which is more a zone than a line, is that commonly it has been abraded in rock, often with a very marked rock terrace. Therefore it probably embraces a considerable period of time with only slight fluctuations in the relative sea-level. At or near the place where the line vanishes proximally there occur important glaci-marine deposits, in some cases with continuation in lateral moraines. Belonging to this sub-stage we have the large Bossekop moraine at the head of Alta-fjord (with "Øskarnes" farther west), and in the Porsangerfjord a moraine at Veines (on the west side) 40 km N of the head of the fjord, and also a corresponding moraine on the east side; furthermore, moraines just south of Laksefjord, at Skipagurra at the Tana river, and in the Kirkenes district in Sørvaranger.

Now it seems evident that the moraines of P_{12}-age of Finnmark (which the writer in 1953 was inclined to refer to a somewhat younger stage) corresponds to the Tromsø—Lyngen substage of Troms. Fur-thermore, as first assumed by Grønlie and later ascertained by C-14 datings, the said substage must, in a general way, be correlated to the Ra substage of southern Norway. Of considerable interest is the dating of *P. arctica,* collected by the author in clay just outside the moraine at Spåkenes in Lyngen (a type locality of the T.-L. substage) to about 10100, a figure which corresponds well with the closing phase of the Younger Dryas period.

Fig. 144. Isobases of "Main line", P_{12} (personal investigations, with some data from J. H. L. Vogt, V. Tanner, O. T. Grønlie, and I. Undås). Full-drawn lines: based on direct observations. Broken lines: scarce and scattered data (or none-sea). Stippled lines: P_{12} not developed because of ice-cover in previously sub-marine areas. The gradient varies in the map area between c. 1,0 and 0,6 m/km.

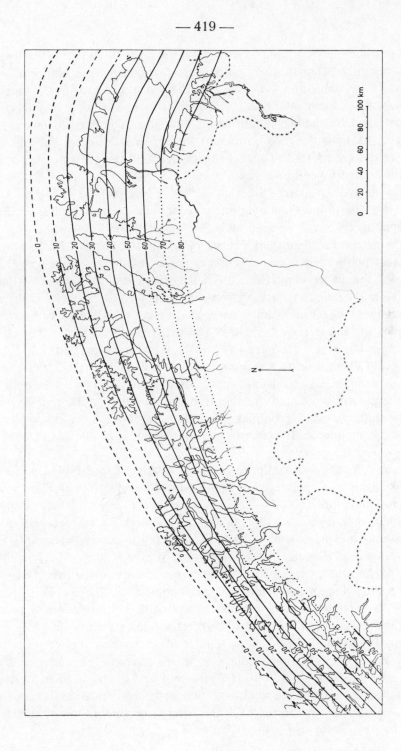

From what has been stated above, it appears that the P_{12} level, a very marked zone in many parts of northern Norway, must represent a *relatively* stable and long phase in the sea-level history of Younger Dryas time. Personal investigations by the author also in parts of Troms and Nordland has confirmed this view. The marked character of the "the main line", as a rock terrace, is a point of great importance for continued efforts to connect shore lines and ice-recession stages in northern and western Norway.

The Tanner relation diagram method is incapable of giving correct results. The proportionality-principle does not seem as valid as first supposed. It cannot be used even for the whole of Finnmark. A comparison between the lines of Tanner's Pl. III (1930) and of Pl. 16 of the present publication shows divergence *transversally* in such a way that the P_{12} line in the inner fjord districts corresponds to Tanner's f-line, in the middle districts to the g-line, while in the outer coastal districts it again approaches f and even comes below this line. Similar conditions are found for other lines. The explanation is that in Tanner's diagram both the "reference line" (b-line), which represents the limit of the "Tapes-Littorina transgression" in fjord- and coast districts, and the other lines, e. g. the f-line, appear as *straight* lines, while the author of the present has found the said transgression limit to be represented by a slightly curved (complex) line. The P_{12} line, on the other hand, no doubt is practically straight, and this seems to be the case also for the other lines.

A more regional comparison of the relation Tapes I—P_{12} and the b—f lines of Tanner's system in the *inner* fjord districts of Finnmark, Troms, and Nordland (For Tapes I-b mainly for the 22—50 m interval) will show a *longitudinal* divergence. In Finnmark P_{12} corresponds to f, in Troms and the northern part of Nordland it corresponds more or less to f_1, in Helgeland and Trøndelag it comes above f_1. For this comparison the writer in some cases has made use of the height of the uppermost pumice accumulation, corresponding to the Tapes I level.

The divergence here pointed to shows that the relation diagram method is not satisfactory for far-distance correlations. Probably no form of relation diagram will be reliable for comparison between northern and southern Fennoscandia. If e. g. we compare the shoreline displacement curves in Bømlo (Fægri) and at Göteborg (Sandegren) with the Finnmark diagram, we find that the shore-level zone of Younger Dryas time in the said areas lies considerably lower than the P_{12} line (or zone) in Finnmark.

The shorelines of Finnmark are all from the last deglaciation period, or younger, and as to the glaciation there can be no doubt but that the ice-sheet of the last glacial period at its maximum has covered all the present land. The falls of marine limits even in the outermost coastal districts, like Ingøy and Syltefjord (on the north side of the Varanger Peninsula), show that we are dealing with a deglaciation period characterized by a marked land uplift, an uplift which most probably also must have taken place in the immediately preceding period, indicating a still wider extent of the ice. Also the outhor's observations on the degree of weathering, occurrence of glacial striae, erratics etc., support the view of a complete ice-cover.

The age of the shorelines of Finnmark is of particular interest with a view to the oldest known history of man in Norway (Bøe and Nummedal 1936). Practically all the finds of implements of the so-called Komsa culture have been made at levels between the P_{12} line and the Tapes limit. If stone-age man has arrived in Finnmark already at the time of the former level, he must have lived in this part of the world as early as 10 000 years ago.

Post-glacial.

After the Tromsø—Lyngen substage there have been some minor stagnation stages in the ice-recession; but, generally speaking, the ice-wasting has no doubt been rapid.

As to the P-lines (below P_{12}) they belong to the first relatively short part of the Post-glacial period only, while the N-lines cover a much longer space of time. The P-lines cannot all be regarded as synchronous levels.

Of the N-lines the four oldest ones (Tapes I—IV) are of special interest, since, from the outer coast to the innermost fjord districts, they represent the so-called Tapes limit. However, the innermost part of this complex of lines, seen in Pl. 16, does not correspond to the uppermost Tapes level, as developed in some fjord districts in other parts of this country or to the higher Littorina level in the northern Baltic region (the Clypeus limit). The corresponding line in Finnmark(with a relatively steep gradient), occurs in the Sørvaranger district only, where in a general way it is represented by Tanner's c-line, inside the 28 m isobase of the Tapes I level. This line included, the Tapes limit here in the east is represented by five metachronous shore levels. The Tapes I—IV levels were, more than two decades ago, largely deduced on the

Fig. 145. Broad (Tapes I) abrasion terrace (in sandstone) 23 m a.s.l., northeast
of Vadsø. (After M. Marthinussen 1945.)

basis of pumice finds. At this time already the existence of several
"Littorina-Tapes" transgressions was known from various southern
districts (the Baltic, S. Sweden, Denmark, SW Norway); but no equidistant diagram had been constructed showing corresponding shore
lines.

Practically everywhere in the fjord- and coast districts of Finnmark the Tapes limit is very distinct, in the outer districts in the form
of shore bars, more centrally as gravel terraces or as abrasion terraces
in rock. Another important feature in this shoreline complex is the
accumulation of pumice found from the innermost part of the fjords
to the outer coast. The author has never found these accumulations at
higher levels (but quite commonly at lower altitudes, inter al. connected with the N_4-level). A critical evaluation of the published finds of
pumice here in the north, points in the same direction. The connection
between the pumice finds and the "Tapes line" was emphasised by
A. M. Hansen (1918). It should be mentioned, however, that Tanner
reports a case from the Fisherman's Peninsula, where pumice is said
to occur as much as 20 m above the Tapes line. It is not stated whether
this refers to actual accumulations or merely to scattered pieces. No

Fig. 146. Shore bars in Stordalen, Laksefjord. To the right transgressive bar complex (max. height a.s.l. 18 m) corresponding to Tapes I—II levels which here come close together. (Phot. M. Marthinussen N.G.)

finds in northern Norway referred to by Tanner, seem to occur above Tapes I—IV. Personal observations in Bindalsfjord southern Nordland, showed accumulations of pumice at a level slightly above 45 m, no doubt corresponding to the said Tapes limit. Higher up no trace of pumice was observed.

Undås (1938) mentions three pumice levels. He emphasises that the upper one cuts the Tapes line at a height of 16 m and he concludes that the latter accumulation zone represents a special pumice drift in the later part of the Tapes period. Judging from personal observations, the author of the present does not believe this to be correct. As previously mentioned, pumice has been found at the Tapes limit even at the heads of the fjords (e. g. Altafjord). On the other hand, the oldest pumice drift may have taken place just after the Tapes I terraces had been formed, but before regression from this level had started.

Previously (1945) the writer has connected the more prominent pumice accumulations with *transgressions,* in part because in Ingøy, at the well marked N_4 pumice level, the accumulation has been found above a layer (about 30 cm thick) of peat. As to the accumulations at Tapes I—IV, it seems reasonable to think that also these belong to transgression phases, at least two, possible three or four such phases. The transgressive character of the Tapes line has, in Finnmark, been pointed to by Tanner, who correlated his II A and b-line respectively, which in a general way corresponds to Tapes I—IV, with the Tapes-Littorina "subsidence" in southern Scandinavia. In 1930 he refers to Nummedal and Rosendahl's find of shore gravel above peat at Tomaselv (just west of Vadsø). In this locality we are dealing with a transgression limit corresponding to the Tapes I level. A C-14 dating of the peat just below the gravel gave c. 7860 y. This is a surprisingly high age for this transgression, which seems to have taken place 6—7000 years ago, according to the present knowledge of the probably corresponding transgression in south Sweden (Sandegren, T. Nilsson et al.). The conditions found are explainable, however, if we assume that a formerly uppermost layer of peat has been removed by marine abrasion, an explanation which it may be useful to have in mind when studying similar transgression profiles in places exposed to vigorous attack by the sea.

Table of C-14 dated samples of peat and driftwood from Finnmark.

Locality	Material	Genus[1]	Corresponding shore-level (event. transgression limit)	Height a. s. l. (m)	Age B. P.
Vadsø (Tomaselv)	Peat below shore bar		Tapes I	25,0	7860±150
Ingøy (Djupdalen)	Drift wood in bog	Pinus	Tapes I—II ? (found at Tapes IV level)	8,9	6350±150
Sørøy (Børfjordbotn I)	— » —	Picea	Tapes II—III	10,5—11,0 (distinct shore l. 11,2—11,7)	5700±150
Sørøy (Børfjordbotn II)	— » —	— » —	— » —	— » —	5500±150
Seiland (Oldervik)	— » —	— » —	Tapes IV	15,0	4820±160
Ingøy (Saraberget)	— » —	Larix	N 4 — level	6,5	4100±100

[1] Identifications of Larix and Picea (Seiland) by Prof. E. Mork, of Pinus and Picea (Sørøy) by Prof. O. A. Høeg.

The dated peat has been pollen-analyzed by K. Egede Larsen, who found: Pinus 51 %, Betula 17½, Alnus 3, Corylus ½, Chenopodiaceae 2½, Cyperaceae 17½, and Gramineae 7½ %, total 99½ % of a total number of pollen of 306. Since pollen-analytical studies of the northernmost part of the country have not yet been carried out, nothing definite can be said regarding the stratigraphical horizon.

Also in another locality of northern Norway a transgression has been demonstrated, most probably corresponding to the one just mentioned from the Vadsø district. It has been described by J. Holmboe (1907) from Ramså in Andøy. Here black mud (gyttja) and peat, which occur 2,5—3,1 m a.s.l. is overlain by a shore bar gravel in thickness 4—6 m. Through a comparison of the diagram, Pl. 16, with a corresponding diagram, constructed by the author from the Ofoten—Andøy district it can be ascertained that the Tapes I level at Ramså is situated c. 4 m a.s.l. This indicates that the initial phase of the transgression can be paralleled with the Tapes I transgression at Vadsø. The thick gravel masses tell of positive shoreline displacement to about 8 m a.s.l., probably corresponding to the Tapes II—IV levels. The very last phase corresponds to the Tapes IV (and to the last pumice accumulations in the so-called Tapes period), the age of which through dating of drift-wood in a bog (Oldervik in Seiland) has been estimated at about 4800 y (see table), and therefore probably must be placed in the first part of Sub-boreal time (Fægri 1943).

As to a possible transgressive character of the Tapes IV level, a transgression described by J. Eggvin from Eggum in Vestvågøy in Lofoten (N.G. p. 744) is of interest. E. Bergstrøm recently arranged a dating of peat collected by him and lying below litoral material in this locality, with the result 5860 ± 100 y (pers. communication). Compared with the age of the Tapes IV level this figure is too high; but possibly an upper peat layer may have been removed also at this locality. In any case, it seems probable that the Tapes shore here (8,5 m), like in other peripheral districts, must represent the maximal transgression in the later part of Tapes time, and it is then difficult to believe that the said transgression at Eggum should be much older than the Tapes IV-line. Thus Tapes I and IV both seem to represent transgressions.

Concerning the Tapes II and III lines, we have as yet no proof of a transgressive character (except for the pumice accumulations). A C-14 dating of driftwood at these levels gave 5500 and 5700 respectively. In the locality in question the two levels are situated so close together that the figures may refer to any of them[1].

---

[1] Also Tapes IV lies here at about the same level, but cannot be considered in this connection. The distinct terraces in this place 11,2—11,7 m, probably represent both Tapes II and III. The locality was sheltered from attack by the sea.

The driftwood from 8,9 m in Ingøy, found at Tapes IV level, was C-14 dated in an attempt to identify rather old drifts (transported upwards) here in the outermost districts. The result, c. 6350 years, so far was satisfactory. The wood may have been deposited while the shore-level was between 2 and 6 m, viz. between the Tapes I and II levels, and have then followed the continued positive shoreline displacement to the Tapes IV level. As to the other driftwood log in Ingøy, at 6,5 m, the N_4 level here, it probably has not drifted ashore till the time of this level, the age of which would be c. 4100 years. Such a figure would fit well with the age 4800 years for Tapes IV.

During the search of driftwood in bogs, the experience was gained that the vertical distribution corresponds to that of the pumice insofar as there are no finds above the Tapes I limit. It should be emphasised that to find *bogs* at the critical altitude in the inner fjord districts is very difficult.

Pumice above the present sea-level is found also farther south in the country (Undås 1942, 1945, Fægri 1943) and along other northern coasts, in Sweden, Denmark, Fisherman's Peninsula (Kola Peninsula), Novaya Zemlya, Spitsbergen, Greenland. Concerning its source, this question has been discussed by various authors (a main contribution from some time ago, is Bäckström 1890).

The problem has more recently regained actuality, partly through a paper by Noe-Nygaard (1951), partly because of recent finds and discussions of pumice at certain levels in the Spitsbergen Archipelago (Donner and West 1957, Kulling 1958, Birkenmajer 1958, Jahn (1959).

Noe-Nygaard has carried out a comparative petrographical investigation of pumice from Denmark (Vendsyssel), Norway (Bømlo, Blomøy), and Greenland[1]. He found that the refractive indices and chemical composition show great correspondence (refringence 1,524—1,525). He also compares the pumice mentioned with material from different volcanic areas, inter al. Iceland, and furthermore he tries to date the pumice drift under consideration and the eventually corresponding volcanic eruption, on the base of a pollen analysis of peat from Denmark (Iversen 1943). His conclusions are the following: 1. "— the pumices examined must belong to the same volcanic focus and, to all

---

[1] Chemical analyses of samples from Spitsbergen (Shoal Pt. Northeast Land) and from the Fisherman's Peninsula are also mentioned.

Some pumice finds in Finnmark and Nordland.

Localities	Colour	Height a. s. l. (m)	Shorelevel	Refractive indices [1]
Nordland: Bindalsfjord (Harangen)..	Brown	45,5	Tapes I	1,522
« « «	Dark brown	«	«	1,525
Finnmark: Alta (Hjemmeluft)	Black	28,0—28,5	«	1,524
« «	Brown	«	«	1,522
« (Bukta)	Dark brown	27,0—27,5	«	1,520
« «	Brown	«	«	1,519
Talvik (Djupvik)	Brown	23,6—23,9	«	1,524
Vadsø (Tomaselv	Brown	ca, 25,0	«	1,517
Måsøy (Revsbotn)	Black	16,3—16,8	Tapes II (also I and perhaps III)	1,525
« «	Dark brown	«	«	1,522
« «	Brown	«	«	1,528
« «	Brown	13,6	Tapes IV	1,524
« «	Black	«	«	1,540
« (Måsøy S.)	Brown	11,5—12,5	«	1,524
North Cape (Girsavaggoppe)	Brown	12,0—12,2	«	1,524
« «	Dark brown	«	«	1,527
Måsøy (Revsbotn)	Dark brown	9,7	N 4—line	1,522
North Cape (Girsavaggoppe)	Brown	7,9	«	1,524
« «	Black	«	«	1,528
« «	Brown	8,3	«	1,524
« «	Dark brown	«	«	1,527
« «	Brown	6,4	N 3—line	1,524
« «	Black	«	«	1,530

[1] Determined by Cand. mag. S. Kollung

appearances, even to the same eruption" (p. 38). 2. "— I consider it to be beyond all doubt that our andesite came from a prehistoric volcanic eruption of Hekla or its immediate neighbourhood in Iceland" (p. 41). 3. The andesitic pumice is found in connection with deposits from the end of the Tapes period (Norway) and from the transition Atlantic—Sub-boreal time (Denmark) respectively (p. 44). 4. The corresponding volcanic eruption of Hekla is thereby placed at the said transition, "in other words, about 4000 years back in time" (p. 44).

Thus Noe-Nygaard finds, based on pollen-analytical data from Vendsyssel and Bømlo (Fægri), that similar pumice accumulations in Denmark and Norway represent a synchronous shore level from about transition time Atlantic—Sub-boreal time[1]. This would correspond to the youngest Tapes level (Tapes IV) of Finnmark. As it is seen from the table p. 427 the refractive indices of pumice from this level, with one exception, have about the same value (1,524—1,527, average 1,525) as those found by Noe-Nygaard. However, in northern Norway there are both older and younger pumice-levels (Tapes I—III, N_4) of very similar petrographic character. The oldest pumice line, Tapes I, is probably 1500—2000 years older than Tapes IV; but the refractive indices of the pumice of the former level (1,517—1,525, average 1,522) also are very similar to those mentioned by Noe-Nygaard. It is the same with the pumice from the N_4-level (1,522—1,528, average 1,525). Between Tapes I and N_4 there is a time interval of probably 2000—2500 years.

From the preceding the following conclusion can be drawn: 1. It does not seem possible to correlate shore levels in the North Atlantic area on the base of pumice finds. 2. Pumice of one and the same type need not belong to one eruption. If one particular district is assumed to be the source area, it looks as if the character of the pumice here has been more or less constant during very long spaces of time. It should be added that the assumption of a more southern origin (cp. Marthinussen 1945) must still considered.

The material discussed by Noe-Nygaard also includes some from Spitsbergen (Shoal Pt., Northeast Land), from 0—2 m a.s.l. The chemical composition corresponds to· that of samples from Norway, Denmark, (and Greenland). According to Donner & West (1957), in the northern part of the Hinlopen Strait there are two distinct pumice-levels with slight inclination toward the northwest (height at Bragenes 13.8 m and 6.4 m respectively). Kulling (1958) refers the above mentioned pumice from Shoal Pt. to the lower of these levels, while Birkenmajer (1958) correlates it to the upper one.

According to Donner & West and Birkenmajer (who use the "whale method") the recent uplift in Spitsbergen has been so fast, that the said pumice levels should not be more than 400 and 200 years old respectively. Jahn (1959) questions these age determinations, and so

---

[1] He also points to the *Purpura lapillus* transgression in Iceland, which Thorarinsson assumes corresponds both to the pumice from W. Norway and the layer of brown pumice directly above the H_4-zone of the tephro-chronological system.

does Feyling-Hanssen (Feyling-Hanssen and Olsson 1960). The latter does not agree with Birkenmajer concerning such very rapid uplift of central Spitsbergen (Billefjord). He points to the fact that a terrace here, 9.7 m a.s.l., which according to Birkenmajer should correspond to a shoreline 350 years old, through C-14 dating of shell material must be more than 5000 years old. It is not improbable that the said terrace in Billefjord more or less may correspond to the upper pumice level at the Hinlopen Strait. In any case, it seems certain that the latter has an age of the same order as those of Norway and Denmark.

Errata

Page numbers are those of the original article.

p.416, l.8	*should read*
early work, and supplemented	early work, supplemented
p.418, l.4	*should read*
an end moraine on Æråsen	an end moraine on Æråsen
p.418, l.8	*should read*
S_4—S_2 fall	S_4—S_2/S_1 fall
p.418, l.11	*should read*
P_{12}, is the "upper line"	P_{12}, corresponds in part to the "upper line"
p.418, l.21	*should read*
a corresponding moraine	a corresponding marginal deposit
p.418, l.22	*should read*
moraines just south	moraines south
p.418, l.27	*should read*
first assumed by Grønli	first assumed inter al. by Gronli
p.418, l.30	*should read*
P. arctica	*Portlandia arctica*
p.418, l.33	*should read*
Dryas period.	Dryas period. Later dating gave about 10,350 yrs. B.P., cf. T-187 (Nydal 1962, p. 162).
p.418, Footnote, l.2	*should read*
from J. H. L. Voght,	from A. Helland, J. H. L. Vogt,
p.420, l.5	*should read*
and Nordland has confirmed	and Nordland have confirmed
p.421, l.9	*should read*
Also the oouthor's	Also the author's
p.421, l.16	*should read*
stone-age man has arrived	stone-age man arrived

p.421, l.18
early as 10,000 years

should read

early as at least 10,000 years

p.423, l.15
writer has connected

should read

writer connected

p.423, l.17
accumulation has been found

should read

accumulation was found

p.423, l.20
two, possible three

should read

two, possibly three

p.424, l.7
gave c. 7860 y. This

should read

gave c. 7860 y.* This

p.424, Table, l.1
Vadsø (Tomaselv) . . . 7860 ± 150

should read

Vadsø (Tomaselv) . . . 7860 ± 150*

p.424 Add footnote to page

should read

*A later dating gave the result 7750 ±
150 years B.P. (Nydal 1962, p. 165).

p.425, l.4
(1907) from Ramså

should read

(1903) from Ramså

p.425, l.18
a possible transgressive

should read

a possibly transgressive

p.426, l.6
levels, and have then

should read

levels, and has then

p.426, l.8
probably has not drifted

should read

probably did not drift

p.426, l.11
search of driftwood

should read

search for driftwood

p.427, l.10
Vendsyssel and Bømlo

should read

Vendsyssel (Iversen) and Bømlo

p.428, l.16
area on the base of

should read

area on the basis of

p.428, l.21
must still considered

should read

must still be considered

p.428, l.33
more than 400

should read

more than some 400

Note: This paper was previously published under the title "Coast- and fjord area of Finnmark."

References

Bergström, E., 1959. Utgjorde Lofoten och Vesteralen ett refugium under sista istiden? Svensk Naturvetenskap 1959, p. 116

Birkenmajer, K., 1958. Remarks on the pumice drift, land-uplift and the recent volcanic activity in the Arctic Basin. Bull. de l'acad. Pol. d. Sci. Chim. etc. No. 8, p. 545.

Bravais, A., 1840. Sur les lignes d'ancien niveau de la mer dans le Finmark. C. R. Ac. Sc., T. X, p. 691.

Bäckström, H., 1890. Über Angeschwemmte Bimssteine und Schlacken der Nordeuropäischen Küsten. Bih. Kgl. Sv. Vetensk.-Akad. Handl. 16. Afd. II. No. 5.

Boe, J. et *Nummedal, A.,* 1936. Le Finnmarkien. Les Origines de la civilisation dans l'extreme-nord de l'Europe. Inst. f. Sammenlign. Kulturforsk. B. 32.

Donner, J. J. and *West, R. G.,* 1957. The Quaternary Geology of Brageneset, Nordaustlandet, Spitsbergen. Norsk Polarinst. Skr. No. 109.

Feyling-Hanssen, R. W. and *Olsson, O.,* 1960. The main trend of the Post-glacial shoreline displacement in central Spitsbergen, five radiocarbon dates. N. Geogr. T. 17 (in print).

Faegri, K., 1943. **Studies on the Pleistocene of Western Norway III. Bømlo.** B. M. 1943. No. 8.

Grønlie, O. T., 1940. On the traces of the ice-ages in Nordland, Troms, and the southwestern part of Finnmark in northern Norway. N.G.T. 20, p. 1.

——— 1951. On the rise of sea and land and the forming of strandflats on the west coast of Fennoscandia. N.G.T. 29, p. 26.

Hansen, A. M., 1918. Isobasesystemet ved slutten av istiden. N.G.T. 4, p. 280.

Helland, A., 1900. Strandliniernes fald. N.G.U , No. 28 (2).

Holmboe, F., 1903. Planterester i norske torvmyrer. Vid Selsk., 1903, No. 2.

Holtedahl, O., 1953. Norges geologi II. N.G.U., 164, p. 575–1118.

Iversen, J., 1943. Et Litorinalprofil ved Dybvad i Vendsyssel. Medd. Dansk Geol. For. 10, p. 324.

Jahn, A., 1959. Postglacial development of Spitsbergen's shores. (Summary of paper in Polish). Czasopismo Geograficzne, 1959, 30, p. 245.

Kulling, O., 1958. Om lavabergarter på havsbotten i Nordpolsområdet och om pimpsten och drivved i höyda strandvallar på kusten av Nordostlandet och angränsande delar av Vastspetsbergen, G.F.F. 80, p. 108.

Marthinussen, M., 1945. **Yngre postglasiale nivåer pa Värangerhaloya.** N.G.T. 25, p. 230.

Nilsson, T., 1959. Aktuella utvecklingslinjer inom allmän svensk kvartärgeologi. G.F.F. 81, p. 127.

Noe-Nygaard, A., 1951. Sub-Fossil Hekla Pumice from Denmark. Medd. Dansk Geol. For. 12, p. 35.

Nydal, R., 1960. Trondheim natural radiocarbon measurements II. Amer. Journ. Sci.: Radiocarbon Supp., 2, p. 82–96.

——— 1962. Trondheim natural radiocarbon measurements III. Amer. Journ. Sci.: Radiocarbon Supp., 4, p. 160–182.

Sandegren, R., 1931. **Beskrivning til kartbladet Goteborg. S.G.U. Ser. Aa.,** No. 173.

——— 1952. Beskrivning til kartbladet Onsala. S.G.U. Ser. Aa., No. 192.

Tanner, V., 1930. Studier över kvartærsystemet i Fennoskandias Nordliga Delar (IV). Fennia 53, No. 1.

Thorarinsson, S., 1951. Laxárgljúfur and Laxárhraun. Geogr. Ann. 32. H. 1–2.

Undas, I., 1938. Kvartæstudier i Vestfinnmark og Vesterålen. N.G.T. 18, p. 81.

——— 1942. On the Late-Quaternary History of Møre and Trøndelag (Norway). K.N.V. Skr. 1943.

——— 1945. Drag av Bergensfeltets kvartærgeologi I. N.G.T. 25, p. 433.

Vogt, J. H. L., 1907. Über die schräge Senkung und die spätere schräge Hebung des Landes im nördlichen Norwegen. N.G.T., Bd. 1, No. 6.

Editor's Comments on Paper 12

Donner: *A Profile Across Fennoscandia of Late Weichselian and Flandrian Shore-lines*

In contrast to the detailed research of Marthinussen, the paper by Donner represents an attempt to derive a regional synthesis of shoreline deformation based on a southeast to northwest transect slightly south of the maximum uplift center over the Gulf of Bothnia (see Donner's Fig. 1).

The importance of radiocarbon dating and pollen analysis in the correlation of strandlines is well demonstrated in this paper. Donner discusses the shoreline evidence from four separate areas with specific attention focused on the problem of determining and dating major strandlines that are correlative throughout the region. The final product is an intriguing picture (Fig. 4) of a broad, dome-shaped uplifted region with a decrease in uplift toward the present (note that in this paper the years are given here in BC notation, not BP). This pattern provides a suggestive link between the shape of Fennoscandia ice cap and the form of crustal unloading upon removal of ice load. The southeast portion of the cross section indicates that crustal bending occurs at the margin of the ice cap, but in the northwest the features are not sufficiently continuous to allow an interpretation. The diagram also indicates the basis on which the relation diagram has been constructed. It is assumed (on good grounds) that there is a proportional relationship between the elevation of the Tapes I shorelines and higher older shorelines.

Donner's tentative reconstruction across the land area suggests that the apparent relaxation time for the central portion of the ice sheet is on the order of 3,000 years.

J. J. Donner was born in Helsinki in 1926. He has Ph.D. degrees from both the University of Helsinki (1951) and Cambridge University (1956). Currently holding the Chair in Geology and Palaeontology at the Universty of Helsinki, he has published extensively on the Quaternary of Scandinavia, Scotland, Spitsbergen, and other areas.

Reprinted from *Comm. Phys.-Math.*, **36**, 1–23 (1969)

12

A profile across Fennoscandia of Late Weichselian and Flandrian shore-lines

J. J. Donner
Department of Geology and Palaeontology
University of Helsinki

Abstract. The altitudes of some Late Weichselian and Flandrian shore-lines were determined in a 1090 km long profile across Fennoscandia from Estonia in the south-east to the Atlantic coast of Norway in the north-west. The shape of the shore-lines suggest that the land uplift during the last 10000 years was dome-like, with the summit of the dome in the eastern part of the Scandinavian mountains. There was possibly a marginal hinge-line or hinge zone in the Gulf of Finland, which bent the shore-lines. No other irregularities in the land uplift could be demonstrated.

INTRODUCTION

The aim of the present study was to determine the altitudes of some Late-glacial and Post-glacial, or, using the division and terminology suggested by WEST (1968) and adopted here, Late Weichselian and Flandrian shore-line surfaces in a narrow belt across the Fennoscandian area of land uplift. In this way a distance diagram of the shore-lines could be constructed, in which the abscissa shows the distance between the sites and the areas studied within the belt, and the ordinate the altitude of the shore-lines above the present sea-level. The area covered by the study and shown in Fig. 1, was comprised of Estonia, south-western Finland, a part of Västerbotten in Sweden and also a part of Nordland in Norway. The total length, including the gaps consisting of the Gulf of Finland, the Gulf of Bothnia and the Scandinavian mountains is 1090 km. The width of the areas from which the observations were included and projected perpendicularly onto the base-line, was in Finland and Sweden 90 km, in Estonia 130 km and in Norway 125 km. When, as here, a straight line is used as a base-line (= abscissa) for the shore-line diagram, it is part of a great circle on the earth's surface. In the map projection used in Fig. 1 it would appear as a slightly curved line, but the curving is negligible in a distance of 1090 km and a straight line was therefore drawn on the map. The angle at which the line crosses the meridians changes, however, from one area to another. The direction

Fig. 1.   Map showing areas studied in Estonia, south-western Finland, Västerbotten in Sweden and Nordland in Norway, and base-line used for shore-line diagrams. Map also gives the outer limit of Weichselian glaciation and the position of the Fennoscandian moraines, where determined, as well as isobases for the recent uplift in mm per year, as compiled by Kukkamäki (1968).

of the base-line for the shore-line diagram should, ideally, be at right angles to the isobases of the land uplift. The only isobases which are sufficiently well determined in the whole area covered by the present study are those for the recent uplift, as determined by repeated precise levellings or by sea-level recordings. The isobases in Fig. 1 for the recent uplift in mm per year were drawn on the basis of the map compiled by Kukkamäki (1968) on the basis of available data from Fennoscandia, the most detailed isobases being those constructed for Finland by Kääriäinen (1966), but in Fig. 1 only given for every mm. From Fig. 1 it can be seen that the base-

line for the shore-line diagram is not everywhere perpendicular to the isobases. If also the isobases for older shore-line surfaces were at an angle to
the base-line, some inaccuracies can be expected in the determination of
the altitudes of the Late Weichselian or Flandrian shore-lines. The effect
of the inaccuracies is lessened by using a narrow belt from which the sites
are chosen, but even then some areas, such as the Norwegian coast, show
that the area must be divided into even more narrow strips before the position of the shore-lines can be determined. Apart from showing the position
of the base-line for the shore-line diagram and the isobases for the recent
uplift, Fig. 1 also shows the outer limit of the Weichselian glaciation (after
WOLDSTEDT 1958 and SEREBRYANNY 1965), within which the area studied
clearly lies, as well as the Fennoscandian moraines, represented by Salpausselkä I and Salpausselkä II in Finland, the Central Swedish moraines, the
Ra moraines in southern Norway and the Tromsö-Lyngen moraines in
northern Norway (see ANDERSEN 1968).

The present paper is a direct continuation of some earlier papers in which
the shore-lines were determined in more restricted areas. Combined with an
evaluation of the various methods which can be used in the determination
of the altitude of the Late Weichselian and Flandrian shore-lines, these
shore-lines were first determined in the area in south-western Finland shown
in Fig. 1 (DONNER 1964, Fig. 11). Later the shore-line diagram thus constructed was extended south-eastwards across the Gulf of Finland to Estonia,
and a shore-line diagram covering a distance of 450 km could be drawn
(DONNER 1966, Fig. 3). South-western Finland and Estonia are therefore
only treated in the present paper as far as corrections and additions to the
former results are concerned. One addition is the more accurately determined
shore-lines determined in the Salpausselkä belt in south-western Finland
with the help of the altitudes of delta plains and their relationship to the
retreat of the ice margin dated with the help of the varve chronology. Thus,
these results (DONNER 1969) could be added to the shore-line diagram constructed earlier for south-western Finland and taken into account in the
construction of the diagram in Fig. 4. The material used in the determination
of the shore-lines in Sweden and Norway was collected from published
papers, but the areas in Sweden and Norway treated in the present investigation were visited in the summer of 1966 so that the field evidence on which
the conclusions in these papers were based could be assessed. In the description of the areas they are treated separately beginning from the south-east.

ESTONIA

The determination of the shore-lines for the Baltic Ice Lake, before its final
drainage, the transgression of the Ancylus Lake, and for the second trans-

gression of the Litorina Sea, L II, as earlier presented in a shore-line diagram (DONNER 1966) on the basis of available data, is in agreement with more recent detailed stratigraphical investigations in Estonia, particularly by KESSEL and RAUKAS (1967). Using their results some additions could be made to the shore-line diagram as well as some corrections in the datings, as new radiocarbon determinations have now been obtained from some sites in Estonia. The additions and changes were all taken into account in constructing the diagram in Fig. 4. The L II (= L II^b) transgression of about 1.5—2 m, which forms the highest Litorina limit in most parts of Estonia dealt with and which reached its peak in 3500 B.C., was preceded by the first transgression of the Litorina Sea, L I, of about 3 m in north-western Estonia (KESSEL and RAUKAS 1967). The position of this shore-line can stratigraphically be determined and, as seen in Fig. 4, it intersects L II in the Tallinn area in north-western Estonia. L I is not, however, traceable with the help of raised beaches (see DONNER 1966) and its altitude is not equally well determined as the shore-line for the upper limit of the Litorina transgression, which is here L II. The age of the earlier trans-gression L I is in Estonia 5000—4500 B.C. A radiocarbon determination of mud, which is synchronous with a peat lens 3 m below the shore-line of L I at Keila-Joa (Site 5, DONNER 1966) gave an age of 7180±270 B.P., 5230 B.C., and is placed at the beginning of the Atlantic period (Mo-223, VINO-GRADOV et al. 1968). The determinations of the pollen stratigraphy at Endla (TA- 85 to TA-98, PUNNING, ILVES and LIIVA 1968), however, indicate that some of the zone boundaries in Estonia are somewhat younger than cal-culated when drawing the earlier shore-line diagram (DONNER 1966). At Endla the zone boundary V/VI, in the division by T. NILSSON (1935) adopted in Estonia, just older than L I, was determined to 6480±70 B.P., 4530 B.C., a date in agreement with the results presented by KESSEL and RAUKAS (1967) in a stratigraphical table also including radiocarbon dates. The two Litorina transgressions in Estonia were correlated with the cor-responding transgressions in Finland and south Sweden by KESSEL and RAUKAS (1967), but the radiocarbon determinations seem to indicate that L I, formed at 5000—4500 B.C. in Estonia, may here be somewhat younger than further north or west in the Baltic area. The dating of L II to 3500 B.C. agrees with the dating of this transgression elsewhere. The transgres-sion of the Ancylus Lake, which exceeded 10 m in northern Estonia reached its maximum between 6500 B.C. and 6000 B.C., as seen from the results presented by KESSEL and RAUKAS (1967) and also confirmed by the radio-carbon determinations from the Endla series of the zone VIII period, cor-responding in age with the Ancylus transgression, which gave an age of 8495±85 B.P., 6545 B.C., and of the zone boundary VII/VIII, following the time of the maximum of the transgression, which gave an age of 7865±

75 B.P., 5915 B.C. (TA-98, TA-96; PUNNING, ILVES and LIIVA 1968).
The date of 6500—6000 B.C. for the maximum of the transgression of the
Ancylus Lake, A, and thus for its shore-line, is younger than the age earlier
calculated for it by the author (DONNER 1966), *i.e.* 7000—6800 B.C. The
age determinations or calculations of the duration of the Ancylus Lake, as
well as for the time at which it reached its highest position in those areas
of the Baltic in which it was transgressive, still vary and only permit a
dating with the accuracy given above, as shown by the review of the study
of the Ancylus Lake by FREDÉN (1967). The uppermost shore-line for
Estonia in Fig. 4 is that for the level of the Baltic Ice Lake before the drainage at Billingen in Sweden. Earlier the date of 8300 B.C. was given for this
shore-line (DONNER 1966), but after the dating of the drainage of the Baltic
Ice Lake to 8213 B.C. in Sweden (E. NILSSON 1968) the shore-line in Estonia
and the corresponding shore-line, B III, in Finland (see DONNER 1969)
can be dated broadly to 8300—8200 B.C.

Of the four shore-lines for Estonia presented in Fig. 4 only two, those
for the Litorina Sea, L I and L II, show the former position of the sea-level,
whereas the shore-line for the transgression of the Ancylus Lake and the
shore-line for the Baltic Ice Lake give the former position of the water-
level at times when it was dammed and stood above sea-level. In Fig. 4
the vertical lines under the shore-lines for the Baltic Ice Lake and the
Ancylus Lake show approximately how much above sea-level the water-
level of these lakes stood. The areas in which there is stratigraphical evidence for a transgression (see KESSEL and RAUKAS 1967) before the water-
level reached the position of a shore-line marked in the diagram, are shown
by arrows pointing upwards, as in the diagram constructed earlier (DONNER
1966). The arrows, however, do not give the total amount of the transgression, as this is often not possible to determine.

SOUTH-WESTERN FINLAND

In the determination of the shore-lines in south-western Finland the Clypeus
limit, at 5500 B.C., was separated from the first transgression of the Litorina Sea, at 5000 B.C., mainly on the basis of a site near the coast where the
transgression could be demonstrated (Site S 17, DONNER 1967). Further,
the first stage of brackish water in the Baltic after the drainage of the
Ancylus Lake, the Mastogloia Sea, was separated from the Litorina Sea on
the basis of the diatom flora containing *Campylodiscus clypeus* in the sediments. As these differences in the diatom flora, especially the occurrence
of *Campylodiscus clypeus*, foremost reflect the conditions, such as water
depth and the presence of lagoons, just before the emergence of the coast
where the basins studied became independent lakes, the limit at which

there was a change from fresh water to brackish water, regardless of diatom flora, was here taken to represent the beginning of the Litorina Sea period (see DONNER 1964, Fig. 7). This limit is somewhat older at higher isobases, about 5500 B.C., with a continuous regression at this time, than near the coast in southern Finland, where the first Litorina transgression, L I at about 5000 B.C., occurred. On the basis of the material used the shore-line, however, is a straight line. If the line is extended towards Estonia it joins the L I shore-line in Estonia, with a slight bend as the younger Litorina shore-line, L II, which it intersects near Tallinn in Estonia. The L I appears to be somewhat younger in Estonia than in southern Finland. Its age decreases towards the marginal parts of uplift and the shore-line is meta-chronous in the way typical for a shore-line formed during a transgression. The dating of the first transgression, is, however, still tentative and, as it is not the uppermost Litorina transgression in Estonia, its dating there is not stratigraphically as clear as in southern Finland where L I is clearly above L II. In the earlier shore-line diagram, in which the shore-lines in Estonia were compared with those in south-western Finland (DONNER 1966, Fig. 3), the Litorina shore-line L I could not be connected across the Gulf of Finland. This is not any more the case when the upper limit for the Li-torina Sea was drawn as one single shore-line, as in Fig. 4.

The shore-line corresponding to the zone boundary V/VI, which was earlier dated to 6300 B.C. (DONNER 1964), can now, when the new radio-carbon dates in south-western Finland are taken into account (ALHONEN 1968, DONNER 1969), be placed at 6000 B.C. At the time corresponding to this shore-line the level of the Ancylus Lake had already dropped to the level of the sea (see DONNER 1964, 1969). As the marked change in the salinity of the coastal waters took place later, at about 5500 B.C., which is the beginning of the Litorina Sea period, there is a transitional period between 6000 B.C. and 5500 B.C. during which salt water had not yet penetrated to the shallow waters of the coast. As it is probable, however, that the Baltic was already connected with the ocean, and that the level of the Ancylus Lake therefore had already dropped, this period of 500 years has in Finland been separated as the Mastogloia Sea period. It is, however, not well defined nor dated in south-western Finland. In those areas where there were rapid land/sea level changes, as in the coastal areas in western Finland with a quick relative uplift of land, the difference between the position of the shore-line for 6000 B.C. and for 5500 B.C., or possibly 5000 B.C., is con-siderable.

The two positions of the shore-line determined for 7000 B.C. and 6000 B.C. are sea-level positions above and below the shore-line for the Ancylus Lake determined in Estonia. On the basis of some sites in southern Finland (DONNER 1964) it could be extended to this area, and it lies immediately below the shore-line determined for 7000 B.C. Its position in Fig. 4 was drawn on the basis of the sites in which the Ancylus transgression in the zone V period was stratigraphically determined (Sites S 38 and S 43, DONNER 1964). Of the shore-lines earlier determined with the help of the pollen stratigraphy that for the zone boundary III/IV was omitted in Fig. 4, because the vegetational change on which this zone boundary was based, is not, in view of radiocarbon determinations, considered to be a synchronous pollen stratigraphical boundary in areas close to the retreating ice margin (DONNER 1967). The oldest shore-lines in Fig. 4 for south-western Finland are those determined in the Salpausselkä belt (DONNER 1969), B I and B III representing levels of the Baltic Ice Lake, the latter the same as the level determined in Estonia, and g and Y I representing positions of the sea-level. g was formed before B I and Y I immediately after the drainage of the Baltic Ice Lake at Billingen, for which the date 8213 B.C. (E. NILSSON 1968) was used here, as also when these shore-lines were determined (DONNER 1969). As 8305 B.C. (see FROMM 1963) was earlier used for the drainage at Billingen, those dates earlier based on the varve chronology become about 100 years younger than before (DONNER 1964). Thus, the new date for the shore-line below Y I is now 7600—7500 B.C., and the positions of the ice margin dated with varve chronology are 8200 B.C., just inside Salpausselkä II, 7900 B.C. and 7500 B.C. The area in which the two Salpausselkä moraines occur are also shown in Fig. 4, their closer relationship to the B levels having been discussed elsewhere (DONNER 1969).

The altitude and tilt of shore-lines B III and g in Fig. 4 is almost exactly the same as for the two highest shore-lines determined earlier (DONNER 1964). There are many raised beaches below B III but above the Y I level. and some inside the position of the ice margin at 8200 B.C. (DONNER 1964, Fig. 2). They may, therefore, represent one or two intermediate levels formed during the drainage of the Baltic Ice Lake from B III to Y I. The marginal delta plains at the levels of B III and Y I are, however, very close to each other (DONNER 1969), which shows that if there were intermediate levels formed during the drainage of the Baltic Ice Lake, they represent a relatively short period of time.

VÄSTERBOTTEN, SWEDEN

When the line for the shore-line diagram in Fig. 4 is continued from Estonia and south-western Finland across the Gulf of Bothnia towards the north-west, it crosses Västerbotten (Fig. 1). The sites used for the determination of the shore-lines are from an area 90 km broad, as in Finland, and 150 km long, in the coastal area. The areas further inland were not affected by land/sea level changes following deglaciation. In Västerbotten the land uplift was still relatively rapid during the time of the Litorina Sea in the Baltic. Therefore, there was no transgression in this area during this time, nor were any clear beaches formed which could be used in determining the Litorina shore-line. The position of the shore-line in the beginning of the Litorina Sea period was in Västerbotten determined with the help of the highest sites in which sediments containing salt-water diatoms occur (GRANLUND 1943). The sites studied by GRANLUND show that below some sites containing only fresh-water diatoms, there is a transitional zone of about 4 m with sediments containing *Mastogloia*-species and below this zone sediments containing *Campylodiscus clypeus*. As the period represented by the *Mastogloia*-species was very short, GRANLUND did not separate it as a Mastogloia Sea period, but used the term salt-water limit for the highest limit to which the influence of salt-water reached at the beginning of the Litorina Sea period (see also J. LUNDQVIST 1965). The salt-water limit thus defined in Västerbotten corresponds to the upper limit of the Litorina Sea in the north-western part of the area studied in Finland. The Mastogloia Sea period as dated and defined in Finland has not been separated in Västerbotten from the Ancylus

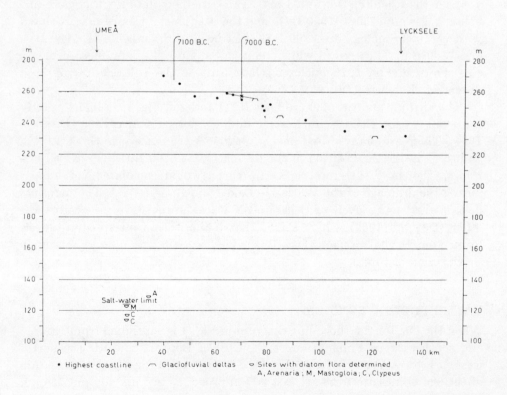

Fig. 2.   Diagram of determined positions of the salt-water limit and the highest coastline in Västerbotten, Sweden. The position of the ice margin in 7100 B.C. and 7000 B.C. is also shown.

Lake period, which here continued until the beginning of the Litorina Sea period (Granlund 1943). As the first penetration of salt-water, marking the beginning of the Litorina period, was determined with the help of pollen analysis of varved clays in Ångermanland to have occurred not earlier than 5400 B.C., and probably at 5100—5000 B.C. (Fromm 1938), the date 5400—5000 B.C. can also be used in the area of Västerbotten dealt with here, lying immediately north-east of Ångermanland. The date of 5500 B.C. used for the salt-water limit in Finland was not dated in the area in which it was used but taken from the earlier determinations in Sweden in general (Florin 1963, Fromm 1963), which may be a few hundred years too old. As the salt-water limit on each side of the Gulf of Bothnia is similar in character, it must be considered to represent an almost synchronous shore-line. Taking into consideration also some younger radiocarbon dates of c. 5000 B.C. (see Fromm 1963) the age of 5500—5000 B.C. can, therefore, be used for this limit on both sides of the Gulf of Bothnia. A more exact date can not be given on the basis of the present material.

The altitude of the salt-water limit in the coastal area of Västerbotten falls from about 126 m in the south-west to just under 120 m in the north-east, as shown in a diagram by Granlund (1934, Fig. 90). Of the sites on which Granlund based his determination of the salt-water limit only four are in the area dealt with here, sites 4 and 5 near Djupsjö, site 6 near Hörnsjö and site 8, earlier studied by Halden, north-west of Umeå (Granlund 1943, p. 107—108). These sites were included in the shore-line diagram in Fig. 2. The sediments of the uppermost site have an *Arenaria*-flora, which shows that it is above the salt-water limit, whereas the others are below it. Site 5 at 123 m is just below it and has *Mastogloia*-forms, whereas the two other sites at lower altitudes also have *Campylodiscus clypeus*. On the basis of these sites the salt-water limit is at about 124 m in the area west of Umeå. As the salt-water limit could only be determined in a small area a line for the limit could not be drawn in the diagram in Fig. 2. The tilt of the salt-water limit along the coast of Västerbotten towards the north-east does not help in determining the position of this limit in the direction of the base-line used in the diagram in Fig. 2 and Fig. 4. In connecting the salt-water limit in the area near Umeå in Västerbotten with the corresponding shore-line on the Finnish side of the Gulf of Bothnia (Fig. 4), it was assumed, without any direct evidence, that the area in which the uplift has been greatest since the time of the formation of the salt-water limit, lies north-west or south-west of the Umeå area, in spite of the isobases for the recent uplift, which rather suggest a centre of uplift in the northern part of the Gulf of Bothnia (Fig. 1). Further away from the coast in Västerbotten, north-west of the area studied, ice-dammed lakes were formed at the time of deglaciation, and in narrow valleys lakes were dammed by endmoraines or eskers (in Swedish called »selsjöar», GRANLUND 1943, p. 82—85). The tilting of the shore-lines of these lakes suggest, according to GRANLUND (1943, p. 85—86), that the area in which the uplift has been greatest is situated north-west of the lakes Malgomaj, Storuman and Storvindeln, near the Norwegian border. The determination of the former shore-line surfaces in northern Sweden is, however, made difficult by the possibility that the centre of uplift has shifted from one area to another during and after the time of deglaciation and, further, that there may even have existed two centres of land uplift (J. LUNDQVIST 1965, p. 174).

The profiles from mires, in which the pollen stratigraphy can be compared with the land/sea level changes by determining the time of the isolation of the basins from the Baltic, are all below the salt-water limit in Västerbotten (GRANLUND 1943) and the older land/sea level changes could not, therefore, be dated in this way. The only level which can be determined in addition to the salt-water limit is the highest coastline, *i.e.* the uppermost limit to which the Baltic Sea reached after the last glaciation. A number of measurements (GRANLUND 1943, Fig. 57) of the altitude of glaciofluvial deltas and of the upper limit of wave action, as shown in Fig. 2, show the position of the highest coastline, falling from an altitude of 270 m near the coast to nearly 230 m further inland. The highest coastline is, however, a metachronous shore-line formed during the relatively rapid regression of the water-level at a time during which the margin of the ice sheet retreated into the Scandinavian mountains. Only by comparing the highest coastline with dated positions of the retreating ice margin, as done in south-western Finland, can the altitude of this line be compared with shore-lines determined in other areas. The recession of the ice margin has not been dated in Västerbotten, but the detailed investigations in Ångermanland (HÖRNSTEN 1964), bordering the area studied in Västerbotten, can be used for this purpose. Using measurements of varved clays HÖRNSTEN dated the retreat of the ice margin in the valleys of Ångermanälven and Moälven and drew recession lines for every 50 years of ice retreat, covering a period of 400 years, the zero year being the youngest (HÖRNSTEN 1964, Fig. 6). The zero year is the same as in the varve counts by BORELL and OFFERBERG (1955), dated to 6923 B.C. (HÖRNSTEN 1964, E. NILSSON 1960, 1968). A radiocarbon measurement of the humus fraction from the varves $+56$ to $+82$ in a varve series from Lugnvik, Ångermanland, gave an age of 9000^{+1400}_{-1200} B.P. (U-215, HÖRNSTEN and OLSSON 1964), the expected age beign 8800 B.P. The determined age may, however, not be the real age, as pointed out by HÖRNSTEN and OLSSON, and the unsoluble portions gave considerably older dates, 30000 years and more. In those varve series where pollen counts have been made, the change-over from a pollen flora dominated by *Betula* to a flora with mainly *Pinus* takes place at about the zero year. The diatoms analyzed show that the water of the Baltic in the year 6823 was fresh in the Ångermanland area but in 7323 B.C. still mesohalobous (HÖRNSTEN 1964). Using the detailed results from Ångermanland the recession lines for 7000 B.C. and 7100 B.C., interpolated from the lines drawn by HÖRNSTEN, were drawn for the bordering area in Västerbotten. The lines were drawn parallel to the tentative lines for the latter area as presented by G. LUNDQVIST (1961), who used the direction of endmoraines in determining the direction of the retreating ice margin. In the shore-line diagram in Fig. 2 the innermost position of the ice margin at 7100 B.C. and 7000 B.C. was drawn on the basis of the recession lines, as earlier done in south-western Finland (DONNER 1964, Fig. 2). The uppermost glaciofluvial deltas and the upper limits of wave action formed at about 7000 B.C. were connected with a line in Fig. 2 and its altitude at the point where the ice margin stood 7000 B.C. is 257 m. About 7100 B.C. the water-level stood at about 265—270 m, when the ice margin was nearer the coast. No shore-lines from the time around 7000 B.C. can, however, be determined in the area.

A connection of the altitude of 257 m for 7000 B.C. in Västerbotten with the shore-line for this period in south-western Finland is possible with a curved line, similar to that for the salt-water limit, assuming the greatest land uplift to have been further inland in Västerbotten also at this time. The shore-line surface for 7000 B.C. was dated in south-western Finland on the basis of the zone boundary IV/V in that area and the corresponding level in the pollen stratigraphy of Ångermanland is also at about 7000 B.C., possibly somewhat younger, as mentioned above. Further, the results from Ångermanland, also applicable in Västerbotten, showed that the Baltic was probably still connected with the ocean at 7000 B.C., thus excluding the possibility that the altitude for the water-level at this time should represent the level of the Ancylus Lake in Västerbotten. The results in south-western Finland agree with this conclusion. There the change-over to the Ancylus Lake period is in the beginning of zone V in the pollen stratigraphy (DONNER 1964).

NORDLAND, NORWAY

The Norwegian sites included in the present study, were all from an area in Nordland. It is 100 km long in the direction of the base-line in Fig. 1 and 125 km broad, stretching from Mo in the south to Fauske and Bodö in the north, the main valley being Saltdalen, with Saltdalsfjorden and Skjerstafjorden outside it. From this area there are a number of levelled altitudes of raised beaches, which were used in the shore-line diagram in Fig. 3. Most of the sites were studied by GRÖNLIE, who after a preliminary paper on the shore-lines of northern Norway (GRÖNLIE 1940) produced a

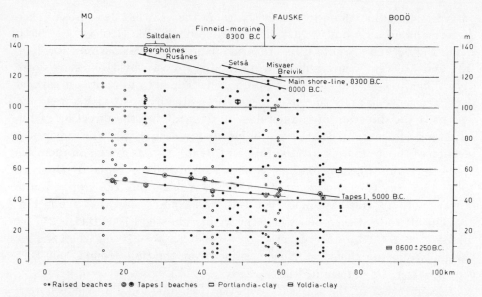

Fig. 3.   Shore-line diagram for Nordland, Norway, with raised beaches and sites with marine clays. Open circles are observations from the south-western half of area studied in Nordland and black dots from the north-eastern half.

more complete list of sites with additions and corrections to the earlier list (GRÖNLIE 1951). Two sites, in which the terraces were levelled by REK-STAD (1910) but not included in the lists published by GRÖNLIE, were included in the diagram. The total number of determined raised beaches included in Fig. 3 was 235. Some sites with marine clays containing shells were also included in Fig. 3 if they were of direct use in dating the shore-lines (the sites used in Fig. 3 are listed in the Appendix).

In representing the shore-lines GRÖNLIE (1940, 1951) used relation diagrams, or epeirogenetic spectra, similar to those introduced by TANNER (1930) for the shore-lines in northern Fennoscandia. GRÖNLIE used, as TANNER had done before, the Tapes line or b-line as a reference level, in relation to which all other altitudes of raised beaches were plotted (the construction of shore-line diagrams has been discussed elsewhere, DONNER 1965). As the uppermost Tapes shore-line is one of the few shore-lines which in northern Norway, especially at lower isobases, can be singled out among the number of raised beaches observed at different altitudes, it was separated in Fig. 3. It was determined with the help of all those sites, listed by GRÖN-LIE (1951), in which a particular raised beach, among a number of raised beaches at different altitudes, falls on the b-line. When all these beaches of the b-line were marked in Fig. 3, it became clear that they represent two separate tilted lines, not one. When the geographical distribution of the sites used is taken into account, this difference can be explained. Those sites which represent the lower, less tilted line, are all in the south-western half of the area, whereas those which represent the higher, slightly more tilted line, are all in the north-eastern half. As the raised beaches representing the b-line are likely to be correctly grouped, the difference must be caused by the direction of the base-line used in the shore-line diagram in Fig. 3. It is probably not perpendicular to the isobases at the Norwegian coast, as suggested by the isobases of the recent uplift of land (Fig. 1) and further supported by the suggested direction for the isobases for the b-line, as presented by GRÖNLIE (1940). Thus, the area 125 km broad along the coast is obviously too broad for an accurate determination of the former shore-lines, as far as they can be separated on observations of raised beaches alone. The area was therefore divided into two halves, the south-western and the north-eastern, and the altitudes of the raised beaches marked differently for each half in the diagram in Fig. 3. Therefore, there is a separate b-line for each area. The determination of the Tapes shore-line to 40—60 m is supported by the occurrence of shell-bearing sediments with a fauna characteristic for Tapes beds. Of the sites described by REKSTAD (1910) the highest is at Moldjord in Beiarn, where a clay with shells is found undernearth terrace sands at 43 m. The Tapes line in Fig. 3 is here at about 47 m.

Assuming that the position of the Tapes shore-line, as drawn in Fig. 3, was correctly determined on the basis of the studies by REKSTAD (1910) and GRÖNLIE (1940, 1951; see also HOLTEDAHL 1953, Pl. 18), it can be correlated with the Tapes I shore-line elsewhere in Norway. At an altitude of 40—60 m, at which it occurs in the area dealt with, it is the highest Tapes shore-line, as seen, for instance, in the diagram for West-Finnmark constructed by MARTHINUSSEN (in HOLTEDAHL 1960). At lower isobases, where the Tapes I shore-line was transgressive, radiocarbon dates have been obtained for peats overlain by beach gravels and sands. At Vadsö in Finnmark a date of 7750±150 B.P. was obtained and another sample from Andöy in Nordland gave a date of 7400±150 B.P. (T-182, T-270, MARTHI-NUSSEN 1962). MARTHINUSSEN gave an age of c. 6600—6500 B.P. for the Tapes shore-line in this area, but, as seen from the dates above, it may be 7000 years old or more, which would be closer to the date obtained elsewhere. In the Oslo area, where the shore-line displacement was studied in great detail by various authors, there were no transgressions at the time during which the Tapes shore-lines were formed in Atlantic time (HAFSTEN 1956, 1959). In a general correlation of the land/sea level changes in the Oslo area with those in other parts of Norway, also with the above-mentioned results by MARTHINUSSEN, the Tapes I shore-line was dated to about 5000 B.C. (FEYLING-HANSSEN 1964, see also ANDERSEN 1965). As the first Tapes

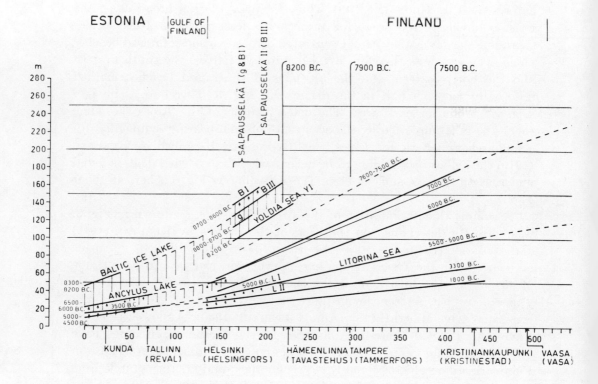

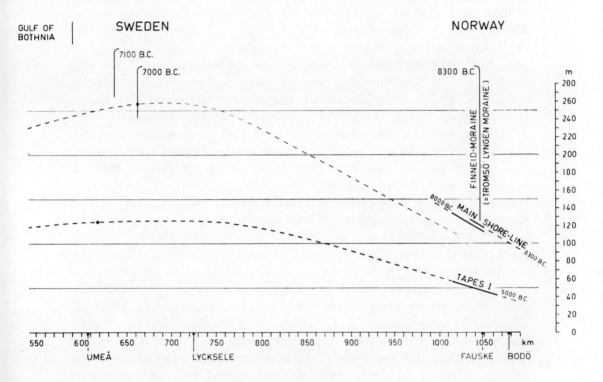

Fig. 4. Combined shore-line diagram for all areas shown in Fig. 1, showing also some positions of the retreating ice margin. Thick lines show determined shore-lines and broken lines show the connection of them. The position of the water-level of the Baltic Ice Lake and the Ancylus Lake above sea-level is shown with vertical lines and transgressions with arrows underneath the levels reached by the transgressions.

225

transgression can be considered to be synchronous with the first Litorina transgression in the Baltic, this date is supported by those obtained in Sweden and Finland. The date for the Tapes I shore-line at higher isobases, where it was not transgressive, may, however, be somewhat older, as was the case in the Baltic.

According to the above-mentioned datings the Tapes I shore-line in the coastal area of Nordland in Norway is of approximately the same age, about 5000 B.C. or slightly more, than the salt-water limit in Västerbotten in Sweden, and therefore these two shore-lines could be joined across the Scandinavian mountains as shown in Fig. 4. The Tapes I shore-line used for Norway was the upper one from Fig. 3, because it represents the north-eastern half of the area dealt with, as do the shore-lines for the highest coastline in the same area. The continuation of the salt-water limit joins the Tapes I limit in Nordland, which slopes down towards the Atlantic, if the curving of the shore-line, already suggested by the results from Väster-botten and south-western Finland, continues across Scandinavia.

The shore-lines above the Tapes line in the area dealt with in Nordland can not be separated on the basis of the altitudes of raised beaches alone, even when the observations were divided into two groups. As seen in Fig. 3, the raised beaches, formed during the relative regression of the sea-level from its highest position, occur irregularly and no grouping along certain tilted shore-lines can be traced. The diagram is similar to the diagram for raised beaches in south-western Finland (DONNER 1964, Fig. 2). The only shore-lines which could be determined above the b-line were those repre-senting the highest coastline. In the coastal areas of Västerbotten and particularly in Finland, where the topographical differences are compara-tively small, the ice margin retreated as a uniform ice front for which re-cession lines can be drawn. In Norway the withdrawal of the ice was more varied because of the broken topography with fjords and deep valleys between high mountains, some of which are still glaciated. Each fjord and valley had its own retreating glacier during the final stages of deglaciation, and in following the position of the highest coastline they have to be treated separately. The north-eastern half of the area dealt with is the only part in which the shore-line for the highest coastline could be reconstructed (Fig. 3). In the inner part of Saltdalen the highest raised beaches at Bergholnes and Rusånes lie on a tilted line, and when the line is extended three observations of raised beaches, i.e. at Setså, Saltdalsfjorden, at Breivik, Skjerstafjorden, and at Misvaer, fall on the same line. At these sites, however, the highest raised beaches are about 5 m above this line, as seen in Fig. 3. The two lines drawn for the highest coastline probably give the approximate altitudes for two shore-lines from the time of the ice recession, the higher shore-line being formed when a glacier was still occupying the lower reaches of Saltdalen, but the fjord was ice-free, and the lower shore-line formed when the glacier had already withdrawn higher up into Saltdalen and the valley was a continuation of the present fjord. Both shore-lines for the highest coastline are more tilted than the Tapes shore-line.

In the north-eastern half of the area dealt with in Nordland, in the area in which the two above-mentioned shore-lines for the highest coastline were determined, there is an assumed position of the ice margin during its retreat at Saltstraumen in the outer part of Skjerstafjorden (GRÖNLIE 1940, HOLTE-DAHL 1953). The two shore-lines for the highest coastline are inside.this line and therefore also younger. South-east of Fauske the well-developed Finneid-moraine blocks Nedrevatnet from Fauskevika (REKSTAD 1910, HOLTEDAHL 1953). It is probable that the glacier in the main valley of Salt-dalen had receded already some way up the valley at the time of the forma-tion of the moraine at Nedrevatnet, formed by a glacier which came down another valley from the east, as indicated by the line drawn by GRÖNLIE (1940, Pl. III) for this moraine stage, which he called M^1. The relationship between the shore-lines for the highest coastline with the moraines show that the upper shore-line was formed approximately at the time of the Finneid-moraine at Fauske, when there was still a glacier in the lower part of Saltdalen, whereas the lower shore-line was formed shortly after the time of the formation of the Finneid-moraine. At Skjerstad, at the mouth of Misvaerfjord, which is a small fjord on the southern side of Skjerstafjorden, there is also a moraine (REKSTAD 1910) which could be of the same age as the Finneid-moraine. This would explain why the highest beach at Misvaer (Fig. 3), which is inside the Skjerstad-moraine, lies below the highest beaches at Breivik and Setså, which are in areas outside both moraines. The above-mentioned age relationship between the shore-lines and the moraines agrees with the conclusions by GRÖNLIE (1951). In his shore-line diagram the Fin-neid-moraine of the M^1 stage corresponds to the d_2 shore-line. The highest coastline in Saltdalen is immediately below the d_2 shore-line, whereas the line in the fjord outside at Setså and Breivik are above d_2, thus being the same age or slightly older than the moraine.

In dating the shore-lines in the area dealt with GRÖNLIE (1940) assumed that the ice-margin was outside the coast at the time of the formation of the Tromsö-Lyngen endmoraine in northern Norway (see Fig. 1), which in Western Troms is shown to have been formed mainly in the Younger Dryas period, partly, however, already in the Older Dryas and Alleröd periods (ANDERSEN 1968). According to these datings the formation of the main part of the Tromsö-Lyngen moraine, formed in the Younger Dryas period, can be correlated with the formation of the Ra moraines in South Norway (Fig. 1). In the shore-line diagram constructed by GRÖNLIE the shore-lines in Fig. 3 lying just below and above the d_2 shore-line would fall, if compared with the shore-lines in the Oslo-Romerike area (GRÖNLIE 1940, Pl. IV), in the zone between the Pholas line and the Litorina line, using the classification by ÖYEN. This would mean that the highest shore-lines in Fig. 3 were formed at about 7500—7000 B.C. (FEYLING-HANSSEN 1964). If a comparison is made on the basis of the altitude of the Tapes I shore-line between the older shore-lines in Fig. 3 and the curve for the land/sea level changes in the Oslo area, determined on the basis of pollen analytically studied sediments in a series of lake basins at different altitudes and formerly connected with the sea (HAFSTEN 1956, 1959), the shore-lines for the highest coastline in Saltdalen and the fjord outside it would correspond to the position of the shore-line at the end of the Pre-Boreal period in the Oslo area, also about 7500—7000 B.C., as dated by HAFSTEN. The comparison of the shore-lines with the Oslo area is, therefore, consistent and places the upper shore-lines in Fig. 3 in the late Pre-Boreal period.

The above-mentioned dating, which was entirely based on the altitudes of the shore-lines and a comparison with the Oslo area is, however, not in agreement with the distribution of the marine clays and their fauna in the area presented in Fig. 3. At Rönvik, 1 km from Bodö, there is a clay at least 8 m thick containing shells of *Portlandia arctica* overlain by younger sediments (REKSTAD 1910, p. 26; HOLTEDAHL 1953, p. 714). A radiocarbon determination of *P. arctica* shells from the clay, which occurs at an altitude of between 7 and 10 m, gave an age of 10550 ± 250 B.P., 8600 B.C. (T-246, MARTHINUSSEN 1962). The presence of *Portlandia arctica*, representing the high-arctic *Yoldia* fauna, at this site and also in Glomfjorden (HOLTEDAHL 1953, p. 714) show that the ice margin must have retreated east of these areas at the time of the formation of the Tromsö-Lyngen moraines in the Younger Dryas period and could thus not have been outside the coast, as assumed by GRÖNLIE (1940). *P. arctica* only occurs in Late Weichselian marine sediments, or in sediments representing the transition to the Flandrian, in North Norway (see ANDERSEN 1968). The radiocarbon date supports the stratigraphical evidence and places the sediment in the Younger Dryas period. Similar dates listed by MARTHINUSSEN (1962, pl. 1) and by ANDERSEN (1968, Table 3) from other areas agree with this dating. Further away from the coast, in the fjords, there are marine clays containing *Portlandia lenticula*, with small stones dropped by icebergs. This clay, called *Portlandia*-clay (REKSTAD 1910, HOLTEDAHL 1953, p. 713), occurs, as shown in Fig. 3, at Nygaard in Beiarn at 60 m (REKSTAD 1910), at Osbak in Beiarn at 105 m (REKSTAD 1910, terraces in same area mentioned by GRÖNLIE 1951, site 369), and at Fauske up to an altitude of about 100 m (REKSTAD 1917, HOLTEDAHL 1953, p. 713). The occurrence of these clays at high altitudes and also close to the Finneid-moraine at Fauske indicate that they were deposited at a time shortly after the formation of this moraine, at a time corresponding to the lower of the two shore-lines for the highest coast-line in Fig. 3, which is about 15 m above the two highest clay-occurrences. The *Portlandia*-clay is younger than the *Yoldia*-clay mentioned above, and it probably represents the time immediately after the formation of the Tromsö-Lyngen moraine, or the Ra moraines in South Norway, *i.e.* the beginning of the Pre-Boreal period, a time about 8000 B.C., as in the Oslo area (FEY-LING-HANSSEN 1964). The marine clays, mentioned above, thus show that the lower shore-line for the highest coastline in Fig. 3 is early Pre-Boreal, from about 8000 B.C. and that the endmoraine or moraines corresponding to the Tromsö-Lyngen moraine is east of Bodö and not outside the coast. This is also supported by the occurrence of *Yoldia*-clay in Glomfjord (HOLTEDAHL 1953, p. 714), which is inside the line at which the ice margin is assumed to have stood when at Saltstraumen (see GRÖNLIE 1940, Pl. III). The well-developed Finneid-moraine, perhaps including the moraine be-tween Nedrevatnet and Övrevatnet just inside the Finneid-moraine, there-fore, most likely corresponds to the Tromsö-Lyngen moraine further north. The upper shore-line for the highest coastline in Fig. 3, corresponding in age approximately to this moraine, would thus be from about 8300 B.C. or somewhat older, using the datings from areas further north (ANDERSEN 1968).

The dating of the uppermost shore-line for the highest coastline in Fig. 3 to correspond to the Tromsö-Lyngen moraine means that this shore-line is at a lower altitude than suggested by GRÖNLIE (1940, 1951), but the dating is in agreement with the new results from Western Troms (ANDERSEN 1968). The relationship between the Tapes I shore-line and the shore-line formed approximately at the time of the Finneid-moraine in the area shown in Fig. 3, is almost exactly the same as between the Tapes shore-line and the Main shore-line, formed at the time of the Tromsö-Lyngen moraine, in Ullsfjord, Balsfjord and Malangenfjord areas in Western Troms if these lines are extended to higher altitudes (ANDERSEN 1968). It is also the same as between the Tapes I shore-line and the P_{12} shore-line, corresponding to the Main shore-line, in West Finnmark (MARTHINUSSEN in HOLTEDAHL 1960). Even if there are some regional variations in the inclination of the Main shore-line in northern Norway (ANDERSEN 1968), its altitude in relation to the Tapes I shore-line seems to be rather consistent in a large area. The close agreement about the age of the shore-lines in Fig. 3 with the results from Western Troms and West Finnmark supports the conclusions made on the basis of the marine clays and their relationship to the raised beaches and moraines, and shows that the comparisons made directly with the Oslo area give somewhat too young ages for the uppermost shore-lines.

As a summary of the dating of the two uppermost shore-lines above Tapes I in the area dealt with in Nordland and shown in Fig. 3, the following table can be given of the relationship between the marine clays, the moraines and the shore-lines.

Marine clays	Moraines	Shore-lines
Portlandia-clay with *Portlandia lenticula*		Highest shore-line in Saltdalen, c. 8000 B.C.
	Finneid-moraine (perhaps also Skjerstad-moraine) = Tromsö-Lyngen moraine	Main shore-line, c. 8300 B.C.
Yoldia-clay with *Portlandia arctica* (C^{14}-date 8600 ± 250 B.C.)		

The two shore-lines in Fig. 3 for the highest coastline are older than the highest position of the shore-line in Västerbotten, in Sweden, from where the ice retreated about 7000 B.C. (Fig. 4). The position of the shore-line for this time is, therefore, in Nordland somewhere below the highest coastline, at approximately the altitude indicated in Fig. 4. It was drawn to show the age relationship between the highest shore-lines in Sweden and Norway but, as for the line for Tapes I joined with the salt-water limit in Sweden, there is no evidence for its true shape in the Scandinavian mountains.

CONCLUSIONS

The shore-lines presented in the 1090 km long profile in Fig. 4 across Fenno-scandia, from Estonia in the south-east to the Atlantic coast of Norway in the north-west, were determined with the help of different methods in different parts and at different altitudes. In the determination of the alti-tudes of the shore-lines raised beaches, glaciofluvial delta plains, pollen analysis and diatoms, Stone Age water-side dwelling places and shellbearing marine clays or beach deposits were used. In dating the shore-lines the varve chronology for Finland and Sweden could be used. Other dates were partly supported by radiocarbon determinations, sometimes from outside the area investigated. Shore-lines in different areas were only connected with each other in those cases where their connection could be supported by stratigraphical evidence. For each area in which the shore-lines were deter-mined, separate more detailed diagrams were first constructed, in which the methods used are mentioned. As some of these diagrams were published earlier, a table is given below in which the figure numbers are listed for all separate shore-line diagrams in earlier papers as well as in the present paper.

	DONNER 1964	1966	1969	Present paper
Estonia		Fig. 2 ⎱ Fig. 3		
South-western Finland	Fig. 11	⎰	Fig. 3	
Västerbotten, Sweden			Fig. 2 ⎱ Fig. 4	
Nordland, Norway			Fig. 3 ⎰	

In Estonia and south-western Finland, in which shore-lines for distances of 80 km and 310 km, perpendicular to the isobases, were determined, no irregularities in their shape could be detected. In joining the shore-lines of Estonia with those in south-western Finland there is, however, a bend in the area of the Gulf of Finland. The shore-lines are clearly more steeply inclined on the Finnish side than on the Estonian side and a hinge-line or hinge zone must be assumed to exist or to have existed in the Gulf of Finland to explain this difference (see DONNER 1966). In addition to this bend, similar bends could possibly exist in the Gulf of Bothnia, the Scandinavian mountains or outside the Norwegian coast, *i.e.* in all those parts of the diagram in Fig. 4 from which there is no and could not be any evidence of the shape of the extended shore-lines. As there is no reason, on the basis of the available material, to assume that the shape of the shore-line surfaces, when extended, is irregular in those areas in which the shore-lines could not be determined, and only regular in those areas where they could, the shore-lines were joined in the way shown in Fig. 4. The uplift was, according to this interpretation, slightly asymmetrically dome-shaped with the summit of the dome in the eastern part of the Scandinavian mountains, in the area in which the last remnants of ice melted. The shore-lines slope more steeply towards the west than the east. It is possible that there is a slight curving of the shore-lines in south-western Finland but it is, however, so slight if it occurs that it was not detected. The shore-lines were, therefore, drawn as straight lines in the diagram. The rather regular dome-shaped uplift of Fennoscandia is also suggested by the isobases for the recent uplift of land, as shown in Fig. 1. The tilt of the shore-lines in Fig. 4 is not always the greatest possible because the base-line used is not everywhere at right angles to the isobases as, for instance, at the Norwegian coast.

In addition to suggesting an asymmetrical uplift in Fennoscandia, with a marginal hinge-line or hinge zone in the Gulf of Finland, the shore-line diagram in Fig. 4 shows how the withdrawal of the ice margin was more rapid on the eastern side of the Scandinavian mountains than at the Atlantic coast and in the mountains themselves. The ice margin stood at the Salpausselkä end-moraines in southern Finland at the same time, just over 8000 B.C., as at the Finneid-moraine in Nordland in Norway, but a thousand years later it had already withdrawn from Finland to Västerbotten in Sweden. The shore-line diagram also shows how the land/sea level changes in the Baltic region were affected by the development of the Baltic Sea. The fluctuations caused by the damming up of the water level of the Baltic Ice Lake and of the Ancylus Lake resulted in shore-line positions above sea-level. The altitude at which the water-level stood above sea-level during these periods is shown in Fig. 4, as well as the areas in which the rise of water-level overtook the land uplift and resulted in transgressions. Two

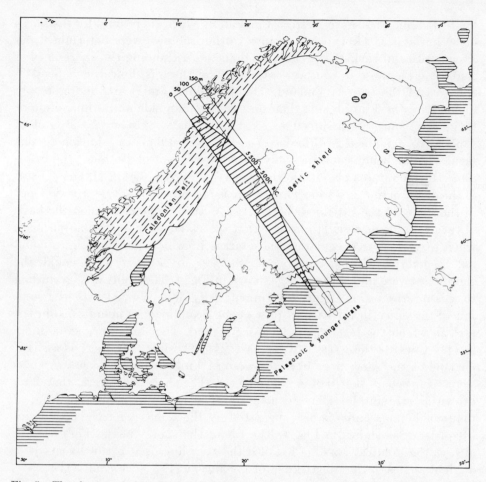

Fig. 5.   The shape and altitude of the Litorina I — Tapes I shore-line from 5500 B.C.—5000 B.C. and the position of the marginal hinge zone in the Gulf of Finland compared with the main tectonic units.

transgressions, that of the Baltic Ice Lake and of the Ancylus Lake, were transgressions in the Baltic when it formed an independent lake, whereas the Litorina transgressions were caused by the eustatic rise overtaking the land uplift.

The relationship of the shore-lines in Fig. 4 to the outer limit of the Weichselian glaciation and to the Fennoscandian moraines, as well as to the isobases of the recent uplift of land, were shown in Fig. 1. In Fig. 5 the general shape of the Litorina I—Tapes I shore-line from 5500—5000 B.C., along the line studied, is shown on the map together with the main tectonic units. In the north-west, along the Atlantic coast, the Caledonian belt borders the Baltic Shield. In the south and south-east there are Palaeozoic and younger strata. Within the Shield there are younger rocks, also marked on the map, as the Devonian intrusions in the Kola Peninsula and the Permian igneous rocks in the Oslo area. No direct relationship can be detected between the shape of the shore-lines and the main tectonic units. The marginal hinge-line or hinge zone.(see Fig. 5) in the Gulf of Finland, if real, is, however, likely to be tectonically controlled.

Appendix

References to raised beaches used in the shore-line diagram in Fig. 3, listed from left to right. GRÖNLIE 1951, sites 381, 380, 384, 378, 394, 367, 389, 386, 383, 354, 377, 375, 372, 369, 363, 360, 337, 366, 339, 334, 358, 350, 341, 320; REKSTAD 1910, Ripnes, Alsvik.

References to sites with *Portlandia*-clay. REKSTAD 1910, Osbak at 105 m, Nygaard at 60 m; REKSTAD 1917, Fauske at 100 m. The site with *Yoldia*-clay is at Bodö at 10 m (MARTHINUSSEN 1962).

References

ALHONEN, P. 1968. Radiocarbon ages from the bottom deposits of Lake Sarkkilanjärvi, southwestern Finland. - Bull. Geol. Soc. Finland 40, p. 65 – 70.
ANDERSEN, B. G. 1965. The Quaternary of Norway (in RANKAMA, *The Quaternary*, Volume 1). — John Wiley & Sons Ltd., London, p. 91 – 138.
 — » — 1968. Glacial geology of Western Troms, North Norway. — Norges Geol. Unders. 256, 160 pp.
BORELL, R. and OFFERBERG, J. 1955. Geokronologiska undersökningar inom Indalsälvens dalgång mellan Bergeforsen och Ragunda (Summary: Geochronological investigations in the Indal River valley between Bergeforsen and Ragunda, N. Sweden). — Sveriges Geol. Unders. Ser. Ca 31, 24 pp.
DONNER, J. J. 1964. The Late-glacial and Post-glacial emergence of south-western Finland. — Soc. Sci. Fennica, Comm. Phys.-Math. 30 (5), 47 pp.
 — » — 1965. Shore-line diagrams in Finnish Quaternary research (Summaries in German and Russian). — Baltica 2, p. 11 – 20.
 — » — 1966. A comparison between the Late-glacial and Post-glacial shore-lines in Estonia and south-western Finland. — Soc. Sci. Fennica, Comm. Phys.-Math. 31 (11), 14 pp.
 — » — 1967. The Late-glacial and early Post-glacial pollen stratigraphy of southern and eastern Finland. — Soc. Sci. Fennica, Comm. Biol. 29 (9), 24 pp.
 — » — 1969. Land/sea level changes in southern Finland during the formation of the Salpausselkä endmoraines. — Bull. Geol. Soc. Finland 41, p. 135 – 150.
FEYLING-HANSSEN, R. W. 1964. A Late Quaternary correlation chart for Norway. — Norges Geol. Unders. 223, p. 67 – 91.
FLORIN, S. 1963. Bodenschwankungen in Schweden während des Spätquartärs (Summaries in English and Russian). — Baltica 1, p. 233 – 264.
FREDÉN, C. 1967. A historical review of the Ancylus Lake and the Svea River. — Geol. Fören. i Stockholm Förh. 89, p. 239 – 267.
FROMM, E. 1938. Geochronologisch datierte Pollendiagramme und Diatomeenanalysen aus Ångermanland. — Geol. Fören. i Stockholm Förh. 60, p. 365 – 381.
 — » — 1963. Absolute chronology of the late Quaternary Baltic. A review of Swedish investigations (Summaries in German and Russian). — Baltica 1, p. 46 – 59.

GRANLUND, E. 1943. Beskrivning till jordartskarta över Västerbottens län nedanför odlings-gränsen. — Sveriges Geol. Unders. Ser. Ca 26, 165 pp.

GRÖNLIE, O. T. 1940. On the traces of the ice ages in Nordland, Troms, and the south-western part of Finnmark in northern Norway. — Norsk Geol. Tidsskrift 20, p. 1—70.

—»— 1951. On the rise of sea and land and the forming of strandflats on the west coast of Fennoscandia. — Norsk Geol. Tidsskrift 29, p. 26—63.

HAFSTEN, U. 1956. Pollen-analytic investigations on the late Quaternary development in the inner Oslofjord area. — Univ. Bergen Årb., Naturv.r. 8, 161 pp.

—»— 1959. De senkvartaere strandlinjeforskyvningene i Oslotrakten belyst ved pollenanaly-tiske undersökelser (Summary: Application of pollen analysis in tracing the late Qua-ternary displacement of shore-lines in the inner Oslofjord area). — Norsk Geogr. Tids-skrift 16, p. 74—99.

HOLTEDAHL, O. 1953. Norges Geologi. — Norges Geol. Unders. 164, 1118 pp.

—»— 1960. Geology of Norway. — Norges Geol. Unders. 208, 540 pp.

HÖRNSTEN, Å. 1964. Ångermanlands kustland under isavsmältningsskedet, preliminärt med-delande. — Geol. Fören. i Stockholm Förh. 86, p. 181—205.

HÖRNSTEN, Å. and OLSSON, I. U. 1964. En C14-datering av glaciallera från Lugnvik, Ånger-manland. — Geol. Fören. i Stockholm Förh. 86, p. 206—210.

KESSEL, H. and RAUKAS, A. 1967. The deposits of the Ancylus Lake and Littorina Sea in Estonia (in Russian with Summary in English). — Eesti NSV Teaduste Akad. Geol. Inst., 134 pp.

KUKKAMÄKI, T. J. 1968. Report on the work of the Fennoscandian Sub-Commission. — Third Symposium of the CRCM in Leningrad, May 22—29, 1968, 4 pp.

KÄÄRIÄINEN, E. 1966. The second levelling of Finland 1935—1955. — Suomen Geod. Lait. Julk. — Veröff. Finn. Geod. Inst. 61, 313 pp.

LUNDQVIST, G. 1961. Beskrivning till karta över landisens avsmältning och högsta kustlinjen i Sverige (Summary: Outline of the Deglaciation in Sweden). — Sveriges Geol. Unders. Ser. Ba 18, 148 pp.

LUNDQVIST, J. 1965. The Quaternary of Sweden (in RANKAMA, The Quaternary, Volume 1). — John Wiley & Sons Ltd., London, p. 139—198.

MARTHINUSSEN, M. 1962. C14-datings referring to shore lines, transgressions, and glacial sub-stages in northern Norway. — Norges Geol. Unders. 215, p. 37—67.

NILSSON, E. 1960. Södra Sverige i senglacial tid. — Geol. Fören. i Stockholm Förh. 82, p. 134—149.

—»— 1968. Södra Sveriges senkvartära historia (Summary: The Late-Quaternary history of southern Sweden). — Kungl. Svenska Vetenskapsakad. Handl. Fjärde Ser. 12:1, 117 pp.

NILSSON, T. 1935. Die pollenanalytische Zonengliederung der spät- und postglazialen Bildun-gen Schonens. — Geol. Fören. i Stockholm Förh. 57, p. 385—562.

PUNNING, J. M., ILVES, E. and LIIVA, A. 1968. Tartu radiocarbon dates II. — Radiocarbon 10, p. 124—130.

REKSTAD, J. 1910. Geologiske iagttagelser fra ytre del av Saltenfjord (Summary in English). — Norges Geol. Unders. Årb. 1910, 3, 67 pp.

—»— 1917. Fjeldströket Fauske-Junkerdalen (Summary in English). — Norges Geol. Unders. Årb. 1917, 4, 70 pp.

SEREBRYANNY, L. R. 1965. Progress of radiocarbon dating in Quaternary Geology (in Russian). — Acad. of Sciences of the U.S.S.R., Inst. of Geogr., Moscow, 269 pp.

TANNER, V. 1930. Studier över kvartärsystemet i Fennoskandias nordliga delar IV (Résumé: Études sur le système quaternaire dans les parties septentrionales de la Fennoscandie IV). — Fennia 53 (1); Bull. Comm. géol. Finlande 88, 594 pp.

VINOGRADOV, A. P. *et al.* 1958. Radiocarbon dating in the Verdandsky institute V. — Radiocarbon 10, p. 454—464.

WEST, R. G. 1968. *Pleistocene geology and biology with especial reference to the British Isles.* — Longmans, London, 377 pp.

WOLDSTEDT, P. 1958. *Das Eiszeitalter*, Band II. — Ferdinand Enke Verlag, Stuttgart, 438 pp.

Societas Scientiarum Fennica
Communicated May 19, 1969
Printed August 1969
Keskuskirjapaino — Centraltryckeriet

Editor's Comments on Paper 13

Schytt, Hoppe, Blake, and Grosswald: *The Extent of the Würm Glaciation in the European Arctic*

The selection of this paper is based on two points; first, the paper uses the relationship between regional isobases and former ice load to suggest that a large ice sheet once covered the area of the Barents Shelf, and second, it considers the question of variations in postglacial uplift at different sites. (In this sense this paper also belongs in the next section.)

Figure 3 of this paper illustrates the isobases on uplift over the last 6500 years. The pumice that Martinussen used as a time-stratigraphic marker in Northern Norway also outcrops on correlative Tapes beaches within the area of the Barents Sea and has been used [together with the postglacial emergence curves (Fig. 2)] to reconstruct the pattern of glacio-isostatic rebound. The regional isobases of Fig. 3 indicate a center of updoming located over the center of the Barents Shelf, where water depths are currently about 200 m. A major problem of the general reconstruction of the Barents Sea ice sheet is the possible effect that it would have on uplift in Northern Norway. Martinussen's paper (Paper 11) indicates that Northern Norway is tilting toward an ice center in central Fennoscandia and there is no evidence from his strandline data of an ice load lying to the north. This raises the question of the age of the Barents Sea ice sheet. Other researchers have suggested that the Barents Sea ice sheet is early Wisconsin in age; this would certainly agree with the general tendency for the early Wisconsin glaciation to be significantly more extensive than younger Wisconsin states. Only two emergence curves from the region (Fig. 2) show the characteristic exponential decrease of uplift—the other four all have remarkable long linear segments with apparent relaxation times of over 3000 years.

The authors of this paper are from Sweden, Canada, and the U.S.S.R. Professors Valter Schytt and Gunnar Hoppe are, respectively, Professor of Glaciology,

Swedish Natural Science Research Council, and Professor of Physical Geography, University of Stockholm. Weston Blake, Jr., is attached to the Geological Survey of Canada, and M. Grosswald is with the Institute of Geography in Moscow. All the authors have had considerable research experience in the European Arctic and Blake has published several significant papers of postglacial rebound in the Canadian Arctic.

Reprinted from *Intern. Assn. Hydrol. Sci., Publ. 79*, 207–216 (1967)

MEDDELANDEN FRÅN
NATURGEOGRAFISKA INSTITUTIONEN
VID STOCKHOLMS UNIVERSITET
Nr A 18 *20*

THE EXTENT OF THE WÜRM GLACIATION IN THE EUROPEAN ARCTIC

A preliminary report about the Stockholm University Svalbard Expedition 1966.

V. SCHYTT([1]), G. HOPPE([1]), W. BLAKE Jr. ([2])
and M.G. GROSSWALD([3])

ABSTRACT

In the light of the increased knowledge of the Antarctic ice sheet it now seems possible that the last ice sheet in northwest Europe was not split up into separate sheets over Scandinavia, the British Isles and the Spitsbergen archipelago but was rather one continuous sheet covering the very shallow North Sea and Barents Sea.

This hypothetical "super ice sheet" has been studied by the Department of Physical Geography, Stockholm University.

Approximately 6500 years ago great quantities of pumice drifted ashore along the northern coasts of Svalbard (on Vestspitsbergen and Nordaustlandet particularly) and the "pumice level" then established has been found to rise from 5 m above sea level at the north end of Hinlopenstretet to just over 20 m at Finn Malmgrenfjorden (on the north coast of Nordaustlandet) and to 28 m at Wilhelmöya (in southern Hinlopen). Systematic studies of the height of the pumice level have given a very reliable map of the land uplift during the last 6500 years, and C^{14}-datings of wood, whale bones and shells from several places in Vestspitsbergen, Nordaustlandet, Kong Karls Land, Barentsöya and Hopen have shown that the rate of uplift in the areas near the edge of the continental shelf was very high about 10 000 years ago and then slowed down very rapidly, whereas this rate has been nearly constant over the last 7000 years in Kong Karls Land and Hopen, far away from the edge.

This is tentatively interpreted as evidence of a thick ice sheet over at least the northern Barents Sea.

RÉSUMÉ

A la lumière des connaissances croissantes de la couche de glace couvrant l'Antarctique, il semble maintenant possible que la dernière couverture de glace de l'Europe du Nord-Ouest n'a pas été déposée en couches distinctes en Scandinavie, dans les îles Britanniques et sur l'archipel du Spitzberg, mais qu'elle constituait plutôt une couverture continue couvrant les mers peu profondes du Nord et de Barents.

Cette hypothétique couverture supérieure de glace a été étudiée par le Département de Géographie Physique de l'Université de Stockholm.

Il y a approximativement 6500 ans, de grandes quantités de pierre ponce furent poussées au rivage des côtes nord de Svalbard (particulièrement sur le Vestspitsbergen et le Nordaustlandet) et la couche de pierre ponce alors établie a été trouvée s'élever à 5 m au-dessus du niveau de la mer à l'extrémité nord du Hinlopenstretet jusqu'à 20 m à Finn Malmgrenfjorden (sur la côte nord de Nordaustlandet) et jusqu'à 28 m à Wilhelmöya (dans le Hinlopen du Sud). Les études systématiques de la hauteur de cette couche de pierre ponce a fourni une carte du relèvement du sol au cours des dernières 6500 années et des déterminations d'âge (à l'aide de C_{14}) de bois, de fanons de baleine et de coquillages de différents endroits du Vestspitsbergen, Nordaustlandet, Kong Karls Land, Barentsöya et Hopen, ont montré que le taux du soulèvement dans les régions près de l'angle de la glace flottante continentale était très élevé il y a environ 10 000 ans, mais se ralentit alors rapidement, tandis que ce taux fut à peu près constant au cours des 7000 dernières années dans le Kong Karls Land et Hopen, très loin de l'angle en question.

Ceci est tentativement interprété comme une preuve de l'existence d'une couche épaisse de glace sur au moins la partie Nord de la mer de Barents.

([1]) Dept. of Physical Geography, University of Stockholm, Stockholm, Sweden.
([2]) Geological Survey of Canada, Ottawa, Canada.
([3]) Institute of Geography, Akademia Nauk, Moscow, U.S.S.R.

The prevalent concept of the maximum extent of the ice sheets in northwestern Europe during the last glaciation shows the Scandinavian Ice Sheet centred over the Baltic Sea, a separate ice sheet over the British Isles and smaller ice caps over Iceland, the Faeroes and the Shetland Islands. Only occasional attention has been paid to the glaciation of Svalbard, Franz Jozef Land and Novaya Zemlya.

Even though most Pleistocene geomorphologists seem to accept that the Scandinavian Ice Sheet covered most of the North Sea area and met the British Ice Sheet during a previous glaciation, only a few scientists have conceived such a large extent during Würm-times. Thus in Scandinavia, the main debate during this century has not been concerned with a possible glaciation of the North Sea, but rather with the possible existence of ice-free refugia in northern and western Norway. That the North Sea outside the Norwegian west coast was ice free was more or less taken for granted. Biologists mapped the distribution of the present flora and fauna, and they maintained that the very distinct distribution patterns of various high-mountain plants, as well as of some insects, could not be explained in any other way than by migration from ice free coastal areas, refugia, —or, possibly from smaller nunataks—where life could have survived throughout a glaciation. However, experts in glacial geology and morphology have found striae and erratics in most of these "ice free areas", and in their opinion these glacial features derive from the Würm glaciation. [1]

Our knowledge of the present-day ice sheets in Antarctica and in Greenland has, however, made more rapid progress during the last 20 years than our knowledge of the Pleistocene ice sheets. Since 1949 seismic methods have been applied to the two major ice sheets of the world-Expeditions Polaires Françaises began a systematic study in Greenland and the Norwegian-British-Swedish Expedition started similar work in the Antarctic. In recent years so much has become known about large parts of the Antarctic ice sheet that it is now feasible to draw reasonably accurate maps of the subglacial topography. In Marie Byrd Land, an area of particular interest to this discussion, an ice thickness of more than 4300 m has been measured at a place where the elevation of the ice surface is 1800 m above sea level. This means that the ice sheet is resting on bedrock more than 2500 m *below* sea level. "The expected isostatic rise of the land surface after removal of the ice and allowing for the weight of overlying water would bring the -2500 m contour up to -1900 m and the present -500 m contour would be the approximate boundary of the waterfilled channel" (Bentley and Ostenso, 1961, pp. 892-893). This suggests a pre-glacial sea with an approximate area of 600,000 km^2; such a sea would be more than 1000 m deep over large areas and it would attain a maximum depth of about 1900 m. It would be slightly more extensive than the North Sea and about half the size of the Barents Sea.

Since both the North Sea and Barents Sea are very shallow and as the general sea level during the maximum stage of the Würm was at least 100 m lower than it is today, it could, in the above context, hardly be the water depth which set a limit to the growth of the ice sheet.

A few scientists have suggested that some marginal seas were glacierized during Würm. Valentin's studies, for example (Valentin, 1957) of the glacial morphology of southeast England and of the bottom topography of the North Sea made him conclude that the British and the Scandinavian ice sheet had coalesced, and De Geer (1900, pp. 430-431) envisaged an extensive ice sheet in the Spitsbergen area which was only separated from the Scandinavian ice by heavily rafted sea ice in the Barents Sea.

---

[1] For a more complete discussion of problems concerned with ice-free refugia see: LÖVE, A. and D. (Editors). *North Atlantic Biota and their History*, Pergamon, London, 1963.

208

Systematic studies of isostatic land rise in Spitsbergen, based on radiocarbon dating of drift wood, whale bones and shells, were begun in 1957 and 1958 by Blake in Nordaustlandet (Blake, 1961 a and 1962,Olsson and Blake, 1962) and by Feyling-Hanssen (Feyling-Hanssen and Olsson, 1960) in Vestspitsbergen. These studies made it clear that Spitsbergen had been covered by a very thick ice sheet which had retreated from the western and northern part of the archipelago more than 10,000 years ago. A rapid isostatic rise 9,000 to 10,000 years ago suggested that the deglaciation of the area began shortly before. Birkenmajer (1958, p. 548) has published an isobase map for the Spitsbergen archipelago. Unfortunately, since the calculations were based upon the erroneous assumption that abundant occurrences of whale bones at a certain level could always be correlated with whaling during the 17th century, excessively high values for recent uplift were obtained (cf. discussion in Blake, 1961 b, and Blake, Olsson and Środoń, 1965). Büdel's expeditions of 1959 and 1960 (Büdel, 1960) contributed further material on land uplift in south-eastern Spitsbergen, whilst Grosswald's (1963) compilation of data from the entire European Arctic, suggested that one continuous ice sheet covered Franz Josef Land, Spitsbergen and the northern parts of Barents Sea. Its maximum thickness was estimated to be nearly 3000 m over Victoria Island. Grosswald did not discuss the southern limits of the large Arctic ice sheet. Corbel (1960), however, had done that a few years earlier and his "Barents ice sheet" covered Svalbard, Franz Josef Land, Novaya Zemlya and the Barents Sea as well as the Kara Sea. Its surface attained almost 3000 m above sea level along a central line from Franz Josef Land to the Kola Peninsula. In our opinion, however, these conclusions were based upon erroneous data as regards Spitsbergen.

In a later paper Corbel (1966) advanced another approach to the calculation of the thickness of the Barents Sea ice sheet. From "known" data on the present rate of uplift he computed the "remaining uplift", which, when added to the elevation of the highest marine limit and corrected for eustatic changes, gave him a value for the total isostatic rebound following removal of the ice load. By multiplying this total change by a factor of seven (see below) Corbel obtained a figure for the former ice thickness over the area.

There are, however, some assumptions in Corbel's calculation that can be questioned. 1). A considerable amount of uplift takes place before the actual deglaciation of any given area, i.e., before the highest marine features can be developed, and this discrepancy cannot be adjusted for in such a simple way by just introducing the factor 7, even if this is obtained from empirical data. 2) The present rate of uplift in northern Nord$_z$ austlandet, as well as in several other areas near the continental slope, is only just balancing the eustatic rise of sea level. Corbel used average rates of uplift over the last several thousand years. 3). Corbel's deduction of the factor, 7, is based on speculations that are completely unconvincing.

When Blake did his work in 1957 and 1958 he was a member of the Swedish Glaciological Expedition organized from the Department of Physical Geography of the University of Stockholm. In 1963 it was decided that a study of the western and northern limits of the Scandinavian Würm ice sheet would be continued as an important part of the Department's research programme. With support from the Swedish Natural Science Research Council three expeditions were sent out in 1964 and 1965. The Shetland Islands were studied by a 6-man party under Hoppe in 1964, and fieldwork suggested a Würm age for the earlier known ice movement from the east. Investigations by a five-man group in May and June of 1965 on Björnöya (Bear Island) indicated that the land uplift had apparently not exceeded the eustatic rise of sea level, and suggested locally a moderate ice thickness close to the continental slope. Later in 1965 a three-man party worked at Russekeila, (near Kapp Linné, Vestspitsbergen) and on Hopen (Hope Island). Some of

209

the results from these investigations, which formed a basis for the more comprehensive efforts in 1966, are discussed below.

THE 1966 EXPEDITION: PURPOSE AND METHODS

The main purpose of the expedition was to collect information on the extent of the last glaciation in northern and north-eastern Svalbard. Striae and erratics in areas not covered by the present ice-caps were to be considered, with the understanding that striae usually date from the final stages of ice retreat and that erratics can have both complex travel-paths and indeterminable ages. The main interest thus centered on studies of the isostatic uplift in as large an area as possible, and a special effort was made to collect samples for radiocarbon dating.

The radiocarbon method has given us a means of obtaining absolute ages from beaches and other marine features at various elevations. Some of the errors inherent in radiocarbon dating of raised beaches have been discussed elsewhere (Blake 1962; Olsson and Blake, 1962), but it should be noted here that our "absolute" dates have to be expressed in radiocarbon years - a result of the long term variations in the initial relative radiocarbon concentration (see e.g. Damon, Long and Grey, 1966). The dates enable us to draw isobase maps and to make comparisons between both the amount and rate of land uplift over large areas. This provides much more interesting and useful data than the mere knowledge of the elevation of the marine limit. This limit is rarely a synchronous feature—it normally develops earlier in peripheral parts than in central parts of an ice-covered region. This is one important reason why it is difficult to use the elevation of the marine limit for quantitative computations of the thickness of the ice sheet.

A radiocarbon age determination, however, cannot be made in the field; one often must wait several months, or even years, after the expedition's return before any dates are obtained. But in Spitsbergen another useful tool for determining the regional pattern of isostatic uplift exists, namely the widespread occurrence of pumice on certain raised beaches. As early as 1827 Parry's expedition recorded the presence of pumice on beaches in the northwestern part of Nordaustlandet, but it was not until the 1955 Oxford University Expedition that Donner and West (1957) recognized the value of the pumice for determining the tilt of raised beaches. They recorded two different levels at which pumice was especially abundant; at Brageneset, on the eastern side of Hinlopenstretet, (see fig. 1) these levels were situated at 13.8 m and 6.4 m a.s.l. They observed that the two pumice levels were found at lower elevations the further north and northwest they worked:, they reported pumice elevations from two sites between Kinnvika and Brageneset and from another two in northern Ny Friesland. Blake continued the survey of the pumice levels in the northwestern parts of Nordaustlandet in 1957 and 1958. He found that the upper, or main pumice level coincided with a broad, well-developed beach which was often cut into bedrock at its inner edge.

From the mouth of Lady Franklinfjorden this beach could be traced along the entire length of the fjord, a distance of 20 km, and it was found to rise from 6.4 m to 10.3 m. It was obvious that a systematic regional study of the upper pumice level along the north coast and along Hinlopenstretet, would provide much information about the gradient of the isobase surfaces and thus also about the direction, in which the ice culmination must have been situated.

For the 1957-1958 expedition yet another method for dating isostatic changes was used. Häggblom recovered a number of cores of bottom sediments from lakes in the inner Murchisonfjorden area at various elevations; from the present sea level and up to about 200 m. All lakes below approximately 50 m proved to have had a marine stage. It was possible to radiocarbon-date the organic material at the transition between marine

210

and fresh-water deposits. This method probably gives the most detailed information about the process of emergence of the land, as long as the core is not contaminated by older or younger material and provided that diatoms are present to indicate the change in environment.

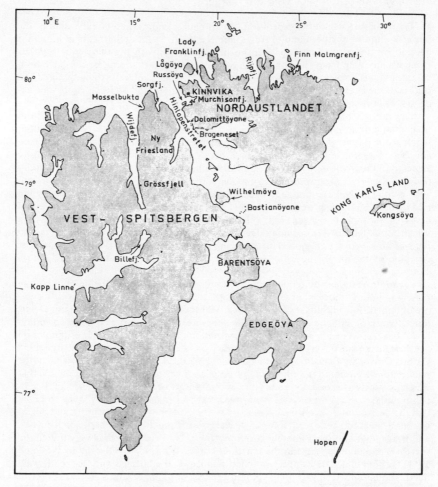

Fig. 1 — General map of Spitsbergen, showing the location of names mentioned in the paper.

PRELIMINARY SCIENTIFIC RESULTS

Striae

Although Österholm and Strömberg had their main interest focused on striae and roches moutonnées, Nordaustlandet and the neighbouring islands proved rather unrewarding in this respect. Ice, snow, beach material, till, solifluction material and badly frost-shattered rocks cover much of the surface area.

211

Österholm's observations on the north coast indicated a general ice movement from the south. In Murchisonfjorden, Blake's observations of 1957, confirmed by numerous observations by the 1966 expedition, showed that the predominant ice motion was westward, out of the fjord, and striae observed by Hoppe at Grössfjell, similarly run parallel to Wijdefjorden.De Geer's observations of striae showing motion from the southeast on the islands in the southern part of Hinlopenstretet were confirmed, while on Bastianöyane, southeast of Wilhelmöya, well developed roches moutonnées with distinct striae indicated an ice movement from the west. However, it must be repeated that all these striations belong to a late stage of deglaciation.

Two weeks of intensive fieldwork by Strömberg's party on Kong Karls Land revealed only one set of striae definitely on bedrock, with the possibility of one other in situ. The striae were probably formed by an ice movement from S-SSE. Since one of the sites was in a col very close to and above a south-facing rock wall, it seems likely that these striae may have been derived from a large ice sheet flowing straight over Kongsöya from a culmination further south.

Lake sediments

The field programme was successful and Hyvärinen observed a change in the diatom flora from marine to lacustrine conditions in several cores; it is of particular interest to note that in a lake only 1 m above sea level, east of Kinnvika, two saltwater and two fresh-water stages were recorded. This suggests that there was a distinct transgression approximately 5000 years ago and that no significant oscillations have taken place during the last few thousand years (cf. Blake, 1961b). The lakes were selected so as to avoid, as far as possible, contamination from both fossil coal and derived graphite.

The pumice level

Nordaustlandet is a most satisfactory area for the study of raised beaches, and since one of the most prominent beaches (the "upper pumice level")'can be identified readily because of the abundance of pumice, the elevation of this synchronous shore feature can be plotted during the course of the field work. This is very valuable as it permits the field scientist to plan his further work according to the results already obtained.

The main pumice level was studied in great detail on the north coast and along the northern shores of Hinlopenstretet; further south pumice becomes scarce. A considerable number of radiocarbon age determinations have been made on drift wood and whale bone found on the same beach as the pumice. As a result of field work in 1957 and 1958 Blake (1961a)reported seven dates of material from the "upper pumice level" within 30 km of Kinnvika. With no exception the dates ranged between 6200 and 7000 years B.P. (using a half-life for C^{14} of 5570 years). This age has been confirmed by dates from material found with the pumice, but at higher elevations in areas far from Kinnvika. On Wilhelmöya, for example, the pumice was found at 28 m a.s.l. and a whale bone at 28 m was dated at 6780 ± 100 years B.P. (St-2293). The pumice, drift wood and whale bones are not necessarily contemporaneous and some variation in ages seems reasonable, since the prominent beach on which the pumice occurs is often cut into bedrock, and thus presumably represents a considerable period of time. Nevertheless, it is clear that the deposition of the pumice can be considered as a synchronous event. Thus, in view of the ages mentioned above it is assumed that the beach on which the main concentration of pumice occurs was in the process of forming some 6500 years ago.

As mentioned above Blake was able to trace the pumice level along most of the heavily indented north coast. He had previously made pumice observations at 4.9 m on Lågöya, at 6.4 m at the mouth of Lady Franklinfjorden and at 10.3 m at its head.

212

Along Murchisonfjorden a distinct tilting was recorded towards the northwest with the highest values in the inner part of the fjord. A steady rise of the pumice level also occurred eastwards, at about 19 m at the head of Rijpfjorden and just over 20 m at Finn Malmgrenfjorden. A clear rise to the south was recorded, for example, along Rijpfjorden. Blake's detailed map (not published here) of the pumice level on the northwest and north coasts suggests that the former ice sheet must have had its culmination in a southeasterly direction.

The tentative conclusion, which can be drawn from these observations alone is in agreement with other pumice observations. Donner and West (1957) reported the upper pumice level at 3.0 m at Mosselbukta, at 5.6 m at Sorgfjorden, at 8.7 m on Russöya, at 12.4 m on Dolomittöyane and at 13.8 m at Brageneset. In southwestern Nordaustlandet no pumice was found, but Knape found abundant pumice at 28 m on Wilhelmöya and at 13 m at Grössfjell, in inner Wijdefjorden (see fig. 1).

The uplift curves (fig. 2), which are now fairly well established for parts of Kong Karls Land, for Kapp Linné, for Edgeöya and for Hopen, indicate approximate elevations at which pumice should be found if it had ever drifted ashore. For Kong Karls Land and for Hopen we get close to 35 m and for Kapp Linné about 10 m. The curve published by Grosswald (1963) gives approximately 20 m for "the pumice level" on Franz Josef Land and those by Feyling-Hanssen and Olsson (1960) give 13 m at Billefjorden. For Edgeöya and Barentsöya (Grosswald et alia 1967, Büdel 1960 and 1962) the information is rather confusing. In an attempt to clarify this problem Knape returned to Spitsbergen in August 1967 for further information about the land uplift in the south and southeast, particularly on Edgeöya. His first available datings from Edgeöya are included in figure 2.

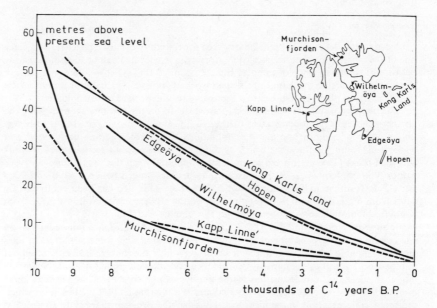

Fig. 2 — Land uplift curves for the Spitsbergen archipelago. The insert map shows where the dated samples have been collected. All curves, except for one, are based on several datings which fall on or very near the plotted lines. The Kapp Linné curve, however, is based on four dates from the period 9000-10,000 years B.P. and only one from 5500 B.P. and is thus not as reliable as the others.

213

In figure 3 the elevation of the "pumice level" is plotted, .e. the strandline which was at sea level some 6500 years ago, according to all the available information *except for Barentsöya and Edgeöya* from where only Knape's datings have been used. The pattern of the resulting isobases indicates the former presence of an ice sheet of considerable extent with its thickest parts over what is now Barents Sea.

The uplift curves

The variations in the rate of land uplift are now quite well known from Hopen (13 radiocarbon dates) back to 9500 years B.P., from Kong Karls Land (6 dates) back to 7000 B.P., from Wilhelmöya and neighbourhood (6 dates) back to 9500 B.P..

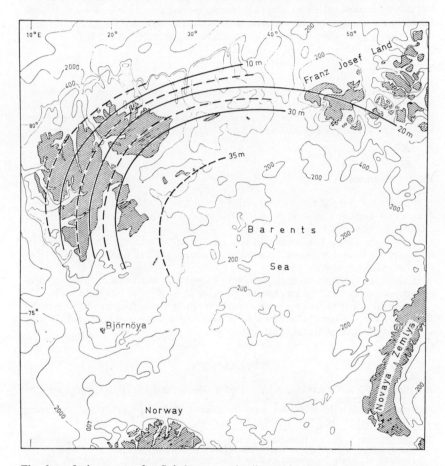

Fig. 3 — Isobase map for Spitsbergen and adjacent islands, showing the present elevation of shore features developed approximately 6500 years ago. The land uplift relative to the present sea level during the last 6500 years has been small near the edge of the continental shelf and considerably larger east and south-east of Spitsbergen. For Nordaustlandet and northeastern Vestspitsbergen the map is based on a great number of observations of the "pumice level" (see text), for other areas data have been taken from the curves in figure 2, as well as from Feyling-Hanssen and Olsson (1960) and from Grosswald (1963).

214

Kapp Linné (Russekeila) with 5 dates back to 10000 years B.P, and from the Murchison Bay area with over 30 datings with a maximum age of 10,700 years. In figure 2 data are included from all six localities. Since more dates are gradually being received a final analysis has not yet been made of these curves. It is, however, interesting to note that the curves from areas near the continental slope show a more rapid rate of uplift 10,000 to 9000 thousand years ago and then a progressively decreasing rate. At a greater distance from the continental slope (and farther from the edge of the ice sheet) uplift seems to have proceeded at a more constant rate over the whole period covered by our data. These differences may be of significance but their implications must be studied more closely. It should also be stressed here that no corrections for eustatic changes have yet been applied to these curves, mainly because we do not feel that the eustatic variations of sea level during the postglacial period are well enough known. Furthermore, eustatic corrections would have similar effects on all curves; they would mainly make them slightly steeper, particularly before 6000 years B.P.

TENTATIVE CONCLUSION

At this early stage, long before all the material has been worked up, all our observations support the idea of a very large Arctic European ice sheet, and even though this idea is not new, we think that the basis for the hypothesis has now progressed from loose speculations to knowledge based on systematic studies and absolute age determinations over a wide area.

ACKNOWLEDGEMENTS

Funds for the expedition were provided by the Swedish Natural Science Research Council, by Konung Gustaf VI Adolfs 70-årsfond and by various Swedish and Finnish scientific foundations.

The success of the field work was to a great extent due to the efficient transportation service offered by Norsk Polarinstitutt and to the valuable advice received from Dr. Tore Gjelsvik and his staff.

Blake's participation was made possible by a grant from the Arctic Institute of North America, under contractual arrangements with the U.S. Office of Naval Research.

Grosswald's participation was made possible through an arrangement between The Royal Swedish Academy of Science and the Academy of Science of the U.S.S.R. with financial support from Svenska Institutet.

REFERENCES

BENTLEY, C.R. and OSTENSO, N.A., (1961): Glacial and subglacial topography of West Antarctica, *J. of Glaciology*, 3, No. 29.
BIRKENMAJER, K., (1958): Remarks on the pumice drift, land uplift and the recent volcanic activity in the Arctic basin, *Bull. Acad. Polon. Sci., Sér. sci. chim., géol. et géogr.*, VI:8, pp. 545-549.
BLAKE, W., Jr., (1961 a): Radiocarbon dating of raised beaches in Nordaustlandet, Spitsbergen, in *Geology of the Arctic*, Univ. of Toronto Press, pp. 133-145.
—, (1961 b): Russian Settlement and land rise in Nordaustlandet, Spitsbergen, *Arctic*, 14:2, pp. 101-111.
—, (1962): Geomorphology and glacial geology in Nordaustlandet, Spitsbergen. Unpubl. Doctor's dissertation. Dept. of Geol. Ohio State Univ., 477 pp. Abstract publ. in *Dissertation Abstracts. Ann. Arbor, Michigan*, Univ. Microfilms, 18, No. 10, 3861 (1963).
—, OLSSON, I. and ŚRODOŃ, A., (1965): A radiocarbon-dated peat deposit near Hornsund, Vestspitsbergen, and its bearing on the problems of land uplift, *Norsk Polarinst*. Årbok 1963, pp. 173-180.

215

Büdel, J., (1960) Die Frostschutt-Zone Südost-Spitsbergens, *Colloqium Geographicum*, Bd 6, 105 p.

—, (1962) Die Abtragungsvorgänge auf Spitzbergen im Umkreis der Barentsinsel, *Deutscher Geographentag Köln*, 1961, Tagungsbericht und wissensch. Abh., pp. 337-373.

Corbel, J., (1960): Le soulèvement des terres autour de la mer de Barentz, *Revue de Geogr. de Lyon*, 35, pp. 253-274.

—, (1966) De l'extension maximale des glaciers dans la zone de la mer de Barentz, in Spitsberg 1964. *C.N.R.S., R.C.P.* 42, pp. 55-70.

Damon, P.E., Long, A. and Grey, D.C., (1966) Fluctuation of atmospheric C^{14} during the last six millennia, *J. Geophys. Res.*, 71, pp. 1055-1063.

De Geer, G., (1900) Om östra Spetsbergens glaciation under istiden. *Geol. fören. förh.*, 22, pp. 427-436.

Donner, J.J. and West, R.G., 1957: The Quaternary geology of Brageneset, Nordaustlandet, Spitsbergen, *Norsk Polarinstitutts Skrifter*, Nr. 109, 29 p.

Feyling-Hanssen, R.W. and Olsson, I.U., (1960) Five radiocarbon datings of Post Glacial shorelines in Central Spitsbergen, *Norsk Geogr. Tidsskr.*, 17, pp. 122-131.

Grosswald, M.G., (1963) The raised shore-lines of the Franz Joseph Land and the late Quaternary history of the archipelago ice sheets, *Glaciological Researches, Articles, IX Section of IGY Program, No. 9. Academy of Sci. USSR.* (In Russian with English summary.)

—, Devirts, A.L., Dobkina, E.I. and Semevsky, D.V., 1967 Earth crust uplift and the age of glaciation stages in the Spitsbergen area, *Geochemistry* (Acad. Sci. USSR), No. 1, pp. 51-56. (In Russian with English summary.)

Löve, A. and D. (Editors), 1963 *North Atlantic biota and their history*. Pergamon, London.

Olsson, I.U. and Blake, W., Jr., (1962) Problems of radiocarbon dating of raised beaches, based on experience in Spitsbergen, *Norsk Geogr. Tidskr.*, 18, pp. 47-64.

Valentin, H., 1957: Glazialmorphologische Untersuchungen in Ostengland, *Abhandl. d. Geogr. Inst. d. freien Univ. Berlin*, pp. 1-86.

DISCUSSION

N.A. Ostenso

Recent geophysical studies in the Greenland and Barents seas show that the steep western boundary of the Svalbard and Murman Rise continental margins is found by the northward extension of the active Mid-Atlantic Ridge. Further, these data show a crustal spreading rate comparable with that calculated for the Mid-Altantic Ridge i.e. 1 cm/yr. Thus Dr. Schytt's assumption of simple isostatic adjustment in response to glacial unloading may be perturbed by such close proximity to a major tectonic element.

O. H. Løken

Have you found any evidence of transgressions in your field area, particularly near the centre of the regenerated ice cap on Spitzbergen?

I should like to draw your attention to work presently going on in the Department of Energy, Mines and Resources in Ottawa, where particularly Dr. Andrews has collected recently and compared a number of uplift curves from Arctic Canada. Curves, corrected for eusostatic sea level changes have been used, and a systematic picture of the postglacial isostatic uplift has appeared. Our work has been based almost exclusively on the basis of shell material but particular attention has been given to the problem of relating shell-deposits to their relevant sea level by detailed stratigraphic studies of beach deposits.

R. F. Black

If glaciers had been regenerated on Spitsbergen after an early deglaciation, would not Hopen Island have been too small to experience the same degree of development of a cap? Do any morphological features on Hopen Island show that its surface has been exposed for a longer time?

216

Editor's Comments on Paper 14

Sissons: *A Re-interpretation of the Literature on Late-glacial Shorelines in Scotland with Particular Reference to the Forth Area*

This paper represents a benchmark in the literature of the United Kingdom on the question of late-glacial shoreline deformation within the area of the last glacial load. Sissons summarizes the orthodox view of the British beach sequence that has been stereotyped into the "100-ft beach" and three lower ones, despite the knowledge that these beaches were differentially uplifted. Sissons gives a very useful and critical review of the early literature and compares the British data with that from Fennoscandia.

In Fig. 1 (not shown) Sissons presents a suggested isostatic uplift curve for Scotland since 12,000 BP. The relaxation time for this curve is close to 4000 years and hence is much slower than estimates related to Fennoscandia that are mentioned in earlier commentaries in this volume.

This paper by Sissons sparked a considerable revival of research into the problems of glacio-isostatic recovery in the United Kingdom, and is thus also an important paper from this viewpoint.

J. Brian Sissons was born in 1926 and educated in Yorkshire. He received his Ph.D. from Cambridge University in 1953. He has been attached to the Geography Department of the University of Edinburgh since 1954, where he is currently a reader.

Reprinted from *Trans. Edinburgh Geol. Soc.*, **19**, 83–86, 95–99 (1962)

A re-interpretation of the Literature on Late-glacial Shorelines in Scotland with Particular reference to the Forth Area

J. B. SISSONS

MS received 1st February 1962

ABSTRACT

Some problems revealed by the literature on Scottish late-glacial raised beaches are indicated and it is suggested that these problems cannot be resolved in terms of the widely-held concept of so-called 100-ft. and 50-ft. beaches. Reasons are given for believing that some late-glacial shorelines in Scotland have considerably steeper gradients than has hitherto been recognised. Particular reference is made to the Forth area, where shoreline altitudes as recorded in the literature reveal anomalies that appear not to be explicable in terms of current ideas. The relation of shorelines to glacial stages is briefly considered and modification of the limit of Simpson's Perth Readvance in the Forth area suggested. Some implications of the hypothesis presented are mentioned.

CONTENTS

I. THE ORTHODOX INTERPRETATION

It has been accepted since before the beginning of this century that there are several distinct raised beaches in Scotland, although

A 2   83

the interpretation that prevails at present was criticised as long ago as 1906 by Jamieson. A 100-ft. beach is repeatedly referred to in the literature and one at 50 ft. or thereabouts is quite frequently mentioned. A 25-ft. beach is widely recognised and in some areas a lower beach, the 15-ft., is also identified. The two higher beaches are interpreted as of late-glacial age and the two lower as of post-glacial age.

It has long been recognised that these beaches are warped owing to unequal isostatic uplift. The 100-ft. beach reaches an altitude of 145 ft. in the Stirling area according to Dinham and Haldane (1932, p. 212) and 135 ft. according to Read (1959, p. 60). Donner (1959) suggested that the area of maximum isostatic uplift of this beach is probably around Callander. The beach is described as sloping outwards from this area and is believed to descend to sea-level in Caithness and to be absent from the Orkneys, Shetlands and Outer Hebrides (e.g. Wright 1937, map p. 369). It is considered to slope southwards in Kintyre and the Firth of Clyde (e.g. Donner 1959) and towards the east or south-east in the Forth area (e.g. Lacaille 1954, map p. 40).

The evidence for the 50-ft. beach is rather indefinite and areal limits for it have not been described in the literature. One of the few areas in which it has been clearly located is Kintyre and Loch Fyne, where it has been described by Donner (1959). The 50-ft. beach is generally considered to mark a halt in the emergence following the formation of the 100-ft. beach.

The 25-ft. beach has a wider distribution than that claimed for the 100-ft. beach. For example, it has been traced along the Lancashire coast (Gresswell 1953, 1958) and is reported in North Wales (Stephens 1957, p. 144). Gresswell has measured its shoreline as 17 ft. O.D. in S.W. Lancashire. It is said to reach 40 ft. or slightly more in the area between Loch Linnhe and the Firth of Clyde (McCallien 1937, p. 198) and 45 to 49 ft. in the Stirling-Lake of Menteith area (Dinham and Haldane 1932). The 25-ft. beach is separated from the two earlier beaches by a period of lower sea-level indicated by peats, tree remains, etc. near or below present sea-level, as Jamieson recognised nearly a century ago (Jamieson 1865).

The 15-ft. beach is less widely mapped than the 25-ft. beach. It is believed to mark a halt in the emergence from the 25-ft. submergence.

This pattern of sea-level change is conveniently summarised in a diagram by Donner (1959, p. 20), which includes a correlation with the British pollen zones and with carbon 14 dates.

II. Discussion of the Orthodox Interpretation

In the following pages the terms ' 100-ft. beach ', ' 25-ft. beach ', etc. will be used as they are normally used in the literature. Thus the term ' 25-ft. beach ' will be used for the feature that is said to be at about 40 ft. in the Loch Linnhe area and 17 ft. in S.W. Lancashire regardless of its local altitude. Altitudes will be expressed in terms of Ordnance Datum.

The 100-ft. beach is remarkable in that it is usually not sharply delimited at its inner margin. This fact is referred to in many of the Geological Survey memoirs and Wright in his summary (1937, p. 372) referred to ' its extremely feeble development, except in certain districts '. Where it is well developed this is usually due to special factors such as contemporaneous supplies of fluvioglacial material or exposure to strong waves, as in parts of western Scotland. In view of the widespread nature of this submergence the normally feeble development of its shoreline seems rather strange, for one might reasonably expect such a major event to have resulted in a marked feature. One should also add that in some places the upper limit of the so-called 100-ft. sea is marked by a thinning out of the marine sands and clays rather than by a distinct erosional feature.

It is also interesting to note that many of the arctic or semi-arctic shell-bearing deposits associated in the literature with the 100-ft. sea are far below this altitude. In part this can be explained by marine clays having been laid down on the floors of drowned river valleys and also by the fact that a considerable number of sections have been revealed by marine erosion along the present shoreline or by excavations near sea-level. Yet often it cannot be proved that the low-level shell-bearing deposits accumulated when the sea-level was at 100 ft. This was recognised by Bailey in his discussion of an arctic shell bed by Loch Sween in Argyllshire. He said that this low-lying deposit, which is covered in part by deposits of the 25ft. beach, has, in its upper part, ' a distinctly littoral facies '. Bailey went on to say: ' It seems likely that the shell bed is of the age of the 100-foot beach, but the connection, as is so often the case, is far from clear ' (Bailey 1911, p. 129). This lack of connection between shell-bearing deposits and the 100-ft. sea is often ignored and in many instances there seems to be no reason why the sea at the time such deposits were accumulated should not have been at a quite different level. One instance where this has been suggested is in F.W. Anderson's (1947) discussion of the Paisley clays.

The altitudinal evidence has also, to some extent, been made to accord with the standard beach pattern of 100, 50, 25 and 15 ft. For example, beaches at altitudes between about 40 and 70 ft. have been quite frequently referred to as ' the 50-ft. beach ', while those between about 80 and 130 ft. have often been called ' the 100-ft. beach '. Wright appears to have had such unsatisfactory correlations in mind when he referred to the feature ' which commonly goes by the name of the " 50-foot " beach ' and said that it is not obvious ' that what has been termed the 50-foot beach is everywhere the same shore-line ' (1937, p. 375). Furthermore, in some parts of Scotland there is published evidence that the simple pattern is not applicable. Thus in the Edinburgh area beaches are said to occur at five levels: 125–130, 100, 75, 50 and 25 ft. (Peach *et al.* 1910, p. 335).

Another problem presented by the higher Scottish raised beaches is their rapid disappearance or poor development in the peripheral areas of their occurrence. For example, in the Edinburgh area beaches are described as occurring up to 130 ft., but some 25 miles to the east at Dunbar there is no clear feature above the 25-ft. beach (Clough *et al.* 1910, p. 183). Farther south, however, the Tweed has definite terraces, the highest of which is said to be 60 to 80 ft. above sea-level (Pringle 1935, p. 93). Beaches up to 100 or 110 ft. occur in Ayrshire (Richey *et al.* 1930, p. 342), Arran (Tyrrell 1928, pp. 264–265) and Kintyre (Donner 1959) and beaches above the 25-ft. continue to Stranraer (A. Geikie 1869, p. 25; A. and J. Geikie 1869, p. 15; A. Geikie 1873, p. 24). Yet along the Solway coast of Scotland there are only scattered occurrences of beaches above the 25-ft. This is rather surprising since the 25-ft. and related fluvial material form extensive flat spreads infilling the heads of bays such as Wigtown Bay, so that one might reasonably expect the higher beaches also to be represented by extensive deposits in these protected localities.

Another perturbing aspect of the interpretation of late-glacial beaches is revealed when one compares the recent work of Charlesworth (1955) and Donner (1959). Charlesworth correlated his glacial stage M, equivalent to the Loch Lomond Readvance of Simpson (1933) with the 100-ft. sea. Donner, however, correlated Simpson's Perth Readvance with the 100-ft. sea and the Loch Lomond Readvance with the 50-ft. sea. Charlesworth correlated his next stage, N, with the 50-ft. sea. Such a major difference between recent correlations causes one to have serious doubts about the basis of such correlations.

[*Editor's note:* The 8 pages omitted from this benchmark paper provide an itemization of research results that are contrary to the standard opinion discussed in the opening pages of the article and reprinted here. Paper 15 in this volume provides a much more recent and comprehensive statement of research results and concepts.]

VII. A Possible Late-Glacial Shoreline in the Forth Area

The preceding discussion explains why the upper ' beach ' line in the Forth area shown in Fig. 2 does not slope more steeply than the lower one. It also shows that true late-glacial shorelines should intersect the upper and lower lines of Fig. 2 at distinct angles. The evidence in the literature is not sufficiently detailed to establish the altitudes of these late-glacial shorelines, except possibly in one instance. In eastern Fife along the coastal stretch between St. Andrews and Kincraig Point Morrison (1961) has recognised a shoreline at 90-100 ft., which he shows to be quite distinct from the maximum submergence limit of 120-130 ft. Morrison measured the altitude of the 90-100 ft. shoreline at 22 points. Some 7 or 8 miles west of Kincraig Point in the East Wemyss-Buckhaven area Allan and Knox stated that ' a distinct notch is cut in boulder clay at a level of about 112 ft. above O.D.' (1932, p. 76). This feature was later described (Knox 1954, p. 119) for a rather larger area, including the East Wemyss-Buckhaven coast and continuing south-westwards almost to Kirkcaldy, as forming a nick at 110-120 ft. Along the northern coast of the Forth southwest from Kirkcaldy a shoreline at 110-120 ft. is not mentioned in Francis's recent description (1961), but one at 120-130 ft. is described. Taken together this evidence suggests the existence of a shoreline that slopes down from 120-130 ft. in the area southwest of Kirkcaldy to 90-100 ft. in the area east of Kincraig Point, a suggestion that is strengthened by the fact that in the accounts cited no reference is made to shorelines at similar altitudes that might permit different correlations.

On the south side of the Forth the evidence available at present is insufficient to permit even tentative correlations to be made. It is perhaps significant, however, that in the Edinburgh area it has been stated that a feature exists west of the city at approximately 100 ft., but that the ' 75- and 50-ft. terraces' occur only east of Granton (Clough et al. 1910, pp. 335-336).

VIII. The Upper Marine Limit and the Perth Readvance in the Stirling Area

One aspect of Fig. 2 that has not so far been referred to is the termination of the high sea-level in the Stirling area. This is described by Dinham (1927, pp. 488-489) as follows: ' while the 100-ft. beach, when followed up the valley of the Carron, merges into the high-level stream terraces of that river, nothing similar occurs at Stirling; the platform comes to a sudden end and cannot be traced beyond Cambusbarron and Bridge of Allan.' Such a marked drop in the upper limit of marine submergence is normally found to occur behind a moraine or other evidence indicating an important halt in glacial retreat or a readvance of the ice. Thus Simpson (1933, pp. 640-641) concluded that a lowering of sea-level of at least 40 ft. occurred when the ice was at the Perth maximum and immediately afterwards. Synge (1956, p. 134) recorded a drop of sea-level of about 75 ft. at this time in the Aberdeen area. Such marked falls of sea-level are due partly to isostatic uplift of the land and partly to a temporary reversal of the eustatic rise in sea-level caused by temporary growth of glaciers in the world as a whole.

One might suggest that the narrow gap in which Stirling lies, itself partly blocked by rocky hills, would result in a retreating ice-margin stabilising itself at this position for a considerable time, during which there was a marked fall of sea-level. This seems to have been Dinham's view, for he suggested that a considerable fall of sea-level occurred ' at this stage in the retreat of the Forth glacier' (1927, p. 489). Thus he did not envisage a readvance of the ice. It seems very probable, however, that such a marked descent of the upper marine limit as occurs in the Stirling area is the result of an important readvance. It also seems very likely that such a readvance would be equivalent to the Perth Readvance of Simpson, especially as nowhere else in the Forth area below Stirling is there evidence of a fall of sea-level of the type that occurred at the time of the Perth Readvance.

Simpson did not suggest this correlation, presumably because he believed the ice in the Forth valley extended much farther down the valley than Stirling at the time of the Perth Readvance. One of Simpson's reasons for this view was the existence of meltwater channels at Glenfarg and Newburgh that carried waters across the Ochils from south to north (1933, p. 638). From this he deduced that the ice stood higher on the southern side of the Ochils than on the northern side and, consequently, that it must have extended much farther east than did the ice in the Tay valley at the same period. It is suggested that this reasoning is not relevant, for there appears to be no evidence that the two meltwater channels referred to were cut at the time of the Perth Readvance.

Simpson's other reason for taking the Perth Readvance far beyond Stirling was that ice from the west flowed up Strath Allan. He argued that this could have occurred only if the region east and south of Stirling was occupied by a still more powerful mass of glacier ice. The writer believes that the flow of ice up Strath Allan is not a serious problem and that the limited width of the gap at Stirling, coupled with the fact that the ice terminus in the Stirling area was situated in the sea and thus subject to melting by sea-water (and possibly to calving), was sufficient to stabilise the ice-margin of the Perth Readvance in the Stirling area.

IX. Some Implications of the Preceding Discussion

If the conclusions reached above concerning late-glacial shore-lines are valid it follows that the terms ' 100-ft. beach ' and ' 50-ft. beach ' as normally used in the literature should be abandoned. The gradients tentatively suggested for some of the late-glacial shorelines also suggest some revision of the current concepts of phases of emergence and submergence may be required. For example, at the time that raised beaches were being formed in some coastal areas sea-level in others may have been lower than it is now. The evidence of the marine deposits with arctic or semi-arctic fauna needs to be reconsidered and the fact that so many of these deposits occur at low levels now has an additional explanation. Furthermore, it has to be recognised that the crossing of shorelines of different ages may occur. At any point the gradients of older shorelines will be steeper than those of younger ones, considerable differences in gradient being expectable between the late-glacial and post-glacial shorelines. Finally, it has to be remembered that the so-called 100-ft. and 50-ft. beaches have been used to correlate glacial stages in different parts of Scotland (Charlesworth 1955; Donner 1958, 1959). Charlesworth (1955, p. 904) stated that, for correlation purposes ' the value of these lines cannot be over-emphasised.' The validity of such correlations is therefore very doubtful.

X. Conclusion

The type of interpretation outlined in the preceding pages is not new, for such interpretations have long prevailed in Scandinavia. It may be that the pattern revealed by detailed study in Scandinavia is quite different from that which will emerge in Scotland. The present writer does not think so, but only research more detailed than that undertaken hitherto in Scotland will reveal the truth.

XI. Acknowledgement

The writer wishes to acknowledge his indebtedness to Mr. F. M. Synge for a valuable discussion of the shorelines of southern Finland.

XII. References

ALLAN, J. K., and KNOX, J., 1933. In *Summ. Prog. geol. Surv. G.B.* 1932, part I.

——, 1934. The economic geology of the Fife coalfields, Area II. *Mem. geol. Surv. Scotland.*

ANDERSON, F. W., 1947. The fauna of the ' 100-feet beach ' clays. *Trans. Edinb. geol. Soc.*, **14**, 220-229.

BAILEY, E. B., 1911, in PEACH, B. N., *et al.* The geology of Knapdale, Jura and North Kintyre. *Mem. geol. Surv. Scotland.*

CHARLESWORTH, J. K., 1955. The late-glacial history of the Highlands and Islands of Scotland. *Trans. roy. Soc. Edinb.*, **62**, 769-928.

——, 1957. *The Quaternary Era.* Edward Arnold, London.

CLOUGH, C. T., *et al.*, 1910. The geology of East Lothian. *Mem. geol. Surv. Scotland.*

DINHAM, C. H., 1927. The Stirling district. *Proc. Geol. Ass.*, **38**, 470-491.

——, and HALDANE, D., 1932. The economic geology of the Stirling and Clackmannan coalfield. *Mem. geol. Surv. Scotland.*

DONNER, J. J., 1958. The geology and vegetation of Late-glacial retreat stages in Scotland. *Trans. roy. Soc. Edinb.*, **63**, 221-264.

——, 1959. The Late- and Post-glacial raised beaches in Scotland. *Ann. Acad. Sci. fenn.*, [A] III Geologica-Geographica, **53**, 25 pp.

DURNO, S. E., 1958. The dating of the Forth valley carse clay : a note. *Scot. geogr. Mag.*, **74**, 47-48.

FAIRBRIDGE, R. W., 1961. Eustatic changes in sea-level. *Physics and Chemistry of the Earth*, **4**, 99-185.

FRANCIS, E. H., 1961. The economic geology of the Fife coalfields, Area II. *Mem. geol. Surv. Scotland.*

GEIKIE, A., 1869. Explanation of sheet 14 (Ayrshire, southern district). *Mem. geol. Surv. Scotland.*

——, 1873. Explanation of sheet 3 (Western Wigtownshire). *Mem. geol. Surv. Scotland.*

——, and GEIKIE, J., 1869. Explanation of sheet 7 (Ayrshire, south-western district). *Mem. geol. Surv. Scotland.*

GRESSWELL, R. K., 1953. *Sandy shores in south Lancashire.* University Press, Liverpool.

——, 1958. The Post-glacial raised beach in Furness and Lyth, north Morecambe Bay. *Trans. and Pap., Inst. brit. Geogr.*, **25**, 79-103.

HAFEMANN, D., 1954. Zur Frage der jungen Niveauveränderungen an den Küsten der Britischen Inseln. *Akad.Wiss. Lit. Abh. Math-Naturwiss.* Klasse 7.

HALDANE, D., and ALLAN, J. K., 1931. The economic geology of the Fife coalfields, Area I. *Mem. geol. Surv. Scotland.*

JAMIESON, T. F., 1865. On the history of the last geological changes in Scotland. *Quart. J. geol. Soc. Lond.*, **21**, 161-203.

——, 1906. On the raised beaches of the Geological Survey of Scotland. *Geol. Mag.*, **53**, 22-25.

KNOX, J., 1954. The economic geology of the Fife coalfields, Area III. *Mem. geol. Surv. Scotland.*

LACAILLE, A. D., 1954. *The Stone Age in Scotland,* Wellcome Historical Medical Museum, New Ser., 6.

McCALLIEN, W. J., 1937. Late-glacial and early Post-glacial Scotland. *Proc. Soc. Antiq. Scot.*, **71**, 174-206.

MACGREGOR, M., and HALDANE, D., 1933. The economic geology of the Central coalfield, Area III. *Mem. geol. Surv. Scotland.*

MORRISON, I. A., 1961. *Former shorelines in eastern Fife.* Unpublished M.A. thesis, Dept. of Geography, Edinburgh Univ.

PEACH, B. N., *et al.*, 1910. The geology of the neighbourhood of Edinburgh. *Mem. geol. Surv. Scotland.*

PRINGLE, J., 1935. *British Regional Geology:* the south of Scotland.

READ, W. A., 1959. The economic geology of the Stirling and Clackmannan coalfield, Scotland: area south of the River Forth. *Coalfield Papers Geol. Surv.*, No. 2, 59-62.

RICE, R. J., 1960. The glacial deposits at St. Fort in north-eastern Fife: a re-examination. *Trans. Edinb. geol. Soc.*, **18**, 113-123.

RICHEY, J. E., *et al.*, 1930. The geology of north Ayrshire. *Mem. geol. Surv. Scotland.*

SIMPSON, J. B., 1933. The late-glacial readvance moraines of the Highland border west of the River Tay. *Trans. roy. Soc. Edinb.*, **57**, 633-646.

STEPHENS, N., 1957. Some observations on the 'interglacial' platform and the early post-glacial raised beach on the east coast of Ireland. *Proc. roy. Irish Acad.*, **58B**, 129-149.

SYNGE, F. M., 1956. The glaciation of north-east Scotland. *Scot. geogr. Mag.*, **72**, 129-143.

TYRRELL, G. W., 1928. The geology of Arran. *Mem. geol. Surv. Scotland.*

VALENTIN, H., 1953. Present vertical movements of the British Isles. *Geogr. J.*, **119**, 299-305.

WRIGHT, W. B., 1937. *The Quaternary Ice Age*, (2nd ed.) Macmillan, London.

Department of Geography
High School Yards, Edinburgh

PRINTED IN GREAT BRITAIN BY OLIVER AND BOYD LTD., EDINBURGH

Editor's Comments on Paper 15

Sissons, Smith, and Cullingford: *Late-glacial and Post-glacial Shorelines in South-east Scotland*

The revival of British interest in glacio-isostatic processes is best illustrated by the publication, in 1966, of a special issue of the Transaction of British Geographers entitled *The Vertical Displacement of Shorelines in Highland Britain.* This paper by Sissons and former graduate students is from that volume.

The paper is basically a summary of the work of Sissons et al. to 1966. It is noted for its details of observations and for not only discussing the late- and post-glacial shorelines, but also for recognizing that buried between the transgressive coarse clays are a number of buried beaches. These have been delimited and studied by the laborious means of handcoring through the overlying clays and peats.

The shoreline diagram (Fig. 1) illustrates a sequence similar to that shown by Marthinussen from Northern Norway. The late-glacial shorelines have gradients between 6.7 and 3.1 ft/mile (1.27 and 0.58 m/km) and ages between about 18,000 (?) and 13,000 BP. The Low Buried Shoreline is about 8800 years old. The transgression that succeeded this is correlative with Tapes I in Norway and the Litorina transgression in Sweden and Finland. As noted previously in this volume, this transgression in Northern Europe was caused by the final rapid deglaciation of the Laurentide Ice Sheet, radiocarbon dated about 8000 BP. The postglacial marine transgression culminated in the Firth of Forth about 5500 BP. Relative sea level then fell to the present, although three additional shorelines were cut into the coarse clays and date from less than 5500 BP. The picture of relative sea level change is complex and the field data well illustrate the varying importance of glacio-isostatic rebound and worldwide sea level changes (e.g., see Wright, Paper 5).

Reprinted from Trans. Papers 1966 Publ. 39, 9–18, 141–145 (1966)

Late-glacial and Post-glacial Shorelines in South-East Scotland

15

J. B. SISSONS, M.A., PH.D.

(*Lecturer in Geography, University of Edinburgh*)

D. E. SMITH, B.A., PH.D.

(*Assistant Professor in Geography, Columbia University in New York*)

AND R. A. CULLINGFORD, M.A.

(*Assistant Lecturer in Geography, University of Exeter*)

MS. received 31 March 1966

THE WRITERS have been studying the Late-glacial and Post-glacial shorelines of south-east Scotland for several years. Some of the results of these studies have been published in various papers dealing with particular shorelines or aspects of the area, but no attempt has so far been made to present a statement of the pattern of sea-level changes in the area as a whole. It is felt that such a statement is now required, and the purpose of this paper is to fulfil this requirement as briefly as possible.

It is not intended to discuss here the field evidence in detail, since this is far too abundant to be considered in a single paper. The nature of the evidence will be apparent, however, from the two papers presented by the authors elsewhere in this volume and from the references to their work at the end of this volume. The main methods of study of the area (which comprises the Forth and Tay lowlands together with the intervening coast of East Fife) and progress to date in their application are as follows.

(i) Mapping on a scale of 1 : 10,560 of all identifiable estuarine and marine landforms and related fluvioglacial landforms. This is completed for the area specified.

(ii) Accurate instrumental levelling of raised shorelines and related outwash deposits and kame terraces at intervals of fifty to eighty yards where practicable. The summits of some kames and the rims of some kettle holes have also been levelled. The measurement of all the principal features and of most of the minor ones is complete, over 10,000 heights having been obtained.

(iii) Study of commercial borehole records, of which about 2000 have been collected so far. The distribution of these boreholes is very variable, some parts of the area having none at all, but one part (in and around Grangemouth) having a very large number. Borehole records continue to be collected as and when they become available and the authors are grateful for the co-operation of the Geological Survey and of many firms and organizations.

(iv) Hand borings to maximum depths of about thirty feet with a strengthened Hiller borer. So far nearly 700 boreholes have been made, some at critical localities scattered through

B

9

the area, but most of them in the western part of the Forth valley. It is planned to put down many more boreholes, and a detailed boring programme linked with laboratory analyses of the sediments has been recently begun by D. E. Kemp.

(v) Pollen analyses of certain deposits are being carried out by W. W. Newey, some of the results being described elsewhere in this volume.

(vi) The relation of the Post-glacial raised beaches to archaeological evidence is being investigated by I. A. Morrison. This work is not yet complete and will not be described here.

(vii) Radiocarbon dating is being carried out by the National Physical Laboratory and by Isotopes Incorporated.

As the above outline indicates, work on sea-level changes in south-east Scotland is still in progress: hence, this paper is a statement of the pattern of changes in the light of evidence so far obtained. It will be noted that no mention has been made of studies of marine organisms. Such studies are needed and it is hoped that specialists in macro- and micro-organisms will turn their attention to the area.

The relations of the various shorelines to one another are best shown by a height-distance diagram. Such a diagram appears as Figure 1 and shows the principal shorelines and some of the minor ones. The diagram has been constructed by projecting shoreline heights into a plane aligned N. 72° W.–S. 72° E. This line was selected as being the one most nearly at right-angles to the isobases so far as they are known at present. It will be seen from the place-names given in Figure 1 that, apart from the East Fife coast between St. Andrews and Fife Ness, the diagram depicts only the Forth valley. This is because the Forth valley extends much farther west than the Tay valley, so that a more widespread and complete sequence of changes can be elucidated here.

It must be pointed out that the term 'raised beach' is used throughout this paper even where the term 'raised mud flat' or 'raised estuarine deposit' would be more appropriate. This usage has been common in Scotland and is adopted here partly for brevity of expression and partly because in specific instances it is often difficult to say precisely where one type of feature ends and the other begins.

Late-glacial Shorelines formed before the Perth Readvance

The earliest Late-glacial shorelines represented in Figure 1 are those of East Fife, for this area, along with the eastern part of East Lothian, was the first to be freed of glacier ice during the period of glacier decay that followed the Aberdeen-Lammermuir Readvance (Sissons, 1965). The six principal shorelines of East Fife are discussed in detail on pages 31–51, so that only brief mention of them is needed here. The relations of the shorelines to glacial outwash and to one another show that, during the period of their formation, the ice-margin was, in general, retreating, thus allowing the sea to extend westward into the Forth area and north-westward along the north coast of Fife towards the site of St. Andrews. The shoreline gradients in East Fife range from 6.7 feet per mile for the oldest to between 3.1 and 3.2 feet per mile for the youngest. It has not yet been found possible to correlate the six Late-glacial shorelines of East Fife with shorelines in East Lothian. This probably results partly from the nature of the local relief in East Lothian and partly from spreads of blown sand in places. The steepest shoreline found in East Lothian has a gradient of slightly over five feet per mile.

Following the formation of the easternmost group of features, the ice continued to decay

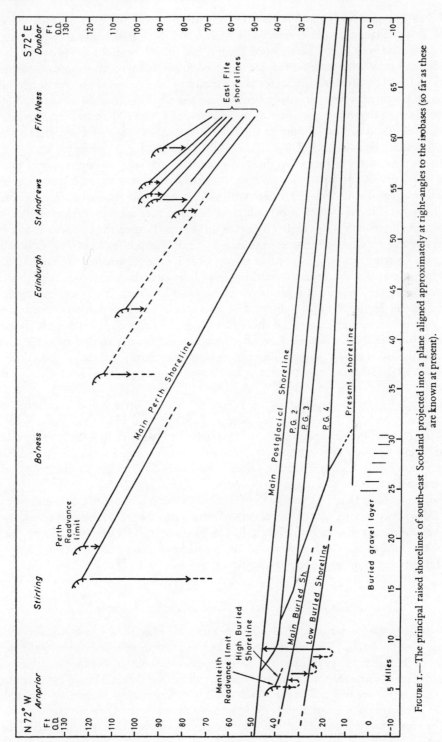

FIGURE I.—The principal raised shorelines of south–east Scotland projected into a plane aligned approximately at right–angles to the isobases (so far as these are known at present).

and the sea extended farther and farther up the Firth of Forth and into the Firth of Tay. It is possible that oscillations of the ice-margin occurred, and there are hints of such oscillations near St. Andrews and west of the Forth Road Bridge, but these have not been proved. The retreat of the ice and its relation to sea-levels are recorded in places by outwash spreads merging into raised beaches. These features are scarce along the Firth of Tay and along the north side of the Forth valley west of Leven, probably because the coastal slopes are often steep and few large river valleys lead down to the coast. On the south side of the Forth, however, a series of rivers enters the sea in and around Edinburgh and here there are large areas of fluvioglacial and raised beach deposits. The shorelines of two of these raised beaches are shown in Figure 1.

The more easterly of these two shorelines occurs in the north-western part of Edinburgh, where its relation to the former ice-margin is clearly seen, owing to the fortunate coincidence of the fluvioglacial and marine features with golf courses and farmlands. The features comprise, from west to east, three morphological elements: (a) a group of kames and dead-ice hollows, leading into (b) a fluvioglacial terrace that descends from 124 to 104–105 feet in three-quarters of a mile, whereupon it merges into (c) a raised beach up to 400 yards broad whose shoreline falls from 104–105 to 101 feet in just over a mile. Thereafter the raised beach continues but its shoreline is lost in the built-up area of the city. Although the raised shoreline has been measured for only a short distance, its minimal gradient, justifying its extension as a broken line in Figure 1, can be inferred in two ways. First, the shoreline must pass below the lowest of the East Fife group of shorelines, since the latter does not appear to have been modified by later marine action. Secondly, near Dunbar, kames and sharp kettles, the latter with rims as low as 44 feet, show no sign of modification by the sea, thus implying that at the time sea-level slightly exceeded 100 feet in north-west Edinburgh it did not exceed 44 feet near Dunbar.

The other shoreline represented on Figure 1 between Bo'ness and Edinburgh occurs a short distance west of the Forth Road Bridge. Here, around Hopetoun House, kames and meltwater channels are succeeded eastward by a large outwash spread, which in turn is followed by a shoreline between 110 and 114 feet. The latter can be traced for only a short distance but its representation in Figure 1 can be justified by lines of reasoning similar to those of the preceding paragraph.

Farther west on the south side of the Forth valley outwash spreads occur intermittently as far as Falkirk but none appears to be associated with a raised beach. A similar situation appears to prevail on the north side of the Forth. It may be that this results from a relative rise of sea-level at the time of the Perth Readvance, the raised beaches associated with the latter having buried those formed during the preceding retreat phase.

Raised Shorelines Associated with and Following the Perth Readvance

Following the formation of the features referred to so far, the ice continued to waste back towards the Highlands. How far it retreated before moving forward again as the Perth Readvance (J. B. Simpson, 1933; Sissons, 1963b, 1964) is unknown. Neither is it known when the Readvance ended, except that this was not later than 12,000 B.P. The ice extended to within a few miles of the site of Perth in the Tay region, to the vicinity of the sites of Kincardine and Plean in the Forth valley, and almost to the site of Larbert in the Carron valley.

The readvance was associated with the formation of very clear raised beaches, of which the highest, which will be referred to as the Main Perth Raised Beach, is usually the most

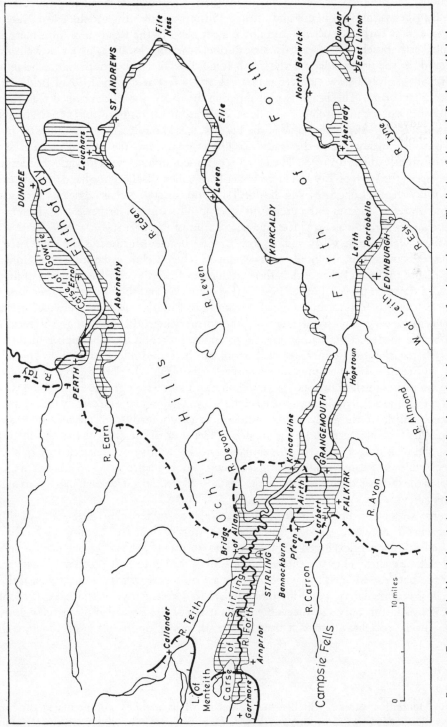

FIGURE 2—Extent of raised beach deposits (small areas omitted) and principal place-names referred to in the text. The limits of the Perth Readvance (dashed line) and the Menteith (Loch Lomond) Readvance are also shown.

conspicuous. These raised beaches have already been described in detail for the part of the
Forth valley between Burntisland and Stirling (Sissons and Smith, 1965b). Near Kincardine
and Plean and at Larbert, outwash laid down at or close to the Readvance limit merges into
the Main Perth Beach. This beach is very distinct in many localities as far down-valley as
Burntisland on the north side and the Forth Road Bridge on the south side. The shoreline
declines from slightly below 125 feet at Plean to 77–78 feet at the Forth Road Bridge and 67
feet near Burntisland. The beach is probably continued by a well-developed feature in the
Aberlady area where the shoreline slopes down eastward from 38–39 to 30–31 feet. The average
gradient of the Main Perth Shoreline in the Forth valley is about 2¼ feet per mile. In the Earn
valley, outwash at or close to the readvance limit merges into the Main Perth Beach at an
altitude of slightly above 100 feet. The beach is conspicuous at various localities bordering the
western part of the Firth of Tay and in the lower Earn valley. Its shoreline has declined to 77 feet
at Errol and 71 feet at Longforgan on the north side and to 70–71 feet at Logie on the south side.
The same shoreline appears to be present in other localities much farther east as, for example,
near Leuchars, where it is at 51–53 feet, and near Carnoustie, where it is at 43 feet.

In the lower Earn and Tay valleys, lower raised beaches, formed as the ice wasted back
from the readvance limit, are locally well developed. They decline eastward with a similar gradient to that of the Main Beach and, in the eastern part of the Carse of Gowrie, one of them has
been traced down to the level of the Post-Glacial carse clay and followed thereafter for a short
distance as a buried feature. In the Forth region near Falkirk, two raised beaches below the
main one form very extensive features on the former intertidal delta of the River Carron
which, before its dissection, had an area of about six square miles. The higher of these two
beaches (that is, the middle member of the trio) can be traced intermittently down-valley to
Bo'ness and can be followed up-valley to near Stirling (Fig. 1). Here the beach merges into
an outwash plain which in turn merges into kame and kettle topography. The beach was thus
formed at the time when the ice terminated in the Stirling gap, and it seems likely that at this stage
there was a pause in the the retreat of the ice-margin or a slight readvance. The maximum altitude
of the raised shoreline near Stirling is about 125 feet, but to the west of the city the highest
raised shoreline is only 73–76 feet. It is thus evident that a fifty-foot relative drop of sea-level
occurred while the ice-margin stood at the Stirling gap. Evidence of a marked relative fall of
sea-level as the ice wasted back following the Perth Readvance is also present in the Carron, Earn
and Tay valleys.

To the west of Stirling, apart from the feature at 73–76 feet close to the city, the highest
raised beaches lie between 65 and 70 feet. Although some of these are clear features locally, they
are, in general, poorly developed and it is not known if they are part of one feature or are of
slightly differing ages. They are therefore not shown in Figure 1. It is certain, however, that
they all pre-date the Menteith moraine for they are quite absent within this twelve mile-long
arcuate loop. Immediately inside the morainic arc (between Arnprior and Lake of Menteith)
kettle holes with rims only a few feet above the surface of the Post-glacial carse clay show that
the Late-Glacial seas have here not been higher than the upper limit of the Post-glacial
seas.

Buried Shorelines

The Menteith moraine, which contains marine shells at many points, marks the limit of
an important readvance (Simpson, 1933). J. J. Donner (1957) concluded from pollen studies

that the readvance occurred in Zone III of the pollen sequence, implying that the moraine was formed about 10,300 years ago. Although there is no reason at present to doubt Donner's pollen dating, it should be noted that it is not conclusively proven.

Relative sea-level changes as recorded by buried shorelines and outwash deposits in the neighbourhood of the Menteith moraine are described in detail on pages 24 to 25 and need be mentioned only briefly here in relation to Figure 1. In this diagram the three buried features that have been identified beneath carse clay and peat are shown. The High Buried Beach was formed while the ice extended to the Menteith moraine and does not occur inside the moraine. As explained on pages 25 to 27, the evidence indicates that, while the ice-margin stood at the moraine, sea-level rose to about 39–40 feet (the altitude of the shoreline near the moraine) and then fell below 35 feet. The shoreline gradient shown on Figure 1 is not proved: the shoreline is depicted as sloping down towards the Main Buried Shoreline since it has not been possible to follow it more than a few miles down the Forth valley from the Menteith moraine.

The Main Buried Beach, formed about 9500 years ago (page 27) and associated with a transgression, has been traced by boring from the area inside the moraine to within a few miles of Stirling and again between Stirling and Bannockburn. Throughout this distance, its shoreline shows a decline in altitude towards the east, falling from about 37 feet in the westernmost locality where it has so far been identified, to about 24 feet near Bannockburn. This gives an average gradient of 0.8 feet per mile, as compared with a gradient of 1.3 feet per mile for the part of the shoreline in the vicinity of the Menteith moraine (page 23). The shoreline is shown with a uniform gradient (0.8 feet per mile) in Figure 1, since it is not yet known if the steeper gradient towards the west is of general significance or is merely a local anomaly (for instance, because of the influence on estuarine sedimentation of the marked constriction caused by the Menteith moraine).

The Low Buried Beach, formed around 8800 years ago, has been traced from the area inside the Menteith moraine to a point a few miles short of Falkirk. Between Stirling and Airth it is a very extensive feature, often a mile broad. Its shoreline has been approximately located at a considerable number of points. Near Bannockburn, for example, it is at about 18 feet, a figure that may be compared with about 27–28 feet for the probable shoreline altitude in the area so far studied within the Menteith moraine. The average gradient of the Low Buried Shoreline is slightly less than that of the Main one, as shown in Figure 1.

Buried raised beaches that may well correlate with those in the Forth valley have been identified by hand-borings in the lower Earn valley. The evidence is not yet sufficient, however, to permit correlation between the two areas.

At this point it is necessary to depart from the sequential treatment of the various shorelines to consider the stratigraphy of the Grangemouth area. In and around this town and extending up the Forth valley to Airth, many hundreds of commercial boreholes have been put down. These reveal a very consistent stratigraphic sequence, which is summarized in Figure 3. The most significant element of this sequence in the present context is a widespread layer of gravel that lies between sediments composed mainly of clay, silt and fine sand. The gravel layer is usually between 2 and 5 feet thick, has many rounded stones, contains many stones of boulder size, includes a variety of rock types ranging from local Carboniferous material to Highland schist, and is mixed with marine shells. The bed slopes regularly towards the Forth, from around Ordnance Datum at its inner margin where in places it meets the steeply rising ground at the back of the carselands, to about 20 feet below O.D. close to the Forth. The gravel layer is known to occur over an area of many square miles and its width at right-angles to the Forth often exceeds a mile. These facts show that the gravel layer represents a buried beach lying at a distinctly lower altitude then the buried features considered so far.

Since the gravel layer is markedly different from the fine sediments above and below it (as well as from most of the other features referred to in this paper) some consideration of its origin is necessary. It is likely that the gravel was in part supplied by the Avon, Carron and smaller streams, for recent artificial excavations near the points where these streams enter the carse plain show spreads of coarse gravel fanning out from the mouths of the valleys. Some of this river gravel was certainly deposited as the carse clay accumulated, for the two deposits interdigitate, but it may well be that the lowest gravel passes into the widespread gravel layer of Figure 3 and contributed to its accumulation. Whether this is so or not, however, it is clear that an important factor in the production of the gravel bed of Figure 3 was erosion by the sea of drift previously deposited. This is shown in several ways. First, the ground at the back of the carse plain often rises steeply, sometimes for a hundred feet, and strongly resembles a marine

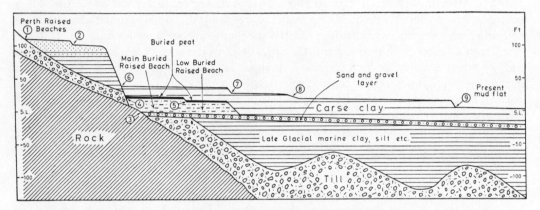

FIGURE 3—Diagrammatic section of the Grangemouth–Airth area to show the main elements of the stratigraphy in relation to the visible and buried morphological features. Circled numbers refer to shorelines in order of formation. The diagram is generalized to enhance clarity; thus, only three visible Post-glacial raised shorelines are shown.

cliff. That this steep slope is in some instances a result of marine erosion is suggested by its being best developed in places that would have been most exposed to former wave attack, whereas in re-entrants a gentle slope often backs the carse plain. Secondly, the till that mantles the solid rock almost everywhere in the Grangemouth area is often missing (beneath the carse clay) in a belt up to several hundred yards wide along the foot of the steep slope, thus suggesting that it has been removed by the sea. Thirdly, farther out from the old cliff the till is sometimes bevelled and on this planed surface rests the buried gravel layer (Fig. 3). In other instances, the higher parts of the strongly undulating ice-moulded buried surface of the till have been truncated and again the gravel layer rests directly on the truncated till. Fourthly, as shown in Figure 3, the marine deposits associated with the Perth Readvance have been considerably eroded to produce a steep bluff in which the horizontally bedded layers of Late-glacial clay, silt and sand crop out.

It thus appears that, while the materials of the gravel layer may have been in part supplied by streams, they were mainly provided by the erosion of till. Some material was probably derived also from the erosion of the Late-glacial marine sediments, for these contain ice-rafted stones. It is relevant to note that the present coast of the Forth eastward from the vicinity of Bo'ness has suffered considerable marine erosion in relatively recent times where it is composed of drift: in places, erosion is in progress today. The result of this erosion is that the present shore is often littered with a residue of stones and boulders of all sizes mixed with sand, mud and shells, this beach deposit closely resembling the buried gravel layer of the Grangemouth area. The close similarity was clearly demonstrated by H. M. Cadell (1883) for he classified in terms of rock type several hundred stones on the shore near Bo'ness, and compared the results with a similar analysis of stones from the buried gravel layer where it was then artificially exposed in Bo'ness. He found that the stone counts were virtually identical, a result that strongly favours the view that the present shore deposits and the buried gravel layer have a common origin.

Commercial borehole data show that near Airth the buried gravel layer passes beneath the deposits of the buried raised beaches already referred to (Figure 3). Thus the layer is older than these beaches, but, from evidence described above, must be younger than the Perth raised beaches. Thus a period of low sea-level, hitherto unrecognized in Scotland, appears to have occurred after the Perth Readvance but not later than the Menteith Readvance, and the erosional evidence suggests this period of low sea-level was of considerable duration. The precise altitude of the shoreline is not yet established and boring is being carried out to determine this. Provisional evidence suggests that the shoreline is approximately at Ordnance Datum south-west of Grangemouth and between six and ten feet below O.D. in Bo'ness (where numerous closely spaced commercial boreholes have been put down). The feature is therefore represented by a band of shading in Figure 1.

Visible Post-glacial Raised Shorelines

Following the formation of the Low Buried Beach, which ended about 8800 years ago, sea-level fell to expose this beach. As pointed out on page 27, the minimal sea-level probably occurred about 8500 years ago. Thereafter sea-level rose relative to the land during the major Post-glacial transgression. The early part of this transgression appears to have been rapid. Whether the relative rise of sea-level was continuous, or was interrupted by temporary regressions, is not at present known. The result of the transgression was that peat accumulating on the now buried beaches was covered by carse clay over extensive areas in the Forth, Earn and Tay valleys. In a few places, however, peat growth managed to keep pace with the rising sea-level, so that occasionally the carse clay is replaced by peat for its full thickness. The two largest areas where this occurred lie near the Menteith moraine and comprise two roughly circular peat bogs, each about a mile across, that continued to accumulate as the carse clay was deposited around them (Sissons and Smith, 1965a).

According to H. Godwin and E. H. Willis (1962), the transgression culminated about five and a half thousand years ago, for wood from a few inches above the base of the peat that extensively overlies the carse clay in the western part of the Forth valley gave a radiocarbon date of 5492 ± 130 years B.P. The associated shoreline (the Main Post-glacial Shoreline) falls in altitude down the Forth valley from 49 feet in the extreme west to 19–20 feet at Dunbar (Fig. 1). Up-valley from Falkirk and in the more sheltered localities farther east, as near Aberlady and in the Tyne estuary, the deposits consist mainly of carse clay (elevated mud flats) and on more exposed coasts they comprise mainly sand and gravel. In the Tay area the carse clay of the Carse of Gowrie was almost entirely deposited at this time. The shoreline at the back of the carselands reaches 36 feet in the Earn valley, falls to 28–29 feet immediately west of Dundee, and has declined to 26 feet in the Eden valley a few miles from St. Andrews. South-eastward from the Eden estuary, the beach is composed of shelly sand and gravel, and the shoreline has been traced intermittently to Fife Ness, where it is at 21–22 feet.

After the formation of the Main Post-Glacial Raised Beach, a series of lower beaches was produced. These are most clearly preserved in the carselands of the Forth valley upstream of Kincardine and Grangemouth, where three features below the main one have been identified. At its western end each shoreline increases slightly in gradient and merges into the one above it, as shown diagrammatically in Figure 1. This is the result of sedimentation during the periods of falling sea-level between the times when sea-level relative to the land was more or less stationary. It is not known if the second and third Post-Glacial shorelines were associated with a slight transgression, but the lowest feature appears to be transgressive since very thin peat is occasionally found beneath the associated carse clay.

For purposes of comparison, the present shoreline as levelled on mud flats is shown in Figure 1. This is at about 5 feet in the Eden estuary. In the Forth valley it declines from 6.6 feet immediately north-west of Grangemouth to about 4.5 feet at the mouth of the Tyne near Dunbar. This faint eastward slope is probably the result of tidal influences, for Admiralty Tide Tables (1964) show that high water mark is 10.3 feet at Alloa, 9.6 feet at Grangemouth and 8.3 feet at Dunbar. It will be noted that the feature measured on the modern mud flats, which is comparable with that measured on the elevated features, is 3 to 4 feet below high water mark, so that one may suggest that the elevated features may have been similarly related to former high water marks.

Conclusion

In the preceding account, the purpose has been to present the pattern of relative sea-level movements in Late-glacial and Post-glacial times in south-east Scotland so far as they are known at present. Discussion of the causes of these movements has intentionally been avoided, as also has discussion of the evidence in relation to sea-level changes in the rest of Scotland and adjacent parts of the British Isles. These subjects are far too large to be dealt with here and it is planned to consider them at length elsewhere.

NOTE

All altitudes are expressed in terms of Ordnance Datum and are related to the second geodetic levelling of the Ordnance Survey.

* * * * * * *

Consolidated Bibliography

ALLAN, J. K. and KNOX, J. (1934) 'The economic geology of the Fife coalfields, Area II', *Mem. geol. Surv. Gt Br.*

ANDERSON, F. W. (1947) 'The fauna of the '100-feet beach' clays', *Trans. Edinburgh geol. Soc.*, 14, 220–9.

BAILEY, E. B. *et al.* (1916) 'The geology of Ben Nevis and Glencoe', *Mem. geol. Surv. Gt Br.*

BAILEY, E. B. *et al.* (1924) 'The Tertiary and post-Tertiary geology of Mull, Loch Aline and Oban', *Mem. geol. Surv. Gt Br.*

BAILEY, E. B. *et al.* (1925) 'The geology of Staffa, Iona and Western Mull', *Mem. geol. Surv. Gt Br.*

BELL, A. (1890) 'Notes on the marine accumulations in Largo Bay, Fife, and at Portrush, County Antrim, North Ireland', *Proc. R. phys. Soc. Edinburgh*, 10, 290–7.

BENNIE, J. and SCOTT, A. (1893) 'The ancient lake of Elie', *Proc. R. phys. Soc. Edinburgh*, 12, 148–70.

BLACKBURN, K. B. (1946) 'On a peat from the Island of Barra, Outer Hebrides. Data for the study of post-glacial history', *New Phytol.*, 45, 44–9.

BLUNDELL, D. (1963) 'Some deep trenches bordering Cardigan Bay', First Irish Sea Colloquium, Univ. of Liverpool.

BREMNER, A. (1925) 'The glacial geology of the Stonehaven District', *Trans. Edinburgh geol. Soc.*, 11, 25–41.

BRINKMANN, P. (1934) 'Zer Geschichte der Moore, Marschen und Walder Nordwestdeutschlands', *bot. Jb.*, 66, 369–445.

BRITISH ASSOCIATION (1893) 'The character of the high-level shell bearing deposits at Clava', *Rep. Br. Ass. Advmt Sci.*

BROWN, T. (1867) 'On the arctic shell-clay of Elie and Errol, viewed in connection with our other glacial and more recent deposits', *Trans. R. Soc. Edinburgh*, 24, 617–33.

BRYANT, R. H. (in the press) 'The 25 ft. "preglacial" raised beach in south-west Ireland'.

CADELL, H. M. (1883) 'Notice of the surface geology of the estuary of the Forth round Borrowstounness', *Trans. Edinburgh geol. Soc.*, 4, 2–33.

CHAMBERS, R. (1843) 'On the existence of raised beaches in the neighbourhood of St. Andrews,' *Edinburgh new philos. J.*, 34, 298–306.

CHAMBERS, R. (1848) *Ancient Sea Margins.*

CHAPMAN, V. J. (1964) *Coastal Vegetation.*

CHARLESWORTH, J. K. (1924) 'The glacial geology of North-west Ireland', *Proc. R. Ir. Acad.*, 36B, 174–314.

CHARLESWORTH, J. K. (1926) 'The readvance marginal kame-moraine of the south of Scotland, and some later stages of retreat,' *Trans. R. Soc. Edinburgh*, 55, 25–50.

CHARLESWORTH, J. K. (1939) 'Some observations on the glaciation of North-east Ireland', *Proc. R. Ir. Acad.* 45B, 255–295.

CHARLESWORTH, J. K. (1956) 'The late-glacial history of the Highlands and Islands of Scotland', *Trans. R. Soc. Edinburgh*, 62, 769–928.

CHARLESWORTH, J. K. (1957) *The Quaternary Era*, 2 vols.

CHARLESWORTH, J. K. (1963) 'Some observations on the Irish Pleistocene,' *Proc. R. Ir. Acad.*, 62B, 295–322.

CLOUGH, C. T. and HARKER, A. (1904) 'The geology of west central Skye', *Mem. geol. Surv. Gt Br.*

CRAMPTON, C. B. (1914) 'The Geology of Caithness', *Mem. geol. Surv. Scotl.*

CUNNINGHAM CRAIG, E. H. *et al.* (1911) 'The geology of Colonsay and Oronsay', *Mem. geol. Surv. Gt Br.*

DAVIES, G. L. (1960) 'Platforms developed in the boulder clay of the coastal margins of Counties Wicklow and Wexford', *Ir. Geogr.*, 4, 107–16.

DONNER, J. J. (1957) 'The geology and vegetation of Late-glacial retreat stages in Scotland', *Trans. R. Soc. Edinburgh*, 63, 221–64.

DONNER, J. J. (1959) 'The late- and post-glacial raised beaches in Scotland', *Ann. Acad. Sci. fenn.*, A III, Geol.-Geogr., 53, 1–25.

DONNER, J. J. (1963) 'The late- and post-glacial raised beaches in Scotland', *Ann. Acad. Sci. fenn.*, A III, Geol.-Geogr., 68, 1–13.

DURNO, S. E. (1956) 'Pollen analysis of peat deposits in Scotland', *Scott. geogr. Mag.*, 72, 177–87.

DURY, G. H. (1953) 'A glacial breach in the North Western Highlands, *Scott. geogr. Mag.*, 69, 106–17.

DWERRYHOUSE, A. R. (1923) 'The glaciation of North-eastern Ireland', *Q. J. geol. Soc.*, 79, 352–421.

ELTON, C. (1938) 'Notes on the ecology and natural history of Pabbay'. *J. Ecol.*, 26, 275–97.

ETHERIDGE, R. (1881) 'Notes on the post-Tertiary deposits of Elie and Largo Bay, Fife', *Proc. R. phys. Soc. Edinburgh*, 6, 105–12.

141

FAEGRI, K. and IVERSEN, J. (1964) *Textbook of Pollen Analysis* (Copenhagen).
FAIRBRIDGE, R. W. (1961) 'Eustatic changes in sea-level', *Physics and Chemistry of the Earth*, 4, 99–185.
FAIRHURST, H. and RITCHIE, W. (1963) *Baleshare Island, North Uist Discovery and Excavation*.
FARRINGTON, A. (1929) 'The preglacial topography of the Liffey basin', *Proc. R. Ir. Acad.*, 38B, 148–70.
FARRINGTON, A. (1934) 'The glaciation of the Wicklow mountains', *Proc. R. Ir. Acad.*, 42B, 173–209.
FARRINGTON, A. (1954) 'A note on the correlation of the Kerry-Cork glaciations with those of the rest of Ireland', *Ir. Geogr.*, 3, 47–53.
FARRINGTON, A. (1959) 'The Lee basin; Part I: Glaciation', *Proc. R. Ir. Acad.*, 60B, 135–66.
FARRINGTON, A. (1960) 'Types of rejuvenated valleys found in Ireland', *Ir. Geogr.*, 4, 117–20.
FARRINGTON, A. (1961) 'The Lee basin; Part II: the drainage pattern', *Proc. R. Ir. Acad.*, 61B, 233–53.
FARRINGTON, A. and STEPHENS, N. (1964) 'The Pleistocene geomorphology of Ireland' in J. A. STEERS (ed.), *Field Studies in the British Isles*, 445–61.
FLEMING, J. (1830) 'Notice of a submarine forest in Largo Bay, in the Firth of Forth', *Q. J. Sci. Lit. Art*, 1, 21–9.
FLINT, R. F. (1961) *Glacial and Pleistocene Geology*.
FLORIN, S. (1944) 'Havsstrandens förskjutningar och bebyggelsseutvecklingen i östra Mellansverige under senkvärtar tid , *geol. Fören. Förh. Stockholm*, 66, 551–624.

GEIKIE, A. (1865) *The Scenery of Scotland*.
GEIKIE, A. (1887) *The Scenery of Scotland*, 2nd edn.
GEIKIE, A. (1902) 'The geology of eastern Fife', *Mem. geol. Surv. Gt Br.*
GEIKIE, A. *et al.* (1903) 'The geology of Buteshire', *Mem. geol. Surv. Gt Br.*
GEIKIE, J. (1878) 'On the glacial phenomena of the Long Island', *Q. J. geol. Soc.*, 34, 819–70.
GEIKIE, J. (1894) *The Great Ice Age*, 3rd edn.
GEORGE, T. N. (1932) 'The Quaternary beaches of Gower', *Proc. geol. Ass.*, 43, 291–324.
GODWIN, H. (1943) 'Coastal peat beds of the British Isles and North Sea', *J. Ecol.*, 31, 199–247.
GODWIN, H. (1956) *The History of the British Flora*.
GODWIN, H., WALKER, D. and WILLIS, E. H. (1957) 'Radiocarbon dating and post-glacial vegetational history: Scaleby Moss', *Proc. R. Soc., B.*, 147, 352–66.
GODWIN, H., SUGGATE, R. P. and WILLIS, E. H. (1958) 'Carbon-14 dating of the eustatic rise in sea level', *Nature*, 181, 1518.
GODWIN, H. (1960) Radiocarbon dating and Quaternary history in Britain, *Proc. R. Soc., B*, 153, 287–320.
GODWIN, H. and WILLIS, E. H. (1961, 1962) 'Cambridge University natural radiocarbon measurements, III and V', *Radiocarbon*, 3, 60–76; 4, 57–70.
GODWIN, H. and WILLIS, E. H. (1965) 'Radiocarbon dates from Roddans Port, Northern Ireland', *philos. Trans. R. Soc., B.*, 249, 249–255.
GRANLUND, E. *et al.* (1949) *Sveriges Geologi Norstedts* (Stockholm).
GRANT WILSON, J. S. (1882) 'Northern Aberdeenshire, Explanation of Sheet 97', *Mem. geol. Surv. Scotl.*
GRESSWELL, R. K. (1953) *Sandy Shores in South Lancashire*.
GRESSWELL, R. K. (1958) 'Hillhouse coastal deposits in South Lancashire', *Liverpool and Manchester geol. J.*, 2, 60–78.
GUILCHER, A. (1958) *Coastal and Submarine Morphology*.
GUILCHER, A. and KING, C. A. M. (1961) 'Spits, tombolos, and tidal marshes in Connemara and west Kerry, Ireland', *Proc. R. Ir. Acad.*, 61B, 283–338.
GUILCHER, A. (1962) 'Morphologie de la Baie de Clew (Comté de Mayo, Irlande)', *Bull. Ass. Géogr. fr.*, 303–4, 53–65.
GUILCHER, A. (1965) 'Drumlin and spit structures in the Kenmare river, south-west Ireland', *Ir. Geogr.*, 5, 7–19.

HAMILTON, W. J. (1835) 'Description of a bed of recent marine shells near Elie, on the southern coast of Fifeshire', *Proc. geol. Ass.*, 2, 180–1.
HARKER, A. (1908) 'The geology of the Small Isles of Inverness-shire', *Mem. geol. Surv. Gt Br.*
HILL, J. B. *et al.* (1905) 'The geology of Mid-Argyll', *Mem. geol. Surv. Gt Br.*
HOLTEDAHL, O. (1953) 'Norges geologi', *Norges geol. Unders.*, No. 164.
HORNE, J. *et al.* (1896) 'The character of the high-level shell-bearing deposits in Kintyre'. *Rep. Br. Ass. Advmt Sci.*, 378–89.
HORNE, J. and HINXMAN L. W. (1914) 'The geology of the country around Beauly and Inverness', *Mem. geol. Surv. Gt Br.*
HOWELL, F. T. (1965) 'The sub-drift surface beneath, and adjacent to, Morecambe and Liverpool Bays, in the Irish Sea', Second Irish Sea Colloquim, Univ. Coll., Dublin.
HULL, E. (1872) 'On the raised beach of the north-east of Ireland', *Rep. Br. Assoc. Advmt Sci.*, 113–14.
HULL, E. (1881) *Explanatory Memoir to Accompany Sheets 60, 61 and part of 71 of the Maps of the Geological Survey of Ireland*, 21.
HYYPPÄ, E. (1963) 'On the Late-Quaternary history of the Baltic Sea', *Fennia*, 89, 37–50.

JAMIESON, T. F. (1865) 'On the history of the last geological changes in Scotland', *Q. J. geol. Soc.*, 21, 161–203.

JAMIESON, T. F. (1882) 'On the causes of the depression and re-elevation of the land during the Glacial Period', *geol. Mag.*, 9, 400–7 and 457–66.

JARDINE, W. G. (1962) 'Post-glacial sediments at Girvan, Ayrshire', *Trans. geol. Soc. Glasgow*, 24, 262–78.

JARDINE, W. G. (1963) 'Pleistocene sediments at Girvan, Ayrshire, *Trans. geol. Soc. Glasgow*, 25, 4–16.

JARDINE, W. G. (1964) 'Post-glacial sea-levels in South-west Scotland', *Scott. geogr. Mag.*, 80. 5–11.

JEHU, T. H. and CRAIG, R. M. (1926) 'Geology of the Outer Hebrides, III', *Trans. R. Soc. Edinburgh*, 54, 467–489.

JESSEN, K. (1949) 'Studies in late Quaternary deposits and flora-history of Ireland', *Proc. R. Ir. Acad.*, 52B, 85–290.

JOHNSON, D. W. (1931) 'The correlation of ancient marine levels. *C. R. Congr. int. de Géogr.,-Paris*, 2, 42–54.

KILROE, J. R. (1908) 'The geology of the country around Londonderry', *Mem. geol. Surv. Ireland*.

KINAHAN, G. H. (1873) 'The estuary of the river Slaney, County Wexford', *J. R. geol. Soc. Ireland*, 14, 60–9.

KIRK, W. and GODWIN, H. (1963) 'A Late-glacial site at Loch Droma, Ross and Cromarty', *Trans. R. Soc. Edinburgh*, 65, 225–49.

KIRK, W., RICE, J. and SYNGE, F. M. (1966) 'Deglaciation and vertical displacement of shorelines in Wester and Easter Ross', *Trans. Inst. Br. Geogr.* 39, 65–68.

KNOX, E. (1954) 'Pollen analysis of a peat at Kingsteps Quarry, Nairn', *Trans. bot. Soc. Edinburgh*, 36, 224–9.

KYNASTON, H. *et al.* (1908) 'The geology of the country round Oban and Dalmally', *Mem. geol. Surv. Gt Br.*

LAMPLUGH, G. W. (1891) 'The drifts of Flamborough Head', *Q. J. geol. Soc.*, 47, 384–431 .

LAMPLUGH, G. W. *et al.* (1904) 'The geology of the country around Belfast', *Mem. geol. Surv. Ireland*.

LAMPLUGH, G, W. *et al.* (1905) 'The geology of the country around around Cork and Cork Harbour', *Mem. geol. Surv. Ireland*.

LAMPLUGH, G. W., *et al.* (1907) 'The geology of the country around Limerick', *Mem. geol. Surv. Ireland*.

LINTON, D. L. (1949) 'Watershed breaching by ice in Scotland', *Trans. Inst. Br. Geogr.*, 15, 1–16.

LINTON, D. L. (1964) 'Tertiary landscape evolution in J. W. WATSON and J. B. SISSONS (eds.), *The British Isles*.

McCALLIEN, W. J. (1937) 'Late-glacial and early Post-glacial Scotland', *Proc. Soc. Antiquaries Scotl.*, 71, 174–206.

McCANN, S. B. (1961a) *The Raised Beaches of Western Scotland*, Unpub. Ph.D dissertation, Univ. of Cambridge.

McCANN, S. B. (1961b) 'Some supposed raised beach deposits at Corran, Loch Linnhe, and Loch Etive', *geol. Mag.*, 98, 131–42.

McCANN, S. B. (1964) 'The raised beaches of north-east Islay and western Jura, Argyll', *Trans. Inst. Br. Geogr.*, 35, 1–16

McCANN, S. B. (1966) 'The limits of the late-glacial highland, or Loch Lomond, readvance along the West Highland seaboard from Oban to Mallaig', *Scott. J. Geol.*, 2, 84–95.

McLEOD, N. K. (1899) *The Churches of Buchan*.

McMILLAN, N. F. (1957) 'Quaternary deposits around Loch Foyle, north Ireland', *Proc. R. Ir. Acad.*, 58B, 185–205.

McMILLAN, N. F. (1964) 'The mollusca of the Wexford gravels (Pleistocene), south-east Ireland', *Proc. R. Ir. Acad.*, 63B, 265–89.

MACRAE, F. (1845) *North Uist: The New Statistical Account*.

MARTHINUSSEN, M. (1945) 'Yngre postglaciale nivåer på Varangerhalvøyal', *Norsk geol. Tidskr.*, 25, 1–230.

MARTIN, C. P. (1930) 'The raised beaches of the east of Ireland', *Sci. Proc. R. Dublin Soc.*, 19, 491–511.

MARTIN, S. (1955) 'Raised beaches and their relation to glacial drifts on the east coast of.Ireland', *Ir. Geogr.*, 3, 87–93.

MITCHELL, G. E. (1948) 'Two interglacial deposits in south-east Ireland', *Proc. R. Ir. Acad.*, 52B, 1–14.

MITCHELL, G. F. (1952) 'Late-glacial deposits of Garscadden Mains, near Glasgow', *New Phytol.*, 50, 277–86.

MITCHELL, G. F. (1956) 'An early kitchen-midden at Sutton, Co. Dublin', *J. R. Soc. Antiq. Ireland*, 86, 1.

MITCHELL, G. F. (1960) 'The Pleistocene history of the Irish Sea', *Advmt Sci.*, 68, 313–25.

MITCHELL, G. F. (1963) 'Morainic ridges on the floor of the Irish Sea', *Ir. Geogr.*, 4, 335–44.

MORRISON, M. E. S. (1965) 'A submerged Late-Quaternary deposit at Roddans Port on the north-east coast of Ireland', *Phil. Trans. R. Soc.*, B, 249, 221–55.

MOVIUS, H. L. (1942) *The Irish Stone Age*.

MOVIUS, H. L. (1953) 'Graphic representation of Post-glacial changes of level in north-east Ireland', *Am. J. Sci.*, 251, 697–740.

MYKURA, W. *et al.* (1962) 'The geology of the neighbourhood of Edinburgh', 3rd edn. *Mem. geol. Surv. Scotl.*

NEWEY, W. W. (1965) *Post-glacial Vegetational and Climatic Changes in Part of South-east Scotland*, Unpub. Ph.D. Thesis, Univ. of Edinburgh.

NICHOLS, H. (1963) *Vegetation Change and Shoreline Displacement History of Western Scotland*, Unpub. Ph.D. Dissertation, Univ. of Leicester.

OGILVIE, A. G. (1923) 'The physiography of the Moray Firth Coast', *Trans. R. Soc. Edinburgh*, 53, 377–404.

OLDFIELD, F. (1960) 'Late Quaternary changes in climate, vegetation and sea-level in lowland Lonsdale', *Trans. Inst. Br. Geogr.*, 28, 99–117.

ORME, A. R. (1962) 'Abandoned and composite seacliffs in Britain and Ireland', *Ir. Geogr.*, 4, 279–91.

ORME, A. R. (1964a) 'Planation surfaces in the Drum Hills, County Waterford, and their wider implications', *Ir. Geogr.*, 5, 48–72.

ORME, A. R. (1964b) 'The geomorphology of southern Dartmoor and the adjacent area', in I. G. SIMMONS (ed.), *Dartmoor Essays*, 31–72.

PAGE, D. (1859) 'On the skeleton of a seal from the Pleistocene clays of Stratheden, in Fifeshire', *Edinburgh new philos. J.*, 9, 149–50.

PEACH, B. N. and HORNE, J. (1892) 'The ice-shed of the N.W. Highlands during the maximum glaciation, *Rep. Brit. Ass. Advmt Sci.*, 720.

PEACH, B. N. *et al.* (1909) 'The geology of the seaboard of Mid-Argyll', *Mem. geol. Surv. Gt Br.*

PEACH, B. N. *et al.* (1910) 'The geology of Glenelg, Lochalsh and the S.E. part of Skye', *Mem. geol. Surv. Gt Br.*

PEACH, B. N. *et al.*, (1911) 'The geology of Knapdale, Jura and N. Kintyre', *Mem. geol. Surv. Gt Br.*

PEACH, B. N. and HORNE, J. (1913) 'Geology of the Fannich Mountains and the country around Upper Loch Maree and Strath Broom (Sheet 92)', *Mem. geol. Surv. Gt Br.*

PEACH, B. N. *et al.* (1913) 'The geology of Central Ross-shire', *Mem. geol. Surv. Gt Br.*

PITCHER, W. S. and CHEESMAN, R. L. (1954) 'Summer field meeting in north-west Ireland with an introductory note on the geology', *Proc. geol. Ass.*, 65, 345–71.

PRIOR, D. B. (1966) 'Late and Post-glacial shorelines in north-east Antrim', *Ir. Geogr.* (in the press).

READ, H. H. *et al.* (1926) 'The geology of Strath Oykell and Lower Loch Shin', *Mem. geol. Surv. Gt Br.*

SIMPSON, J. B. (1933) 'The late-glacial readvance moraines of the Highland border west of the River Tay', *Trans. R. Soc. Edinburgh*, 57, 633–45.

SIMPSON, J. B. (1949) 'Geology of Central Ayrshire', *Mem. geol. Surv. Scotland,*

SISSONS, J. B. (1962) 'A re-interpretation of the literature on late-glacial shorelines in Scotland with particular reference to the Forth area', *Trans. Edinburgh geol. Soc.*, 19, 83–99.

SISSONS, J. B. (1963a) 'Scottish raised shoreline heights with particular reference to the Forth valley', *geogr. Annlr*, 45, 180–5.

SISSONS, J. B. (1963b, 1964) 'The Perth re-advance in central Scotland', *Scott. geogr. Mag.*, 79, 151–63; 80, 28–36.

SISSONS, J. B. (1965) 'Quaternary' in G. Y. CRAIG (ed.), *The Geology of Scotland*, 467–503.

SISSONS, J. B., CULLINGFORD, R. A., and SMITH, D. E. (1965),'Some pre-carse valleys in the Forth and Tay basins', *Scott. geogr. Mag.*, 81, 115–24.

SISSONS, J. B. and SMITH, D. E. (1965a) 'Peat bogs in a post-glacial sea and a buried raised beach in the western part of the Carse of Stirling', *Scott. J. Geol.*, 1, 247–55.

SISSONS, J. B. and SMITH, D. E. (1965) 'Raised shorelines associated with the Perth readvance in the Forth valley and their relation to glacial isostasy', *Trans. R. Soc. Edinburgh*, 66, 143–68.

SMITH, A. G. (1965) 'Problems of inertia and threshold related to post-glacial habitat changes', *Proc. R. Soc.*, B, 161, 331–41.

SMITH, D. E. (1965) '*Late and postglacial changes of shoreline on the north side of the Forth valley and estuary*', Unpub. Ph.D. thesis, Univ. of Edinburgh.

STEPHENS, N. (1957) 'Some observations on the "interglacial" platform and the early post-glacial raised beach on the east coast of Ireland', *Proc. R. Ir. Acad.*, 58B, 129–49.

STEPHENS, N. (1958) 'The evolution of the coastline of north-east Ireland', *Advmt Sci.*, 56, 389–91.

STEPHENS, N. and SYNGE, F. M. (1958) 'A Quaternary succession at Sutton, Co. Dublin', *Proc. R. Ir. Acad.*, 59B, 19–27.

STEPHENS, N. (1963) 'Late-glacial sea levels in north-east Ireland', *Ir. Geogr.*, 4, 345–59.

STEPHENS, N. and SYNGE, F. M. (1965) 'Late Pleistocene shorelines and drift limits in north Donegal', *Proc. R. Ir. Acad.*, 64B, 131–53.

STEPHENS, N. (1966) 'Late-glacial and post-glacial shorelines in Ireland and south-west Scotland', Special Geol. Soc. Am. papers, Int. Stud. Quaternary, VII Congress Int. Ass. Quaternary Res., Boulder, Colorado, U.S.A. (in the press).

STEPHENS, N. and SYNGE, F. M. (1966) 'Pleistocene shorelines', in G. H. DURY (ed.), Geomorphological Essays, 1–51.

SVENSSON, H. (1956) 'Method for exact characterizing of denudation surfaces, especially peneplains, as to the position in space', *Lund. Stud. Geogr., phys. Geogr.*, 8.

SWEETING, M. M. (1955) 'The landforms of north-west County Clare, Ireland', *Trans. Inst. Br. Geogr.*, 21, 33–49.

SYNGE, F. M. and STEPHENS, N. (1960) 'The Quaternary period in Ireland—an assessment, 1960', *Ir. Geogr.*, 4, 121–30.

SYNGE, F. M. (1964) 'Some problems concerned with the glacial succession in south-east Ireland', *Ir. Geogr.*, 5, 73–82.

SYNGE, F. M. (1966a) 'The Würm ice-limit in the West of Ireland', Special Geol. Soc. Am. papers, Int. Stud. Quaternary, VII Congress Int. Ass. Quaternary Res., Boulder, Colorado, U.S.A. (in the press).

SYNGE, F. M. (1966b) 'The relationship of the raised strandlines and main end-moraines on the Isle of Mull, and in the district of Lorn, Scotland', *Proc. geol. Ass.* (in the press).

TANNER, V. (1930) 'Studier över kvartärsystemet i Fennoskandian nordliga delar. IV'. *Bull. Comm. geol. Finland*, 88, 1–594.

TRICART, J. and CAILLEUX, A. (1963) 'Initiation à l'étude des sables et des galets', Tome I, *Cent. Documn Univ. Paris*.

TING, S. (1937) 'The coastal configuration of West Scotland', *geogr. Annlr.*, 19, 62–83.

VALENTIN, H. (1953) 'Present vertical movements of the British Isles', *geogr. J.*, 119, 299–305.

VON POST, L. (1947) 'Hallands marina fornstränder', *Geol. Fören. Stockh. Förh.*, 69, 293–320.

VON POST, L. (1956) 'The Ancient sea fiord of the Viskan valley: Chapter II: The shore marks', *Medd. Stockholms Högsk. geol. Inst. No.* 113, 101–121.

WALKER, R. (1863) 'On the skeleton of a seal (*Phoca groenlandica?*), and the cranium of a duck, from the Pliocene beds, Fifeshire', *Ann. Mag. nat. Hist.*, 12, 382–8.

WALKER, R. (1876) 'On clays containing *Ophiolepis gracilis* and other organic remains, with notes on recent geological formations near St. Andrews', *Scott. Nat.*, 3, 41–6.

WALTON, K. (1956) 'Rattray; a study in coastal evolution', *Scott. geogr. Mag.*, 72, 85–96.

WALTON, K. (1959) 'Ancient elements in the coast-line of North-East Scotland' in *Geographical Essays in Memory o Alan G. Ogilvie*, 93–109.

WATSON, R. B. (1864) 'On the great drift beds with shells in the south of Arran', *Trans. R. Soc. Edinburgh*, 23, 523–46.

WATSON, J. W. with SISSONS, J. B. (1964) *The British Isles*.

WATTS, W. A. (1959) 'Interglacial deposits at Kilbeg and Newtown, Co. Waterford', *Proc. R. Ir. Acad.*, 60B, 79–134.

WILKINSON, S. B. *et al.* (1907) 'The geology of Islay', *Mem. geol. Surv. Gt Br*.

WOOD, W. (1887) *The East Neuk of Fife*.

WRIGHT, W. B. and MUFF, H. B. (1904) 'The pre-glacial raised beach of the south coat of Ireland', *Sci. Proc. R. Dublin Soc.*, 10, 250–324.

WRIGHT, W. B. (1911) 'On a pre-glacial shoreline in the Western Isles of Scotland', *geol. Mag.*, 48 (1911), 97–109.

WRIGHT, W. B. (1914) *The Quaternary Ice Age*.

WRIGHT, W. B. (1927) 'The geology of Killarney and Kenmare', *Mem. geol. Surv. Ireland*.

WRIGHT, W. B. (1928) 'The raised beaches of the British Isles', *1st Rep. Commn Pliocene and Pleistocene Terraces, Int. geogr. Un.*, 99–106.

WRIGHT, W. B. (1937) *The Quaternary Ice Age*, 2nd edn.

ZEUNER, F. E. (1959) *The Pleistocene Period*, 2nd edn.

Editor's Comments on Paper 16

Farrand: *Postglacial Uplift in North America*

Farrand's paper is a classic in North American studies of postglacial uplift. The advent of radiocarbon dating had already been utilized by researchers working on Spitsbergen to construct time/elevation curves of postglacial relative sea level changes (e.g., Feyling-Hanssen and Olsson, 1959, *Norsk Geog. Tids.*) but no one had yet attempted a continental survey of the available radiocarbon dates and dated shoreline elevations (see, however, Fairbridge, 1961). Farrand's paper (1) shows the form of postglacial uplift at 11 sites stretching from the Great Lakes northward to northern Ellesmere Island, and (2) uses these curves to construct the form of glacio-isostatic recovery along a cross section from the Great Lakes to the ice divide (Fig. 11). This figure should be compared to the cross section of Fennoscandia shown by Donner (Paper 12). A marked flexing of the crust is noticeable in the vicinity of Cape Rich.

The postglacial uplift curves presented by Farrand are based on radiocarbon dated marine shells, wood, and archaeological materials. The curves are similar in geometry and show a rapid deceleration of uplift toward the present. The curves are corrected for eustatic sea levels at the time of the death or emplacement of the dated samples. Postglacial uplift curves of similar form had been presented in the late 1920s and 1930s by Nansen and Liden in Scandinavia, but the papers by Feyling-Hanssen and Olsson, Olsson and Blake, and Farrand indicated that a detailed elevation and age survey of raised marine beaches would provide unique information concerning the relaxation properties of the earth in response to loads in excess of 1000 km diameter, of 3000 m maximum thickness, and of 10^3-10^4 yr time duration. This information will provide essential data and supplement existing knowledge of the elastic properties of the earth resulting from seismological research.

William R. Farrand was born in 1931 and received his B.Sc. from Ohio State University in 1955, and his Ph.D. from the University of Michigan in 1960. Currently head of the Quaternary Research Laboratory at the University of Michigan, he has worked in Greenland, the area of the Great Lakes, and the Middle East on a variety of topics in Quaternary geology.

Reprinted from Amer. J. Sci., **260**, 181–199 (1962)

16

POSTGLACIAL UPLIFT IN NORTH AMERICA*

WILLIAM R. FARRAND

Lamont Geological Observatory, Columbia University, Palisades, New York

ABSTRACT. Tilted beaches around the Great Lakes and raised marine features throughout Arctic Canada which have been dated by radiocarbon analysis furnish sufficient data for the construction of curves of postglacial uplift vs. time for eleven areas. These curves show (1) a uniform pattern of strongly decreasing rate of uplift from the time of deglaciation to the present and (2) a time-displacement which seems to correlate with the time of deglaciation in a given area. These systematic relationships are additional evidence that postglacial and recent uplift around the Canadian Shield (as in Scandinavia) is a result of glacial loading.

Furthermore, these curves indicate that far northern areas such as northwestern Victoria Island and northern Ellesmere Island probably had an ice cover comparable to that over the Great Lakes. Thus the Wisconsin ice border lay considerably north or northwest of these areas. Also, if we may assume that the Wisconsin ice sheet achieved isostasy, then the major part of isostatic rebound occurred prior to complete ice removal in any given place.

INTRODUCTION

The succession of glacial lakes in the Upper Great Lakes is now well enough known and dated by the radiocarbon method that curves of uplift versus time for the past 10,500 years can be drawn for several areas. Such curves for four localities around the Lake Huron basin show the regularity of the uplift phenomenon and a systematic relationship to the Wisconsin-age ice cover. The mean pattern of uplift with respect to time at any given locality is that of a strongly exponential decrease from approximately the time of ice removal in that locality to the present day. Curves for different localities are quite similar although displaced in time, and this displacement correlates with the time of deglaciation of each locality. Therefore, these uplift curves should bear a distinct relationship to the perimeter of the Wisconsin ice sheet, if one may assume that shrinkage of the ice sheet took place uniformly, in broad terms, along all radii. Such an assumption appears to be justified by present knowledge of the southern border of the Wisconsin Laurentide ice sheet.

This relationship of uplift to deglaciation in the Great Lakes area is applied to areas of Arctic Canada in an attempt to learn more about the limits of Wisconsin glaciation in that area. Numerous raised marine features have been observed throughout the Canadian Arctic Archipelago and mainland, and recently some of them have been dated by radiocarbon. Curves constructed from these data, although fragmentary, show a systematic relationship to the Great Lakes curves and suggest that the Wisconsin ice border lay perhaps farther out onto the Arctic Archipelago than is interpreted from geomorphic studies.

This study was initiated in connection with the writer's research on the history of the Great Lakes (Farrand, 1959, 1960). At first radiocarbon-dated sequences of beach features in the Arctic were considered simply as a means of long-distance correlation of such features, which is nearly impossible by any other means at the present time. The similarity of uplift curves from the Great Lakes and the Arctic was immediately apparent, and a genetic relationship suggested itself when the curves were compared to the extent of glaciation.

* Lamont Geological Observatory Contribution No. 526.

181

ACKNOWLEDGMENTS

This paper was made possible by a grant from the U. S. Steel Foundation to Lamont Geological Observatory, Columbia University and from Air Force Cambridge Research Laboratories, Office of Aerospace Research, Contract AF 19(604)7442. Valuable criticism was given by William L. Donn and other colleagues at Lamont. Wallace S. Broecker kindly read the manuscript.

UPLIFT AROUND LAKE HURON

Lake Huron, in the center of the Great Lakes system, has a longer and better known record of glacial lake succession than any other basin. Glacial Lakes Maumee through Algonquin, as well as the postglacial Nipissing Great Lakes and Lake Algoma, have left their shorelines around this basin. In addition, the very important outlet channels at Port Huron, Michigan, and North Bay, Ontario, are integral to the Lake Huron basin. For these several reasons four localities within the Huron basin are considered in this paper: Port Huron, Sault Ste. Marie, Cape Rich, and North Bay (fig. 1).

Fig. 1. Location map for stations discussed in text.

For each locality the total uplift since a given time has been plotted for several lake stages. In order to plot these points the following data for each locality are necessary: (a) the present elevation of the shoreline of the lake stage being considered: this elevation has been observed in the field in most cases, but has to be obtained by extrapolation in one instance; (b) the original elevation of that shoreline before uplift, which is known from the still horizontal portions of that same beach in the southern parts of Lakes Huron and Erie; and (c) the age of the shoreline as determined directly or indirectly from radiocarbon dating of associated organic materials. Then, (a) minus (b)

Table 1

Glacial Lake chronology of Lake Huron Basin

	Lake stage original elevation (after Hough, 1958)		Time of maximum water level	Source of date
	Lake Huron	580		
	Algoma	595	3200 B.P.	(M-659, Crane & Griffin, 1960)
	Nipissing	605	4200	this report & Dreimanis (1958)
Post-Algonquin low stages	Stanley	180	>9500	Terasmae & Hughes (1960)
	Korah	390(?)		
	Sheguiandah	?		
	Payette	465(?)		
	Cedar Point	493(?)		
	Penetang	510(?)	>9560	GRO-1926, Terasmae & Hughes (1960)
	Wyebridge	540(?)		
	Algonquin	605	10,500	this report, estimated
	Lundy	620		
	Grassmere	640		
	Warren III	675	10,800	Valders glacial maximum (Flint,
	Wayne	655		1957)
	Warren II	682		
	I	690		
	Whittlesey	738	12,500	based on S-31 (below)
	III	695		
	Arkona II	700	12,660	S-31, McCallum (1955)
	I	710		
	III	790		
	Maumee II	760	ca. 14,000(?)	estimated, this report (cf. W-198, Rubin & Suess, 1955)
	I	800		

is the amount of uplift since the time (c) when that lake stage existed. Table 1 gives the lake succession in the Lake Huron basin and the absolute ages of certain stages as they are now known. Table 2 lists the total uplift at each locality since several different lake stages. The lake succession and amounts of uplift are mainly from Hough (1958). The data given in these two tables form the basis for uplift curves in figure 2.

Some explanation of these data is necessary, however, before proceeding farther. First, the time at which the Nipissing Great Lakes reached their maximum elevation has been determined by calculating the time necessary for water which covered C-14 dated logs at Blackwell, Ontario, to rise an additional 19 feet. This method was used by Dreimanis (1958) and is modified slightly by me to accord with more recent data. The date thus obtained is 4200 ± 270 B.P.[1]

[1] B.P.—before present.

TABLE 2

Uplift in Lake Huron basin

Lake Stage		Total uplift since given lake stage (feet)		
	Port Huron	Cape Rich	Sault	North Bay
Algoma	0	?	25	?
Nipissing	0	27	45	95
Stanley	0	?	?	520
Payette	0	159	320(?)	?
Penetang	0	174	370	?
Algonquin	0	199	410	960*
Warren (I?)	17			
Whittlesey	27			
Maumee III	60			
Maumee I	60			

* Extrapolated, not observed.
Data assembled from Leverett and Taylor (1915), Stanley (1936, 1937), and Hough (1958).

The Algoma stage of the Great Lakes has been dated recently in connection with an excavation at Saginaw, Michigan, by University of Michigan archeologists and the writer. Here an Indian burial in the crest of an Algoma beach deposit has been dated 3170 ± 300 B.P. (M-659, Crane and Griffin, 1960). Unpublished radiation absorption studies by Lewis Binford of the University of Michigan indicate that this burial has not been submerged. There-

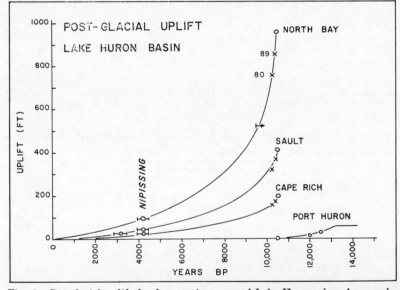

Fig. 2. Postglacial uplift for four stations around Lake Huron plotted as total uplift since a given time. Circles indicate primary control points. One standard deviation is shown by horizontal bracket through each circle controlled by radiocarbon dating. Other circles are estimated datings. X's indicate secondary control as explained in the text and derived from table 3.

fore, Lake Algoma waters must have abandoned the 595-foot shoreline prior to 3200 B.P.

Glacial Lake Algonquin has not been directly dated, but there is abundant indirect chronological information. Algonquin shorelines cut across Valders substage till in the Lake Michigan basin and are, therefore, younger than the Valders maximum, dated about 10,800 B.P. (Flint, 1957). Radiocarbon dates from Manitoulin Island (northern Lake Huron) and from near North Bay, Ontario, show that not only Lake Algonquin but also several post-Algonquin lake stages (Wyebridge through Stanley) terminated prior to 9500 B.P. The most critical of these dates relates to a peat bog, 9500 years old, near North Bay (Terasmae and Hughes, 1960). This bog must postdate deglaciation of the North Bay outlet channel, and it was this deglaciation that initiated the Stanley low-water stage. Terasmae and Hughes (1960) calculated three possible dates for this deglaciation: 9275, 9715, or 10,970 B.P. The middle date accords best with other events; the oldest date is considerably too old because it permits no time for the Valders ice advance.

To allow for at least six or seven separate lake stages between 10,800 and 9500 B.P., I have estimated the date of Glacial Lake Algonquin to be about 10,500 B.P., which is considerably older than previous suggestions (Hough, 1958).

Determination of the amount of uplift is straightforward, as explained above, except for North Bay. The Lake Algonquin shoreline did not extend as far north as North Bay, so its elevation there is imaginary and must be determined by extrapolation in order to complete the North Bay curve. Chapman (1954) has traced the Algonquin beach along the east shore of Georgian Bay to Bernard Lake, only 40 miles south of North Bay, where it lies about 1240 feet above sealevel. By projecting Chapman's curve, the elevation of the Algonquin beach at North Bay would be about 1565 feet, which is 960 feet above the original level of Glacial Lake Algonquin.

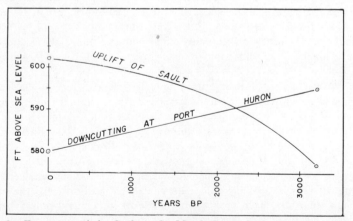

Fig. 3. Emergence of the Sault in St. Mary's River, Sault Ste. Marie, Ontario. The uplift curve is taken from figure 2, and the downcutting at Port Huron since the Algoma stage (3200 B.P.) is plotted linearly against time. The curves intersect at 591 feet above sealevel and about 2200 B.P. (250 B.C.).

The amount of uplift of the Lake Stanley beach can be determined only at North Bay. The North Bay outlet that drained Lake Stanley now is 700 feet above sealevel. The original level of the Stanley low stage, as determined from lake bottom topography in the Mackinac Straits area (Hough, 1958), was about 180 feet above sealevel. Therefore, North Bay has risen about 520 feet since the inception of Lake Stanley.[2]

Further control for the Sault and Cape Rich curves in figure 2 was derived from Stanley's (1936, 1937) work on the post-Algonquin lake stages. He showed a slight decrease in slope of successively younger water planes in the Algonquin to Payette series. Table 3, which is assembled from Stanley's data and additional information from Hough (1958, p. 231-232), shows that 93 percent of total post-Algonquin uplift has occurred since the Wyebridge stage, 89 percent since the Penetang, and 80 percent since the Payette stage. After having drawn the North Bay curve (fig. 2) from data presented in table 2, one can read from that curve the times since which 93, 89, and 80 percent of post-Algonquin uplift occurred. By this means the dates of the Wyebridge, Penetang, and Payette stages are obtained (only the latter two have been plotted) and, coupled with uplift data for these stages in table 2, additional control is available for important segments of the Sault and Cape Rich curves.

Uplift at Port Huron is less well known: the record is shorter, uplift terminated much earlier than at the other stations, and absolute dating is less satisfactory. However, a portion of the uplift curve can be drawn. The earliest shorelines recorded at Port Huron are the three Lake Maumee beaches which have been uplifted 60 feet and show no differential uplift from earliest to latest Maumee times. The Maumee lakes came into existence upon retreat of the ice sheet from the Fort Wayne moraine (Cary subage) and persisted through the advance to and retreat from the subsequent Defiance moraine. The Fort Wayne moraine must be younger than the Wabash moraine which is dated 14,300 ± 450 years (W-198, Rubin and Suess, 1955). Furthermore, the Maumee lakes must be older than Glacial Lake Arkona, which is dated 12,660 ± 440 B.P. (S-31, McCallum, 1955). Therefore, Lakes Maumee I, II, and III can be roughly bracketed between 14,000 and 13,000 B.P.

Glacial Lake Whittlesey is the next stage younger than Lake Arkona, perhaps dating about 12,500 B.P., and its shoreline at Port Huron has been uplifted 27 feet.

The Warren beach at Port Huron now lies at 707 feet above sealevel, but it is not clear which of the three Warren stages (original elevations: 690, 684, and 675 feet) is represented. If one assumes the 707-foot beach to be Warren I, which probably occurred about 12,000 B.P. (post-Whittlesey, pre-Two Creeks), a smooth curve can be drawn for Port Huron (fig. 2). The Grassmere, Lundy, and Algonquin beaches suffered no uplift here.

Thus, we see that the rate of postglacial uplift (rebound) was very great at the time of deglaciation but that it decelerated rapidly toward the present day. Because there are large gaps among some of the control points, the possibility exists that this uplift was, in fact, not a continuous phenomenon as the

[2] Melhorn's (1959) estimate of 152 feet for the original level of Lake Stanley is not valid because he assumes no uplift whatsoever occurred between Lake Algonquin and Lake Stanley time.

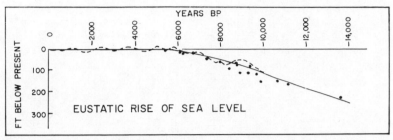

Fig. 4. Eustatic rise of sealevel in late glacial and postglacial time as interpreted by Godwin, Suggate, and Willis (1958, solid dots) and Fairbridge (1958, 1961, dashed curve). The present writer has used only the mean (solid) curve for this paper.

curves in figure 2 imply. Step-wise uplift movements are possible, as suggested by Flint (1957, p. 254-255), although the necessity for stable periods are by no means proven. For example, Hough (1958, p. 261) explains a mechanism by which the strong Nipissing beach—evidence of a stable period for Flint— could have been produced during continuing uplift. In either case, these curves apparently depict the *mean course* of postglacial rebound around the Lake Huron basin during the past 10,500 years.[3]

GREAT LAKES CHRONOLOGY

In light of these uplift curves certain other aspects of the history of the Great Lakes can be reconstructed. As previously discussed the extremely low Chippewa-Stanley stages are now known to have occurred about 9600 B.P. I have discussed elsewhere (Farrand, 1960) evidence for the contemporaneous Hougton low stage in the Lake Superior basin, but without the uplift curve for the controlling outlet at Sault Ste. Marie it was not possible to determine the original water level of the Houghton stage. Now, figure 2 shows that the Sault was about 260 feet lower, or about 340 feet above sealevel, at 9600 B.P. There- fore, the Houghton low stage was originally about 340 feet above sealevel and was separated from Lake Stanley in the Huron basin by the ancestral St. Mary's River which descended 160 feet, probably through a series of falls and rapids.

As North Bay was uplifted, lake waters rose and again flooded the Sault area, bringing the lakes in the Superior and Huron basins to a confluent level. The final emergence of the bedrock barrier at Sault Ste. Marie has occurred since the Algoma stage (3200 B.P.). The Algoma water level was controlled by the outlet at Port Huron at about 595 feet above sealevel. Since 3200 B.P. the

[3] Confirmatory evidence of the shape of these curves comes from the Mackinac City, Michigan, area where logs have been buried by shore sediments during the rise of water level to the Nipissing beach. Their average date is 5460 ± 425 B.P. (M-855, J. B. Griffin, personal communication). The logs were found 16 feet above present lake level (or 597 feet above sealevel) and the Nipissing beach at Mackinac City is 629 feet above sealevel. The apparent 32-foot rise in water level is not the total rise, however, because the land was also rising at this place. Mackinac City lies nearly on the same isobase as Cape Rich, where about 14 feet of uplift occurred between 5460 and 4200 B.P. (fig. 2). Therefore, the total rise was 32 plus 14 feet, or 46 feet. This total rise of water level should be equal to the uplift of the controlling outlet at North Bay. And, from the North Bay curve (fig. 2) we see that about 48 feet of uplift occurred there between 5460 and 4200 B.P.

TABLE 3

Comparative uplift of Post-Algonquin beaches

Area	Lake Algonquin orig. 605 ft		Wyebridge Stage orig. 540 ft			Penetang Stage orig. 510 ft			Payette Stage orig. 465 ft		
	now	uplift	now	uplift	% post-Algonquin uplift	now	uplift	% post-Algonquin uplift	now	uplift	% post-Algonquin uplift
Giants' Tomb Island	875	270	785	245	91	748	238	88	686	221	82
Mackinac Island	812	207	722	182	88*	—	—	—	—	—	—
Sault Ste. Marie	1015	410	935	395	96	880	370	90	785	320	78
Little Current (Manitoulin I.)	1013	408	918	378	93	870?	360	88	782	317	78
Cape Rich	801	196	718	178	92	684	174	89	624	159	81
Averages	—		—	—	93	—	—	89	—	—	80

* Perhaps miscorrelated by Hough (1958) and should apply to Penetang stage.
Data are from Hough (1958) and Stanley (1936, 1937).

Port Huron outlet has been cut down to 580 feet, its present level. During this same period of time the Sault has risen from 577 to 602 feet (fig. 2). If the rate of downcutting at Port Huron is plotted linearly against the uplift curve for Sault Ste. Marie (taken from fig. 2)—this is obviously a simplification— the curves (fig. 3) cross at about 2200 B.P. (250 B.C.), which is the approximate date of final separation of Lakes Superior and Huron. Lake level was about 591 feet above sealevel at the moment of separation.

The rapid rate of uplift between 10,500 and 9600 B.P. implies that the several lake stages which occurred in that interval were quite short-lived. Six or more such stages are known from the Lake Huron basin (Hough, 1958) and sixteen to eighteen shorelines occur between the Lake Duluth and Houghton stages in the Superior basin (Farrand, 1960). The average duration of these stages apparently was between 50 and 150 years, and, as would be expected, these shorelines are rarely strongly developed and entirely absent in many places.

UPLIFT IN ARCTIC CANADA

In Arctic Canada many emerged marine features have long been recognized as indicators of relative changes of land and sea. However, until the advent of radiocarbon dating the age (interglacial, glacial, or postglacial) and nature (eustatic or isostatic) of the features were debated. Washburn (1947, p. 58-59) presents a strong case for postglacial isostatic emergence to account for raised beaches 200 or 300 feet above present sealevel throughout the northwestern part of the Arctic archipelago, but he does not suggest the age of this glaciation. Craig and Fyles (1960) suggest that raised marine features on the westernmost Queen Elizabeth Islands are possibly pre-Wisconsin (?). It is now obvious that most, perhaps all, of these marine features are late-Wisconsin or post-Wisconsin in age and, therefore, related to isostatic movements, because eustatic changes of the sea to elevations several hundred feet above the present are unknown along these stable coasts in post-Wisconsin time.

These features—marine strandlines and terraces, marine fossils, limits of wave action as determined by perched boulders, etc.—occur as high as 875 feet above sealevel and are commonly 500 to 700 feet in elevation. They have long defied correlation (but see Craig and Fyles, 1960, fig. 5) because they lack diagnostic fossils, and the relatively few observations are widely scattered through a vast and little explored territory. Elevation alone is not a sufficient criterion because of past sealevel fluctuations and because different areas were deglaciated at different times. However, with C-14 dates, we can begin gross correlation and, furthermore, analysis of uplift rates in a manner similar to that applied to the Great Lakes sheds additional light on the nature of these features and their relationship to the Wisconsin ice sheet.

A number of radiocarbon dates are now available from the Arctic (Craig and Fyles, 1960; Terasmae and Hughes, 1960), but there are relatively few localities from which we have a *series* of dates on marine features. Seven series of from two to eight dates each have been considered here and the results are plotted in figures 5 through 9.

TABLE 4

Dated marine features in Arctic Canada

For locations refer to Figure 1.

Location	Feature	C-14 Date	Strand Line Elev. (ft)			Lab No. and Reference
			Observed	S.L. Corr'n	Corr. Elev.	
A. James Bay						
Missinaibi R.	marine shells	7875±200	400-500	45	445-545	I(GSC)14	Lee, 1960
Opasatika R.	marine shells	7280±80	400-500	30	430-530	GRO-1698	"
Gt. Whale R.	wood in marine silt	3150±50	90	0	90	L-441A	Lee, Eade, & Heywood, 1959
Ft. George	wood in stony silt	3700±130	175-190	0	175-190	L-433A	
B. Keewatin Ice Divide						
Carr Lake	marine shells	6975±250	560	22	582	I(GSC)8	Lee, 1959
C. Southampton Island						
Southampton Island	marine strands	5600±300	170	0	170	S-12	McCallum, 1955
	marine strands	3670±290	105	0	105	S-13	"
	eskimo site	2060±200	45-60	0	45-60	P-62	Collins, 1956
	burned bone	2508±130	70	0	70	P-75	Rainey & Ralph (1959)
	burned bone	2191±120	40	0	40	P-77	"
	burned bone	2632±128	70	0	70	P-76	"
	burned bone	2183±122	70	0	70	P-74	"
D. Somerset & Prince of Wales Islands						
Somerset I.	marine shells	7150±350	100	26	126	L-571A	J. B. Bird, unpub.
Prince of Wales I.	marine shells	9200±160	370	85	455	L-571B	"

TABLE 4 (Continued)

Location	Feature	C-14 Date	Strand Line Elev. (ft)			Lab No. and Reference
			Observed	S.L. Corr'n	Corr. Elev.	
E. Melville Peninsula						
Igloolik	eskimo site	3700±150	172	0	172	K-505	Tauber, 1960
	eskimo site	600±150	26	0	26	K-504	"
	ivory	3958±168	165	0	165	P-207	Rainey & Ralph (1959)
	antler	3560±123	165	0	165	P-208	"
	ivory	3906±133	165	0	165	P-209	"
	antler	2898±136	145	0	145	P-210	"
	antler	2354±135	80	0	80	P-211	"
	antler	2404±137	70	0	73	P-212	"
	ivory	2910±129	70	0	73	P-213	"
F. Coronation Gulf						
	marine shells	9100±180	495+	80	575+	I(GSC)16	Craig & Fyles, 1960
	shore deposits	8290±330	320	54	374	I(GSC)13	"
G. Northwest Victoria Island						
	marine shells	12,400±320	230	200	430	I(GSC)18	Craig & Fyles, 1960
	marine shells	8895±220	25	75	100	I(GSC)20	"
H. Ellesmere Island						
northern coast	spruce wood	980±100	20	0	20	L261A	Crary, 1960
	spruce wood	2190±150	23	0	23	L261B	"
	tamarack wood	6050±200	100	5	105	L261C	"
	marine shells	7200±200	125	26	151	L248A	"
	marine shells	7200±250	200	26	226	L248B	"

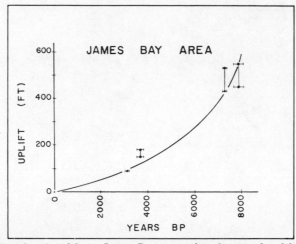

Fig. 5. Postglacial uplift in James Bay area plotted as total uplift since a given time. Horizontal brackets indicate one standard deviation for C-14 datings. Vertical brackets indicate uncertainty of former sealevel position relative to the dated feature. For source of data, see table 4.

The Arctic uplift curves are based on radiocarbon-dated marine features (table 4), some of which involve assumptions in order to approximate the sea level at the time of their formation or deposition. In all cases, I have taken the sealevel value assumed by the original investigator.

Eustatic changes of sealevel that were contemporaneous with this post-glacial uplift must also be considered. The dated marine features used in this analysis must obviously be related to sealevel of the time at which they formed (not present-day sealevel) if we are to deduce the total uplift. Recent compilations of sealevel rise by Godwin, Suggate, and Willis (1958) and by Fairbridge (1958, 1961) have been plotted in figure 4. Both interpretations agree very well in generalities, although the second-order fluctuations suggested by Fairbridge (dashed curve) seem perhaps too exact for this stage in our knowledge. For the purpose of this paper I have drawn a mean line (solid)

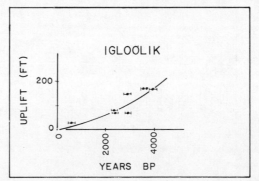
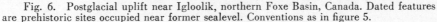

Fig. 6. Postglacial uplift near Igloolik, northern Foxe Basin, Canada. Dated features are prehistoric sites occupied near former sealevel. Conventions as in figure 5.

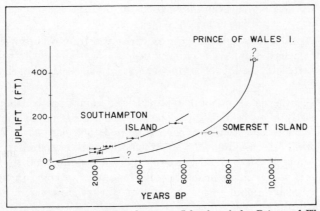

Fig. 7. Postglacial uplift of Southampton Island and the Prince of Wales-Somerset Island area. Conventions as in figure 5.

through Godwin's and Fairbridge's data for the period prior to 5500 B.P. For the period since 5500 B.P. I have used no eustatic correction. Thus, the present elevation of a raised marine feature *plus* the eustatic correction from figure 4 gives the total uplift of that feature since it formed.

A short description is necessary for the individual curves in figures 5 through 9. The basic data for these curves is given in table 4 and the locations are shown on figure 1. The limits of one standard deviation for the C-14 date are shown by the short horizontal lines; the vertical range of uncertainty in estimation of sealevel as given in table 4 is shown by short vertical lines.

The four features from the James Bay area (fig. 5) have a wider geographic scatter than the data of any other curve, but they lie in a central area with respect to the former ice sheet and were probably deglaciated essentially at one time. Only the wood from stony silt near Fort George does not lie on

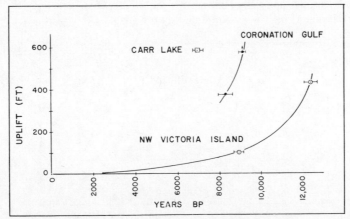

Fig. 8. Postglacial uplift at Carr Lake on Keewatin Ice Divide; Coronation Gulf, N.W.T.; and northwestern Victoria Island. Conventions as in figure 5.

the curve as drawn; perhaps the implications of that specimen need to be re-examined.

The information from Igloolik (fig. 6) is derived from prehistoric Eskimo sites which are believed to have been very near the seashore of their time, but above high tide (Meldgaard, 1960). Only one point out of eight lies significantly below the mean curve.

In figure 7, the data for Southampton Island comes partly from marine strands and partly from Eskimo sites. The internal consistency is quite good. For Prince of Wales and Somerset Islands there is one date each. Although these neighboring islands should be expected to have very similar rebound histories, the simplest curve drawn through these two points and the zero co-ordinate is slightly discordant in detail with the curves grouped in figure 10, from which it has been omitted for the sake of clarity. However, it does definitely agree with the general pattern.

As only two points on a rather steep portion of the Coronation Gulf curve (fig. 8) are available, it is probably not wise to continue that curve downwards at the present time. Older marine shells from Coronation Gulf dated 10,215 and 10,530 B.P. occur in bottom clay related to a much higher but unspecified strand (Craig, 1960). The record from northwestern Victoria Island is the longest from either Arctic Canada or the Great Lakes; yet it gives the same pattern of decreasing uplift. Only one dated marine feature is available from the Carr Lake area on the Keewatin Ice Divide (Craig and Fyles, 1960). It is included here simply to indicate the order of magnitude of the uplift in that central region.

Northernmost Ellesmere Island (fig. 9) is far removed from the main "centers" of the Wisconsin Laurentide ice sheet, but three of the four dated features there give a curve very similar to those from lower latitudes. The two 7200 B.P. dates for strandlines 75 feet apart vertically cannot obviously both be correct. The 125-foot strand dated 7200 B.P. is favored because it gives a curve more in accord with all other curves presented in this paper. Shells of younger age might have been blown by strong arctic winds onto the upper (200-foot) beach after it was abandoned by the sea.

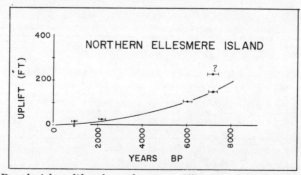

Fig. 9. Postglacial uplift of northernmost Ellesmere Island. Conventions as in figure 5. Of the two features which dated 7200 B.P. the lower one, 125 feet uplift, appears more conformable with the other data.

INTERPRETATION

The individual curves of figures 5 through 9 have been grouped for comparison with the Great Lakes curves in figure 10. Only the Prince of Wales-Somerset Island curve (fig. 6) has been omitted. The fact that all these curves nest together conformably seems to indicate a systematic relationship. Although the phenomenon of glacio-isostatic rebound is now widely accepted (Flint, 1957, chap. 14), the relationships shown by these curves strengthen it further.

In the first place, areas more centrally located, with respect to the Wisconsin continental ice cover in North America, show their most intensive uplift later than do the more peripheral areas. This most intense uplift is nearly coincident with the time of deglaciation in a given place. For example, at North Bay the uplift curve (A in fig. 10) is nearly vertical between 10,500 and 10,000 B.P., immediately *prior* to the actual moment of removal of the ice sheet (ca. 9600 B.P.). In the Coronation Gulf area, deglaciation must have taken place prior to 10,530 B.P., perhaps immediately prior to that date; and the uplift curve (no. 5 in fig. 10), although limited in length, shows rapid uplift between 10,000 and 9000 B.P. We know that the James Bay area was ice-free by 7875 B.P. (I-GSC-14, Terasmae and Hughes, 1960), a date from marine shells along the Missinaibi River, but not before about 9000 B.P., an estimate relative to the opening of the North Bay outlet at about 9600 B.P. The steepest part of the James Bay curve (no. 3 in fig. 10) would lie about 8500 to 8000 B.P., essentially coincident with deglaciation. Although only one date is avail-

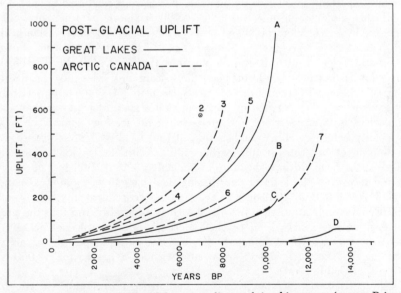

Fig. 10. Composite of all uplift curves discussed in this paper (except Prince of Wales-Somerset Island). Control is not shown for the sake of clarity. Solid curves are from Lake Huron (fig. 2); dashed curves are from Arctic Canada (figs. 5 through 9). A—North Bay, B—Sault Ste. Marie, C—Cape Rich, D—Port Huron, 1—Igloolik, 2—Carr Lake, 3—James Bay, 4—Southampton Island, 5—Coronation Gulf, 6—northern Ellesmere Island, 7—northwest Victoria Island.

able from the Keewatin Ice Divide (Carr Lake), its position (no. 2) in figure 10 suggests intensive uplift around 7000 B.P., the approximate time at which that area was freed from ice (Lee, 1959). Moreover, the historic rate of uplift at Churchill, Manitoba, near the Keewatin Ice Divide, is approximately 3 feet per century (Bird, 1959; not plotted in figure 10) which nearly coincides with the Igloolik curve and might also be considered a logical extension of the Carr Lake date.

The second relationship that appears in figure 10 is that all of these curves, in so far as they are known, can be drawn with essentially the same form,[4] but they are simply displaced in time as a function of time of deglaciation. In the Great Lakes area we know that the three localities best documented lay at least 250 miles inside the Wisconsin border. The overlying thickness of Wisconsin ice was probably nearly the same at Cape Rich, Sault Ste. Marie, and North Bay and was perhaps 10,000 feet thick—both conclusions by analogy with the present Greenland and Antarctic ice sheets in light of Nye's (1959) theoretical conclusion that the thickness of an ice sheet is only very slightly dependent on its accumulation (i.e., climatic) regime. It is plausible that a similar thickness of ice overlay James Bay, the Keewatin Ice Divide, Southampton Island, Igloolik, and Coronation Gulf. Therefore, it is unlikely that the curves in figure 10 reflect differential ice thickness at the various localities. Rather, the curves are interpreted as showing nearly identical isostatic response to nearly identical ice loads, but—as mentioned before—displaced in time.

Some important corollaries follow from this second relationship. (1) Ice loads of this same magnitude (around 10,000 ft. more or less) probably covered northwestern Victoria Island and northernmost Ellesmere Island, although both of these areas are very close to the "Wisconsin" ice border shown by Craig and Fyles (1960). It is quite possible, therefore, that the Wisconsin ice cover lay farther north and northwest than Craig and Fyles have drawn it, perhaps coinciding with their area of "pre-Wisconsin (?)" glaciation. (2) If the Wisconsin ice sheet reached or approximated a state of isostasy, then the major part of isostatic recovery upon removal of the ice load took place in a given locality before complete deglaciation of that locality. Unfortunately we know nothing of the rate of this earlier uplift. With an ice load of about 10,000 feet, there should have been approximately 3000 to 3300 feet of subsidence and recovery—the ratio of densities of glacier ice to average rock being about 1:3. Yet less than 1000 feet of recovery is recorded at any place, and the uplift curves indicate that uplift has slowed considerably and that the pre-glacial position has nearly been reached. Perhaps isostasy was not achieved by the Wisconsin ice sheet, although the ice sheet covered peripheral areas such as the St. Lawrence valley, apparently without a break, for 40,000 to 50,000 years (Gadd, 1960).

Isostatic recovery may be plotted as in figure 11 to show in a rough, schematic way the nature of crustal rebound. The dates and amounts of uplift are taken directly from figure 10, and the geographic (horizontal) axis has

[4] Uplift curves of the same general form, i.e., showing a strong decrease in rate of uplift with time, have been constructed for central Spitsbergen (Feyling-Hanssen and Olsson, 1959) and for the coast of Maine (Bloom, 1959, 1960).

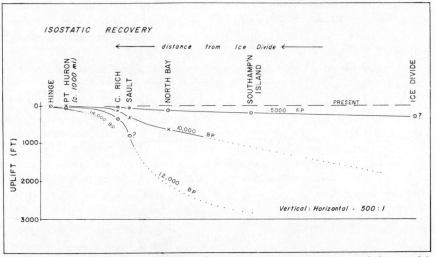

Fig. 11. Schematic representation of isostatic recovery as interpreted from uplift curves in figure 10, assuming the Wisconsin ice sheet was in complete isostatic adjustment and depressed the crust approximately 3000 feet. The approximate horizontal scale is distance from a given station to the nearest ice divide, either in Labrador (Schefferville) or in Keewatin. The dotted portions of each curve have no control whatsoever.

been arranged with respect to distance from a line connecting the Keewatin (Lee, 1959) and Labrador (Ives, 1960) ice divides across the center of Hudson Bay. (The Hudson Bay area holds the record for the greatest recorded post-Wisconsin rebound, about 900 feet at Richmond Gulf, and is perhaps the true "center" of the Wisconsin ice sheet. The Keewatin and Labrador ice divides mark the position of the last remnants of the Wisconsin ice sheet and are interpreted here as roughly the limits of this Hudson Bay ice center.) The distances between points around Lake Huron were measured orthogonally to known isobases in that region. The solid lines have a factual basis, but the dotted extrapolations have no control. This construction assumes complete isostatic compensation of an ice sheet 10,000 feet thick, resulting in about 3000 feet of subsidence.

It should be noted that the horizontal scale does not extend to the margin of the Wisconsin ice sheet, rather it stops some 200 miles short of the margin, acknowledging the fact that we know almost nothing of subsidence or rebound in this extreme peripheral fringe. Therefore, figure 11 shows a radial line through the central essentially uniformly thick portion of the Wisconsin ice sheet. Although this scheme for isostatic recovery is strictly a first approximation, it supports the idealized concept of Flint (1957, fig. 14-7) and differs from Lougee's concept (1953, fig. 3) by the absence of sharp "hinge lines."

SUMMARY

In conclusion, the radiocarbon-dated uplift curves presented in this paper show:

1) a rapidly decreasing rate of postglacial uplift in a given locality immediately following deglaciation of that locality;

2) similar curves from all localities, but these curves are displaced in time. This displacement is correlated with the pattern of deglaciation;

3) an ice cover over Victoria Island and northern Ellesmere Island comparable in thickness to that over the Great Lakes; and

4) that the major part of total isostatic adjustment occurred before complete disappearance of the ice sheet from a given locality.

The scheme presented here requires additional testing which should be forthcoming soon. Numerous geolgical and geographical expeditions in Arctic Canada are, among other things, observing features of marine submergence and collecting materials for radiocarbon dating. These additional data will either verify this interpretation of uplift or show that the data presented in this paper are not a fair sample. However, it seems significant that all the data available at the time of this writing are concordant with the interpretation offered here.

References

Bird, J. B., 1959, Recent contributions to the physiography of northern Canada: Zeitschr. Geomorphologie, v. 3, p. 151-174.

Bloom, A. L., 1959, Late Pleistocene changes of sealevel in southwestern Maine: Office of Naval Research, Final report, project no. NR 388-040, 143 p.

———— 1960, Pleistocene crustal and sea-level movements in Maine [abs.]: Geol. Soc. America Bull., v. 71, p. 1828.

Chapman, L. J., 1954, An outlet of Lake Algonquin at Fossmill, Ontario: Geol. Assoc. Canada Proc., v. 6, p. 61-68.

Collins, H. B., 1956, T1 Site at Native Point, Southampton Island, NWT: Alaska Univ. Papers, v. 4, p. 63-89.

Craig, B. G., 1960, Surficial geology of north-central District of MacKenzie, Northwest Territories: Canada Geol. Survey Paper 60-18, 8 p.

Craig, B. G., and Fyles, J. T., 1960, Pleistocene geology of the Canadian Arctic: Canada Geol. Survey Paper 60-10, 21 p.

Crane, H. R., and Griffin, J. B., 1960, University of Michigan radiocarbon dates V: Am. Jour. Sci. Radioc. Supp., v. 2, p. 31-48.

Crary, A. P., 1960, Arctic ice island and ice shelf studies, pt. II: Arctic, v. 13, p. 32-50.

Dreimanis, Aleksis, 1958, Beginning of the Nipissing phase of Lake Huron: Jour. Geology, v. 66, p. 591-594.

Fairbridge, R. W., 1958, Dating the latest movements of the Quaternary sea level: New York Acad. Sci. Trans., ser. 2, v. 20, p. 471-482.

———— 1961, Eustatic changes in sea level, *in* Ahrens, L. H., et al., eds., Physics and chemistry of the Earth, v. 4: Oxford, Pergamon Press, p. 99-185.

Farrand, W. R., 1959, Pleistocene beaches along the north shore of Lake Superior [abs.]: Geol. Soc. America Bull., v. 70, p. 1600.

———— 1960, Former shorelines in western and northern Lake Superior basin: Unpub. Ph.D. dissertation, Univ. of Michigan, Ann Arbor, 226 p.

Feyling-Hanssen, R. and Olsson, Ingrid, 1959, Five radiocarbon datings of Post-glacial shorelines in central Spitsbergen: Norsk geog. tidsskr., v. 17, p. 122-131.

Flint, R. F., 1957, Glacial and Pleistocene geology: New York, John Wiley and Sons, 553 p.

Gadd, N. R., 1960, Surficial geology of the Becancour map-area, Quebec, 31 I/8: Canada Geol. Survey Paper 59-8, 34 p.

Godwin, H., Suggate, R. P., and Willis, E. H., 1958, Radiocarbon dating of the eustatic rise in ocean-level: Nature, v. 181, p. 1518-1519.

Hough, J. L., 1958, Geology of the Great Lakes: Urbana, University of Illinois Press, 313 p.

Ives, J. D., 1960, The deglaciation of Labrador-Ungava—an outline: Cahiers de Géographie de Québec, no. 8, p. 323-343.

Lee, H. A., 1959, Surficial geology of southern District of Keewatin and the Keewatin Ice Divide, Northwest Territories: Canada Geol. Survey Bull. 51, 42 p.

———— 1960, Late glacial and postglacial Hudson Bay sea episode: Science, v. 131, p. 1609-1611.

Lee, H. A., Eade, K. E., and Heywood, W. W., 1959, Surficial geology Sakami Lake map area: Canada Geol. Survey Map 52-1959.

Leverett, Frank, and Taylor, F. B., 1915, The Pleistocene of Indiana and Michigan and the history of the Great Lakes: U. S. Geol. Survey Mon. 53, 529 p.

Lougee, R. J., 1953, A chronology of postglacial time in eastern North America: Sci. Monthly, v. 76, p. 259-276.

McCallum, K. J., 1955, Carbon-14 age determinations at the University of Saskatchewan: Royal Soc. Canada Trans., ser. 3, v. 49, sec. 4, p. 31-35.

Meldgaard, Jørgen, 1960, Origin and evolution of Eskimo cultures in the eastern Arctic: Canad. Geog. Jour., v. 60, p. 64-75.

Melhorn, W. N., 1959, Geology of Mackinac Straits in relation to Mackinac Bridge: Jour. Geology, v. 67, p. 403-416.

Nye, John, 1959, The motion of ice sheets and glaciers: Jour. Glaciology, v. 3, p. 493-507.

Rainey, Froelich, and Ralph, Elizabeth, 1959, Radiocarbon dating in the Arctic: Am. Antiquity, v. 24, p. 365-374.

Rubin, Meyer, and Suess, H. E., 1955, U. S. Geological Survey radiocarbon dates II: Science, v. 121, p. 481-488.

Stanley, G. M., 1936, Lower Algonquin beaches of Penetanguishene Peninsula: Geol. Soc. America Bull., v. 47, p. 1933-1960.

———— 1937, Lower Algonquin beaches of Cape Rich, Georgian Bay: Geol. Soc. America Bull., v. 48, p. 1665-1686.

Tauber, Henrik, 1960, Copenhagen radiocarbon dates IV: Am. Jour. Sci. Radioc. Supp., v. 2, p. 12-25.

Terasmae, Jaan, and Hughes, O. L., 1960, Glacial retreat in the North Bay area, Ontario: Science, v. 131, p. 1444-1446.

Washburn, A. L., 1947, Reconnaissance geology of portions of Victoria Island and adjacent regions, Arctic Canada: Geol. Soc. America Mem. 22, 142 p.

Editor's Comments on Paper 17

Washburn and Stuiver: *Radiocarbon-dated Postglacial Delevelling in Northeast Greenland and Its Implications*

By a strange coincidence the year 1962 saw the publication of two papers other than Farrand's on the form of the postglacial uplift curve—that by Lee (*Biult. Perig.*) and the one reprinted here. Furthermore, Paper 18 by Bloom was published the same year and involves a discussion of the complex interactions betwen eustatic sea level changes and glacio-isostatic recovery of the crust in Maine. Prior to the advent of ^{14}C dating, vertical crustal movements within glaciated areas had been studied primarily by the lateral tracing of one or more major strandlines (see Papers 8 through 15). Now, however, the ability to date material in raised, or drowned, marine deposits enabled the study of vertical crustal movement to be made at a single site.

The note by Washburn and Stuiver discusses the construction of a postglacial uplift curve from Meters Vig, Northeast Greenland. The radiocarbon dates are on a variety of shell species and on driftwood. The eustatically corrected uplift curve (Fig. 3) shows that uplift was extremely rapid with about 100 m of recovery occurring in approximately 4000 years. Relaxation time for this curve is only about 1200 years, and it offers a marked contrast to the curves from Arctic Canada (compare with illustrations in Paper 16). The very rapid relaxation of the Greenland curve could imply a relatively low viscosity for the asthenosphere, or it might indicate primary elastic recovery of a site near the inland ice margin, where the elastic properties of the crust become important in determining the deformation.

This Greenland curve has been used extensively to model relaxation processes (e.g., Broecker, 1966, Paper 26) of the earth. It must be noted, however, that the curve has few documented counterparts from the area of the Laurentide Ice Cap, although it is typical of postglacial uplift on Greenland (Lasca, *Arctic*, 1966, and Ten Brink, Ph.D. Thesis, University of Washington, 1971). Figure 4 indicates that the postglacial uplift is well described by an exponential decay equation (i.e., data plot as a straight line on semilog paper).

The stratigraphic relationships of the dates to sea level is not too firmly established in some instances (e.g., Y-704); nevertheless, the points fall close to the straight line on Fig. 4. This paper had a major impact on research and has been frequently quoted in, and used for, other studies.

A. L. Washburn was born in 1911 and received his A.B. from Dartmouth and his Ph.D. from Yale in 1942. He has been involved in Arctic research much of his career; he was Director of the Arctic Institute of North America 1950–1951 and Director of SIPRE 1952–1953. He is currently Director of the Quaternary Research Center, University of Washington, Seattle.

M. Stuiver was born in 1929 in Groningen, where he received his Ph.D. in Biophysics in 1958. He was Director of the Yale University Radiocarbon Laboratory from 1962–1969 and he is now Professor of the Geological Sciences and Zoology, University of Washington. He has published many papers connected with problems of radiocarbon chronology and calibration.

Reprinted from *Arctic*, 15, 66–72 (1962)

Short Papers and Notes 17

RADIOCARBON-DATED POSTGLACIAL DELEVELLING IN NORTHEAST GREENLAND AND ITS IMPLICATIONS

Mesters Vig, Northeast Greenland lies at 72°N. 24°W. on the south side of Kong Oscars Fjord, about 50 km. from the entrance (Fig. 1.). As part of a geomorphic study of the district it was necessary to reconstruct the local history of postglacial delevelling. The following preliminary report summarizes some of the results; a detailed report is planned for publication in *Meddelelser om Grønland*. The radiocarbon dating was carried out by Stuiver, the field work by Washburn.

A number of studies of emerged strandlines have been made in the fiord region of Northeast Greenland (ref. 1, p. 204-22; ref. 2, p. 162-92), but information on absolute dating of emergence has been lacking. Radiocarbon dating of driftwood and shells from emerged marine deposits was therefore essential. A related problem was to determine whether a widespread and locally till-like deposit, containing in places numerous well-preserved shells, had been transported by a glacier, or whether it was an emerged fiord-bottom deposit. A till-like aspect of a fiord-bottom deposit could be due to deposition from debris-loaded icebergs as they floated past sites colonized by molluscs, and solifluction following emergence could have contributed to it. Lithologic criteria were useless, for material carried and deposited by a glacier could have been picked up from the fiord bottom. Even the presence of uncrushed, paired mollusc valves need not be diagnostic in some situations (ref. 3, p. 25-6). However, if the till-like deposit was laid down during a major glacier advance it should be (1) significantly older than fossiliferous deltaic beds deposited after the ice retreated, and (2) much more

nearly of one age and lack any systematic correlation between age and altitude. Radiocarbon dating of the shells should therefore cast light on the origin of the deposit as well as on emergence, without involving a circular argument in fitting the deposit into the delevelling history on the basis of the same dates.

Shells were collected from a number of places and were dated at the Yale Geochronometric Laboratory; a few driftwood specimens were included. The localities, shell identifications, and radiocarbon ages are summarized in Table 1. The ages are plotted against altitude in Figs. 2-4, together with suggested curves.

Several points deserve comment: (1) the plotted altitudes in Fig. 2 are field altitudes (i.e., altitudes uncorrected for eustatic rise of sea-level). The altitudes in Figs. 3 and 4 are adjusted to allow for eustatic rise of sea-level of 0.9 m. per 100 years prior to 6,000 B.P. (ref. 4, Fig. 14, p. 156; ref. 5, pp. 556-7; cf. also ref. 6; ref. 7, Fig. 1, p. 31). Table 1 and Fig. 2 will facilitate application of alternative corrections, based on different interpretations of the rate of rise of sea-level (cf. ref. 8). Comparison of Fig. 2 with Fig. 3 demonstrates that the break in the curve at about 6,000 B.P. is not significantly affected by the correction for rise of sea-level in Fig. 3. (2) The shell dates have been corrected for an apparent age of 550 years, based on modern shells (Y 606) collected in the district and used as a standard. (3) As with all shell dates, there is uncertainty as to the depth at which a mollusc died, and a correlation of the date with a former sea-level involves a possible error of several metres. (4) Field altitudes were determined largely by Paulin aneroid, corrected for changes of pressure and temperature. Since the tidal range is of the order of only 1 m. no attempt was made to distinguish between high-tide and mean-tide levels

66

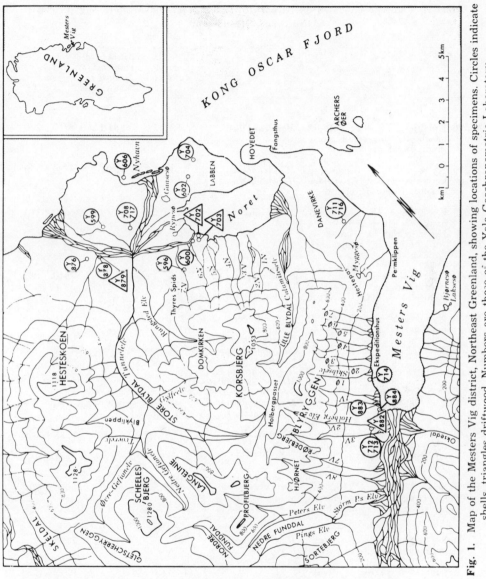

Fig. 1. Map of the Mesters Vig district, Northeast Greenland, showing locations of specimens. Circles indicate shells, triangles driftwood. Numbers are those of the Yale Geochronometric Laboratory.

in computing altitude. Accuracy of measurement for altitudes less than 5 m. is estimated to be ±0.5 m.; for those between 5 and 25 m. ±1 m.; and for those above 25 m. ±2 m. (5) The vertical sides of the squares associated with the specimens as plotted in the figures represent the altitude range (corrected in Figs. 3 and 4 for eustatic rise of sea-level) within which the specimens were collected; the

horizontal sides represent the statistical error of the age. (6) Four dates are derived from driftwood, but only two of these (Y 702 and Y 703) represent driftwood clearly related to emerged strandlines. In these two cases the wood was from logs lying at, and parallel to, the inner ridge of the strandlines. However, the strandlines were low nips in unconsolidated material, and because they may have been associated

Table 1. Radiocarbon-dated shells and driftwood from the Mesters Vig district, Northeast Greenland.

No.	Locality	Species	Field altitude m.	Altitude (m.) corrected for eustatic rise of sea-level	C-14 age years B.P.†
Y 596	Korsbjerg, NE slope, deltaic* bench. At surface of stony sand, abundant	*Mya truncata* L.	59 ± 2	84 ± 2	8760 ± 250
Y 599	Nyhavn hills, NW side trap knob ms* 112 m. At surface of till-like deposit, abundant	*Mya truncata* L. *Hiatella arctica* L.	66-69 ± 2	91-94 ± 2	8780 ± 250
Y 600	Korsbjerg, NE slope, cut bank of emerged delta at Noret outlet of Tunnelelv. *In situ* in silt, shells with both valves	*Mya truncata* L.	2-4 ± 0.5	8-10 ± 0.5	6690 ± 210
Y 602	Labben hills, cut bank of stream adjacent to experimental site 5. In stony silt	*Hiatella arctica* L.	7-8 ± 1	21-22 ± 1	7540 ± 180
Y 606	Nyhavn, S cove. Dredged, modern shells used for standard	*Astarte borealis* Schumacker *Astarte crenata* Gray *Cardium ciliatum* Fabricius *Hiatella arctica* L. *Margarites undalata* Sowerby *Mya arenaria* L.* *Mya truncata* L.	−2 to −15 (estimated)		0 ± 70
Y 702	Korsbjerg, NE slope, inner edge of emerged strandline on delta at Noret outlet of Tunnelelv	(Driftwood)	3 ± 0.5	3 ± 0.5	735 ± 110
Y 703	Same locality as Y 702 but inner edge of another emerged strandline 1 m. above Y 702	(Driftwood)	4 ± 0.5	4 ± 0.5	2980 ± 120
Y 704	Labben hills, adjacent to experimental site 3. In clayey silt containing stones, abundant shell fragments (frost-worked*) in patterned ground	*Mya truncata* L.	31 ± 2	47 ± 2	7730 ± 210
Y 708	Nyhavn hills, cut in deltaic beds at SW base trap knob ms 78 m. At surface	*Mytilus edilus* L.	16-20 ± 1	22-26 ± 1	6670 ± 250
Y 711	Danevirke hills, cut in deltaic beds 0.5 km. SE of trap knob ms 125 m. At surface of sand	*Hiatella arctica* L. *Mya truncata* L.	67-76 ± 2	90-99 ± 2	8500 ± 250
Y 712	Blyryggen, SE slope, cut bank of emerged delta, SW side Holberg Elv. At surface of stony sand, abundant, some shells with both valves	*Mya truncata* L.	45-52 ± 2	67-74 ± 2	8480 ± 140

Y 713	Same locality as Y 712. Abundant, upper limit 3 m. below delta tread.	*Mya truncata* L.	52-57 ± 2	73-78 ± 2	8360 ± 140
Y 714	Blyryggen, SE slope, cut bank of bench on which Expeditionshus is located. At surface of till-like stony sand and silt, abundant	*Hiatella arctica* L. *Mya truncata* L. *Serripes groenlandica* Bruguiere	7-10 ± 1	15-18 ± 1	6910 ± 200
Y 716	Same locality as Y 711. Part of same collection but independent C-14 check	*Hiatella arctica* L. *Mya truncata* L.	67-76 ± 2	92-101 ± 2	8780 ± 210
Y 717	Same locality as Y 708. At surface, profuse	*Mya truncata* L.	16-20 ± 1	22-26 ± 1	6650 ± 200
Y 876	Hesteskoen, NE slope, cut bank in deltaic beds at E base trap knob ms 90 m. *In situ* in sand, shells with both valves	*Hiatella arctica* L. *Mya truncata* L.	29 ± 2	47 ± 2	8000 ± 160
Y 878	Hesteskoen, NE slope, cut bank at N tip 2nd large delta sector NW of Tunnelelv. *In situ* in sand, shells with both valves	*Hiatella arctica* L. *Mya truncata* L.	19-20 ± 1	28-29 ± 1	6950 ± 150
Y 879	As Y 878. At surface	(Driftwood)	20 ± 1	33 ± 1	7460 ± 130
Y 882	Blyryggen, SE slope, channel of 1 V Elv. Partly in till-like deposit	(Driftwood)	4 ± 0.5	4 ± 0.5	5590 ± 140
Y 883	Same locality as Y 882. At surface of till-like deposit, abundant	*Astarte borealis* Schumacker *Clinocardium ciliatum* Fabricius *Hiatella arctica* L. *Mya truncata* L. *Mytilus edulis* L.	3-4 ± 0.5	11-12 ± 0.5	6840 ± 210
Y 884	Blyryggen, SE slope, cut bank by first small stream SW of 1 V Elv. *In situ* in sand, shells with both valves	*Astarte borealis* Schumacker	1-3 ± 0.5	1-3 ± 0.5	4960 ± 320

*ms (map summit) identifies by altitude knobs and hills lacking a name in Fig. 1.
†The radiocarbon half life used for calculation is 5570 years.

with storms, they may have been formed a little above the high-tide level of the time. This may account for the slightly anomalous position of Y 702 and Y 703 above the curve in Figs. 2-4. In Fig. 4 their position is grossly exaggerated by the use of the logarithmic altitude scale. The other two wood specimens (Y 879 and Y 882) may have been derived from somewhat higher altitude by mass-wasting.

The following conclusions can be drawn from Figs. 2-4:

(1) The fossiliferous till-like material is an emerged fiord-bottom deposit. The points of the curve line up too well, and are internally too consistent between the altitudes and associated ages of the till-like material on the one hand and of fossiliferous deltaic deposits of similar age and altitude on the other to permit the interpretation that the till-like deposit was laid down by an advancing glacier prior to deglaciation.

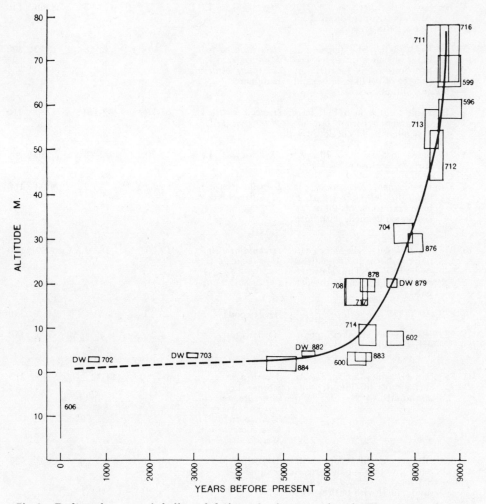

Fig 2. Radiocarbon age of shells and driftwood relative to altitude, Mesters Vig district, Northeast Greenland. Altitudes not corrected for eustatic rise of sea-level. [Y] 884 — Yale Geochronometric Laboratory Number. DW — Driftwood. All other specimens are shells.

(2) The Mesters Vig district was open to the sea and, therefore, deglaciated at least in part by 9,000-8,500 B.P.

(3) The Mesters Vig District has remained largely free of glaciers since about 8,500 B.P. This is significant in view of the fact that valley glaciers, fringed by old moraines, occur nearby today. A sizeable glacier near the head of Mesters Vig bay is only about 8 km. from an emerged delta with shells dated 8480 ± 140 B.P. (Y 712). Therefore, the

climate since about 8,500 B.P. could not have been very much more conducive to glaciation that at present.

(4) Deglaciation of the Mesters Vig district is closely related in time and effect to the Hypsithermal[9].

(5) It follows from (4) that emergence in the Mesters Vig district is probably primarily related to ice thinning and deglaciation and is, therefore, probably due to isostatic adjustment.

(6) The rate of emergence was high

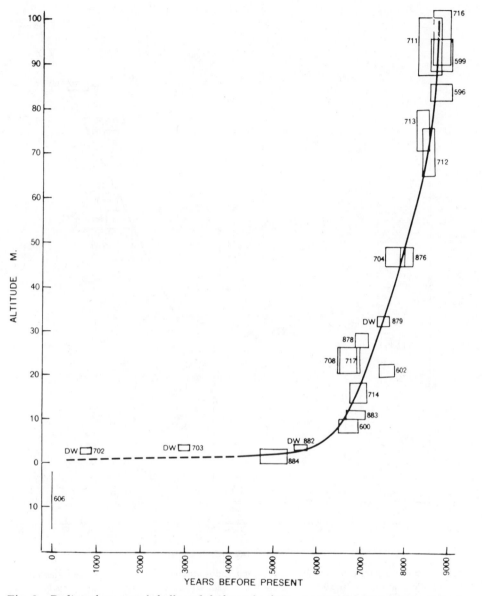

Fig. 3. Radiocarbon age of shells and driftwood relative to altitude, Mesters Vig district, Northeast Greenland. Altitudes corrected for eustatic rise of sea-level; 0.9 m/100 years prior to 6,000 B.P. [Y] 884 — Yale Geochronometric Laboratory Number. DW — Driftwood. All other specimens are shells.

initially, of the order of 9 m. 100 years, and decreased approximately exponentially to about 0.6 m. 100 years, for the interval 9,000 B.P. to 6,000 B.P. It should be emphasized that the absolute value of rate of emergence for the exponential part of the curve depends on the altitude, and comparison with other curves should take this into consideration. Compared with curves of other

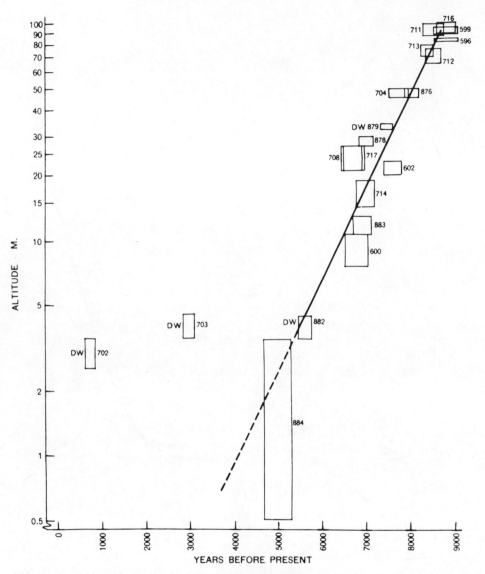

Fig. 4. Semi-logarithmic graph of Fig. 3. Altitudes corrected for eustatic rise of sea-level; 0.9 m./100 years prior to 6,000 B.P. [Y] 884 — Yale Geochronometric Laboratory Number. DW — Driftwood. All other specimens are shells.

regions showing postglacial delevelling, the Mesters Vig district has one of the highest rates of emergence. For the interval 6,000 B.P. to the present the exponential character of the curve disappears, with rates of emergence of perhaps as little as 7 cm./100 years. However, this rate must be regarded as

highly tentative pending further investigations.

(7) The general aspect of the curve in Figs. 2 and 3 is very similar to that deduced for northern Canada[10], and for Spitsbergen (ref. 11, Fig. 1, p. 123; ref. 12, Fig. 9, p. 143).

Editor's Comments on Paper 18

Bloom: *Late-Pleistocene Fluctuations of Sealevel and Postglacial Crustal Rebound in Coastal Maine*

Many papers dealing with postglacial crustal movements in northern North America indicate that postglacial changes of relative sea level were exclusively related to coastal emergence (e.g., Farrand, Paper 16). However, in coastal areas near the margin of the former Laurentide Ice Sheet which were deglaciated fairly early, the relative sea level movements were considerably more complex and involved periods of marine regression and transgression. Bloom's paper from coastal Maine documents such a complex record of sea level changes and presents a series of crustal response models based on (1) the inferred history of relative sea level movements along this stretch of Maine coast from 15,000 BP to the present, and (2) three different eustatic sea level curves. Combining the relative sea level (R) changes with the three eustatic sea level curves (S) results in three separate models of isostatic response (I) based on the relationship $I = R - E$, all measured with respect to present sea level as datum. This general model has been used by Mörner in slightly different form to develop a eustatic sea level curve (see Paper 21).

Bloom's paper is an excellent example of the possible variations in relative sea level caused by variations in the sign and rate of isostatic rebound and eustatic sea level fluctuations. The paper clearly illustrates the importance of determining the "correct" eustatic sea level curve if the true nature of isostatic rebound is to be determined.

A. L. Bloom was born in 1928 and received his B.A. in 1950 from Miami University, Ohio; his M.A. from Victoria University (New Zealand) in 1952; and his Ph.D. from Yale University in 1959. He has been associated with Cornell University's Geology Department since 1960 and is an Associate Professor at that institution. He has written extensively on the question of eustatic sea level changes and hydro-isostatic responses of coasts based on field work in the United States and the Pacific Oecan region.

303

Reprinted from *Amer. J. Sci.*, **261**, 862–879 (1963)

18

LATE-PLEISTOCENE FLUCTUATIONS OF SEALEVEL AND POSTGLACIAL CRUSTAL REBOUND IN COASTAL MAINE

ARTHUR L. BLOOM

Department of Geology, Cornell University, Ithaca, New York

ABSTRACT. . Deglaciation and postglacial crustal rebound near Portland, Maine took place during a time of generally rising sealevel. The relative positions of land and sea-level at several points in time are indicated in part by: (1) glacial-fluvial deposits buried by marine silty clay to depths of 70 feet below present sealevel, (2) radiocarbon-dated shells from emerged marine beds 160 to 250 feet above present sealevel, (3) pollen-stratigraphic profiles of coastal sites underlain by marine sediment, which lack the late-glacial and early postglacial portion of the record, and (4) submerged tree stumps dating back to 4200 years B.P. The position of a reference point in Maine with respect to chang-ing sealevel is graphically shown, using several alternative interpretations of late-Pleisto-cene eustatic rise of sealevel as base lines. Only one of the constructions can be reasonably interpreted in terms of postglacial crustal rebound, demonstrating that areas of known crustal deformation can be useful in interpreting eustatic sealevel fluctuations.

INTRODUCTION

A strip of land a few tens of miles wide (fig. 1) inland from the coastline of southwestern Maine displays evidence of both emergence and submergence during late Pleistocene time. The region was described by Davis (1912, p. 532) as an unusual type of emerged coastal plain because it has only a dis-continuous cover of late-glacial marine sediment partially filling valleys and lapping up on highlands to the altitude of maximum postglacial marine sub-mergence. In contradiction to such obvious evidence of emergence, topographic maps of the area, especially the Boothbay Quadrangle, are used in many geol-ogy classes to illustrate the characteristics of a submerged coast.

The late-glacial and postglacial history of coastal Maine can be sum-marized as follows: the weight of glacier ice isostatically depressed the region, then postglacial eustatic rise of sealevel submerged a fringe of land extending inland from the present shoreline to about the 200- or 300-foot contour, and finally, lagging postglacial crustal rebound caused re-emergence. Marine sedi-mentation was not sufficient to blanket completely the crystalline basement rocks and thus did not convert the coastal region into the monotonous lowland that is envisioned by the elementary definition of a coastal plain.

The present report is an attempt to fill in the details of the late-glacial and postglacial history, with emphasis on the interrelation of glacial-eustatic and glacial-isostatic movements. Work was supported during the 1957 and 1958 field seasons by the Office of Naval Research and during 1959 by the Maine Geological Survey. The stimulating discussion of the Friends of the Pleistocene during their annual meeting in May, 1961 is gratefully acknowl-edged.

The plan of attack is as follows: relative movements of land and sealevel in the vicinity of Portland, Maine are reviewed; then models are constructed, showing the position of a hypothetical bench mark relative to sealevel during late Pleistocene time; finally, the difficulties of the constructions are discussed and points of interest noted. The model presented as the most satisfactory is only one of an infinite number of possible interpretations, but it appears to be in best accord with the facts now available. The intent is to illustrate a method of analyzing interrelated land and sealevel movements as well as to elucidate

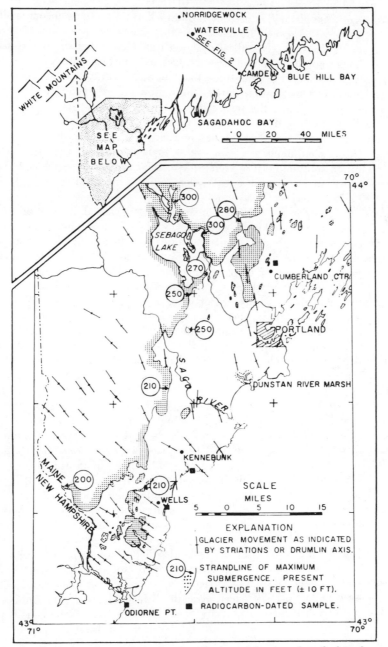

Fig. 1. Map of southern Maine showing places and features described in the text.

the history of a region. For more detailed description of the late-Pleistocene geology of the region, the reader is referred to Bloom (1960).

The term "late glacial" is used here in an informal sense as encompassing that part of late Pleistocene time when southwestern Maine was being deglaciated for the last time. Not all of the area was deglaciated simultaneously, but even though one locality had been deglaciated and was thus in its "postglacial" environment at the same time that another locality was still covered with glacier ice, as long as some part of the region was characterized by some glacier-controlled regimen, the term "late glacial" is appropriate. The term "postglacial" is used to describe events that followed regional deglaciation. Such terms avoid the problem of defining an artificial Pleistocene-Recent boundary in the continuous series of events to be described. Because events in Maine form a continuum, it is preferable to refer the entire sequence to late Pleistocene time and to use the term "recent" in the informal sense proposed by Flint (1957, p. 283).

Mode of deglaciation.—Till is not abundantly exposed in southwestern Maine. It forms drumlins, locally abundant, and both downglacier tails and upglacier ramps against bedrock hills. In addition, clusters of closely spaced till ridges, or washboard moraines, cross several areas (Bloom, 1960, p. 28). The predominant amount of drift has been stratified or sorted by running water, and takes the form of kames, kame terraces, eskers, ice-contact deltas, and outwash fans. In the vicinity of Portland, many deltaic deposits show ice-contact stratigraphy on their north or northwest sides grading into deltaic foreset beds of sand dipping south or southeast. The surface form of these ice-contact deltas has been completely masked or destroyed by subsequent marine erosion or deposition, but abundant artificial exposures indicate that ice-contact deltas are regionally abundant forms.

The significance of the regional abundance of stratified drift lies in the fact that both the stratification and sorting indicate deposition in meltwater streams or ice-dammed lakes. Through-flowing drainage enabled most of the silt and clay-size sediment to be flushed from the drift, leaving a profusion of sand and gravel deposits. The fact that finer-sized sediment constituted a part of the glacier load is evidenced by exposures of compact fissile till, rich in silt and clay. The general lack of fines in the stratified drift, except for silt lenses that mark local meltwater ponds, emphasizes the through-flowing nature of the meltwater streams. The sea, the ultimate base-level of stream flow as well as the ultimate settling basin for flocculated glacial rock flour, must have been lower than the lowest-known deposit of coarse, well sorted stratified drift. The log of an engineering testhole near Portland (Bloom, 1960, p. 58) reported sand at 64 to 68 feet below present sea level in a buried valley, overlain by gray silty clay. Another testhole log reported, "coarse sand and gravel; hole caves" between −62.2 and −70.2 feet, beneath 63 feet of marine fine sand, silt, and clay (Maine Turnpike Authority, boring 2148+44, Nonesuch River Bridge, unpublished). Although a subaerial origin for these deeply buried sand and gravel beds can not be proved, it seems unlikely that well sorted sand or gravel would be deposited in a sea that was receiving great quantities of silt- and clay-sized rock flour from a melting ice sheet. Marine reworking of previously deposited drift, well shown in figure 3 by a lens of mixed sand, silt, and clay underlying marine silty clay and thinning over the crests of buried hills, is additional evidence that valley floors now well below sealevel were sub-aerially exposed during deglaciation, prior to marine submergence. Sealevel, relative to the land, must have been at least 70 feet below present during some part of late-glacial time.

Marine submergence.—Stratified glacial drift and till in the vicinity of Portland are unconformably overlain by gray sandy, silty clay of marine origin. The marine sediment has been named the Presumpscot Formation (Bloom, 1960, p. 55) from exposures in the valley of the Presumpscot River near Portland. Fossils from the type area were described as early as 1842 (Mighels and Adams, 1842) and the marine deposit, referred to by various terms such as "the Portland clays" (Hitchcock, 1861, p. 275), "the Leda clay" (Katz and Keith, 1917, p. 29), and "clay of the Champlain submergence" (Meserve, 1919, p. 207), has been well known ever since. A formation name has been given the unit in order to prevent unwarranted assumptions of its stratigraphic equivalence (as was done in referring it the Champlain sub-mergence) and also to define a mappable unit of regional extent.

The Presumpscot Formation is composed of poorly sorted, glacially abraded sediment deposited in a transgressing sea of late-Pleistocene age. The average grain size of 43 samples (Goldthwait, 1951, p. 25) was 23.5 percent sand, 37.5 percent silt, and 39 percent clay, although extreme variations are common. Fossils are locally abundant, converting some beds to veritable coquina. All 30 species of macrofossils collected (Bloom, 1960, fig. 3) from the formation are extant today. Most species are described in reference works as circumpolar or arctic forms, although eleven species are found in deep water as far south as Cape Hatteras, and several inhabit coastal waters to southern Georgia.

Two of the most common species of pelecypods in the Presumpscot Forma-tion, *Yoldia arctica* and *Nucula expansa*, are not found today south of Cape Breton Island. Another less common species, *Astarte arctica*, is not found south of the Labrador coast. The general aspect of the fauna suggests that the en-vironment of deposition of the Presumpscot Formation was comparable to present conditions about seven degrees latitude farther north, in southern Labrador. The faunal evidence of a colder climate is compatible with abundant ice-rafted erratics and ice-thrust features in the formation that confirm a peri-glacial marine origin.

In the vicinity of Portland the Presumpscot Formation overlies with regional unconformity stratified and sorted glacial drift. As is to be expected, the angle-of-rest slopes of sand and gravel that formed part of the newly de-glaciated landscape were unstable when submerged, and reworking and mixing of subaerially deposited drift into the base of the Presumpscot Formation was common. Gravel pits opened in stratified drift on the coastal plain commonly show sandy basal beds of the Presumpscot Formation dipping radially off buried former hills at angles approaching 30°. No buried soil profile or other evidence of postglacial, pre-submergence weathering has been found on the surface of buried glacial deposits.

Near the inland limit of marine submergence the Presumpscot Formation intertongues with, rather than unconformably overlies, subaerially deposited drift. Delta fans, each several miles long on radii that converge on major gaps in the foothills at the back of the coastal plain, trend in a line northeastward across southwestern Maine (Bloom, 1960, p. 42). At their heads, the delta fans are composed of poorly sorted, but usually well-rounded, cobble gravel that passes transitionally inland into ice-contact stratified drift. Several of the delta fans head at eskers that follow sinuous courses northwestward into the foothills of the White Mountains.

At their distal margins the delta fans are composed of fine sand, commonly rhythmically interbedded with silt and clay typical of the Presumpscot Formation. Thus, while deglaciation was apparently pre-marine on the coastal plain near Portland, at the inland limit of marine submergence deglaciation and the maximum marine transgression were synchronous. The late glacial transgressing sea apparently "caught up" with deglaciation.

Radiocarbon dates on the marine submergence not only measure the transgression of the Presumpscot sea, but also give maximum dates for the final deglaciation of southwestern Maine. Three dates are now available, two from the vicinity of Waterville, Maine, 75 miles north of Portland, and one from Cumberland Center, 10 miles north of Portland. All three dates were determined on shells of shallow water or even intertidal mollusks (*Mytilus edulis, Panomya arctica*). They are as follows, listed with sample number, locality, height above sealevel, and reference:

W-737	$11,800 \pm 240$	Waterville	ca. 186 ft.	(Rubin and Alexander, 1960, p. 130)
W-947	$11,950 \pm 350$	Norridgewock	ca. 250 ft.	(Rubin, 1961, written communication)
L-678	$12,100 \pm 300$	Cumberland Center	ca. 160 ft.	(Broecker, 1961, written communication)

The limit of marine submergence in the Waterville-Norridgewock area is about 320 feet (C. F. Hickox, 1959, written communication), and near Cumberland Center it is about 260 to 280 feet. Thus the shallow-water fauna of all three radiocarbon-dated sites shows that the dates are older than the time of maximum marine transgression and final deglaciation.

All the dates are from localities north of Portland that have been uplifted more than Portland by postglacial isostatic recovery. The limitations of attempting to project former water levels can be illustrated by an example: The maximum amount of submergence measurable at Portland is 160 feet, the height of the highest hills in the city. Actual submergence was probably greater. If 12,100 years ago the sea was about 100 feet below the level of maximum submergence at Cumberland Center (the difference between the present altitude of the marine limit there and the present altitude of the dated shell bed) then at the same time Portland must have been submerged more than 60 feet (100 feet less than the measurable submergence), but how much more is indeterminate. The extent of submergence at Portland at the times indicated by the more northerly sites cannot even be estimated because of the uncertainty of the amount of differential uplift over a distance of 75 miles.

POSTGLACIAL HISTORY

Re-emergence of the coastal plain.—It would be hard to improve upon the description by Davis (1912, p. 532-539) of the morphological changes that followed re-emergence of the coastal plain. The newly exposed Presumpscot Formation had partially filled the valleys and reduced the local relief considerably. Wave erosion of the previously submerged landscape during the re-emergence through the surf zone was minimized by the blanket of marine sediment. Nevertheless, the newly emerged, fine-grained Presumpscot Formation was much more susceptible to subaerial erosion than were the permeable sand and gravel deposits of the upland valleys, and a fine-textured drainage pattern quickly developed on the clayey lowlands. Major streams generally followed the axes of partially filled valleys across the coastal plain. A marked change in drainage texture marks the inland edge of marine deposition.

Marine submergence and subsequent re-emergence involved a transgressive and a regressive phase. The Presumpscot Formation was deposited during the transgressive phase in the proximity of melting ice, as witnessed by: (1) its basic character as a deposit of glacial rock flour, (2) its cold-water marine fauna, (3) abundance of ice-rafted erratics in it, and (4) intertonguing of the formation with outwash along the inland edge of submergence. The maximum transgression coincided with regional deglaciation. Thus re-emergence must have been a postglacial event. This inference is supported by the pollen stratigraphy of two ponds in Maine that are floored by the Presumpscot Formation. In both ponds (Potzger and Friesner, 1948, p. 181, and Deevey, 1951, p. 192) blue-gray silty clay of the Presumpscot Formation is overlain by freshwater organic sediments that bear a dominant percentage of pine pollen. The late-glacial and early postglacial sequence of tundra and spruce-birch pollen that typically precedes the pine maximum in the pollen stratigraphy of New England (Deevey, 1958) is missing on the coastal plain of Maine. Therefore, re-emergence apparently did not take place until pine dominated the pollen accumulation in swamps and ponds. The pine zone (zone B of pollen stratigraphers) is not radiocarbon dated in southern Maine, but zone B in south-central New England, subdivided into B1 and B2, dates from 6000 to 8500 B.P., and in northern Maine and Nova Scotia the undifferentiated zone B dates from 6000 to 7000 years B.P. (Deevey, 1958, p. 34). Therefore, a tentative date of 7000 to 8000 years B.P. for the re-emergence of southern Maine is proposed.

As is shown by figure 2, the ponds used in the pollen-stratigraphic studies now stand 250 feet and 140 feet above sealevel. To the degree of accuracy with

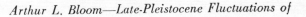

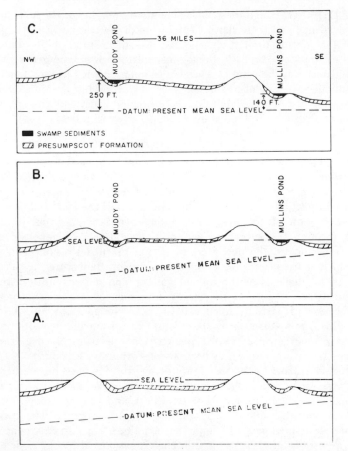

Fig. 2. Episodes in the differential postglacial upwarping of south-central Maine. Diagrammatic only; vertical scale greatly exaggerated. Note that earliest episode A is at the bottom. A. Maximum marine submergence, <11,800, >7000 to 8000 years ago. B. Re-emergence of Muddy Pond and Mullins Pond, 7000 to 8000 years ago. C. Present altitudes.

which the pollen stratigraphy permits an estimate of simultaneous emergence, in the same length of postglacial time Muddy Pond, near Waterville (Deevey, 1951, p. 192) has been elevated 110 feet more than Mullins Pond, near Camden (Potzger and Friesner, 1948, p. 181). Thus differential upwarping of about 3 feet per mile toward the northwest has accompanied re-emergence during the last 7000 to 8000 years. The differential uplift, greatest in the direction of thickest glacier ice as inferred from directions of ice movement, supports the idea that re-emergence was caused by postglacial crustal rebound. The total amount of postglacial rebound is unknown, as only the last 7000 to 8000 years are subject to measurement. Furthermore, the northwestern component of 3 feet per mile is not necessarily the maximum component of uplift.

Recent submergence.—Renewed submergence of at least 35 feet has followed the postglacial re-emergence. Evidence of this recent submergence consists of: (1) subaerially weathered sediments now submerged, (2) the well known "drowned forests" of coastal Maine and adjacent regions, and (3) tide-gauge records.

The following quotation is from an engineering report (Gray, 1954a, p. 1) on the problem of roadbed subsidence of U. S. Route 1 across the Dunstan River tidal marsh, 7 miles southwest of Portland. It illustrates the economic importance of a significant geologic feature—a weathered zone on the Presumpscot Formation now below sealevel.

> There exists over the greater part of the marsh a surficial layer of up to 20 feet in thickness which is composed essentially of organic material. This organic substratum does contain variable quantities of sand, but in spite of this it is very soft and unstable . . . Underlying the organic stratum in numerous localities is a layer of firm weathered silty clay. It is characterized by its consistency and frequently by a brownish color. At the southwest end of the section this clay appears at the ground surface. Elsewhere it is buried at depths varying from 10 to 20 or more feet. It appears to represent the weathered surface zone of a much thicker stratum [Presumpscot Formation] which was formerly exposed to atmospheric conditions . . .

Figure 3 (generalized from Gray, 1954b, pl. II) is a cross section constructed from highway borings 2.5 miles southeast of U. S. Route 1 on the Pine Point Road, across another branch of the Dunstan River marsh. Note that tidal-marsh peat has accumulated to a thickness of 15 feet. Moreover, stiff, mottled brown and gray clay, interpreted by highway engineers as a weathered zone on the Presumpscot Formation, extends 35 feet below the high-tide marsh surface. Thus, at some time subsequent to the deposition of the Presumpscot Formation, the Portland area must have been emerged at least

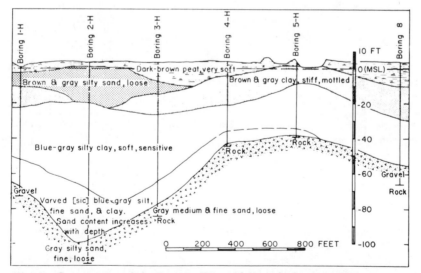

Fig. 3.   Cross section of the Dunstan River tidal marsh, Scarboro, Maine, along Pine Point Road. Stiff, mottled brown and gray clay, interpreted by highway engineers as a weathered zone on the Presumpscot Formation, extends to 35 feet below high-tide marsh surface. Interbedded silt, sand, and clay in the buried valley is interpreted as marine-reworked glacial drift.

35 feet more than present, for it seems doubtful that the Presumpscot Formation could be discolored and partly cemented with oxidized iron in the reducing environment of a tidal marsh. The testhole logs of this section recorded an abrupt decrease in the number of blows required for a foot of penetration from 64 to 88 in "partly weathered firm brown and blue clay and silt" to 10 to 20 in the underlying blue-gray silty clay.

Additional evidence of recent submergence is to be found in radiocarbon dates from rooted tree stumps exposed on tidal flats or in the banks of tidal-marsh channels. The following list of radiocarbon-dated stumps gives the minimum amount of submergence indicated, on the assumption that all trees grew with their roots above the high-tide level.

W-396	2980 ± 180	Wells Marsh (2 to 3 ft)	(Hussey, 1959)
W-508	2810 ± 200	Wells Beach (6 ft)	(Hussey, 1959)
W-509	1280 ± 200	Kennebunk Beach West (10 to 13 ft)	(Hussey, 1959)
W-510	3250 ± 200	Kennebunk Beach East (7 to 13 ft)	(Hussey, 1959)
L-118	4150 ± 200	Sagadahoc Bay (9 ft)	(Bradley, 1953)
Y-156	4190 ± 70	Odiorne Point, N. H. (7.6 ft)	(Deevey, Gralenski, and Hoffren, 1959, p. 145)
Y-773	2970 ± 140	Blue Hill Bay (13.5 ft)	(Stuiver and Deevey, 1961, p. 127)

The dates show no consistent relationship of age and amount of submergence. Some of the stumps are rooted in peat that has been overridden and compacted by the landward retreat of a barrier beach; the amount of submergence indicated by such stumps is excessive. Nevertheless the total evidence of the stumps is for submergence, perhaps of different amounts at different localities, and perhaps proceeding at an irregular rate, during the last 4200 years. The maximum amount of submergence indicated by an intertidal stump, 13.5 feet at Blue Hill Bay, is less than half the amount indicated by the engineering data of marsh borings. However, lobstermen's tales of other stumps farther offshore suggest that the intertidal stumps are only the visible part of a more extensive occurrence.

Submergence has continued to the present decade on the Atlantic coast. Disney (1955) summarized tide-gauge records and concluded that submergence at a mean annual rate of 0.007 feet had been in progress at Portland since 1912, when the record began. The rate of submergence indicated at Portland by the tide-gauge record is only about half that indicated by tide gauges in other Atlantic Coast cities from Boston southward.

RELATIVE MOVEMENTS OF LAND AND SEALEVEL

A summary of the preceding description of late-glacial and postglacial events in southwestern Maine contains the following points of interest to the problem of relative movements of land and sealevel at Portland:

1. During deglaciation the area around Portland was emerged at least 70 feet greater than present, because stratified drift was deposited in valleys now that far below sealevel.

2. Submergence was in progress 12,100 years ago, with rapid marine transgression over a newly deglaciated landscape. The result was a transgressive wedge of glacial-marine, fine-grained sediment, the Presumpscot Formation. By 12,100 years ago Portland was submerged at least 60 feet greater than present.

3. Some time less than 11,800 years ago the maximum submergence took place. Late-glacial conditions still persisted along the inland edge of marine submergence. Sealevel at Portland was at least 160 feet higher than present relative to the land at the time of maximum submergence.

4. Postglacial re-emergence, accompanied by differential crustal upwarping toward the northwest, was in progress 7000 to 8000 years ago. Ponds now 140 and 250 feet above sealevel re-emerged then, and localities now at lower altitudes, including the Portland area, probably re-emerged late in this interval.

5. Postglacial re-emergence apparently continued until areas seaward of the present shoreline were emerged, producing a relative sealevel at least 35 feet lower than present in the vicinity of Portland. The weathered zone of the Presumpscot Formation below present sealevel and the submerged rooted stumps along the Maine coast are the record of this greater emergence. Radiocarbon-dated stumps suggest that the greater emergence dates back to at least 4200 years ago.

6. Since at least 4200 years ago, sealevel relative to the coast of Maine has been rising. Radiocarbon-dated stumps and tide-gauge records indicate that submergence has continued to the present decade.

A graphic model.—The time span of the events summarized above covers an interval when, according to widely accepted theory and substantial evidence, eustatic (worldwide) sealevel was rising rapidly. Sealevel during the maximum Wisconsin Glaciation was at least 300 feet below present, and deglaciation must have resulted in a rapid rise as water was returned to the sea. Several graphic interpretations of successive positions of sealevel during late Pleistocene time have been published (for example, Shepard, 1960; McFarlan, 1961; Fairbridge, 1961). The data supporting all these graphs are inadequate, and caution is invariably urged in their use. It is instructive, nevertheless, to apply the information on fluctuations of relative level in Maine to the various interpretations of glacial-eustatic sealevel.

In figure 4, the locus of a point now at mean sea level near the city of Portland, Maine has been plotted with respect to the sealevel curve proposed by Shepard (1960). Shepard's curve is based primarily on studies in the northwestern part of the Gulf of Mexico (Curray, 1960) an area considered to be tectonically stable. The curve is not in conflict with radiocarbon-dated sealevel positions reported from other areas, but it is not a "best fitted" curve through a collection of reference points from many areas. Maxima and minima on the curve are based on local stratigraphic and topographic evidence in the northwest Gulf of Mexico.

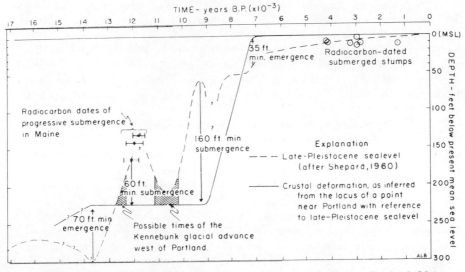

Fig. 4. Position of a reference point now at mean sea level near Portland, Maine, relative to late-Pleistocene sealevel as interpreted by Shepard. Arrows indicate the amount of emergence or submergence of the reference point at various times.

The major consideration in drawing figure 4 was to fit a curve of crustal deformation at Portland, Maine, to the curve of eustatic sealevel in such a way that at least 70 feet of late-glacial emergence and subsequently at least 160 feet of late-glacial submergence could be reasonably accounted for prior to 7000 to 8000 years ago. In other words, relative vertical displacement of coastal Maine and the sea amounted to a minimum of 230 feet in the late-glacial interval. Based on Shepard's curve, 70 feet of emergence a little less than 14,000 years ago and 160 feet of submergence a little more than 9000 years ago can be accounted for by drawing a horizontal line about 225 feet below present sealevel.

There is no evidence for the age of deglaciation in the Portland area. The 14,000 year date suggested in figure 4 is used only because 70 feet of emergence at that time does not require downwarping of the crust to account for a subsequent 160 feet of submergence. However, deglaciation 14,000 years ago is not inappropriate for the Portland area. Southern Connecticut was deglaciated more than 13,500 years ago, judging by radiocarbon ages of peat samples from near the bottom of bogs; similar evidence suggests that central Massachusetts was deglaciated an estimated 13,000 years ago (Deevey, 1958, p. 34-36.) A minimum age for deglaciation of the Portland area is 12,100 years, the radiocarbon age of marine shells from Cumberland Center (L-678).

A horizontal curve of crustal deformation fixed by the two points defined above is compatible with two other features of the late-Pleistocene history of Maine. First, the radiocarbon date of 12,100 years B.P. on shells from Cumberland Center has been shown in a previous part of this paper to represent at least 60 feet of submergence at Portland. Shepard's sealevel curve is almost exactly 60 feet above the curve of crustal deformation at 12,100 years B.P. The two other radiocarbon dates of the marine transgression in Maine also fall on a steeply ascending limb of Shepard's curve.

Second, there is evidence of a late advance of glacier ice 15 to 20 miles west of Portland that entered the sea (Bloom, 1959). Kennebunk, the type locality for this ice advance, was submerged about 40 feet at the time, because marine sediment now that far above sealevel was deformed by ice movement. Kennebunk is southwest of Portland and has probably been uplifted about the same amount by differential postglacial rebound. Thus at the time of the Kennebunk advance, Portland also would have been submerged about 40 feet. Shepard's sealevel curve lies less than 40 feet above the curve of crustal deformation at only three times on figure 4, and the youngest of these is in postglacial time. Therefore 12,300 to 12,800 and 10,200 to 11,200 years ago are the times of the Kennebunk advance permitted by the construction. Both sets of dates narrowly bracket times of significant ice readvances in the north-central United States, the earlier dates bracketing the Port Huron advance, and the later the Valders advance. The lesser age is more likely because of the evidence that late-glacial conditions persisted at the inland limit of marine transgression until the time of maximum submergence, which would have been about 9,300 years ago by the construction.

The suggested correlation of the Kennebunk advance in Maine with the Valders readvance in Wisconsin raises a larger question of significance to New England Pleistocene geology. If, as is now believed (Karrow, Clark, and Terasmae, 1961) the St. Lawrence lowland was ice-free during Valders time, then a Kennebunk advance of Valders age would have been from an independent ice cap over the White Mountains. Ice of the Kennebunk advance did indeed move over southwestern Maine from a northwesterly direction (the direction of the White Mountains) in contrast to the southward-moving ice sheet that covered the Portland area prior to marine submergence (fig. 1).

The fact that figure 4 permits an ice advance in Maine at times of known glacier expansion in central United States is permissive evidence for the validity of the construction. It is emphasized that this placement of the Kennebunk advance is not a basis for the construction of figure 4, but the result of the construction.

No evidence reported from Maine supports the fall of sealevel indicated by Shepard as culminating about 10,700 years ago. However, some visitors to a field conference in Maine (Friends of the Pleistocene, 1961) argued that the sand overlying highly fossiliferous Presumpscot Formation at the Dowdy Gravel Pit, near Cumberland Center (Bloom, 1961, p. 5) represents a regressive episode. Shells from that pit were subsequently radiocarbon dated as 12,100 years old (L-678, p. 866). Thus it is possible that a fall of sealevel coinciding with the Valders (= Kennebunk?) readvance produced a temporary marine regression in parts of coastal Maine that had already been submerged. Regional evidence for or against the hypothesized regressive episode has been destroyed or buried by subsequent more extensive submergence.

Southwestern Maine records little evidence of final re-emergence. Re-emergence apparently began very shortly after the time of maximum submergence, for the erosional and depositional forms along the strandline of maximum submergence are few and poorly developed. Furthermore, re-emergence must have been caused primarily by crustal rebound, as evidenced by the differential upwarping described on a previous page.

The segment of the crustal deformation curve in figure 4 between 9300 and 7000 years ago is drawn as a straight line intersecting the sealevel curve at 7500 years B.P., midway in the thousand-year period of probable re-emergence. Between 9300 and 8000 years ago Shepard's curve, based on data from Curray (1960, p. 257) shows a substantial drop to a low stand of at least –90 feet (–15 fathoms) and possibly –126 feet (–21 fathoms). Curray (1960, p. 263) followed Fairbridge in attributing this temporary lowering of sealevel to the "Cochrane and Bothnian ice return". If sealevel fell only to –90 feet 8700 years ago, then by figure 4, from 9300 to about 7000 years ago re-emergence in Maine would have been continuous although at a variable rate. However, if sealevel reached the extreme low level of –126 feet suggested for this period by Curray, then from 8700 to about 8000 years ago crustal rebound in Maine and rise of sealevel would have been nearly synchronous, and a relative still-stand of the sea in Maine, about 75 to 90 feet above present sealevel, would have resulted. No strandline features from this altitude have been found in Maine. Therefore, on the basis of the model being considered, the evidence in Maine favors the higher alternative stand of the sea shown by Shepard at 8700 years B.P.

Re-emergence of the reference point about 7500 years ago permits adequate space above Shepard's curve for the 35 feet of greater-than-present emergence recorded by the buried submarsh weathered zone on the Presumpscot Formation. If crustal deformation ceased about 7000 years ago, and sealevel continued its slow rise to the present as interpreted by Shepard, then radiocarbon-dated submerged stumps ought to plot on figure 4 along Shepard's curve, with younger stumps showing progressively less submergence. That they do not raises the possibility of either minor fluctuations in one or both of the variables or, equally likely, differential post-depositional compaction of the submerged stumps. The substratum of only one of the stumps (W-396, Hussey, 1959) has escaped probable compaction by a migrating barrier beach. Some of the dated stumps may not be in place; there is considerable room for uncertainty concerning the validity of criteria used to demonstrate that stumps are in an undisturbed growth position. About all that can be said for submerged stumps is that they demonstrate submergence; how much and at what rate remain open questions.

In summary, the following points concerning figure 4 can be noted:

1. A simple curve of crustal rebound, horizontal between 14,000 and 9000 years ago, then rising 225 feet in about 2000 years (about 0.1 ft/yr.), then horizontal from 7000 years ago to the present, when superimposed on Shepard's curve of late-Pleistocene eustatic sealevel, accounts for all of the known late-glacial and postglacial relative fluctuations of land and sealevel in southwestern Maine.

2. Deglaciation of the Portland area by about 14,000 years ago is required by the construction.

316

3. The model permits the Kennebunk ice advance to have occurred at either of two favorable times. The younger of these times, correlative with the Valders advance in the north-central states, is favored because of the evidence that melting glacier ice persisted until the time of maximum submergence.

4. Re-emergence caused by rapid crustal uplift began immediately after regional deglaciation, and abruptly reversed the previous submergent trend.

5. The evidence of submergence during the last 4200 years and probably longer is accounted for by crustal stability and a eustatic rise of sealevel. Minor fluctuations in either variable could account for departures from the linear trend.

6. Crustal rebound near Portland of 225 feet could be caused by postglacial isostatic recovery from an ice sheet three times as thick, or about 675 feet. The latter value is too small by an order of magnitude for the total thickness of New England glacier ice, which is known to have covered Mt. Washington, with a height above the surrounding country of 5000 feet. However, the ice of the Kennebunk advance west of Portland was thin, and during deglaciation the Kennebunk ice must have ceased movement and melted in place when its thickness no longer exceeded the local relief of 1000 feet. Regional stagnation of Kennebunk ice over a zone at least 25 miles wide in southwestern Maine is indicated by the profusion of ice-contact stratified drift in the valleys. It is possible that the model shows only the rebound at Portland attributable to the removal of about 1000 feet of Kennebunk ice from a region 15 to 40 miles to the west, earlier rebound having accompanied deglaciation of the Portland area by 14,000 years ago.

Other possible models.—It is obvious that an infinite number of sets of intertwining curves can be drawn to explain the relative movements of land

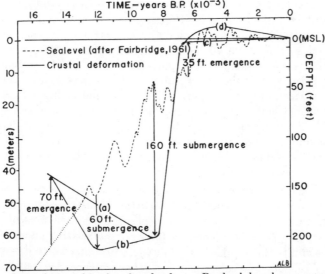

Fig. 5.   Interrelation of land and sealevel near Portland, based on a curve of eustatic sealevel by Fairbridge. The data and method of construction are the same as used in figure 4.

and sealevel in Maine. Most of them would be unrelated to the meager data we possess concerning the absolute position of either variable with time. However, it is possible to use the data from Maine to test hypotheses of eustatic sealevel fluctuation. Figure 5 illustrates one such test, applied to a curve by Fairbridge (1961, p. 156). The method of construction and the data are the same as those used in figure 4.

Alternative constructions are made for parts of the crustal deformation curve in figure 5. Alternative (a) represents a straight line between an early episode of 70 feet of emergence at Portland and a subsequent episode of 160 feet of submergence. Even though these values are minimal, the construction requires a negative slope of the crustal deformation curve, indicating continued crustal downwarping until the time of maximum submergence, which has been shown to coincide with final regional deglaciation. The negative slope could only be eliminated by assuming an unreasonably early deglaciation of the Portland region. Furthermore, although alternative (a) would provide for progressive submergence during the 12,100 to 11,800 year interval dated by radiocarbon, the magnitude of the submergence would be far too small to reach the altitudes of any of the dated sites. Alternative (a) would date the advance of Kennebunk ice into the shallow transgressing sea at 11,300 to 13,000 years ago; while in terms of glacial chronology this is an admissable time range, it is hard to visualize how the thin ice of the advance could have persisted in the region for several thousand years until the time of maximum submergence. Furthermore, if the maximum advance of the Kennebunk ice were some 12,000 years old, and the ice subsequently lingered stagnant in the foothills, what force can be invoked to explain continued crustal subsidence during the entire late-glacial interval?.

Alternative (b) is an attempt to remove some of the difficulties presented above. Specifically, the locus of the Portland reference point is set 60 feet below sealevel at 12,100 years B.P. in order to allow for submergence of the radiocarbon-dated sites to the north and at higher altitudes. This difficulty being removed, the shallow submergence correlated with the Kennebunk advance must be pushed still farther back to more than 13,000 years ago, and the problems of accounting for lingering ice cover and continued crustal subsidence during deglaciation become even more serious. The major criticism of Fairbridge's eustatic sealevel curve in this time range is that it does not have sufficient amplitude to account for the minimum estimate of 230 feet of relative movement in Maine without requiring an unexplained negative slope for the curve of crustal deformation.

Re-emergence between 7,000 and 8,000 years B.P. is easily shown by a curve of crustal rebound intersecting Fairbridge's eustatic sealevel curve at the appropriate time. The 35 feet of emergence greater than present and subsequent submergence can be accounted for by hypothesizing either (alternative c) crustal stability during the last 6800 years or (alternative d) overcompensated glacial rebound and subsequent crustal subsidence. Alternative (c) would predict numerous emerged shoreline features along the Maine coast correlated with the remarkably detailed fluctuations Fairbridge described for this interval.

No emerged marine features of this young age have been found. Therefore, alternative (c) seems highly unlikely.

The lack of emerged shoreline features can be accounted for by arching the curve of crustal rebound over Fairbridge's emergent peaks. A curve of this shape recalls the now discredited idea of marginal superelevation, but it could be explained by hypothesizing crustal subsidence not related to glacial isostasy superimposed on the curve of postglacial recovery. However, the magnitude and rate of the submergence seem excessive for a non-glacial tectonic phenomena, when one considers the geologic terrane involved.

Another construction is attempted in figure 6. As a basis for this model, a eustatic sealevel curve published by McFarlan (1961, p. 145) is used. McFarlan drew two curves, (1) a mean based on all plotted sealevel positions in the Mississippi delta region, many of which were corrected for structural movement, and (2) a curve based on the fewer points provided by "hingeline" samples that required no structural adjustment. The hinge-line curve is used here because the mean curve does not have sufficient amplitude to accommodate 230 feet of relative movement in Maine prior to 7000 to 8000 years ago. Broecker (1961, p. 160) in discussing McFarlan's data, also preferred the hinge-line curve in spite of the fact that it is based on fewer points and cannot be expected to show much detail.

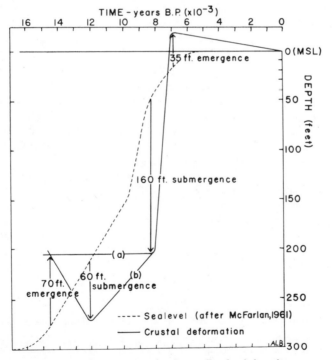

Fig. 6. Interrelation of land and sealevel near Portland, based on a curve of eustatic sealevel by McFarlan. The data and method of construction are the same as used in figure 4.

It is possible to draw in figure 6 a line of crustal deformation that does not have a negative slope between 14,000 and 9000 years ago. Alternative interpretation (a) shows that line, 70 feet above sealevel 14,600 or more years ago and 160 feet below sealevel about 8400 years ago. By this construction the sealevel reference point at Portland would not have been submerged until less than 12,000 years ago, whereas radiocarbon dates show submergence in progress to much higher altitudes by that time. Therefore, even though the interpretation would date the Kennebunk advance into the shallow sea at a reasonable time of about 10,000 to 12,000 years ago, the construction must be rejected as not satisfying the known facts. Alternative curve (b) provides for the submergence indicated by radiocarbon dates about 12,000 years ago, but requires either excessively early deglaciation at Portland, or an unexplained crustal downwarping during the time of deglaciation.

Re-emergence of the Portland area 7000 to 8000 years ago, 35 feet of emergence greater than present, and subsequent slow submergence are illustrated as in the preceding models. McFarlan's curve requires postglacial rebound to superelevate the reference point. However, McFarlan noted in an addendum (1961, p. 148) new evidence that suggests sealevel rose from 17 feet below its present level 5650 years ago to its present level about 3000 years ago. If his curve of eustatic sealevel based on hinge-line samples were to be revised in accordance with this new evidence, no postglacial crustal superelevation and subsequent downwarping would be required to account for relative changes of sealevel in Maine during the last 6000 to 7000 years.

CONCLUSIONS

A method of using relative changes of land and sealevel in a glaciated region to test the validity of various interpretations of postglacial eustatic fluctuations of sealevel is demonstrated. A simple curve of postglacial isostatic rebound in southwestern Maine, compatible with the time of deglaciation and estimated ice thickness, is superimposed on the curve of late-Pleistocene eustatic rise of sealevel proposed by Shepard. The result provides a remarkably complete explanation of late-Pleistocene events in southwestern Maine, restricts the possible times of an otherwise undated glacier advance, and accounts for all known relative changes of level.

Two other models are constructed to illustrate the method of analysis, but neither of them can be made to fit the observed facts. Although only one of three models is found acceptable, this is not to imply that it is the ultimate solution. Data from mobile areas, especially glaciated regions where a late-Pleistocene chronology can be established, can be useful in studying the difficult problem of glacial-eustatic sealevel fluctuations.

REFERENCES

Bloom, A. L., 1959, Late Pleistocene events in southwestern Maine [abs.]: Geol. Soc. America Bull., v. 70, p. 1571.
———— 1960, Late Pleistocene changes of sealevel in southwestern Maine: Augusta, Maine, Dept. of Econ. Devel., Maine Geol. Survey, 143 p.
———— 1961, Late Pleistocene stratigraphy and history of southwestern Maine, Eastern Friends of the Pleistocene Guidebook 24th Ann. Reunion, Portland, Maine: Ithaca, N. Y., Cornell Univ., 17 p.

Bradley, W. H., 1953, Age of intertidal tree stumps at Robinhood, Maine: Am. Jour. Sci., v. 251, p. 543-546.

Broecker, Wallace, 1961, Radiocarbon dating of late Quaternary deposits, South Louisiana, a discussion: Geol. Soc. America Bull., v. 72, p. 159-162.

Curray, J. R., 1960, Sediments and history of Holocene transgression, continental shelf, northwest Gulf of Mexico, *in* Shepard, F. P., Phleger, F. B, and van Andel, Tj. H., eds, Recent sediments, Northwest Gulf of Mexico: Am. Assoc. Petroleum Geologists, p. 221-266.

Davis, W. M., 1912, Die erklärende Beschreibung der Landformen: Leipzig, B. G. Teubner, 565 p.

Deevey, E. S., 1951, Late-glacial and postglacial pollen diagrams from Maine: Am. Jour. Sci., v. 249, p. 177-207.

————— 1958, Radiocarbon-dated pollen sequences in eastern North America, *in* Verhandlungen der vierten Internationalen Tagung der Quartarbotaniker 1957: Geobot. Inst. Rübel, Zurich, Veröffentl., no. 34, p. 30-37.

Deevey, E. S., Gralenski, L. J., and Hoffren, Väinö, 1959, Yale natural radiocarbon measurements IV: Radiocarbon, v. 1, p. 144-172.

Disney, L. P., 1955, Tide heights along the coasts of the United States: Am. Soc. Civil Engineers Proc., v. 81, no. 666, 9 p.

Fairbridge, R. W., 1961, Eustatic changes in sea level, *in* Ahrens, L. H., ed., Physics and chemistry of the Earth: New York, Pergamon Press, v. 4, p. 99-185.

Flint, R. F., 1957, Glacial and Pleistocene geology: New York, Wiley, 553 p.

Goldthwait, Lawrence, 1951, Marine clay of the Portland—Sebago, Maine region: Maine Devel. Comm., Rept. State Geologist 1949-1950, p. 24-34.

Gray, Hamilton, 1954a, Soils report, Scarboro—Cumberland County, Dunstan River Crossing: Soils Laboratory, Maine Highway Comm., Orono, Maine, open file report.

————— 1954b, Soils report, Scarboro—Cumberland County, F-01-1 (13) Rt. 9: Soils Laboratory, Maine Highway Comm., Orono, Maine, open file report.

Hitchcock, C. H., 1861, General report upon the geology of Maine: Maine Board of Agriculture, 6th Ann. Rept., 1861, p. 146-328.

Hussey, A. M., 1959, Age of intertidal tree stumps at Wells Beach and Kennebunk Beach, Maine: Jour. Sed. Petrology, v. 29, p. 464-465.

Karrow, P. F., Clark, J. R., and Terasmae, Jaan, 1961, Age of Lake Iroquois and Lake Ontario: Jour. Geology, v. 69, p. 659-667.

Katz, F. J., and Keith, Arthur, 1917, The Newington moraine, Maine, New Hampshire, and Massachusetts: U. S. Geol. Survey Prof. Paper 108-B, p. 11-29.

McFarlan, E. Jr., 1961, Radiocarbon dating of late Quaternary deposits, South Louisiana: Geol. Soc. America Bull., v. 72, p. 129-158.

Meserve, P. W., 1919, Note on the depth of the Champlain submergence along the Maine coast: Am. Jour. Sci., 4th ser., v. 48, p. 207-208.

Mighels, J. W., and Adams, C. B., 1842, Description of fossil shells (*Nucula* and *Bulla*) occurring at Westbrook, Maine: Boston Soc. Nat. History Jour., v. 4, p. 53-54.

Potzger, J. E., and Friesner, R. C., 1948, Forests of the past along the coast of southern Maine: Butler Univ. Botan. Studies, v. 8, p. 178-203.

Rubin, Meyer, and Alexander, Corrinne, 1960, U. S. Geol. Survey radiocarbon dates V: Radiocarbon, v. 2, p. 129-185.

Shepard, F. P., 1960, Rise of sea level along northwest Gulf of Mexico, *in* Shepard, F. P., Phleger, F. B, and van Andel, Tj. H., eds., Recent sediments, Northwest Gulf of Mexico: Am. Assoc. Petroleum Geologists, p. 338-344.

Stuiver, Minze, and Deevey, E. S., 1961, Yale natural radiocarbon measurements VI: Radiocarbon, v. 3, p. 126-140.

Editor's Comments on Paper 19

Løken: *Postglacial Emergence at the South End of Inugsuin Fiord, Baffin Island, N.W.T.*

As Løken notes in his first paragraph, a number of studies on postglacial uplift had been reported in the early 1960's from Spitzbergen, Greenland, and the Canadian Arctic. The paper by Løken is one of the most detailed studies of postglacial emergence (i.e., no correction has been made to the elevation for eustatic sea level changes), with nine dates on marine shells, eight of which come from the same fiord-head delta. The shells in the glacio-marine delta were collected from foreset beds which were traced to the surface of the delta and then precisely surveyed for elevation by level and staff. Løken presents detailed profiles of the emerged submarine sections of the delta. A noticeable change in gradient occurs at about +9 m.a.s.l., which is ^{14}C-dated between 2990 and 3520 BP (see Fig. 3). The break in slope at +9 m may represent a correlative strandline of one of the younger Tapes waterplanes in Scandinavia.

The postglacial emergence curve from Inugsuin Fiord (Fig. 7) has two major straight segments where relative emergence was constant with time—a break is inferred at about 2900 BP in order to bring the curve down to present sea level today. It is possible, of course, that the emergence continued linearly and ceased about 1700 BP. The addition of the eustatic sea level correction indicates a relaxation time of nearly 3000 years—a remarkable contrast with the data from Northeast Greenland based on a comparable situation. The main difference between the curve presented by Washburn and Stuiver (Paper 17) and the one illustrated by Løken is the presence of the long "tail" on the recovery. From the point of view of load it might be critical that the Greenland ice retreated toward the present margin but has not been much smaller in size than we see today; conversely, in Baffin Island the northeastern margin of the Laurentide Ice Cap commenced retreat about 8000 BP and was largely gone by 2000 BP. The "tail" on the Inugsuin curve might then represent the true isostatic component of uplift; this would be missing in Greenland because of the continued presence of the ice load that would be felt tens of kilometers from the ice edge, owing to the rigidity of the crust.

O. H. Løken was born in Norway in 1931; he received his early degrees from the University of Oslo (cand. real in 1954) and his Ph.D. from McGill University in 1962. He has worked extensively in Norway and the eastern Canadian Arctic and has published several papers on shoreline deformation. Since 1967 he has been head of the Glaciology Division, Department of Environment, Canada.

Reprinted from *Geogr. Bull.*, **7**, 243–258 (1965)

POSTGLACIAL EMERGENCE AT THE SOUTH END OF INUGSUIN FIORD, BAFFIN ISLAND, N.W.T.

Olav H. Løken

ABSTRACT

A beach deposit from Inugsuin Fiord, Baffin Island, is described and its mode of formation outlined. It was possible to relate a number of shell samples to the sea levels prevailing when the shells lived. An accurate emergence curve has been drawn which is similar in form to curves obtained from other areas. The early part of postglacial time was characterized by a rate of emergence much smaller than observed in other areas and there are distinct differences between the pattern of emergence on the west and the east coasts of Baffin Island.

A pronounced bench on the beach deposit is discussed and is believed to be associated with a postglacial halt in the process of emergence. This is possibly a parallel to a postglacial transgression in northern Labrador and of late Tapes age. Fossil evidence suggests a climate in early postglacial time warmer than the present.

RÉSUMÉ

L'auteur décrit un dépôt de plage du fjord Inugsuin, Île Baffin (T. N.-O.) et expose son mode de formation. Il a été possible de rattacher un certain nombre de coquillages fossiles à différents niveaux de la mer à l'époque de leur vie. L'auteur a tracé une courbe précise d'émersion, d'une allure semblable aux courbes d'autres régions. Le début des temps post-glaciaires a été caractérisé par une vitesse d'émersion beaucoup plus faible que celles qui ont été observées ailleurs. L'allure générale de l'émersion sur la côte ouest de l'Île Baffin diffère nettement de celle de l'émersion sur la côte est.

L'auteur étudie l'existence d'une terrasse bien marquée qu'il croit causée par un arrêt dans le mouvement d'émersion postglaciaire. Cette terrasse semble correspondre à une transgression marine postglaciaire, d'âge Tapes récent, dans le nord du Labrador. Les indices fournis par les fossiles font croire que le climat du début des temps postglaciaires était plus chaud qu'actuellement.

INTRODUCTION

The postglacial isostatic recovery of glaciated areas has been studied in many localities and several curves to show the vertical displacement of the shoreline as a function of time have recently been published (Blake, 1961; Farrand, 1962; Feyling-Hanssen and Olsson, 1959; Henoch, 1964; Ives, 1964; Washburn and Stuiver, 1962.) These curves have been based on radiocarbon dates of marine mollusc shells, whale bones, driftwood, peat deposits and from archaeological sites.

Ives (1964) lists some of the errors that might be inherent in such curves, i.e., errors in the dating process, contamination and possible redeposition

MS submitted May, 1965.

of the shells, the difficulties of assessing the depth at which the molluscs originally lived and finally the inaccuracy introduced by the usually limited number of observations on which a given curve is based. Although the discussion was restricted to marine molluscs, the same applies to other organic deposits such as driftwood, whale bones and peat deposits. An additional error is introduced because dates on which some curves have been constructed were derived from widely separated localities without proof that a comparison could be made.

This paper presents the results of a detailed study of a raised shore feature in Inugsuin Fiord, Baffin Island (Figure 1) carried out during the summer of 1964. Radiocarbon dates have been obtained from shell samples and their faunal

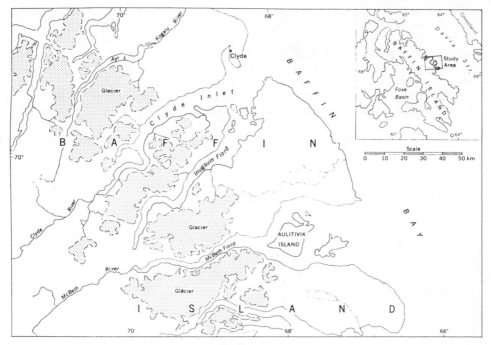

FIGURE 1. Location map.

composition provides information about ecological conditions in postglacial time. (In this paper, postglacial time means the time since deglaciation of the locality in question.) The height of sea level at the time the shells lived has been accurately determined from their stratigraphic position and all samples are from a small area; none of the sampling sites is more than 700 metres from another.

On this basis, a curve showing time versus height of sea level above present sea level has been drawn. This curve shows the time when any point below the marine limit emerged above sea level and the slope of the curve shows the rate at which the land emerged from the sea.

Such a curve will be referred to as an emergence curve, a better term than the commonly used "uplift curve". It is proposed to reserve the latter for a curve showing crustal uplift, that is the time change of geocentric distance for a point on the earth's crust. An uplift curve can be obtained from the emergence curve by taking into account the eustatic changes of sea level that have occurred.

GENERAL DESCRIPTION

Inugsuin Fiord lies on the east coast of Baffin Island and extends inland toward the southwest for about 100 km (Figure 1); the locality under consideration is at the head of the fiord.

The fiord occupies a typical fiord basin; it has steep sides and is long and narrow. The steepest slopes are found along the middle part of the fiord at the sharp bend, while gentler slopes prevail farther southwest. At the head of the fiord the surrounding mountains rise to between 600 and 1,000 metres except where a valley leads toward a series of lakes farther west (Figure 2). The lower lakes are dammed by a series of terminal moraines lying between the fiord and the second lake. Apart from this, the only substantial area of low-lying land along the inner part of the fiord is at its head where a large drift deposit has been formed under the steep cliffs that rise approximately 1 km south of the present shore.

During the last glaciation, outlet glaciers from the ice sheet to the southwest flowed toward

Baffin Bay along the valleys and fiords that cross the high mountains of the coastal area (Ives and Andrews, 1963). The ice that filled Inugsuin Fiord came from the southwest and the high level lateral moraines and terraces show a uniform slope down toward the northeast. At a later phase, ice movement was largely determined by the local topography, and the last ice stream to reach the inner part of the fiord flowed along the valley that ends at base camp (Figure 2). Several

readvances occurred shortly before final deglaciation of the fiord head, and the trend of the lateral moraines shows a re-entrant of the main outlet glacier flowed into the embayment at the end of the fiord. This is significant for the interpretation of the deposits in that locality.

The area surrounding the head of the fiord was deglaciated more than 7,900 years ago (Figure 2 and Table 1) at a time when sea level relative to the land was some 65 metres higher than today.

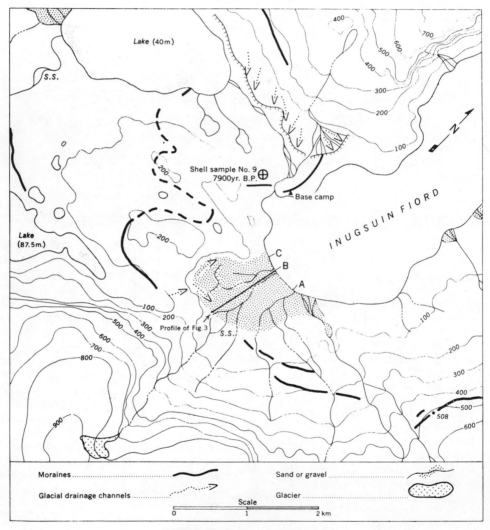

FIGURE 2. Area at head of Inugsuin Fiord.

O. H. LØKEN 245

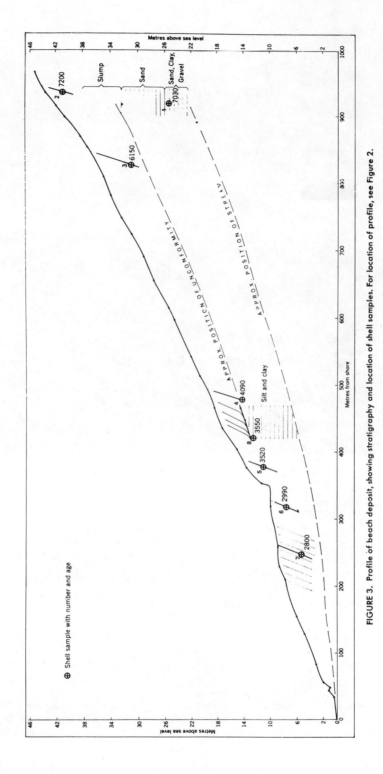

FIGURE 3. Profile of beach deposit, showing stratigraphy and location of shell samples. For location of profile, see Figure 2.

FIGURE 4. The stratigraphy under the highest part of the deposit. Men are collecting shell sample 1.

THE DEPOSIT AT THE HEAD OF THE FIORD

The drift deposit has the appearance of a gently sloping surface extending from sea level to an altitude of 44 metres over a horizontal distance of approximately 1,000 metres. Detailed study shows the apparently even surface is irregular and numerous ridges, roughly parallel to the present beach and diminishing in size away from it, cross the surface. A profile levelled (by Zeiss N2 level) across the deposit approximately perpendicular to the trend of the ridges is shown in Figure 3.

A stream enters the fiord from the south and has eroded a deep cut in the drift deposit following its emergence above sea level. The cut runs generally parallel to the profile in Figure 3 and exposes the material in steep banks up to 20 metres in height (Figure 4). The material ranges in grain size from gravel to silt and clay, with coarse sand being most common in the upper half of the river bank. At lower levels, silt and clay predominate except in the seaward section where the coarser material extends from the top to the bottom of the profile.

The stratigraphy of the deposit as it is exposed along the river bank is shown in Figure 3. The most conspicuous feature of the profile is a series of seaward-dipping strata each of almost uniform thickness which strike at about 90 degrees to the direction of the profile. The layers, varying in thickness between one and approximately 25 cm consist of well-sorted material varying in size from fine silts to coarse sand or fine gravel. The

O. H. LØKEN 247

328

dip of the beds is between 18 and 33 degrees, commonly around 24 degrees, and was determined by laying an Abney level or a Brunton compass on the beds. The strata are all but absent from the middle part of the profile, where the slope of the river bank is gentler and covered by vegetation. The permafrost level is high, and it was not possible to make excavations deep enough to expose the actual strata, although their existence can be assumed.

Near the present shore the dipping strata extend from stream level almost to the top of the profile. Farther inland, however, an unconformity separates them from lower, almost flat-lying strata of fine sands, silts and clays. This angular unconformity is accentuated by the change in material and increases in altitude, as shown toward the right of Figure 3. Where the contact is exposed, the lowest 20 cm of the dipping strata become almost horizontal before disappearing.

Figure 3 shows that the dipping strata do not reach the surface of the deposit. In all localities where excavations were made, except near sample 7, the stratification fades toward the top and the uppermost metre shows no stratification. As the distinct stratification disappears, the material becomes poorly sorted.

Well-preserved shells were embedded in many of the layers and samples were collected as indicated in Figure 3. The shells were usually found as complete bi-valves and in only a few cases were the single shells broken prior to excavation. Where occurring in silts and clays, the tubes of some of the shells were preserved.

The altitude and position of the shell samples were determined by T12 theodolite and tachymetry based on known points established along the main profile. The ridges on the surface of the deposit form angles between 80 and 106 degrees with the main profile, and the sample sites shown in Figure 3 have been projected onto the profile along the strike of the ridges.

Terrestrial organic matter was commonly embedded in the sand and silt especially in the strata between samples 4 and 7. Usually the material was scattered among the mineral matter but in some localities, near sample 4 for instance, layers of organic matter one to two cm thick occurred. A sample has been studied by Dr. J. Terasmae of the Geological Survey of Canada, who describes the material as "a detrital accumulation consisting of abundant remains of moss . . . of several species. Small twig fragments of *Ericaceae* and

leaf fragments were found . . . but no pollen or diatoms."

Of particular interest are the fossils found under the highest part of the deposit, sample 1. Due to slumping and vegetation the top part of the section through this point could not be observed but it is assumed that it consists of about 7 metres of steeply dipping strata. Below this at least 12 metres of almost flat-lying beds, predominantly of fine sand, are exposed. The lower beds show less uniformity in grain size and several layers of clay occur. At the top of one of them a shell-bearing gravelly layer was found, and sample 1 was collected from it. Unlike the other samples, it consisted mainly of shell fragments and a complete valve was rarely found. Such whole valves were small strong shells of *Hiatella arctica* and in one case of *Chlamys islandicus*. Evidently these shells were deposited under conditions different from those prevailing when the others were laid down. The shells were possibly broken by re-deposition. At the top of the shell-bearing stratum, a large tussock of grass was found. It has been identified as *Puccinellia* (W. G. Dore, personal communication 1965), a species common in the area today. Driftwood was not found in or on the deposit except along the present shore.

On the basis of these observations the feature is interpreted as a littoral deposit that started to form immediately after deglaciation when the horizontal beds (shown at the extreme right of Figure 3) were deposited in relatively deep water. As the land rose relative to sea level, the material became coarser; and the steeply dipping strata of the highest part of the feature were deposited when the land was submerged to about 44 metres above present sea level. With the continuing emergence of the land, stratum was added to stratum and beach ridge to beach ridge at gradually lower levels and the seaward-sloping ruffled surface was formed. The unsorted material immediately below the surface was deposited as beach ridges above sea level, while the dipping beds were deposited in the sea, close to the shore where wave action is less and the long shore current more significant. These strata overlie the silts and clays deposited earlier in deeper but steadily shallowing water. Cryoturbation could have influenced the top part of the deposit, but the well-preserved beach ridges suggest that this process has not been important.

It is assumed that the processes that formed the beach deposit were similar to those active along the present shore. For comparison with past conditions, three profiles A, B and C were surveyed off the present beach to determine the submarine topography. The approximate position of the profiles is shown in Figure 2 and the profiles themselves in Figure 5. The depth soundings were taken by line from a boat and they were located by taking readings with a level to a stadia rod in the boat.

The three profiles are similar in form and consist of three parts. The upper part shows a very gentle slope, the middle part is relatively steep, while a gentler slope prevails in the third and lowest part. The profiles depict a depositional landform, and it can be assumed that this will be preserved during further deposition. Unsorted material will be deposited in the highest parts where wave action is significant. At greater depth better sorted strata conforming to the profile will be deposited under influence of the longshore current.

The same three elements occur in the raised deposit. The steeply dipping strata correspond to the steep middle sections shown in Figure 5; the dip of the beds is comparable. The unsorted, unstratified coarser material of the surface layer shown in Figure 3 corresponds to the upper section close to the shore. Here the greater energy of the waves provides for the greater variation in grain size while more constant forces act at greater depths. Finally, the lowest parts of profiles A, B, and C correspond to the lower flat-lying sediments seen in Figure 3. There, in the deepest water, the finest sediments are deposited.

The survey of the offshore zone thus supports the statement about the mode of formation of the drift deposit although the offshore stratigraphy has only been inferred and is not known from core samples or drilling. Furthermore, it is clear from the study of the stratigraphy and the shells themselves that, with the possible exception of sample 1, the shells collected were found in living position as the younger strata have buried the older ones without disturbing them.

The material for the deposit is brought by the steep stream that enters the fiord at its eastern end. The size of the boulders on its bed testifies to the stream's great geomorphic activity; the beach deposit extends up to the south side of this stream, but not beyond. A check of the material along the present beach shows the grain size decreases away from the river mouth, thus indicating transport towards the south. The two streams entering the fiord from the south have supplied less although redeposition of material eroded from the already emerged part of the deposit clearly occurs.

The material is carried southward by the longshore current, driven by the prevailing up-fiord wind. During the summer of 1964, 35 per cent of all wind observations taken at base camp between June 27 and August 25, recorded up-fiord wind. Although the observations cover a short period, eolian features in the area show that northeast is the prevailing wind direction. Furthermore, these observations cover a substantial part of the ice-free period, the season when the littoral processes are most active. Break-up at the head of the fiord takes place in early July, while the date of freeze-up is not known. Mean freeze-up date for the bay at Clyde River is October 15 (Allen, 1964), but the fresher surface water at the fiord head probably freezes earlier.

Given a constant supply of material, the rate of prograding would depend only on the length of the beach over which the material was distributed. The length of the beach does not change significantly with altitude, except in its highest part. The slope of the profile in Figure 3 will depend on the rate of both prograding and land emergence. For a certain rate of emergence the steepness of the profile will increase with a decreasing rate of prograding. Similarly, with a constant rate of prograding the steepness of the profile will increase with an increasing rate of emergence. Any break of slope on the profile in Figure 3, therefore, indicates (1) a change in the rate of supply of material, (2) a change in the rate of emergence or (3) a change in both factors.

The upper part of the profile is steeper despite a shorter length of beach at this than at any later stage. The short beach length should indicate the prograding would have been faster at that time than later, and should result in a gentle slope. Thus, the steep slope could be created only by assuming very rapid emergence shortly after deglaciation (*see* below).

The most significant break of slope occurs at an altitude of about 9 metres, where the profile is very gentle. This can be explained by assuming a great increase in the amount of material supplied to the beach, with a constant rate of emergence or, alternatively, by assuming a constant supply of material and a virtual halt

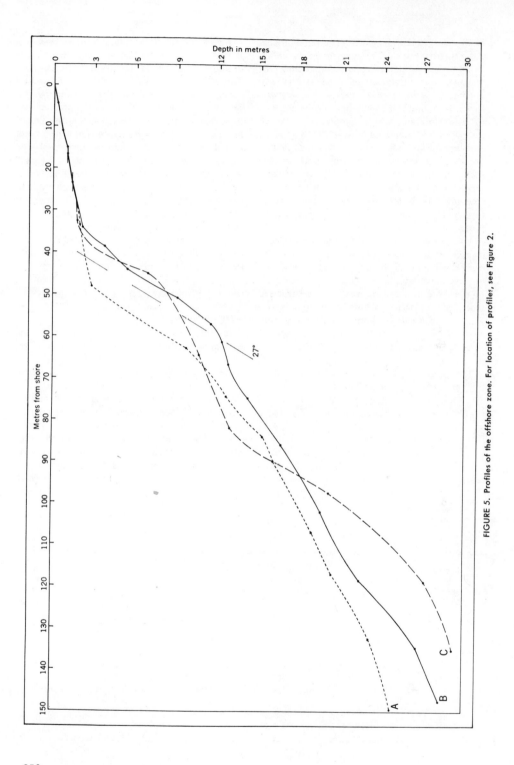

FIGURE 5. Profiles of the offshore zone. For location of profiles, see Figure 2.

in the postglacial emergence. (The latter implies there was no vertical displacement of the shoreline.) It is worth noting that following the irregularity at an altitude of 9 metres, "normal conditions" were re-established, and it is therefore very doubtful if the change can be ascribed to any variation in the supply of material. Support for this conclusion was found in inner Sam Ford Fiord where similar although smaller breaks of slope were observed at approximately the same altitude. The geomorphological evidence seems to imply that a temporary halt in the land emergence occurred when sea level relative to the land was approximately 9 metres above present.

The author has shown (1962) that a postglacial transgression is associated with a strandline in northern Labrador; he suggested that evidence of the same distinct sea level occurs in other parts of the eastern Arctic. It was further suggested that this transgression was contemporaneous with one of the Tapes transgressions in Scandinavia. The almost horizontal part of the deposit at Inugsuin might correspond to the Labrador strandline.

The preceding paragraphs have emphasized the very gently sloping section at an altitude of 9 metres, but an equally important part of the profile is the steep section immediately above 10 metres.

In this context, the altitude of the unconformity (see above) is important. It was measured in three' places between samples 4 and 8, and a minimum altitude recorded above sample 1. A suggested trend of the unconformity has been drawn in Figure 3, and its convergence with the surface profile to the left of sample 8 is noteworthy. The unconformity was not seen in the section through sample 6. A significant change in the stratigraphy therefore must take place between these two samples, and the steepening of the surface profile supports this suggestion. The change is considered to be related to the topography 'on which the beach deposit was formed. Figure 2 shows the lateral and terminal moraine at the head of the fiord and reference has been made to the re-entrant that flowed into the embayment here. The steep part of the profile (Figure 3) is believed to lie at the proximal side of the re-entrant moraine which is associated with the terminal moraine approximately three km northeast of the fiord head.

THE EMERGENCE CURVE

Radiocarbon dates

In order to study the development of the deposit and the postglacial emergence, eight shell samples were dated. Table 1 lists these and a ninth relevant date. The location and number of each sample are indicated in Figures 2 or 3.

There is an apparent discrepancy between the two oldest samples; sample 1, however, is from the bottom of a section approximately 20 metres deep, whereas sample 2 was collected at the top; there can be little doubt about their relative age. The two radiocarbon dates overlap and do not necessarily contradict this statement. A shell

Table 1

Radiocarbon dates

Sample no.	Geog. Br. lab. no.	Altitude of sample (metres)	Isotopes Inc. lab. no.	Age in years	Related sea level (metres above present sea level)
1	OHL–64– 9S	25.6	I–1554	7030 ± 190	>44.5
2	–14S	41.0	I–1598	7200 ± 150	44.5
3	–11S	31.2	I–1596	6150 ± 170	37.0
4	–12S	14.4	I–1597	4090 ± 150	19.5
5	–18S	11.1	I–1600	3520 ± 230	14.5
6	17S	7.7	I–1599	2990 ± 140	9.0
7	–10S	5.4	I–1555	2800 ± 140	8.0
8	–21S	12.8	I–1601	3550 ± 130	?
9	DMB–64– 3S	32.6	I–1602	7900 ± 210	>32.6

sample collected at base camp (by D. M. Barnett) was found to be 7,900 years old, thus showing that the area was ice-free much earlier than indicated by sample 2. While whole valves made up the bulk of the other samples, sample 1 consisted of small fragments and its condition might have influenced the result of the dating. Another possibility is that old shells from higher levels might have been mixed into sample 2, thus giving an exaggerated age. The whole sample in each case was submitted for dating and all dates are based on the inner 80 per cent of each shell sample. None of the dates has been adjusted for the apparent age of the sea water. As none of the species present would bury more than approximately 30 cm into the bottom sediments (F. J. E. Wagner, personal communication), it is therefore assumed that the age of the shell samples is the age of the layer where they were collected.

The samples from the deposit have been arranged in chronological order of decreasing age according to the mode of formation presented above. Sample 8 cannot be fitted into this relative chronology as it was collected in the fine sediments below the dipping strata.

Determination of sea level

In order to establish an emergence curve, it is necessary to relate each dated shell sample to its relevant sea level. This is possible by studying Figure 3.

The existence of the beach deposit shows that considerable accumulation has taken place, and as the land emerged, successive beach ridges have been deposited at gradually decreasing levels. Depending on the rate of emergence and prograding of the shore, abrasion of the top part of the steeply dipping strata might have occurred before emergence. Figure 5 shows that the average slope, approximately 4.5:100 is gentler than the shore profile immediately below high tide level which has a slope of approximately 6:100. Providing the average height of the beach ridges has not changed, this is evidence that the prograding has proceeded faster than the rise of the land and no abrasion of the steep strata has taken place except during storms.

During periods of exceptionally intense wave action, abrasion in the littoral zone extends to greater depths and parts of the dipping strata

may disappear. When "normal" wave conditions are restored, prograding continues and any "scars" created are covered. Evidence of one such episode was found close to sample 7 where the stratigraphy is as shown in Figure 6. The middle bed was truncated under storm conditions and later covered by the younger layer. As this occurrence was unique, it may be concluded that abrasion of the strata was not common. The reason is probably the limited fetch, as approximately 20 km from its head, the fiord makes two sharp bends. Furthermore, it is ice-covered during the winter period when the strongest winds occur.

In order to determine the sea level at which each dipping bed was formed, the strata of Figure 3 were extended upward until they intersected the present surface of the deposit. The level of this intersection is considered as the former sea level. A similar method has been used by Feyling-Hanssen and Olsson (1959-60). The accuracy of the method depends on the slope of the top profile. As seen from Figure 3, the slope is almost constant, and the result is believed to be correct to one metre. Special consideration must be given to the beds of samples 6 and 7 as the surface slope here is much gentler. The sea levels at which these strata were formed were determined by subtracting 1 metre from the altitude of the intersection between the extended strata and the surface. The value of 1 metre is derived from Figure 3 where it will be seen that the present beach ridge rises approximately 1 metre above high tide. In this case also sea level is determined within an accuracy of one metre. It should be noted that the sea level associated with sample 7 is derived in this way, although the locality differs from the others as the dipping strata here extend to the surface of the deposit (*see* above). The implication is that a part of the deposit has been removed thus giving a sea level too low for this sample.

Corresponding dates and altitudes have been plotted in Figure 7, and an emergence curve drawn. The lack of dates from the last 2,800 years should be noted. A constant rate of emergence has been assumed between all points, although additional dates might result in a modified curve. The curve covers the last two-thirds of the postglacial emergence at Inugsuin Fiord; no reliable information is available on the first third.

FIGURE 6. Exposed strata near sample 7. Truncated bed ends at bend near centre of photo.

DISCUSSION

The curve consists essentially of two sections of constant rate of emergence separated by a sharp break approximately 3,000 years ago. Between 7,200 and 3,000 years ago the land emerged at an average rate of approximately 0.8 metre per century; this was reduced later to approximately 0.3 metre per century. The older section is well documented while no date is available in the last 2,800 years.

The emergence of about 21 metres that occurred prior to 7,200 years ago was completed in a period of more than 700 years as indicated by the date of deglaciation of the fiord head. Thus, the rate of emergence did not exceed 3 metres per century. Although this is a maximum value it can be assumed that the emergence prior to 7,200 years ago, was faster than the more recent

emergence. This agrees with the interpretation of the geomorphological evidence.

The bend in the curve appears at the same altitude as the notch on Figure 3, but the curve shows no indication of a halt in the emergence as suggested on the basis of geomorphological evidence. However, there are strong indications (*see* above) that sample 7 is associated with a relative sea level higher than 8 metres which would make a section in the curve in Figure 7 horizontal indicating a halt in the emergence.

Table 1 shows that the pause in the emergence took place some 2,900 years ago. This was at the end of the postglacial warm period and it supports the suggestion of Tapes age made above.

The general form of the curve is similar to that of other published curves, with a steep older and a gently sloping younger part. Detailed comparison with other curves is difficult because of great

O. H. LØKEN 253

97309—7

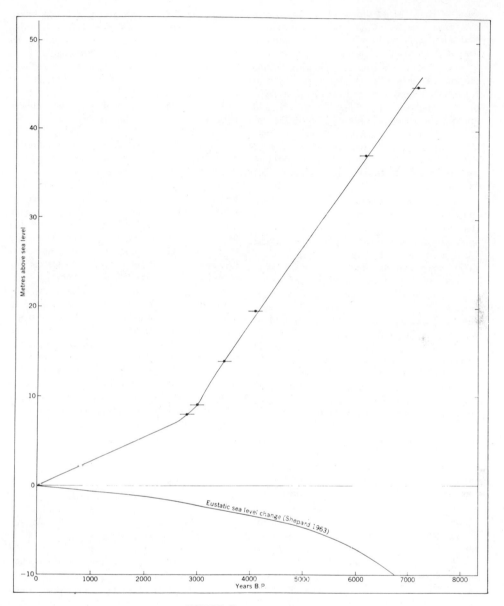

FIGURE 7. The emergence curve.

variations in reliability and differences in glacial history between various localities. A comparison can be made however of the middle third (in terms of total postglacial emergence) of various curves.

On Spitzbergen emergence at an average rate of 2.9 metres per century has been derived from the curve presented by Olsson and Blake (1961-62) and a rate of 2.6 metres per century on the basis of a similar curve published by Feyling-Hanssen and Olsson (1959-60). An emergence curve from East Greenland (Washburn and Stuiver, 1962) shows an average rate of 2.4 metres per century, and Ives' curve (1964, Figure 3) from the west coast of Baffin Island shows an average rate of emergence of 4 metres per century. It is evident that the Inugsuin uplift was very slow relative to that of other areas.

When the average rate of emergence during the last 3,000 years is considered, the uplift at Inugsuin of approximately 30 cm per century is rapid compared with East Greenland and Spitzbergen where averages of 7 to 17 cm per century have been derived from the published curves. Ives' results from the west coast of Baffin Island, however, show an emergence at an average rate of 50 cm per century in the last 3,000 years.

The difference between the east and west coasts of Baffin Island is significant. On the west coast the oldest shells date from about 6,700 years ago and the period of rapid uplift ended almost 5,000 years ago, that is, in the course of some 2,000 years after deglaciation. At Inugsuin Fiord, the oldest shells are 7,900 years old and the period of rapid uplift was still in progress some 5,000 years later. The accuracy of the uplift curves is not the same, and later results might lead to modifications. The difference, however, is so great that it is doubtful whether any major change will be necessary, especially as the oldest shells on the west coast were collected only 16.5 metres below the marine limit and the chances for finding much older shells are, therefore, small. The reason for this difference is not known but it may be related to the more peripheral positions of the Inugsuin area in relation to the centre of the continental glaciation. Another possibility is the presence of local glaciers along the east coast, while the west coast was influenced only by the continental ice and the Barnes Ice Cap. Further detailed work is needed.

Figure 7 shows only the emergence of the land. To indicate crustal uplift the curve must be adjusted for the eustatic changes. Postglacial eustatic changes are still a matter of controversy, but the curve presented by Shepard (1963) has been plotted on Figure 7, and gives a picture of the adjustments that must be made. An adjusted curve will be similar to the emergence curve in form, but both sections will be steeper.

The change in rate of crustal uplift approximately 3,000 years ago coincides with the transition from the sub-Boreal to the sub-Atlantic period. In early sub-Atlantic time glaciers re-advanced in Norway(Liestøl, 1960) and a similar development has been suggested in Canada

(Terasmae, 1961). More recently, lichenometric studies by Andrews and Webber (1964) show that a major moraine system around the Barnes Ice Cap is of sub-Atlantic age. Presumably, a re-advance of the ice cap would be associated with similar increases of the glaciers along the east coast of Baffin Island. The resulting increased load would retard the crustal uplift that followed the disappearance of the continental ice sheet, whereas the isostatic readjustment on the west coast of Baffin Island would not have been affected by any local ice load. This may explain the difference between the emergence curves.

An interesting fact appears from Table 1; all shell samples except no. 1 were collected at shallow and fairly constant depths of between three and six metres below their related sea level. The samples were always taken where the shells were most abundant and it appears that the ecological conditions for the species involved are most favourable at this depth.

Assuming that the deposit was formed as outlined above, it is possible to make an estimate of the rate of deposition by using the radiocarbon dates and the perpendicular distance between the dated strata. From the currently forming steep bed in the offshore zone to the layer of sample 2, there lies a sediment pack some 750 metres thick. As this has been deposited in the course of 7,200 years, the average rate of deposition has been between 10 and 11 cm per year. This corresponds to an average rate of prograding of 13 cm per year.

THE FOSSIL EVIDENCE

The shell samples have been examined by Dr. F. J. E. Wagner of the Geological Survey of Canada. The results are summarized in Table II which lists the samples from top to bottom according to increasing relative age.

When collecting the shells efforts were made to obtain representative samples, but obviously many species might have been excluded, giving an incomplete picture of the fauna. However, some information about the ecological conditions can be derived from Table II. This is particularly true when the argument is based mainly on the species which are present rather than on those which are absent. Dr. Wagner remarks that, "These are all shallow water assemblages indicating depths from zero to probably no more than 50 metres. Water temperatures were probably similar to those prevailing in the area at present."

O. H. LØKEN   255

97309—8

Table II

Macro Fossils

Sample No.	Mya truncata Linné	Hiatella arctica (Linné)	Macoma balthica (Linné)	Mytilus edulis Linné	Clinocardium ciliatum (Fabricius)	Chlamys islandicus (Müller)	Mya pseudo- arenaria Schlesch	Age in years
7		X	X	X				2800 ± 140
6	X	C	C					2990 ± 140
5	R		C					3520 ± 230
4	R	R	A	R				4090 ± 150
3	C	R	R					6150 ± 170
2	X	C			R			7200 ± 150
1	X	X		X	X	X		7030 ± 190
8	X	X	X	X			X	3550 ± 130

A – abundant
C – common
R – rare
X – species present—no particular comment made

In northern Canada, only limited work has been done on the mollusc communities of raised beach deposits (see however, Wagner, in Craig, 1961 and 1964). The most important attempt to deduce information about past climatic conditions on the basis of such communities is still that of Laursen (1946) who examined the shell collection from northern Baffin Island and the west coast of Foxe Basin made more than two decades earlier by the Fifth Thule Expedition. Emphasis since then has been on the collection of shells for radiocarbon dating in order to determine the date of deglaciation and to assess the postglacial emergence. Little or no consideration has been given to the species collected or to the zoogeographical and ecological significance of the faunal composition. Ellis (1955 and 1960) and Ellis and Wilce (1961) studied the present mollusc fauna of eastern arctic Canada, but the extensive works of Jensen (1905), Laursen (1946 and 1950) and Ockelmann (1958) on the past and present mollusc fauna of Greenland have no parallel in arctic Canada.

Of the species collected in the raised beach deposits at Inugsuin, Mya truncata, Hiatella arctica and Clinocardium ciliatum are widely found in arctic areas today. Mya pseudoarenaria and Macoma balthica are also common in arctic waters (F. J. E. Wagner, personal communication).

The two most interesting species are Chlamys islandicus and Mytilus edulis. The former is living in the Disco Bay area of West Greenland (Ellis,

1960) but has not been reported on the west and colder side of Baffin Bay. The species was common in the collections of the Fifth Thule Expedition in samples taken from between 30 and 50 metres above sea level (Laursen, 1946). In the Inugsuin area, the species occurred in sample 7 and also in a sample collected on the proximal side of the large terminal moraine at base camp (Figure 2), at an altitude of 33 metres. The radiocarbon age of this latter sample is 7,900 ± 210 years (Table I). The species has not been found living north of 64°N in eastern Canada (Laursen, 1946) and this suggests that climates warmer than the present prevailed between 7,000 and 8,000 years ago. Laursen's observations are compatible with those made at Inugsuin Fiord where Chlamys islandicus appears only in the oldest deposits. This warmer interval falls in the early part of the postglacial warm period and is thus in agreement with the evidence for postglacial climatic change. Why the species occurred only in the earliest part of the warm period and then apparently disappeared is not known.

Mytilus edulis lives along the west coast of Greenland as far north as Thule (Ockelmann, 1958) and on the east coast of Baffin Island as far north as Padloping Island at 67°N (Ellis, 1955). At Pond Inlet (approx. 73°N) Ellis and Wilce (1961) and B. G. Craig (personal communication) report accumulation of Mytilus shells along the shore. "Some of these contained undecomposed tissues, but attempts to find living colonies were unsuccessful but there is a distinct possibility of

an intertidal population of *M. edulis* in that area"
(Ellis and Wilce 1961, p. 233).

Mytilus edulis is a common species in the deposit at Inugsuin, and is represented by some remarkably large specimens. A shell from a sea level of 19.5 metres above present was more than 80 mm long. This is larger than any of the specimens reported from Spitzbergen by Feyling-Hanssen (1957), and from West Greenland, Ockelmann (1958) reports only one shell longer than this. It follows, therefore, that this species in the past found very favourable ecological conditions in an area that today is on the fringe of its geographical distribution. This lends further support to the suggested warmer climates in the past.

The common occurrence of *Mytilus edulis* at several levels in the raised deposit is noteworthy because this species has been reported from only a limited number of localities in the Canadian arctic. Washburn (1947) reports *Mytilus edulis* in a raised deposit along the southwest coast of Victoria Island. Farther east Blake (1963) found shells of this species in Bathurst Inlet, and Craig (1961 and 1964) reported it from localities west of Pelly Bay and along the east coast of Boothia Peninsula. Along the shores of the Hudson Strait and Bay, only fragments of *Mytilus edulis* have been found: by the Fifth Thule Expedition at Baker Lake (Laursen, 1946) and in the Belcher Islands–James Bay area (F. J. E. Wagner, personal communication). It thus appears that *Mytilus edulis* in the Canadian arctic has been restricted to the Baffin Bay–Labrador Sea coast south of Lancaster Sound and to the channels immediately to the north of the mainland coast from the Mackenzie Delta area and eastward to Boothia Peninsula.

Laursen (1946) quoted *Astarte montagii* var. *striata* (Leach) as evidence of warmer climates in the past, and the occurrence of this species (F. J. E. Wagner, personal communication) in a sample (OHL–64–15S) collected at Inugsuin Fiord lends further support to the conclusion reached above. The shells were collected at 35.3 metres above sea level, and are stratigraphically slightly younger than the sample dated 7,900 years old. The sub-species has not been reported living in this area or farther north on Baffin Island.

The same species also occurs in a sample collected at an altitude of 30 metres in Ikpik Bay on the west coast of Baffin Island (J. T. Andrews,

personal communication). The emergence curve from that area shows that a point at this altitude emerged above sea level some 5,000 years ago, and the shells must predate this event.

The existence of strata of organic material in the beach deposit was referred to above and it was pointed out that they were particularly common between samples 4 and 5 (near the large .*Mytilus* shell) and therefore deposited 3,500 to 4,000 years ago. Similar occurrences of twigs and plant material have been found in marine deposits along the west coast of Baffin Island (J. T. Andrews, personal communication) and along the south shore of Hudson Strait (B. Matthews, personal communication). In both cases, the beds were between 3,500 and 4,000 years old. Such accumulations can be explained by assuming an abundant vegetation in the local area, or by a longer ice-free period thus allowing floating matter more time to drift in along the shore. Either explanation requires a more favourable climate.

In summary, the fossil evidence shows that mollusc fauna of the raised beach deposit is not very different from that found in the area at present. The presence of *Chlamys islandicus*, *Mytilus edulis* and *Astarte montagii* var. *striata* (Leach) indicates that ecological conditions have been less severe in the past as two of these species have not been found living in the area; a warmer climate can therefore be inferred.

ACKNOWLEDGMENTS

The writer is indebted to T. W. Fielding and B. B. Smithson for assistance in the field, and to Dr. F. J. E. Wagner, Geological Survey of Canada, who identified the shells. Dr. W. Blake Jr., Geological Survey of Canada and Drs. J. T. Andrews, M. J. J. Bik and J. D. Ives, Geographical Branch, read the manuscript and contributed many valuable comments.

REFERENCES

Allen, W. T. R.
    1964 : Break-up and freeze-up dates in Canada. Met. Br., Dept. Trans., Circ. 4116, 201 p. mimeo.
Andrews, J. T. and Webber, P. J.
    1964 : A lichenometrical study of the northwestern margin of the Barnes Ice Cap: a geomorphological technique. *Geog. Bull.*, no. 22, 80-104.

O. H. LØKEN 257

Blake, W. Jr.
1961 : Radiocarbon dating of raised beaches in Nordaustlandet, Spitsbergen. Raasch, G. O. (ed.), *Geology of the Arctic*, v. 1, 133-145. Univ. of Toronto Press, Toronto.
1963 : Notes on glacial geology, northeastern District of Mackenzie. Geol. Surv. Can., Paper 63-28, 12 p.

Craig, B. G.
1961 : Surficial geology of northern District of Keewatin, Northwest Territories. Geol. Surv. Can., Paper 61-5, 8 p.
1964 : Surficial geology of Boothia Peninsula and Somerset, King William and Prince of Wales islands, District of Franklin. Geol. Surv. Can., Paper 63-44, 10 p.

Ellis, D. V.
1955 : Some observations on the shore fauna of Baffin Island. *Arctic*, v. 8, no. 4, 224-236.
1960 : Marine infaunal benthos in Arctic North America. Arctic Inst. of N. A., Tech. Paper no. 5, 53 p.

Ellis, D. V. and Wilce, R. T.
1961 : Arctic and subarctic examples of intertidal zonation. *Arctic*, v. 14, no. 4, 224-235.

Farrand, W. R.
1962 : Postglacial uplift in North America. *Am. J. Sci.*, v. 260, no. 3, 181-199.

Feyling-Hanssen, R. W.
1955 : Stratigraphy of the marine late Pleistocene of Bille-fjorden, Vestspitsbergen Norsk Polarinstitutt, Skrifter no. 107, 186 p.

Feyling-Hanssen, R. W. and Olsson, I.
1959 : Five radiocarbon datings of postglacial shorelines in central Spitsbergen. *Norsk Geog. Tids.*, bind 17, h. 1-4, 122-131.

Henoch, W. E. S.
1964 : Postglacial marine submergence and emergence of Melville Island, N.W.T. *Geog. Bull.*, no. 22, 105-126.

Ives, J. D.
1964 : Deglaciation and land emergence in northeastern Foxe Basin, N.W.T. *Geog. Bull.*, no. 21, 54-65.

Ives, J. D. and Andrews, J. T.
1963 : Studies in the physical geography of north-central Baffin Island, N.W.T. *Geog. Bull.*, no. 19, 5-48.

Jensen, Ad. S.
1905 : On the mollusca of East Grönland I Lamellibranchiata. *Medd. om Grönland*, v. 29, no. 9, 287-362.

Laursen, D.
1946 : Quaternary shells collected by the Fifth Thule expedition 1921-24. Report on The Fifth Thule expedition, v. I, no. 7, 59 p.
1950 : The stratigraphy of the marine Quaternary deposits in West Grönland. *Medd. om Grönland*, v. 151, no. 1, 152 p.

Liestøl, O.
1960 : Glaciers of the present day. Holtedahl, O. (ed.), *Geology of Norway*, Norges Geol. Unders, no. 208, 482-490.

Løken, O. H.
1962 : The late-glacial and postglacial emergence and the deglaciation of northernmost Labrador. *Geog. Bull.*, no. 17, 23-56.

Ockelmann, W. K.
1958 : The zoology of East Greenland. Marine Lamellibranchiata. *Medd. om Grönland*, v. 122, no. 4, 256 p.

Olsson, I. U., and Blake, W. Jr.
1961-62 : Problems of radiocarbon dating of raised beaches, based on experience in Spitsbergen. *Norsk Geog. Tids.*, bind 18, h. 1-2, 47-64.

Shepard, F. P.
1963 : Thirty-five thousand years of sea level. Clements, T. (ed.), *Essays in marine geology in honour of K. O. Emery*, 1-10.

Terasmae, J.
1961 : Notes on late Quaternary climatic changes in Canada. *Ann. N.Y. Acad. Sc.*, v. 95, 658-675.

Washburn, A. L.
1947 : Reconnaissance geology of portions of Victoria Island and adjacent regions, Arctic Canada. Geol. Soc. Am. Mem. 22, 142 p.

Washburn, A. L. and Stuiver, M.
1962 : Radiocarbon-dated postglacial delevelling in northeast Greenland, and its implications. *Arctic*, v. 15, no. 1, 66-73.

Editor's Comments on Paper 20

Fillon: *Possible Causes of the Variability of Postglacial Uplift in North America*

Investigations of postglacial uplift by several authors has led to the suggestion that the decay constant, k, in the exponential equation is constant at least within the error limits of field and ^{14}C measurements. This means, of course, that the relaxation time for uplift is constant. Such a constant is implicitly or explicitly assumed in many geomorphological studies of postglacial uplift which utilize the elevation of one strandline of known age as a means of predicting the ages of higher (older) waterplanes. Schofield (1964, *N.Z.J. Geol. Geophys.*), Andrews (1968, *Can. J. Earth Sci.*), and Ten Brink (1971, Ph.D. Thesis, Univ. Washington) have proposed that postglacial curves from Fennoscandia, Arctic Canada, and Greenland, respectively, have similar decay constants *within* each major region, although relaxation times vary *among* the three regions. Interpretations to the contrary have been reported by McConnell (Paper 25).

Fillon draws on the literature and his own field observations to plot various decay constants against distance from the former ice margin (D) and the date of deglaciation (A). Andrews (1970, *Inst. Br. Geog. Spec. Publ. 2*) has used basically the same data and has suggested that k appears to be independent of both A and D and was constant with a sample mean of $0.44 \pm 0.07 \times 10^{-3}$ yr^{-1} (i.e., relaxation time of 1/0.44). Fillon replots the values using the exponential uplift rate as the dependent variable and interprets the resulting graph (Fig. 2) as reflecting variations in postglacial uplift as a function of variable isostatic adjustments in a layered earth. It also indicates variations in the pattern of deglaciation. Based on Fig. 2, Fillon suggests the existence of five major deformation regions at distances of about 1–20, 20–400, 400–1100, 1100–1500, and > 1500 km from the former ice edge that have responses controlled by mantle properties and deglacial history.

Richard H. Fillon received his Ph.D. from the University of Rhode Island in 1973 in the field of marine geology. He was born in 1944 and received his B.S. from Rennsselaer Polytechnic Institute in 1966. He worked for Chevron Oil Company prior to attending the University of Vermont, where he obtained his M.S. in 1970.

Copyright © 1972 by the University of Washington, Seattle

Reprinted from *Quaternary Res.*, **1**, 522–531 (1972)

20

Possible Causes of the Variability of Postglacial Uplift in North America

RICHARD H. FILLON [1]

Received February 1, 1971

Postglacial uplift data from 33 sites in northeastern North America reveal that during the period from 11,000 years B.P. to 7000 years B.P., glacio-isostatic uplift rates varied in a consistent manner with distance from the former margin of the Laurentide Ice Sheet. The consistent trends of these uplift rate variations with distance from the former ice sheet margin suggest that they were not the result of changes in the rate of ice sheet retreat or local tectonic activity. They instead may have resulted from rebound affected significantly by the earth's viscosity at a depth approximately equal to the wavelength of isostatic deformation [McConnell, R. K., Jr., *Journal of Geophysical Research* **70**, 5171 (1965)]. Extremely high viscosities below 600 km, however, probably provide the lower limit for this relationship.

INTRODUCTION

Extensive investigations of glacio-isostatic uplift in Canada (Andrews, 1968a,b) have indicated that relationships are not entirely predictable between the amount of uplift (U_p), rate of rebound $(dU_p)/dt)$, approximate date of deglaciation (A), and distance from the former maximum limit of the Laurentide Ice Sheet (D). An attempt to empirically describe uplift by means of a quadratic polynominal in which $U_p = f(D,A)$ has been carried out by Andrews (1968b). However, in the same paper it was shown that the coefficient of determination for the equation is 80% for uplift data from Arctic Canada. Andrews regarded that some of the remaining uplift variability may be due to the effect of possible local tectonic activity on rebound, in addition to some inherent imprecision in the data.

Broecker (1966) found that field data from the Lake Huron area deviate from his own theoretically predicted curve of uplift versus distance, which was based on a constant rate of ice margin retreat. He proposed that these differences resulted from fluctuations in the rate of ice margin retreat. Brotchie and Silvester (1969) considered a linear increase in rate of ice margin retreat but found similar discrepancies between predicted and field values for uplift. Variability in postglacial uplift was also discussed by Mörner (1970), who observed that, in general, curves representing isostatic uplift versus time since deglaciation in Europe and North America fluctuate for the period from deglaciation to 7000 years B.P. but are smooth frm 7000 years B.P. to the present day. Mörner attributes these fluctuations to response of the crust to changes in ice load inland and water load in the oceans during deglaciation.

It is reasonable to expect that a portion of this variability of glacio-isostatic uplift is

[1] Graduate School of Oceanography, Narragansett Marine Laboratory, University of Rhode Island, Kingston, Rhode Island 02881.

due to the imprecision in ^{14}C age dating and in the estimation of shoreline elevations and eustatic sea level changes. However, by dealing with uplift data from a variety of locations that differ in time of deglaciation, it is hoped that the effect of experimental and other errors in the field data can be minimized.

The object of this study is to contribute to the understanding of the variability of postglacial uplift rates in North America. A possible explanation of the areal variability of uplift rates will also be offered.

SOURCES OF DATA AND METHODS OF ANALYSIS

The data used in this paper (Table 1, Fig. 1) have been obtained from 28 sites in the Canadian Arctic (Andrews, 1968a,b; Webber et al., 1970; Andrews and Falconer, 1969; Andrews et al., 1970), 4 sites in southern Canada (Farrand, 1962), 1 site in northern Vermont (Fillon, 1969), and 1 site in Utah (Crittendon, 1963). For each of his sites, Andrews (1968a,b) solved the equation,

$$U_p = a + r \ln t, \qquad (1)$$

where a and r (which is equal to $\Delta U_p / \Delta \ln t$) are constants, and t represents time since deglaciation. This equation provides a close fit to glacio-isostatic uplift data for the period after $t = 1000$ years. Andrews also presented the times of deglaciation and distances along assumed flow lines from the former ice sheet margin. Webber et al. (1970), Farrand (1962), Andrews and Falconer (1969), and Andrews et al. (1970) each gave approximate dates of deglaciation and presented uplift versus time curves from which $\Delta U_p / \Delta \ln t$ (here defined as the exponential uplift rate) could be calculated. Distances from the former ice sheet margin were determined by measuring along ice movement flow lines for the Woodfordian Substage inferred from the "Glacial Map of the United States East of the Rocky

Mountains" (Flint, 1959). The Vermont site (Fillon, 1969) yielded an additional value of exponential uplift rate. Distance from this site to the Long Island morainic system was measured parallel to the axis of the Champlain–Hudson Valley which controlled the direction of ice movement (Connally and Sirkin, 1969).

To test relations between exponential uplift rate (r), distance from the former ice sheet margin (D), and date of deglaciation (A), r was plotted against D and A (Fig. 2). The shape of the resulting three-dimensional surface $[f(r,D,A)]$ is represented by isochrons of date of deglaciation.

DISCUSSION

Possible Mechanism

It is apparent (Fig. 2) that the deviations of the exponential uplift rate about the least squares straight line are not random, but are cyclic with distance from the former maximum extent of the ice sheet. For a given distance, the rate of uplift decreased as the age of deglaciation increased.

The cyclic variations of exponential uplift rates indicate that during deglaciation in North America, there existed, at constant distances (D) from the ice sheet margin (see isolines of D, Fig. 1), alternating concentric bands of relatively high and low rates of uplift

$$\left(\frac{dU_p}{dt_{t>1000 \text{ yr}}} \right),$$

which is defined as the rate of uplift for a site at time t after deglaciation at that site. $dU_p/dt_{\ t>1000 \text{ yr}}$ variations have been determined from exponential uplift rates according to the relationship,

$$\frac{dU_p}{dt} = \frac{r}{t}$$

To test the uniqueness of Fig. 2 and to investigate the possibility of extending the inferences drawn from Fig. 2 to the period

TABLE 1

DATA USED AS BASIS FOR INVESTIGATION OF VARIABILITY OF POSTGLACIAL UPLIFT IN NORTH AMERICA

Site	Date of deglaciation (A) (yr B.P.)	Distance from former ice margin (D) (km)	Decay constant (k) (yr^{-1} $\times$ 10^3)	Exponential uplift rate (r) (m/ln yr)	Data source
1	7300	1500	0.50	51.8	
2	8400	850	0.53	51.6	
3	10,200	600	0.43	47.5	
4	9200	700	0.37	49.2	
5	8800	1000	0.42	44.9	
6	11,300	150	0.54	15.5	
7	9200	150	0.56	26.7	
8	8800	450	0.45	33.0	
9	6800	1300	0.40	42.4	
10	7000	1000	0.43	59.5	
11	6900	680	0.50	65.6	Andrews (1968b)
12	8000	180	0.49	18.5	
13	8400	115	0.40	32.6	
14	9000	70	0.40	27.7	
15	7900	160	0.43	27.9	
16	6500	150	0.48	23.4	
17	8000	160	0.38	25.8	
18	8000	770	0.34	63.8	
19	7900	1500	0.41	77.6	
20	6800	350	0.45	35.0	
21	6200	330	0.46	29.1	
22	10,500	800	0.32	33.0	
23	10,000	550	0.35	72.1	Farrand (1962)
24	10,500	400	0.36	13.2	
25	14,000	450	0.19	8.8	
26	11,700	500	0.34	41.7	Fillon (1969)
27	7500	1500	0.43	117.0	Webber et al. (1970)
28	7400	1500	0.44	47.5	Andrews and Falconer (1969)
29	9250	60	0.31	22.0	
30	8750	75	0.36	26.6	
31	8250	60	0.36	24.1	Andrews et al. (1970)
32	9750	50	0.28	38.8	
33	9250	75	0.33	33.2	
34	14,000 [a]	120 [b]	0.17	18.6	Crittendon (1963)

[a] Date of the unloading of Lake Bonneville is about 20,000 B.P.; however, according to Crittendon, >99% of uplift was completed from 4000 to 6000 years B.P.

[b] Location, Bonneville pluvial lake in Utah; D, distance from former shoreline to lake center.

between deglaciation and 1000 years after deglaciation, two additional parameters were studied. First, changes in k, the decay constant (Table 1), from the equation

$$U_p = U_{p\ max} (1 - e^{-kt}) \qquad (2)$$

were investigated with respect to distance from the ice sheet margin (D). Variation of the decay constant with distance appears to

be generally similar to, but not nearly as uniform as, variation of the exponential uplift rate with distance (Fig. 2). This lack of correspondence may be due to the inverse proportionality of the decay constant to the relaxation time (τ),[2] which is not only de-

[2] Relaxation time is defined as the time required for the amplitude of deformation to be reduced by a factor of $1/e$.

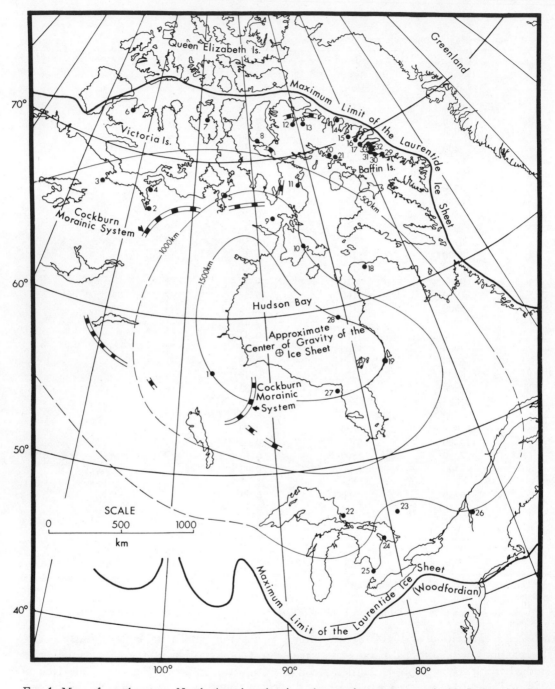

Fig. 1. Map of northeastern North America showing the maximum extent of the Laurentide Ice Sheet (Andrews, 1968b; Flint, 1959) and the Cockburn morainic limit (Falconer *et al.*, 1965). The approximate center of gravity of the ice sheet is from Andrews (1968b). Site numbers (shown in Fig. 2) are indicated by arabic numerals. Contour lines connect points of equal distance from the maximum limit of the ice sheet margin as measured along ice flow lines.

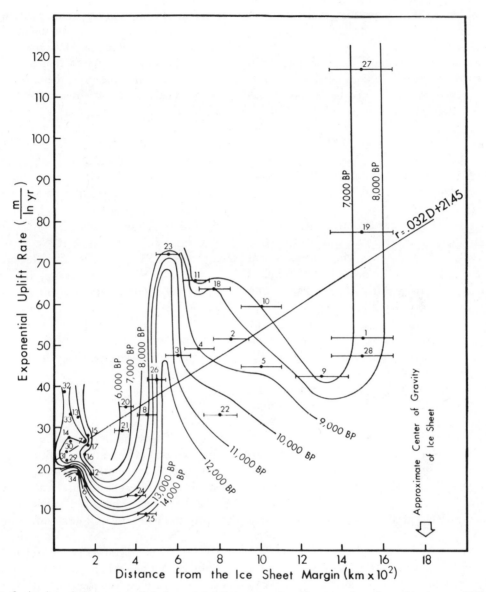

FIG. 2. A plot of the exponential slope (r) of the equation $U_p = a + r \ln t$ (Andrews, 1968a,b) versus distance from the furthest extent of the former ice sheet margin measured along ice flow lines (D). Dates of deglaciation (A) are indicated by isochrons. The center of gravity of the ice sheet is placed at 1800 km from the maximum former extent of the ice sheet (Andrews, 1968b, p. 406, Fig. 1). Site numbers (Table 1; Fig. 1) are indicated by arabic numerals. $r = 0.032\ D + 21.45$ is the least squares straight line through the data points.

pendent on the rate of uplift but also on the maximum amount of depression $(U_{p\ \max})$. Thus, it may be that prior to 1000 years after deglaciation in areas subjected to greater than the expected amounts of isostatic depression (Brotchie and Silvester, 1969) uplift rates were not high enough to compensate. This has resulted in relatively low values of k. In constrast, r is unaffected by variations in U_p before 1000 years after deglaciation.

Second, the variability of uplift rates with D for the period less than 1000 years after deglaciation was examined using the derivative of Eq. (2), which is

$$\frac{dU_p}{dt} = - U_{p \, max} \, ke^{-kt}. \qquad (3)$$

It can be shown according to this equation that the initial uplift rate (at $t - 0$ years) is $-U_{p \, max}k$. Therefore, as t approaches zero, the trend of uplift rate variation with distance appears to depart significantly from the relationship r/t (which is proportional to r in Fig. 2) and approaches the trend of k with D. Thus, uplift rates calculated from Eq. (3) for the period preceding 1000 years after deglaciation are dependent on the same assumptions about the relation between $U_{p \, max}$ and uplift rate that are implicit in the determination of k.

In summary, it has been previously assummed that Eq. (2) gives a true representation of uplift for all times since deglaciation. Instead, lack of correspondence between the decay constant and the exponential uplift rate seems to indicate that various amounts of crustal recovery occurring beneath the ice before deglaciation and differences in ice sheet retreat rates combined during the first 1000 years after deglaciation to produce nonexponential rebound.

Because, however, the widespread uplift rate fluctuations for t greater than 1000 years (Fig. 2) have remained generally similar in form during a period of 8000 years, it is unlikely that these fluctuations could have been caused by local tectonic activity (Andrews, 1968b) or variations in the speed of ice sheet retreat (Broecker, 1966). Linear ice sheet volume reduction since 19,000 years B.P. demonstrated by Farrand (1965) futher minimizes the possibility of extensive variations in rate of ice sheet retreat. A possible explanation for the ob-

served uplift rate variation can be found in the theory of isostatic adjustments in a layered earth presented by McConnell (1965, 1968). McConnell suggests that a less viscous layer at a depth (d) in the earth would produce a relatively short relaxation time for surface deformation having a wavelength (λ) approximately 1–6 times d (Fig. 3).[3] He supports this theory by presenting a 1400-km deep rheological profile that closely fits the observed relaxation times for Fennoscandia. It is probable, however, that the effects of glacio-isostatic loading are insignificant below a discontinuity at 600 km where viscosities as high as 10^{26} cm²/sec are postulated (Knopoff, 1967). McConnell's relationship may be directly applied to the data treated in this paper within the limits imposed by viscosity distributions (as defined by gravity and seismic evidence and the equatorial bulge) because an increase or decrease in τ results respectively in a decrease or increase in r.

Estimation of Relative Viscosities from Uplift Rates

Figure 2 may be interpreted as being the sum of several deformational units, each having its own wavelength of deformation. The first (λ_1) consists of sites near the periphery of the ice sheet where the value for λ_1 is roughly the width of ice lobes and reentrants into the ice margin. These are on the order of 10 km (Champlain–Hudson Valley lobe) to 200 km (Lake Michigan lobe) in width and are controlled by viscosities at depths of less than 50 km (Fig. 3). This zone is restricted to within 200 km of the ice sheet margin and is represented by relatively *high exponential uplift rates* (Fig. 2) perhaps due to an elastic layer (McConnell, 1968). The second deformational unit (λ_2) extends from about 200 to 400 km from the ice sheet

---

[3] The term *wavelength* is used because glacio-isostatic deformation may be described as the sum of several superimposed depressions having the shape characteristics of sine wave segments.

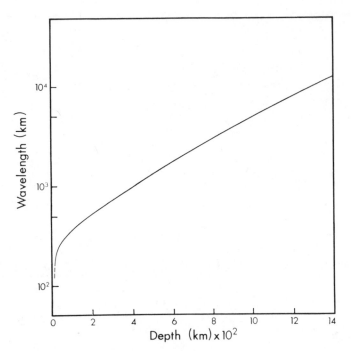

Fig. 3. A generalized curve showing the relationship between wavelength of isostatic deformation (λ) and the depth (d) of greatest movement of mantle material during recovery. This curve has been inferred from plots of wavelengths versus relaxation time and viscosity versus depth presented by McConnell (1968).

margin. It is a region of relatively *low r values*. In this case, the ice margin retreated along a broad front (cf. Fig. 1 and 4). The value of λ_2 appears to have been about 1000–3000 km as measured along chords connecting points of significant change of curvature on the isochrons of deglaciation, indicating control of rebound by a relatively viscous layer at depths from 400 to 800 km. This agrees well with McConnell's (1968) model (No. 66-22), which is designed to fit viscosity data derived from gravity anomalies and the equatorial bulge of the earth.

The third deformational unit (λ_3) extends from approximately 400 to 1100 km from the ice sheet margin and includes sites having relatively *high r values* (Fig. 2). In this region, dates of deglaciation (Fig. 4) indicate that the retreating ice sheet margin was embayed. The widths of the embayments and remaining ice lobes in this region were about 500–1000 km, which implies (Fig. 3)

control of rebound by a relatively low viscosity layer at a depth of 200–400 km. This also agrees with McConnell's (1968) model (No. 66-22). The fourth deformational unit from about 1100 to 1500 km (Figs. 1 and 2) is represented by only three stations but may indicate another zone of ice retreat along a broad front. The fourth deformational unit grades into a fifth from 1500 to 1800 km, and the increase in *r* values probably reflects the decrease in size of the remaining ice sheet and wavelength of depression as total deglaciation drew near.

The procedure used to determine viscosity distribution is admittedly based on several assumptions and approximations, and the results are not to be considered conclusive. However, agreement between the relative viscosity distribution as inferred from North American uplift data and viscosity profiles derived from other data (McConnell, 1968) tends to support the idea that wavelength of

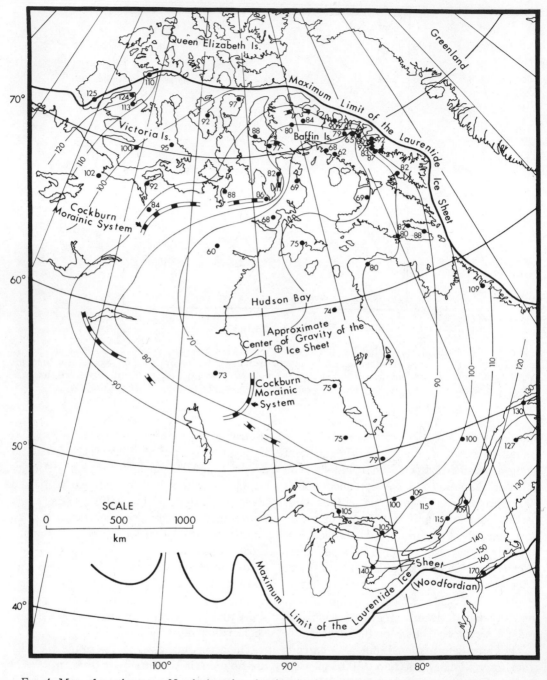

FIG. 4. Map of northeastern North America showing isochrons of date of deglaciation based on data from Table 1 (sites 1–33) and from Andrews (1970). Arabic numerals are dates of deglaciation in years $\times 10^2$.

deformation and viscosity distribution may contribute significantly to the observed variability in uplift data. Fluctuations of uplift rates with distance from the ice margin (Fig. 2), and fluctuations of uplift–time relations (Mörner, 1970) and uplift–distance curves (Broecker, 1966), may thus be attributed in part to variations of deformational wavelength and the resulting effect on uplift of regions of differing viscosities in the crust and upper mantle. On maps of uplift rates for various times (Andrews, 1970), the influence of changing deformational wavelengths is masked by asymmetrical ice sheet retreat.

Causes of Noncyclical Exponential Uplift Rate Variation

A gradual increase of uplift rate with distance from the ice sheet margin, indicated by the slope of the least squares straight line through the data (Fig. 2), is a result of the decrease of maximum isostatic depression with distance from the center of the former ice cap (Farrand and Gajda, 1962; Andrews, 1968b). The nearly linear increase in rate of uplift with decrease in time since deglaciation, at a given distance from the ice sheet margin, is considered by the writer to be most reasonably explained by accelerated ice margin retreat that probably occurred as the ice sheet area was depleted, thus creating a high margin to area ratio. This explanation is consistent with the evidence of linear ice sheet volume reduction reported by Farrand (1965) that also implies accelerated ice margin retreat.

Discussion of Irregularities in Exponential Uplift Rate Variation and Data Errors

Nearer than 200 km to the former ice sheet margin, the isochrons deviate from the smoothly fluctuating curve. This may be attributed in part to the effect of sub-ice topography on the amount of isostatic depression along the thinner periphery of the ice sheet. The error involved in measuring D is probably no greater than about $\pm 10\%$, so that in all but the extreme central portion of the glaciated area the effect of this error on the form of the empirical curve is minimal. Errors involved in determination of the date of deglaciation include imprecision in

the ^{14}C age dating technique as well as in estimation of the relationship of dated material to the maximum marine limit. A conservative estimate of the total deviation due to dating errors at most locations is probably ± 500 years, an amount that is not sufficient to substantially affect the general trend of the isochrons (Fig. 2). Calculation of the exponential uplift rate requires the fitting of a straight line to uplift–time data on semilogarithmic graph paper and the calculation of the slope of that line. The fit obtained has been consistently excellent for the period greater than 1000 years since deglaciation. The slopes determined, therefore, probably do not vary by more than $\pm 2\%$. This amount of deviation also has only a negligible effect on the trend of exponential uplift rate variation.

CONCLUSIONS

1. Exponential rate of uplift varies approximately cyclically as a function of date of deglaciation, and distance from the maximum extent of the Laurentide Ice Sheet. Therefore, the suggestion by Andrews (1968a) that the decay constant (k) is uniform across North America is correct only as a first approximation.

2. Fluctuations in uplift rate have possibly been caused by varying wavelengths of isostatic deformation. This proposal uses the premise (McConnell, 1965) that relaxation times (and therefore uplift velocities) associated with a deformational wavelength of λ are influenced by the earth's viscosity at a depth of one to six times λ.

This model should provide a basis for further quantitative studies on the prediction of glacio-isostatic uplift. Uplift curve segments with differing slopes or decay constants may be the result of changing wavelengths of deformation in an area.

ACKNOWLEDGMENTS

I would like to thank James P. Kennett, W. Phillip Wagner, and Norman D. Watkins for their critical appraisal of the manuscript and suggestions for its improvement. During the study, the author was supported at the University of Vermont by a Water Resources Research grant to Allen S. Hunt, and at the University of Rhode Island by NSF grants GA 27092 and GA 21485 to James P. Kennett. I also thank Lianne Armstrong for drafting the figures.

REFERENCES

ANDREWS, J. T. (1968a). Postglacial rebound: similarity and prediction of uplift curves. *Canadian Journal of Earth Sciences* 5, 39–47.

ANDREWS, J. T. (1968b). Pattern and cause of variability of Postglacial uplift and rate of uplift in Arctic Canada. *Journal of Geology* 76, 404–425.

ANDREWS, J. T. (1970). Present and Postglacial rates of uplift for glaciated northern and eastern North America derived from Postglacial uplift curves. *Canadian Journal of Earth Sciences* 7, 703.

ANDREWS, J. T., and FALCONER, G. (1969). Late Glacial and Postglacial history and emergence of the Ottawa Islands, Hudson Bay, Northwest Territories: Evidence on the deglaciation of Hudson Bay. *Canadian Journal of Earth Sciences* 6, 1263–1276.

ANDREWS, J. T., BUCKLEY, JANE T., and ENGLAND, J. H. (1970). Late-glacial chronology and glacio-isostatic recovery, Home Bay, East Baffin Island, Canada. *Geological Society of America Bulletin* 81, 1123–1148.

BROECKER, W. S. (1966). Glacial rebound and the deformation of the shorelines of proglacial lakes. *Journal of Geophysical Research* 71, 4777–4783.

BROTCHIE, J. F., and SILVESTER, R. (1969). On crustal flexure. *Journal of Geophysical Research* 74, 5240–5252.

CONNALLY, G. G., and SIRKIN, L. A. (1969). Deglacial events in the Hudson–Champlain Valley and their possible equivalents in New England. Abstracts for 1969, North East Section of the Geological Society of America, Part I.

CRITTENDON, M. D. (1963). Effective viscosity of earth derived from isostatic loading of Pleisto-cene Lake Bonneville. *Journal of Geophysical Research* 6, 5517.

FALCONER, G., ANDREWS, J. T., and IVES, J. D. (1965). Late Wisconsin end moraines in Northern Canada. *Science* 147, 608–610.

FARRAND, W. R. (1962). Postglacial uplift in North America. *American Journal of Science* 260, 181–199.

FARRAND, W. R. (1965). The deglacial homicycle. *Geologische Rundschau* 54, 385–398.

FARRAND, W. R., and GAJDA, R. T. (1962). Iso-bases on the marine limit in Canada. *Geographic Bulletin* 5–22.

FILLON, R. H. (1969). "Sedimentation and Recent Geologic History of the Missisquoi Delta, Lake Champlain, Vermont." Unpublished Master's thesis, University of Vermont.

FLINT, R. F. (1959). "Glacial Map of the United States East of the Rocky Mountains," published by the Geological Society of America, New York, compiled and edited by a Committee of the Division of Earth Sciences, National Research Council, Washington, D.C.

KNOPOFF, L. (1967). In "The Earth's Mantle" (T. F. Gaskell, ed.), Academic Press, New York.

McCONNELL, R. K., JR. (1965). Isostatic adjustments in a layered earth. *Journal of Geophysical Research* 70, 5171–5188.

McCONNELL, R. K., JR. (1968). Viscosity of the mantle from relaxation time spectra of isostatic adjustment. *Journal of Geophysical Research* 73, 7089–7105.

MÖRNER, NILS-AXEL. (1970). Isostasy and eustasy, Late Quaternary isostatic changes in southern Scandinavia and general isostatic changes of the world. Abstracts of Papers presented at the International Symposium on Recent Crustal Movements and Associated Seismicity, Royal Society of New Zealand.

WEBBER, P. J., RICHARDSON, J. W., and ANDREWS, J. T. (1970). Postglacial uplift and substrate age at Cape Henrietta Maria, Southeastern Hudson Bay, Canada. *Canadian Journal of Earth Sciences* 7, 317–325.

Erratum: p. 349, 2nd col., 8th line of Conclusion 2, should read: "a depth of one to one-sixth times λ."

Editor's Comments on Paper 21

Mörner: *Eustatic Changes During the Last 20,000 Years and a Method of Separating the Isostatic and Eustatic Factors in an Uplifted Area*

Workers study postglacial uplift (e.g., Washburn and Stuiver, Løken, Andrews) have corrected the present elevation of dated marine deposits by using a eustatic sea level curve derived from another region that was not glaciated and ideally is supposed to be "stable." In this paper Mörner effectively argues that the most detailed eustatic sea level curve can only be reconstructed from a glacio-isostatically rising area where the detailed stratigraphy of minor relative sea level changes can be detected within the raised deposits; such an area is southern Sweden. In 1964 Schofield (*N.Z.J. Geol. Geophys.*) has outlined a mathematical method of determining the eustatic sea level changes from postglacial emergence curves, but the method was quite different from that used here by Mörner.

Postglacial uplift curves from Greenland, Arctic Canada, and parts of Fennoscandia have all been portrayed as smoothly decelerating, but in contrast Mörner indicates that in southern Sweden the early part of the uplift was marked by fluctuations caused by variations in both the ice and water loads. These variations lead to an irregular relationship between strandline age and gradient (Fig. 4), although in central Sweden gradient as a function of age appears to relax very smoothly. Mörner attempts to match previously published eustatic sea level curves to his shoreline diagram but finds that the results are unacceptable and hence develops the relationships among isostatic factors, shoreline displacement, and eustatic sea level changes to present his own eustatic sea level curve (Fig. 12). The isostatic curves on Fig. 11 should be contrasted with those in earlier papers in this volume. Mörner's results are certainly significant, but other areas will have to reveal the same complex reactions to ice/water loads before this degree of crustal sensitivity can be fully accepted. By implication, Mörner is saying that the rigid part of the crust responds rapidly (elastically) to fluctuations in load. As world coastlines tilt in response to water tides of amplitude 10 m or so, such a rapid response is possible, but why has it not been recognized in other areas? The answer is (1) it did not occur in other areas, or (2) no one has done the detailed work to establish the field relationships.

Nils-Axil Mörner was born in Sweden and studied in the University of Stockholm, where he received his Swedish Doctorate in 1969. He is a member of the INQUA Shoreline Commission and is active in the area of Quaternary chronology and climatic change. He has published a significant number of papers on these topics.

Reprinted from *Palaeogeogr. Palaeoclimat. Palaeocol.*, **9**, 153–181 (1971)

21

EUSTATIC CHANGES DURING THE LAST 20,000 YEARS AND A METHOD OF SEPARATING THE ISOSTATIC AND EUSTATIC FACTORS IN AN UPLIFTED AREA

NILS-AXEL MÖRNER

University of Stockholm, Stockholm (Sweden) and University of Western Ontario, London, Ont. (Canada)

(Received April 3, 1970)
(Resubmitted November 25, 1970)

ABSTRACT

MÖRNER, N.-A., 1971. Eustatic changes during the last 20,000 years and a method of separating the isostatic and eustatic factors in an uplifted area. *Palaeogeography, Palaeoclimatol., Palaeoecol.*, 9: 153–181.

The present paper (presented at the VIII INQUA Congress in Paris 1969) deals with the eustatic changes since the maximum lowering of the ocean level during the glaciation maximum about 18,000 B.C. (20,000 B.P.). The eustatic curve is believed also to depict the climatic changes quite closely. For the period 11,750 B.C.–200 B.C., the eustatic curve is derived from southern Scandinavia by a calculation of the eustatic and isostatic factors, respectively, in shorelevel displacement curves from several points in the direction of tilting from strongly uplifted to slightly subsided isobase latitudes (MÖRNER, 1969a). The eustatic curve obtained is highly fluctuating, though with a small amplitude. The present position was reached 1,700–1,400 B.C. (PTM-7 at + 0.4 m). With information from Vendsyssel (in comparison with other parts of the Kattegatt area), the Gulf of Mexico and the shelf off northeast America (excluding the isostatic factor) the eustatic curve was extended back in time to the glaciation maximum about 18,000 B.C., where a eustatic sea level of about − 85–90 m was found. At about 13,000–12,800 B.C. a 10 m drop in sea level was found, corresponding to the second glaciation maximum during the Late Würm–Weichsel–Wisconsin.

INTRODUCTION

It has often been stated that a eustatic curve can only be determined in a so-called "stable area". However, coastal stability is questioned. Furthermore subaquatic studies can only give approximate information. The details must be studied in a suitable uplifted area. But this requires a well-fixed and closely dated shoreline diagram covering a large area, from which shorelevel displacement curves can be drawn for numerous points in the direction of tilting and the isostatic and eustatic changes calculated and checked.

The intensity of land upheaval, the direction of tilting, the distribution of sediments, the negligible tide, the good pollen chronology, etc., all make the

Swedish west coast and the Kattegatt Sea uniquely suitable for a detailed investigation of shorelevel displacement.

The isostatic changes were reconstructed in detail and the land upheaval was shown to have had a very high glacio-isostatic and hydro-isostatic sensitivity. The deduction of hydro-isostatic changes is important, as it means that the transgressed shelf areas of the world most probably have sunk hydro-isostatically.

The eustatic curve is highly fluctuating. It is the only one that can in a reliable way express the shorelevel displacement and isostasy of the area investigated. The regressions correspond to climatic deteriorations and the transgressions to climatic ameliorations. The eustatic curve was compared with other results from all over the world (MÖRNER, 1969a, pp.426–448).

At 7,330 B.C. the eustatic curve starts to rise very rapidly from −38 m. This is supposed to give the most logical boundary between the Würm (Weichsel) Glacial Age and Flandrian Interglacial Age as well as between the Pleistocene Epoch and Holocene Epoch (MÖRNER, 1969b, fig.8; 1970a, fig.7).

The author has earlier discussed the Late Quaternary eustatic-climatic changes (MÖRNER, 1969a, b, c, 1970a, c), as well as the isostatic changes (MÖRNER, 1969a, 1970a, d). In this paper the stress is laid on the method of separating the eustatic and isostatic factors from the shorelevel displacement found.

INVESTIGATION AREA AND BACKGROUND

I have investigated (MÖRNER, 1969a) the shorelevel displacement in southern Scandinavia, i.e., the margin of the Fennoscandian shield which, during the glaciation, was pressed down by about one-third of the thickness of the covering ice and later, with the disappearance of the ice, it rose again. This means that all shorelines in this area are tilted as shown in Fig.1. Every eustatic sea level (E) is tilted (S) by

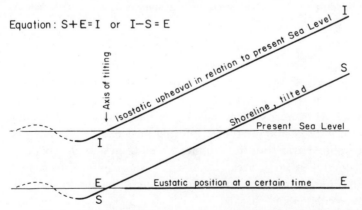

Fig.1. All the shorelines in southern Scandinavia are tilted. This figure illustrates the relations between a eustatic sea level (E), its tilted shoreline in southern Scandinavia (S) and the amount of isostatic upheaval in relation to the present sea level (I).

Palaeogeography, Palaeoclimatol., Palaeoecol., 9(1971) 153–181

the isostatic upheaval (I), the intensity of which decreases from northeast towards southwest where it changes over into subsidence. The amount of isostatic upheaval in relation to the present sea level is the difference between the shoreline found (S) and the present sea level (above = + and below = −) + the difference between the eustatic position (E) and the present sea level (above = − and below = +). This gives the following equation, which was used for the calculation of the eustatic (E) and isostatic (I) changes, respectively (MÖRNER, 1969a, b, 1970a):

$$S + E = I \text{ or } I - S = E$$

The advantage of this area as compared to other areas of the world is partly that the negligible tide allows very close determinations of the mean sea level (in contrast to, for example, Holland) and partly that we know that a land upheaval is going on (in contrast to most other areas (often called stable) which might have gone up—e.g., parts of Australia, South America and Africa—and/or gone down, e.g., Bermuda, Florida and Micronesia). For this area the intensity of upheaval, could be exactly reconstructed by the time-gradient curves (see Fig.4; MÖRNER, 1969a, b, 1970a, d). However, every calculation of the eustatic and isostatic changes from uplifted areas would be more or less senseless if the isobases and direction of tilting were not closely determined, and if the shorelines were not very closely fixed in age and altitude and extended over a large area giving numerous shorelevel displacement curves from which the isostatic and eustatic changes could be calculated and checked.

The isobases and direction of tilting were closely fixed in the investigation area (MÖRNER, 1969a, p.406). Bohuslän, Vendsyssel and southeastern Sweden were shown to have had a different isostatic history. For the rest of the area, no shift whatsoever of the direction of tilting and the isobases was found from the time of deglaciation up to the last shoreline determined over a large area, viz., PTM-7 at about 1,500 B.C. This is a very important and fundamental thing for the whole investigation.

The shorelines are in Mörners opinion very closely fixed in age and altitude. In total 111 ^{14}C datings and numerous pollen diagrams were made—together with numerous pollen diagrams and ^{14}C dates made by others—and these give the investigation a very closely fixed chronology. The ALV-1 shoreline (the regression maximum), i.e., the shoreline of the transition period between Late Glacial regression and Postglacial transgression, now submerged and usually buried by sediments, was extremely well preserved and could easily be followed over the whole area investigated (i.e., about 350 km in the direction of tilting). It was ^{14}C dated at various places along its whole extension. The PTM shorelines were all formed by transgressions (Postglacial Transgression Maxima) and their respective morphological elements are thus definitely synchronal. They were followed over the whole area (MÖRNER, 1969a, plate 6). In the northeast the PTM shorelines lie one below the other (regression) and in the southwest one above the other (transgression). All the PTM shorelines were ^{14}C and pollen dated at various places along the diagram. The regressions (PR) between the transgressions (PTM) were not seen as shorelines in the field (except for ALV-3). They could, however, later be calculated from the eustatic and isostatic curves, since the upheaval was linear between the PTM (cf., below), if the amount of regression was only closely determined at some points, i.e., in some of the shorelevel displacement curves. The number of shorelines and transgression sequences found, clearly show that all PTM/PR changes are true and that none of them are just misinterpretations due to variations of the atmospheric ^{14}C activity (MÖRNER, 1969a, fig.41; 1969b, fig.2). The Late Glacial shorelines older than Alleröd (i.e., 11,750–9,800 B.C.) were determined by the Marine Limit (ML) at synchronal ice-marginal positions (MÖRNER, 1969a, plate 4). Especially the shorelines of the distinct Fjärås, Moslätt and Berghem ice marginal lines (terminal moraines) were very closely fixed and also dated by pollen and ^{14}C in basins isolated from the sea at the same time. There are abundant Late Glacial shore marks in the area investigated. However, these are not believed to represent synchronous Late Glacial raised beaches (MÖRNER, 1969a, p.113). For the Alleröd and Younger Dryas periods the shorelines were based upon time-elevation indications from basins at various altitudes in the direction of tilting, the isolation of which from the sea was dated by ^{14}C and pollen. This means that the gradients are not so well fixed for these periods as for the other, and minor changes are possible.

When starting this investigation, the author thought that the land upheaval would give quite smooth isostatic curves at least for the older part (i.e., the time of deglaciation) with strong intensity of upheaval, but possibly undulating curves for the youngest part where local changes could have influenced the slow upheaval of that time. However, this idea was totally revised, as in fact the opposite was found by this investigation. The Late Glacial part of the isostatic curves is fluctuating, whereas the same curves are quite smooth for the last 7,000 years. This might not be so surprising, as the upheaval during the time of deglaciation acted as a balance responding very sensitively to changes in the load of ice (glacio-isostasy) inland, and in the load of water (hydro-isostasy) seawards (MÖRNER, 1969a, pp.408–420). Also there were changes of the centre of upheaval (MÖRNER, 1969a, fig.146; 1970d, fig.3). Fig.2 illustrates the three types of isostatic changes found.

(*1*) (the common type). The removal of ice load gives strong upheaval and, usually peripheral, subsidence.

(*2*) The adding of ice load (readvances, etc.) gives a retarded upheaval or even subsidence.

(*3*) The adding of water load (eustatic transgressions), gives peripheral subsidence (sometimes quite strong) and upheaval further inland—the axis of tilting is moved inland.

The relation between the eustatic curve (E) and shorelevel displacement curves (S) in isostatically uplifted ($I+$) and sunken ($I-$) area respectively are

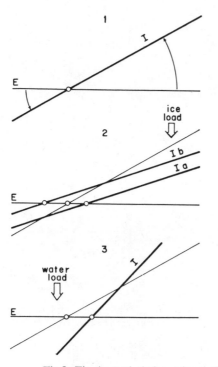

Fig.2. The isostatic balance in southern Scandinavia. E = eustatic level; I = isostatic uplift. 1 = Normal uplift due to removed ice load; 2 = Retarded upheaval (compared to 1) due to increased ice load; 3 = Accelerated upheaval (compared to 1) due to increased water load.

illustrated in Fig.3 (cf. MÖRNER, 1969a, b). From this relationship the following conclusions are reached:

(*1*) The shorelevel displacement curves in areas of land subsidence (e.g., the coast of The Netherlands) are too smooth, as eustatic regressions are here decreased or absent.

(*2a*) The shorelevel displacement curves in areas with a low intensity of land upheaval (e.g., the areas studied by the author) are too fluctuating, as eustatic regressions are increased and a flattening out of the eustatic rise may be enough to cause a transgression + regression of the relative sea level.

(*2b*) The shorelevel displacement curves in areas with a high intensity of land upheaval (e.g., northern Sweden) are stepped, since eustatic regressions are increased and eustatic transgressions are transformed to retardations of a continuous regression.

Going back to the equation $S + E = I$ or $I - S = E$, there is one factor (S) which is known in detail and two factors (I and E) which are usually both unknown. However, the isostatic factor (I) is, in fact, well known here from the time-gradient curve in Fig.4 (MÖRNER, 1969a, fig.142; 1969b, 1970a, d), i.e., the curve where the age is plotted against the gradient of the 33 shorelines identified.

Palaeogeography, Palaeoclimatol., Palaeoecol., 9(1971) 153–181

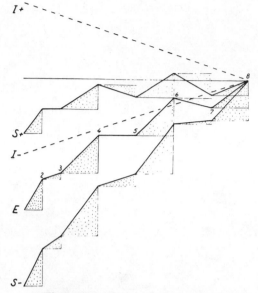

Fig.3. Theoretical model showing the relation between eustatic changes (*E*) and the shore-level displacement (*S*+ and *S*−) in uplifted (*I*+) and sunken (*I*−) areas. The height of the dotted areas gives shorelevel displacement between two points.

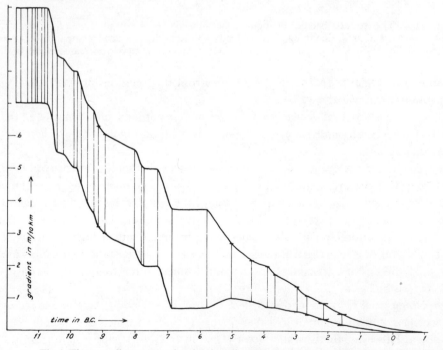

Fig.4. Time-gradient curves for the shorelines (vertical lines) in southern Scandinavia. These curves directly give the intensity of the isostatic upheaval. The curves fluctuate up to 5,000 B.C. after which they decrease more or less continuously to the present.

Palaeogeography, Palaeoclimatol., Palaeoecol., 9(1971) 153–181

The curve fluctuates up to 5,000 B.C., showing the influence of the changes in the load of water and ice, and of the changes of the centre of upheaval. Consequently, the isostatic curves calculated must also show all these changes in intensity of upheaval. However, there are no changes indicated for the period after 5,000 B.C., which makes this part very suitable for the first calculation of the isostatic and eustatic factors.

Factor E (the eustatic curve) was the unknown unit, which had to be calculated. No earlier-published "eustatic" curve can be characterized as giving more than an approximation and they also all gave quite unrealistic isostatic curves if added to the author's shorelevel displacement curves. However, some important facts were derived from these curves. JELGERSMA's curve (1961, 1966) represents a sunken area and must thus lie below the true eustatic curve. Almost all "eustatic" curves lie at about -10 m at 5,000 B.C. The Florida curve (SCHOLL, 1964; SCHOLL and STUIVER, 1967) was originally supposed to give the best picture of the eustatic rise during the last 4,000 years. However, this supposition had to be revised, as the Florida curve gave unrealistic isostatic curves and probably represents a slowly subsiding area.

CALCULATION OF THE ISOSTATIC AND EUSTATIC CHANGES

Since the time the author made the calculation separating the eustatic and isostatic factors from the shorelevel displacement curves (MÖRNER, 1969a), some new "eustatic" curves have been published or presented (at the VIII INQUA Congress in Paris, 1969), e.g., the ones from Florida (SCHOLL, CRAIGHEAD and STUIVER, 1969), Bermuda (NEUMANN, 1969) and Micronesia (BLOOM, 1969). However, none of them have changed Mörner's earlier results.

The last 7,000 years

This was the best part with which to start the calculations. The shorelevel displacement (S) was known in detail over a large area. The time-gradient curve (Fig.4) showed a tilting uniformly decreasing with time and distance towards the periphery. Several "eustatic" curves were published for this part (Fig.5). JELGERSMA's (1961, 1966) sea level curve for the subsiding Dutch coast must lie below the eustatic curve.

For the first calculation all the oscillations, definitely established, were ignored and only the eight PTM shorelines were used (MÖRNER, 1969a, plate 6). The corresponding depths of different "eustatic" curves were simply added to the altitude of these PTM levels at every 50 km in the direction of tilting, e.g., added to the levels in Fig.6. Four main "eustatic" curves are used here and the resulting isostatic curves (Fig.7) are discussed.

(*1*) FAIRBRIDGE's curve (1961): this is a synthesis of information from the whole world. This curve was a priori believed to show excessively large and rapid

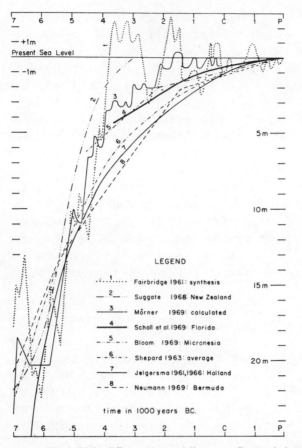

Fig.5. Eight different "eustatic" curves. Curves *1*, *3*, *4* and *5* converge at about 4,000–4,500 B.C. Curves *1*, *(2)*, *3*, *6*, *7* and *8* converge at about 5,000–5,500 B.C. This converging is the real Holocene eustatic sea level problem.

oscillations. Also, some of the basic material (e.g., that from Sweden, New Zealand and Florida) has later been revised (SCHOLL, 1964; SUGGATE, 1968; MÖRNER, 1969a). The points obtained using Fairbridge's curve cannot be combined to form reliable isostatic curves (Fig.7A), although the most favourable correlations (for the F:s curve) were chosen. The isostatic curves fluctuate too much even at this stage when Fairbridge's large oscillations are excluded.

(*2*) SHEPARD's curve (1963a): this is an average curve of nine "relatively stable areas". A priori this curve was believed to lie too close to the Dutch (sub-sided) curve. Also a smooth curve seems incompatible with the changing Late Quaternary climate of the world (MÖRNER, 1970c). The points obtained (Fig.7B) using this "eustatic" curve form isostatic curves that in the older part look fairly good but after 2,500 B.C. give quite impossible curves. At −250 km in the direction of tilting—i.e., at a position at or just outside the Danish 0-isobase according to

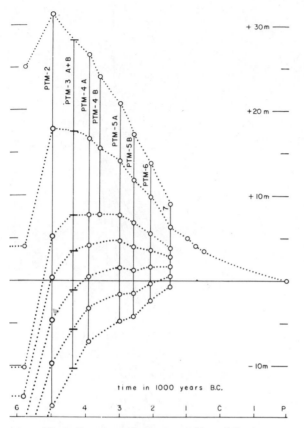

Fig.6. Shorelevel displacement (dotted lines) at every 50 km in the direction of tilting. The oscillations, obviously found are excluded in this figure (cf., Fig.10). Vertical lines = PTM shorelines.

MERTZ (1924; cf. HANSEN, 1965)—a subsidence of 1.8 mm/year from 2,500 to 1,500 B.C. is followed by an upheaval of about 0.5 mm/year. Furthermore, Shepard's curve cannot explain the oscillations found.

(3) Scholl's curve from Florida (SCHOLL, 1964; SCHOLL and STUIVER, 1967; SCHOLL et al., 1969) and Bloom's curve from Micronesia (BLOOM, 1969): the Florida curve was a priori believed to be the best and possibly the true eustatic one because it was determined from one single area and based upon numerous ^{14}C dates. The smoothness of the Florida curve was not too great a problem as SCHOLL and STUIVER (1967, p.446) write: "the smooth nature of the submergence curve . . . should not be constructed to indicate the lack of possible high-frequency low-amplitude oscillations in sea level during the last 4,000 years". The points obtained using these curves (those of SCHOLL et al., 1969 and BLOOM, 1969 were not available at the time of the first calculation) form isostatic curves that are unrealistic in the outer part (Fig.7C). The curve for −250 km shows a subsidence of 2.0 mm/

Palaeogeography, Palaeoclimatol., Palaeoecol., 9(1971) 153–181

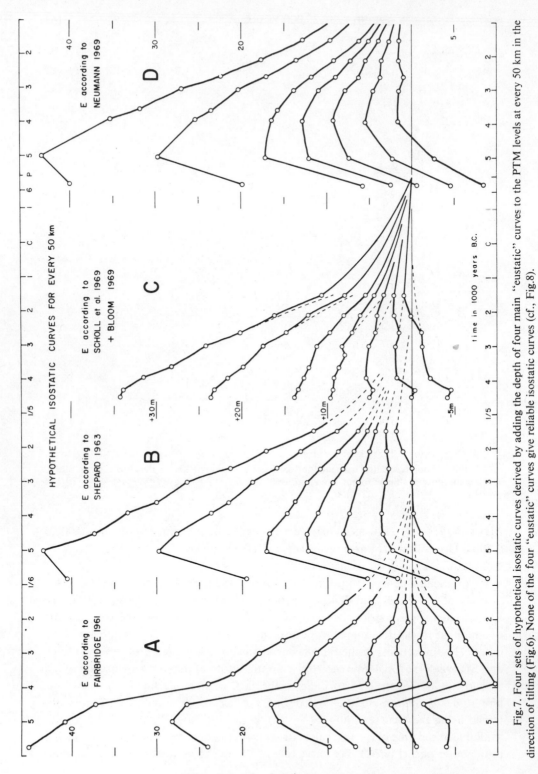

Fig.7. Four sets of hypothetical isostatic curves derived by adding the depth of four main "eustatic" curves to the PTM levels at every 50 km in the direction of tilting (Fig.6). None of the four "eustatic" curves give reliable isostatic curves (cf., Fig.8).

year prior to 1,500 B.C., where it changes into an upheaval of 0.26 mm/year (0.6 mm/year at −200 km). This does not at all agree with historical data indicating the 0-isobase at about −200 km, nor is it indicated by the curve of isostatic intensity (Fig.4). As is shown by the dashed lines in Fig.7C, smooth isostatic curves can be obtained if the two youngest points are ignored. However, it is incorrect to do so, as this means a correction of Scholl's curve from −1.6 m to −0.4 m at the PTM-7 level at about 3,500 B.P. (1,500 B.C.), which is the best fixed part of his curve. (It would also give an axis of tilting moving seawards). Thus the Florida curve is supposed to represent a slowly subsided area (MÖRNER, 1969a, fig.156). In order to test whether the Micronesia–Florida curve could after all be the best "eustatic curve", these calculations were made all over again in October 1969. However, the same result as previously (MÖRNER, 1969a) was obtained.

(4) NEUMANN's curve from Bermuda (1969): this is a well-dated (quite new) curve. It is drawn as a smooth curve, though the time-depth indications may allow minor oscillations. Neumann states that "there is some evidence of sea-level reversal between 8,500 and 9,000 B.P.", i.e., at the time of Mörner's regression after PTM-1 (MÖRNER 1969a). The Bermuda curve lies close to the Dutch curve (Fig.5); above it after 2,500 B.C., below it between 2,500 and 5,000 B.C. and again above it prior to 5,000 B.C. This may suggest that Bermuda is slowly subsiding. The points obtained (Fig.7D) form fairly nice isostatic curves, though the 0-isobase would lie decidedly too far out (Table I). Consequently, the Bermuda curve is believed to show a slow subsidence (agreeing with Neumann's own opinion).

TABLE I

MEAN ANNUAL ISOSTATIC CHANGES DURING THE LAST 3,500 YEARS AT THE ISOBASES −200 KM AND −250 KM AS DERIVED FROM DIFFERENT "EUSTATIC" CURVES

	−200 km (mm)	−250 km (mm)
Fairbridge's curve	−0.43	−0.50
Mörner's curve	±0	−0.32
Scholl's curve	+0.60	+0.23
Neumann's curve	+1.12	+0.32
Shepard's curve	+1.23	+0.54

As shown by Fig.7 none of the four "eustatic" curves used here can (even excluding the oscillations found) express the isostatic changes in a reliable way. The mean annual isostatic changes during the last 3,500 years, derived from the different "eustatic" curves (Table I), show that only the eustatic curve calculated by the author gives isostatic curves (Fig.8) that harmonize with the historical and

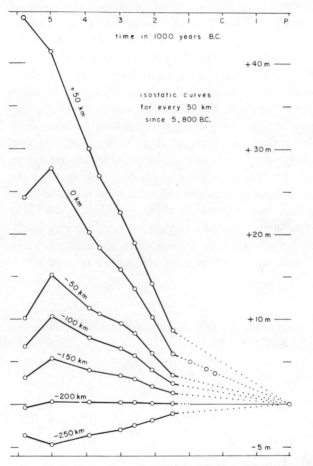

Fig.8. Isostatic curves derived by adding the eustatic curve (Fig.12) to the shorelevel displacement curves (Fig.6 and 9). These curves are the only ones that can express the isostatic upheaval in a reliable way (cf., Fig.7). The curve for −200 km in the direction of tilting almost exactly gives the eustatic changes and gives the 0-isobase or the axis of tilting for the last 7,000 years. The extension to the present (dotted lines) should probably be more curved lines (cf., MÖR-NER, 1969a, fig.148; 1969b, fig.7B). With a calendrian chronology instead of the radiocarbon chronology, these curves would probably form parabolic curves.

geological information from this area. Fairbridge's curve gives too much sub-sidence and the other curves too much upheaval.

Palaeogeography, Palaeoclimatol., Palaeoecol., 9(1971) 153–181

The tide-gauges give zero figures for points lying about -150–100 km in the direction of tilting (LENNON, 1966, fig.4–5). However, the true 0-isobase lies further out, as the tide-gauges give too little uplift due partly to eustatic rise (e.g., FAIR-BRIDGE, 1961, fig.1; WAALEWIJN, 1966, fig.2; VAN VEEN, 1954) and partly to local subsidence due to compaction (many tide-gauges outside the area of archean bed-rock are located on harbour constructions resting upon sediments). Some minor irregularities of the isobases have also to be considered (MÖRNER, 1969a, p.407). A comparison between the isobases from the repeated Finnish geometrical levellings and those from the Baltic tide-gauges (JAKUBOVSKY, 1966, fig.2) shows a difference of about 0.5 m (Mörner, in preparation). These calculations bring the 0-isobase out to about -200 km, i.e., in full agreement with Mörner's isostatic curves (Fig.8), but far too much inland to agree with the isostatic curves derived by using the "eustatic" curves by Shepard, Scholl, Bloom or Neumann (Fig.7; Table I).

The next step was to find out what changes were necessary to obtain a eustatic curve, which gave logical isostatic curves. For the first part of these calculations all the oscillations found were ignored and only the eight shorelines identified were used. All of these shorelines were known to be formed by transgressions. There were two shortcomings: PTM-4B was not identified outside -50 km and PTM-3 was known to consist of two morphologically indistinguishable transgression maxima. The shorelevel displacement at every 50 km in the direction of tilting (from $+50$ km to -250 km) was used (see Fig.6). In order to be able to draw corresponding isostatic curves, a more or less hypothetic eustatic curve had to be added, about which the following was known: (*1*) it must lie above the Dutch curve; (*2*) it must lie above the Florida curve (at least) at PTM-7; and (*3*) it must lie at about -10 m at 5,000 B.C.

The next step was detailed comparison and alteration until a reliable isostatic picture was obtained—the only one that could express the shorelevel displacement curves, time-gradient curves and all other information in a satisfying and logical way (the slightest change leads to the opposite). In this way isostatic curves could be drawn for every 50 km in the direction of tilting (Fig.8; cf., MÖRNER, 1969a, fig.148; 1969b, fig.7B) and also a generalized eustatic curve.

Having established the isostatic curves and a main eustatic curve, the rest was a fairly easy calculation. As the isostatic curves must be straight between PTM-2 and PTM-4, it was now possible to separate the two PTM-3 maxima and test the result by field observations (see MÖRNER, 1969a, e.g., pp.368 and 400). Now it was also time to include all the sea level oscillations definitely found.

As the isostatic curves form smooth lines, it was easy to calculate the eustatic oscillations. The magnitude of every relative sea level (shorelevel displacement) oscillation had only to be known in some places in order to be calculable along the whole diagram. And the magnitude of all the oscillations was well established at least in some places (see MÖRNER 1969a, pp.302–403). After having included the oscillations in the shorelevel displacement curves (Fig.9; cf., MÖRNER, 1969a, fig.147; 1969b, fig.7A), these were closely checked with and compared to the field evidence (MÖRNER 1969a, e.g., fig.131) showing (e.g., isolated and/or transgressed basins) or not showing (e.g., not isolated and/or not transgressed basins) these oscillations. The oscillations in Fig.9 agree completely with the field evidence (MÖRNER, 1969a).

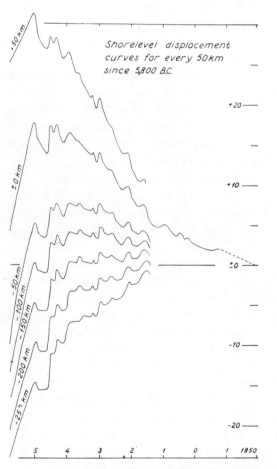

Fig.9. Complete shorelevel displacement curves including all the oscillations found. The transgression maxima (PTM) were determined as shorelines while the regressions (PR) were calculated from the field information and the isostatic curves.

The eustatic curve obtained is shown in Fig.4 and 12. It is important to remember that this curve refers to mean sea-level changes, since the tide is negligible in the Kattegatt Sea, whilst most other curves refer to areas having a tidal range of several or at least one or two meters. PTM-7 is supposed to give the real

Holocene eustatic maximum at +0.4 m and also the point where a generally rising sea level changes to a generally horizontal one, which corresponds to the climatic boundary between the period of optimum climate (Period II or Megathermal) and the first major deterioration of climate (Period III or Katathermal) at about 1,400 B.C. or 3,450 B.P. (MÖRNER, 1969b, fig.8; 1970a, fig.7).

The final part of the calculations of the eustatic changes during the last 7,000 years, was to compare my curve with the results from other areas. These comparisons have been described earlier (MÖRNER, 1969a, pp.426–448; 1969b, 1970a, b). The agreement is in many cases striking, even in detail, and, what is equally important, the disagreements are in no case too serious.

One of the main questions about the Holocene eustatic sea level changes has always been the presence or absence of oscillations. Mörner's eustatic curve is highly oscillating, though with a small amplitude. As the Holocene climate was oscillating and sea-level oscillations are well recorded in all areas studied in sufficient detail, I (Mörner) think a low-amplitude oscillating eustatic curve must—sooner or later—be generally accepted (cf., MÖRNER, 1969a, e.g., fig.138; 1969b, e.g., fig.3; 1970a, b, c).

Another main question about the Holocene eustatic sea level changes has always been whether the ocean level ever exceeded the present one (type Fairbridge's curve) or whether it was more or less steadily rising (type Shepard's curve). Mörner's eustatic curve rises to a maximum level at +0.4 m at PTM-7 dated at 1,750–1,450 B.C. This means some sort of a mean value between that of Fairbridge and that of Shepard and many others (cf., Fig.5). If the field evidence is to be trusted, which the author thinks must be done at least in most cases, several areas (e.g., Florida, Bermuda and parts of Micronesia) clearly show a more or less steadily rising sea level, while other areas (e.g., parts of Australia, South America and Africa) show sea levels of Middle Flandrian (Holocene) age well above the present one. The sea level stands at 5,000 B.P. (3,000 B.C.) and 3,500 B.P. (1,500 B.C.), according to different authors, are compared in Table II and III (left column). The three columns to the right give the difference in elevation (in meters) to a eustatic position (E) according to three different alternatives (Fairbridge, Mörner and Scholl, respectively) and the corresponding mean rate of upheaval or subsidence to the present in mm per year. However, the present upheaval or subsidence is probably much smaller than the mean values given in Table II and III, as the rate of upheaval or subsidence usually decreased with time.

If Fairbridge is correct, areas such as Florida (SCHOLL, 1964; SCHOLL and STUIVER, 1967; SCHOLL et al., 1969), Bermuda (NEUMANN, 1969) and The Netherlands (JELGERSMA, 1961; WAALEWIJN, 1966; HAGEMAN, 1969) would be subsiding too rapidly. If, on the other hand, Scholl is correct, areas such as Australia (FAIRBRIDGE, 1961; HOPLEY, 1969) would be so strongly uplifted that this would probably also be recorded by various historical data and by the tide-gauges. Finally, if Mörner's curve is correct, this does not require too strong an upheaval

TABLE II

SEA LEVEL STAND AT 5,000 B.P. (3,000 B.C.)

Sea level at 5,000 B.P.	Fairbridge correct	Mörner correct	Scholl correct
+3–5 m	+1.5–3.5 m	+4.8–6.8 m	+6.5–8.5 m
some areas	0.3–0.7 mm/y	0.96–1.36 mm/y	1.3–1.7 mm/y
+1.8 m	(E)	+3.6 m	+5.3 m
Hopley		0.70 mm/y	1.06 mm/y
+1.5 (2.5) m	E	+3.2 m	+5.0 m
Fairbridge		0.64 mm/y	1.0 mm/y
±0 m	−1.5 m	+1.8 m	+3.5 m
Suggate	0.30 mm/y	0.36 mm/y	0.70 mm/y
−1.8 m	−3.2 m	E	+1.7 m
Mörner	0.64 mm/y		0.34 mm/y
−3.5 m	−5.0 m	−1.7 m	E
Scholl	1.0 mm/y	0.34 mm/y	
−4.8 m	−6.3 m	−3.0 m	−1.3 m
Shepard	1.27 mm/y	0.60 mm/y	0.27 mm/y
−5.7 m	−7.2 m	−3.9 m	−2.2 m
Jelgersma	1.44 mm/y	0.78 mm/y	0.44 mm/y
−6.3 m	−7.8 m	−4.5 m	−2.8 m
Neumann	1.56 mm/y	0.90 mm/y	0.56 mm/y

TABLE III

SEA LEVEL STAND AT 3,500 B.P. (1,500 B.C.)

Sea level at 3,500 B.P.	Fairbridge correct	Mörner correct	Scholl correct
+4.8 m	+2.8 m	+4.4 m	+6.4 m
Hopley	0.80 mm/y	1.25 mm/y	1.83 mm/y
+2.0 (2.7) m	E	+1.6 m	+3.6 m
Fairbridge		0.46 mm/y	1.0 mm/y
+0.8 m	−1.2 m	(E)	+2.4 m
Guilcher	0.34 mm/y		0.70 mm/y
+0.4 m	−1.6 m	E	+2.0 m
Mörner	0.46 mm/y		0.57 mm/y
−1.6 m	−3.6 m	−2.0 m	E
Schóll	1.0 mm/y	0.57 mm/y	
−2.2 m	−4.2 m	−2.6 m	−0.8 m
Neumann	1.2 mm/y	0.74 mm/y	0.20 mm/y
−2.6 m	−4.6 m	−3.0 m	−1.0 m
Shepard	1.3 mm/y	0.86 mm/y	0.30 mm/y
−3.2 m	−5.2 m	−3.6 m	−1.6 m
Jelgersma	1.50 mm/y	1.00 mm/y	0.46 mm/y

Palaeogeography, Palaeoclimatol., Palaeoecol., 9(1971) 153–181

of areas such as Australia or subsidence of areas such as Florida, Bermuda and The Netherlands. A similar result is given in Table I.

However, different isostatic changes are not sufficient to explain the differences in sea level. The author has earlier called attention to and discussed his so-called "real Holocene eustatic sea level problem" (MÖRNER, 1970a, b, d), viz., that all the different sea-level curves converge back in time (see Fig.5) instead of diverging. The explanation of this convergence seems either to be that the ocean surface (for the interjacent period) was different from and unequal to that of today and that of 5,000–4,500 B.C. (7,000–6,500 B.P.), i.e., eustatically different; or that there was a time of generally changed isostatic movements (e.g., by a hydro-isostatic deformation of a coast with a compensational upheaval inside the axis of subsidence: MÖRNER 1970a, fig.3; 1970d, fig.4).

The period 5,000 B.C. to 11,750 B.C.

The separation of the isostatic and eustatic factors from the shorelevel displacement curves is harder for the part prior to 5,000 B.C., since the intensity of upheaval was then irregular. The final isostatic curves must, however, contain all the details shown by the time-gradient curves (see Fig.4). The difficulties especially concern the real magnitude of the two major transgressions starting 7,330 B.C. (causing peripheral subsidence) and 5,800 B.C. (causing the bend at −43.4 km, where subsidence took place). There were two periods of syngression (i.e., no shorelevel displacement either from isostatic or eustatic causes), viz., at 7,750–7,330 B.C. and at 6,550–5,800 B.C. These periods of syngression were caused by a more or less stable ocean level and a temporary change of the centre of upheaval and the axis and margin of tilting (MÖRNER, 1969a, pp.415–420; 1970d, fig.3). Another complication is the obvious irregularities in the atmospheric ^{14}C content occurring during the period 8,000–5,000 B.C. (MÖRNER, 1969a, pp.179–181; 1970e), making the periods of syngression too long and the time of the very big transgression to PTM-2 too short, and also causing stratigraphical irregularities of dates (common at about 7,400–6,800 B.C. and at about 6,300–5,500 B.C.). The upheaval during the period 7,800–10,700 B.C. was suitable for these calculations.

At 5,800 B.C. the sea level started to rise rapidly. This transgression most probably brought it directly up to the PTM-2 level at 5,000 B.C. At the same time— and hydro-isostatically initiated—the centre of upheaval temporary moved back to the highland of Småland causing such a rapid tilting outside that the bend at −43.4 km was created, where the land sunk in relation to the area inside as well as outside. The eustatic magnitude of the PTM-2 transgression must be less than 15.6 m, which is the amount of shorelevel displacement at the subsided bend. Minimum figures of about 7.5 m (BERGLUND, 1964, fig.23) or 6.5 m (MÖRNER, 1969a, fig.149) are obtained from southeast Sweden, and of about 7–8 m from northern Norway (MARTHINUSSEN, 1962, plate 2; MÖRNER, 1969a, fig.150). A rise of about 9.8 m or somewhat more (from −22.5 m to −12.7 m) is indicated from

The Netherlands (MÖRNER, 1969a, p.393). SUGGATE (1968) found a rapid transgression of about 13.9 m between − 18.5 m at 6,000 B.C. and −4.6 m at 4,500 B.C. (i.e., between the author's ALV-3 and PTM-3A shorelines). Mörner's eustatic curve gives a corresponding transgression of 13.7 m from − 20.3 m at 5,800 B.C. to − 6.6 m at 4,500 B.C. (PTM-3A) via a PTM-2 level at − 10.0 m at 5,000 B.C. The calculated eustatic rise from 5,800 to 5,000 B.C. is 10.3 m.

The period 5,800–6,550 B.C. was a time of syngression without traceable eustatic changes. However, there might have been minor fluctuations as suggested by some results (MÖRNER, 1969a, p.247; TERS and PINOT, 1969, fig.11; MÖRNER, 1970a).

Between 6,850 B.C. (PTM-1) and 6,550 B.C. there was a regression of about 1.8 m. This figure was taken directly from the shoreline diagram, since the axis and the margin of tilting were then located further inland. This regression is also found in Ireland (MÖRNER, 1969a, p.436) and on Bermuda (NEUMANN, 1969; "there is some evidence of a sea-level reversal between 8,500 and 9,000 years B.P.").

At 7,330 B.C. the sea level started to rise very rapidly, bringing the sea level up to PTM-1 reached at about 6,900 B.C. (^{14}C age). This is evident from the whole of southern Scandinavia, from Ireland, from South America and indirectly from several other areas (MÖRNER, 1969a, 1970a). The hydro-isostatic load made the outer area subside, causing the centre of upheaval temporary to move back to the highland of Småland and the axis of tilting further southwestward (MÖRNER, 1969a, p.417; 1970d, fig.3). This makes an exact calculation of the corresponding eustatic rise difficult. From South America, AUER (1959) reported a transgression of 12–14 m, which, however, only is a eustatic minimum figure, as this area evidently has gone up (decreasing the real eustatic magnitude of transgression). In the North Sea this transgression started from a level close below − 46 m, as is evident from BEHRE's (1969) description of a submarine core from the Dogger Bank. Outside Brittany this transgression most probably started from a − 37 m shorelevel (see below). By the Würmian regression, the Black Sea was isolated from the oceans at about − 40 m (e.g., ZENKOVICH, 1969) as the neoeuxinic state. The change from this lake stage to the marine Old Black Sea stage (the transgression of the − 40 m level) has been ^{14}C dated at 7,400 ± 220 B.C. (Mo-287; VINOGRADOV et al., 1966, p.313). Mörner's calculations gave a eustatic transgression of 19.5 m from − 38.0 m to − 18.5 m.

Between 7,330 B.C. and 7,700–7,750 B.C. there was a period of syngression with a eustatic sea level at − 38.0–38.5 m. This level was supposed (MÖRNER, 1969a, p.434, 1970a) to correspond to Pinot's "excellently" preserved shore at about − 37 m outside Brittany (PINOT, 1965, 1966, 1968). Pinot thinks that the − 37 m level was followed by a distinct regression. However, the "excellent" preservation of the shore features can hardly agree with a subsequent regression; only with a rapid transgression. Furthermore, Ters reports (PINOT, 1968, p.216) a submarine peat layer at − 35–40 m at Pas-de-Calais, ^{14}C dated at 8,005 ± 100 B.C.

Palaeogeography, Palaeoclimatol., Palaeoecol., 9(1971) 153–181

at the top and 8,615 ± 100 B.C. at the bottom, and DELIBRAIS (1969) reports ^{14}C dates from −26.7–27.7 m in Le Havre, giving about 7,390–7,950 B.C. and from −37.9 m at Cherbourg, giving about 7,520 B.C. The morphology of Pinot's shore requires a sea-level stillstand for its formation, which is in full agreement with the syngression here discussed. See also Table V.

Between about 7,700 and 7,800 B.C. there must have been a small eustatic regression, as the sea level dropped almost equally and by about 4 m throughout the whole diagram. A eustatic regression of 2.5 m was calculated. It corresponds to the Preboreal deterioration of the climate (MÖRNER, 1969a, pp.423–424) and a distinct change from dry to wet climate, at least in northwest Europe.

From 7,800 to 10,700 B.C. the calculations are easier again, since the land upheaval was unbroken (though oscillating), with the centre the whole time in the highland of Småland. However, the upheaval was a balance, which very sensitively responded to changes in the load of ice and water (cf., above, Fig.3), as is seen in the shorelevel-displacement and time-gradient curves (Fig.4 and 10). We must speak

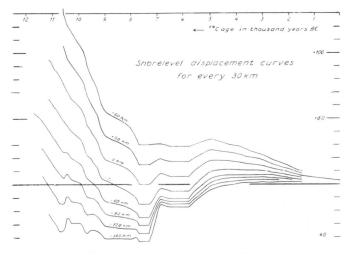

Fig.10. Shorelevel displacement curves (*S*) for the whole period studied. The oscillations during the last 7,000 years are not shown.

of a very high "crustal sensitivity" (MÖRNER, 1969a, 1970a, d). Table IV shows the relation between climatic division (pollen and molluscs) and the changes in ice load, changes in sea level and changes in tilting. These changes in the intensity of upheaval, both from glacio-isostatic and from hydro-isostatic causes, have all to be expressed in the isostatic curves (the glacio-isostatic influence increasing inland and the hydro-isostatic influence increasing seawards). The isostatic curves calculated (Fig.11) contain all the details suggested by the above-mentioned information. The eustatic curve obtained (Fig.12) is highly fluctuating (Table IV,

Palaeogeography, Palaeoclimatol., Palaeoecol., 9(1971) 153–181

TABLE IV

THE CLIMATIC SUBDIVISION OF THE PERIOD 10,700–7,700 B.C. AND THE CHANGES IN THE ICE LOAD, WATER LOAD, TILTING AND OCEAN LEVEL

Climatic division stadials/interstadials	Pollen zones	Age in B.C.	Ice load	Sea level	Tilting	E Calculated (m)
		10,700				−56.0
Ågård Interstadial			−	+	rapid	
		10,350				−47.0
Fjärås Stadial	Ia		+	−	slow	
		10,300				−49.0
Bölling Interstadial	Ib		−	+	rapid	
		9,950				−42.0
Older Dryas (Berghem) Stadial	Ic		+	−	no	
		9.800				−46.0
Alleröd Interstadial	II		−	+	rapid	
		9,000				−42.0
Younger Dryas (Ra-Billingen-Salpausselkä) Stadial	III		+	−	slow	−45.0
		8,050		+		−42.0
Early Preboreal (Friesland) Interstadial	IVa		−	+	rapid	
		7,800				−36.0
Preboreal (Ejdfjord) Stadial	IVb		−			
		7,700				−38.5

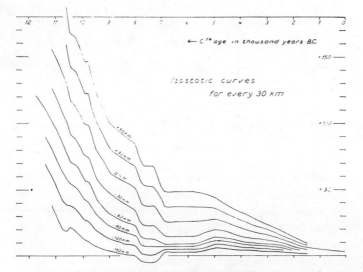

Fig.11. Isostatic curves (*I*) derived by adding the eustatic curve (Fig.12) to the curves in Fig.10. These curves include all the details shown by the time-gradient curves (Fig.4) and the reconstruction of the geographic changes of centre and margin of upheaval.

Palaeogeography, Palaeoclimatol., Palaeoecol., 9(1971) 153–181

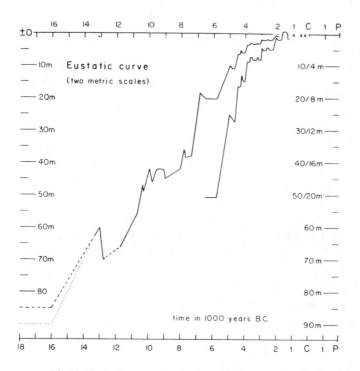

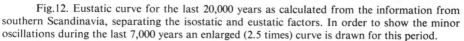

Fig.12. Eustatic curve for the last 20,000 years as calculated from the information from southern Scandinavia, separating the isostatic and eustatic factors. In order to show the minor oscillations during the last 7,000 years an enlarged (2.5 times) curve is drawn for this period.

right column), though with a low amplitude. Changes in the eustatic curve would immediately make the isostatic curves illogical. There is an exact correlation between climatic changes (as indicated by pollen and molluscs), changes of the ice margin and changes in the eustatic sea level—no retardation at all is indicated.

For the period 10,700–11,750 B.C. the upheaval was quite different, since the area investigated then rose up as a block. No tilting at all was found (further out there must, of course, have been tilting). This means that the calculations are not at all as well founded as those for the younger periods. But, as the time is short, it is only a question of detail. The rate of shorelevel displacement was about 3 m/ 100 year. Assuming an unbroken rate of upheaval, the eustatic rise was calculated at 10 m from −66 m to −56 m. No eustatic oscillation was found corresponding to the drastic climatic change from the Vintapper Interstadial to the Low Baltic Stadial at about 11,150 B.C. (13,100 B.P.), though this was supposed to correspond to a global event (MÖRNER, 1969a, p.94). However, this might well be the case, if the rate of upheaval was just slightly irregular. A change from 4 m to 2 m isostatic upheaval per 100 year between 11,250 B.C. and 11,150 B.C. is sufficient to cause an eustatic regression of 1 m.

Palaeogeography, Palaeoclimatol., Palaeoecol., 9(1971) 153–181

The period 11,750 B.C. to 18,000 B.C.

As the shoreline observations stop at 11,750 B.C., due to the complete cover by ice, the continuous calculations of the eustatic changes must also stop there. However, the eustatic curve can be extended back to 18,000 B.C. by indirect calculations (MÖRNER, 1969a, 1970a).

From Vendsyssel on the northwest side of the Kattegatt Sea, there are indications of a eustatic drop in sea level of 10 m (MÖRNER, 1969a, p.430, fig.151) in connection with the formation of the East Jylland ice marginal line or the D line, the age of which was calculated at about 13,000 B.C. (MÖRNER, 1969a; recently nicely supported by the results of CHEBOTAREVA, 1969, fig.3).

From the northwest part of the Gulf of Mexico, CURRAY (1960) and SHEPARD (1960) have given time-depth graphs for the last 20,000 years based upon ^{14}C-dated shells, the distribution and character of sediments, and the morphology. CURRAY (1960) gave 12 ^{14}C dates of shells of types now found in brackish shallow water, adding (p.254): "There is always the possibility, however, that the shells were transported either up or down the shelf after the death of the organisms". These dates, ranging from 15,000 to 6,000 B.C., all fall close—above or below— to Mörner's eustatic curve (Fig.13), suggesting that this area has been more or less

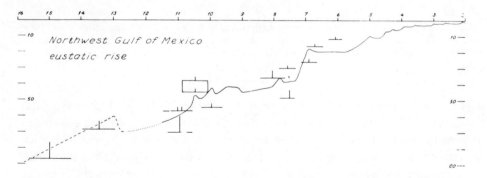

Fig.13. Curray's time-depth indications from the northwest Gulf of Mexico compared to Mörner's eustatic curve. Time scale in 1000 years B.C. The figure shows that this area has not been subsided and has probably been stable during the last 18,000 years (cf., MÖRNER 1970d, fig.7). The two oldest dates consequently give a continuation of Mörner's eustatic curve (dashed line), agreeing well with the information from Vendsyssel.

stable during the period in question. Consequently, the two oldest dates may give a good idea about the continuation of Mörner's eustatic curve back in time. They also fit very well with an extrapolation of Mörner's eustatic curve and with the indication from Vendsyssel of a regression by 10 m at about 13,000 B.C. A 10 m eustatic drop from − 60 m to − 70 m was calculated, which gave quite logical isostatic curves for Vendsyssel (MÖRNER, 1969a, fig.151). This drop occurred at about 13,000–12,800 B.C. (15,000–14,800 B.P.) and corresponds to the distinct

Palaeogeography, Palaeoclimatol., Palaeoecol., 9(1971) 153–181

second glaciation maximum of the Main Wisconsin–Weichselian–Würm–Valdaj, which evidently was a global event (MÖRNER, 1969a, b, 1970a, b, f).

The eustatic changes between 12,800 B.C. (or somewhat later) and 11,750 B.C. are not known in detail; only a rise from −70 m to −66 m. Minor oscillations corresponding to the oscillating climate are expected.

From the Atlantic shelf off northeast North America, GARRISON and McMASTER (1966) reported four shore cut terraces. The lowermost one lies at −145 m and corresponds to the Nicholls Shore (VEACH and SMITH, 1939) and a distinct sub-bottom erosion surface (EWING et al., 1963). This terrace was most probably formed during the main Wisconsin Glaciation maximum at about 18,000–15,000 B.C. (MÖRNER, 1969a, p.438). Time-depth indications from this shelf (e.g., EMERY and GARRISON, 1967) show a position of the relative sea level during the second main Wisconsin Glaciation maximum (about 13,000 B.C.) at −130 m (MÖRNER, 1969a, fig.155). This means that the difference between the eustatic sea-level position at the two glaciation maxima at 18,000–15,000 B.C. and 13,000–12,800 B.C. was small or about 15 m, which is in full agreement with the volume of the continental ice (MÖRNER, 1969a, pp.436–438, 1970f). Such a large eustatic difference between these two maxima as CURRAY (1960, 1965) gives, for example, is quite incompatible with the extension of the continental ice sheets (MÖRNER, 1969a, 1970a, d). Consequently, the eustatic sea level should have been at about −85 m (−85 m to −90 m) during the first (main) maximum. EMERY and GARRISON (1967) also gave a date of 11,470 ±210 B.C. for a −120 m terrace, giving a 25 m difference to the −145 m terrace. As Mörner's eustatic curve lies at −60 m at the suggested time of the formation of the −120 m terrace, it again means that the eustatic position at 18,000–15,000 B.C. can be calculated at −85 m.

At the time of the preparation of the manuscript (1967–1968) of my first paper on this subject (MÖRNER, 1969a), I (Mörner) could only note that a maximum lowering of the eustatic sea level during the main Wisconsin–Weichselian–Würm to −85 m is much smaller than the general North American view of −118–125 m, but it fits better with the general European view of −80–100 m (MÖRNER, 1969a, p.328). However, new papers (e.g., MILLIMAN and EMERY, 1968; PINOT, 1968; FAURE, 1969; FLINT, 1969) have later provided material for an extended discussion (MÖRNER, 1970a).

FLINT (1969) suggested that the ocean floor would have been deformed by about one-third of the thickness of the water added and removed during the interglacial/glacial changes. He calculated the "ocean volume changes" (terminology from MÖRNER, 1970a) at about 132 m for a maximum glaciation and a 10% smaller figure for the Last Glacial Age. This would give a corresponding eustatic rise and lowering of 88 m and 80 m, respectively. However, there is—as is closely discussed in another paper (MÖRNER, 1970d)—no ground to expect an upheaval of the ocean floors by the same amount (as assumed by FLINT, 1969) during a drop of sea level in connection with the building up of the glaciers, as the amount of the subsidence during the rise of sea level after a glaciation period.

SHEPARD (1963b) showed that the average shelf break of the world lies at −132 m. MILLIMAN and EMERY (1968) supposed that 35 dates from 11 different shelf areas of the world (other than the Atlantic) show good agreement with their Atlantic sea-level curve, suggesting that the latter "defines a eustatic curve". However, the Atlantic shelf off northeast North America is most probably strongly subsided (MÖRNER, 1969a; NEWMAN and MARCH, 1968), indicating that the other shelves are most probably also subsided (MÖRNER, 1970d).

FAURE (1969) gave a sea-level curve from the slightly uplifted Senegal area, showing a sea level at −75 m at 18,000 B.C. and at −50 m at 13,000 B.C.

Outside the more or less stable coast of Brittany, submarine shorelines have been found at −89 m and −92 m (PINOT, 1968). They were formed during a periglacial climate and were supposed to be 22,000–18,000 years old. Below these levels there are traces of a shore at −106 m. However, valleys were found down to the −92 m shore, but not down to the −106 m level. This indicates that the −89–92 m levels were formed during the maximum of the last glacial age, while the fairly indistinct −106 m level is much older and probably of Riss Glacial age (MÖRNER, 1970a). The −89–92 m levels fit nicely with Mörner's calculated eustatic level at −85 m (or −85–90 m). This indicates—together with the above-mentioned −37 m level—that this area has been stable (or very close to stable). The −106 m level would, with one-third isostatic deformation of the ocean floor, correspond to an "ocean volume" of 159 m, which might correspond to the buried −165 m terrace on the Atlantic shelf off northeast North America with cold-water molluscs dated at more than 35,000 years (EWING et al., 1963). Table V gives a reinterpretation of the shorelines outside Brittany (cf., Fig.12 and PINOT, 1968).

TABLE V

REINTERPRETATION OF THE SHORELINES OUTSIDE BRITTANY, FRANCE

Shoreline (m)	Interpretation
−106	Riss Glacial; maximum lowering
+1.5	Late Eemian Interglacial or, more probable, the Brörup Interstadial
−52	some part of the Weichselian
−92	maximum Late Weichselian lowering at about 18,000–15,000 B.C.
−89	later part of same stage (16,000–15,000 B.C.?)
−55	Late Glacial, at onset of rapid transgression 10,700 B.C.
−37	Postglacial, 7,700–7,330 B.C., before onset of rapid transgression
−21	Postglacial, 6,800–5,800 B.C., before onset of rapid transgression

BLANC's (e.g., 1937, fig.6) classical section from Bassa Versilia in Italy clearly shows a regression down to −90–85 m, where a thick layer of continental sediments at −91–75 m reaches far out in the marine sequence (Fig.14). Repeated levellings in this area show a relative subsidence of more than 5 cm and somewhat less than 10 cm during the 60 years from 1897 to 1957 (i.e., 0.83–1.66 mm/year). Probably this area is more or less stable—at least it is not uplifted.

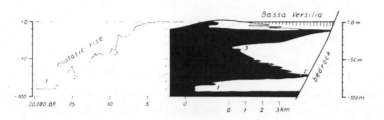

Fig.14. To the right Blanc's classical section from Bassa Versilia in Italy. To the left the Mörner eustatic curve. Black areas in the section = marine sediments, and white areas = continental sediments. This section show a maximum sea level lowering to about −90 m. The following interpretation was found: Regression maximum with continental sedimentation (*1*); transgression with marine sedimentation up to −70 m; regression with continental sedimentation at about −70 m (*2*); oscillating transgression between −70 m (*2*) and −35 m (*3*); rapid transgression with marine sedimentation from −35 m (*3*) up to about −12 m; slow oscillatory rise with final building out of continental sediments.

Mörner's eustatic curve (Fig.12) is calculated in such a way that it shall give true eustatic changes. The disagreement about the amount of eustatic lowering during the Last Glacial Age between Mörner's calculation and the information from e.g., Brittany and Bassa Versilia on one hand and Flint's calculation of the "ocean volume changes" and the information from different shelves of the world on the other hand (Table VI), strongly speaks for a hydro-isostatic deformation of the ocean floor (MÖRNER, 1970a; 1970d). Assuming a hydro-isostatic deformation by

TABLE VI

SUMMARY OF THE INFORMATION ABOUT THE MAXIMUM LATE WEICHSELIAN-WISCONSIN SEA LEVEL LOWERING

The MÖRNER curve	ocean level	−85–90 m		
Brittany (PINOT, 1968)	shoreline	−90–89 m	stable	true eustatic lowering
Bassa Versilia (BLANC, 1937)	sediment	−90 m		
FLINT (1969)	ice volume	−132–120 m		"ocean volume" changes
GARRISON/MCMASTER (1966)	terrace	−145 m		
SHEPARD (1963a)	shelf break	−132 m	subsided	eustatic + hydro-
MILLIMAN/EMERY (1968)	sea level	−130 m		isostatic lowering

Palaeogeography, Palaeoclimatol., Palaeoecol., 9(1971) 153–181

one-third of the water thickness, the following relation applies for the "ocean volume" and eustatic changes:

"ocean volume" increase (decrease?)	30 m	60 m	90 m	120 m	150 m	180 m
eustatic rise (drop?)	20 m	40 m	60 m	80 m	100 m	120 m

However, this relationship must also be influenced by several other factors; e.g., local upheaval or subsidence of other causes.

CONCLUSIONS

The author's investigation in southern Scandinavia (MÖRNER, 1969a) revealed a very high "crustal sensitivity" to changes in the load of ice (glacio-iso-stasy) and water (hydro-isostasy). This means that a transgressed shelf most probably was hydro-isostatically deformed. If there was a compensational upheaval inside the axis of subsidence (MÖRNER, 1970a, b, d), the deformation must have been complex, changing from the undeformed stage to upheaval and later to subsidence. Furthermore, there are strong arguments for a hydro-isostatic deformation of the ocean floor in general also (KAITERA, 1966; BLOOM, 1967; FLINT, 1969; MÖRNER, 1970d). Finally, there are several other factors that may change the ocean level in general or locally; e.g., changes in the tide, water density changes, changes of major currents, changes of the earth's rotation, changes in major wind pattern and air pressure.

All these things mean that it must be very hard to find a really "stable area", where the sea-level changes are true eustatic changes. It also means that a detailed study of an uplifted area, where the isostatic factor can be closely determined and the eustatic and isostatic factors separated, is definitely one of the best ways to determine the eustatic changes. For such studies, southern Scandinavia is uniquely suitable (MÖRNER, 1969a, p.8).

Mörner's eustatic curve is calculated with a very high precision (for the last 12,700 years) and future changes or additions will probably only concern minor details. However, we must remember that it gives the eustatic changes in the northwest Atlantic (55–58°N lat. and 10–13°E long.), as the ocean level has not necessarily been equally distributed back in time.

ACKNOWLEDGEMENTS

I am indebted to Professor Alexis Dreimanis for critical reading of the manuscript. Final drafting and manuscript work were financed by the Department of Geology, University of Western Ontario, London, Ontario, Canada.

NOTE ADDED IN PROOF

Concerning the maximum Late Weichselian-Wisconsin sea level lowering, it is interesting to note that the detailed mapping by MCMASTER et al. (1970) of the shelf morphology off West Africa did not reveal any shore features below −90 m, where remnants of a well-developed shoreline were found, and that a coral from −103–111 m (*Porites benardi*) was ^{14}C dated at 16,800 ± 350 B.C., i.e., the time of the First Glaciation maximum.

By submarine coring off southeast Devon, England, CLARKE (1970) was able to reconstruct the shore morphology of a distinct sea level stand at about −40 m (CLARKE, 1970, Fig.11). Pollen-analytically, this level could be dated at about the Preboreal/Boreal boundary. Obviously, this is another example of distinct shore features formed in connection with the eustatic level at −38.0–38.5 m during the period 7,330–7,700 B.C.

REFERENCES

AUER, V., 1959. The Pleistocene of Fuego-Patagonia. III. Shoreline displacement. *Ann. Acad. Sci. Fennicae, A. III*, 60: 1–247.
BEHRE, K.-E., 1969. Pollenanalytische Untersuchungen an einem Bohrkern der südlichen Dogger-bank. *Beitr. Meereskunde*, H.24–25: 122–130.
BERGLUND, B., 1964. The Post-Glacial shore displacement in eastern Blekinge, southeastern Sweden. *Sveriges Geol. Undersökn., Ser. C*, 599: 1–47.
BLANC, A. C., 1937. Low levels of the Mediterranean Sea during the Pleistocene glaciation. *Quart. J. Geol. Soc. London*, 93: 621–651.
BLOOM, A., 1967. Pleistocene shorelines: a new test of isostasy. *Geol. Soc. Am. Bull.*, 78(2): 1477–1494.
BLOOM, A., 1969. Holocene submergence in Micronesia as the standard for eustatic sea-level changes. Lecture (multigraphed), *Symposium on the Evolution of Shorelines and Continental Shelves during the Quaternary, 1969*. U.N.E.S.C.O., Paris (in press).
CHEBOTAREVA, N. S., 1969. Recession of the last glaciation in northeastern European U.S.S.R. In: H. E. WRIGHT (Editor), *Quaternary Geology and Climate—Intern. Congr. Quaternary, 7th, 16*. Natl. Acad. Sci. Publ., Washington D.C. 1701, pp.79–83.
CLARKE, R. H., 1970. Quaternary sediments off south-east Devon. *Quart. J. Geol. Soc. London*, 125(3): 277–318.
CURRAY, J. R., 1960. Sediments and history of Holocene transgression, continental shelf, North-west Gulf of Mexico. In: F. P. SHEPARD et al. (Editors), *Recent Sediments, Northwest Gulf of Mexico*. Am. Assoc. Petrol. Geologists, Tulsa, Okla., pp.221–266.
CURRAY, J. R., 1965. Late Quaternary history, continental shelves of the United States. In: H. E. WRIGHT and D. J. FREY (Editors), *The Quaternary of the United States*, Princeton Univ. Press, Princeton, N.J., pp.723–735.
DELIBRAIS, G., 1969. Niveau de la mer sur la côte atlantique et la manche depuis 10,000 ans par le C[14]. *Abstracts, Intern. Congr. Quaternary, 8th, Paris, 1969*, p.347.
EMERY, K. O. and GARRISON, L. E., 1967. Sea level 7,000 to 20,000 years ago. *Science*, 157: 684–687.
EWING, J., LE PICHON, X. and EWING, M., 1963. Upper stratification of Hudson Apron region. *J. Geophys. Res.*, 68(23): 6303–6316.
FAIRBRIDGE, R. W., 1961. Eustatic changes in sea level. *Phys. Chem. Earth*, 4: 99–185.
FAURE, H., 1969. Neotectonique et lignes de rivage en Afrique. *Abstracts, Intern. Congr. Quaternary, 8th, Paris, 1969*, p.301.

FLINT, R. F., 1969. Position of sealevel in a Glacial Age. Lecture (multigraphed), *Intern. Congr. Quaternary, 8th, Paris, 1969* (in press).

GARRISON, L. E. and McMASTER, R. L., 1966. Sediments and geomorphology of the continental shelf off southern New England. *Marine Geol.*, 4: 273–289.

HAGEMAN, B. P., 1969. The western part of The Netherlands during the Holocene. *Geol. en Mijnbouw*, 48(2): 373–388.

HANSEN, S., 1965. The Quaternary of Denmark. In: K. Rankama (Editor). *The Geologic Systems. The Quaternary, Vol. 1.* Wiley, New York, N.Y., pp.1–90.

HOPLEY, D., 1969. World sea levels during the past 11,000 years—evidence from Australia and New Zealand. *Abstracts, Intern. Congr. Quaternary, 8th, Paris, 1969*, p.260.

JAKUBOVSKY, O., 1966. Vertical movements of the earth's crust on the coasts of the Baltic Sea. *Ann. Acad. Sci. Fennicae, A. III.*, 90: 479–488.

JELGERSMA, S., 1961. Holocene sea-level changes in The Netherlands. *Mededel. Geol. Sticht., Ser. C*, VI(7): 1–101.

JELGERSMA, S., 1966. Sea-level changes during the last 10,000 years. In: *Proc. Intern. Symp.— World climate from 8000 to 0 B.C., 1966.* Roy. Meteorol. Soc., London, pp.54–71.

KAITERA, P., 1966. Sea pressure as a cause of crustal movements. *Ann. Acad. Sci. Fennicae, A. III.*, 90: 191–200.

LENNON, G. W., 1966. An investigation of secular variations of sea level in European waters. *Ann. Acad. Sci. Fennicae, A. III.*, 90: 225–236.

MARTHINUSSEN, M., 1962. C^{14}-datings referring to shore lines, transgressions, and glacial stages in northern Norway. *Norg. Geol. Undersökelse*, 215: 37–67.

McMASTER, R. L., LACHANCE, T. P. and ASHRAF, A., 1970. Continental shelf geomorphic features off Portuguese Guinea, Guinea, and Sierra Leone (West Africa). *Marine Geol.*, 9: 203–213.

MERTZ, E. L., 1924. Oversigt over De sen- og postglaciale Niveauforandringer i Danmark. *Danmarks Geol. Undersögelse, II R.*, 41: 1–50.

MILLIMAN, J. D. and EMERY, K. O., 1968. Sea levels during the past 35,000 years. *Science*, 162: 1121–1123.

MÖRNER, N.-A., 1969a. The Late Quaternary history of the Kattegatt Sea and the Swedish West Coast: deglaciation, shorelevel displacement, chronology, isostasy and eustasy. *Sveriges Geol. Undersökn., Ser. C*, 640: 1–487.

MÖRNER, N.-A., 1969b. Climatic and eustatic changes during the last 15,000 years. *Geol. en Mijnbouw*, 48(4): 389–399.

MÖRNER, N.-A., 1969c. Eustatic and climatic changes during the last 20,000 years. *Abstracts, Intern. Congr. Quaternary, 8th, Paris, 1969*, p.226.

MÖRNER, N.-A., 1970a. Late Quaternary isostatic, eustatic and climatic changes. *Quaternaria* (in press).

MÖRNER, N.-A., 1970b. The Holocene eustatic sea level problem. *Geol. en Mijnbouw*, 50 (in press).

MÖRNER, N.-A., 1970c. Eustatic and climatic oscillations. *Arctic Alpine Res.*, 3(2) (in press).

MÖRNER, N.-A., 1970d. Isostasy and eustasy. Late Quaternary isostatic changes in southern Scandinavia and general isostatic changes of the world. In: B. W. COLLINS (Editor), *Proc. Intern. Symp.—Recent Crustal Movements and Associated Seismicity.* Royal Society of New Zealand (in press) and Abstracts Symp. 1970, Victoria Univ., Wellington, pp.25-27.

MÖRNER, N.-A., 1970e. Comparisons between the varve and C^{14} chronologies. In: I. OLSSON (Editor), *Proc. Nobel Symp.*, 12, *Uppsala*, 1969. Almqvist and Wiksell, Stockholm, New York, pp.225–229.

MÖRNER, N.-A., 1970f. Comparison between Late Weichselian and Late Wisconsin ice marginal charges. *Eiszeitaltes Gegenwart*, 21, (in press).

NEUMANN, A. C., 1969. Quaternary sea-level data from Bermuda. *Abstracts, Intern. Congr. Quaternary, 8th, Paris, 1969*, pp.228–229.

NEWMAN, W. S. and MARCH, S., 1968. Littoral of the northeastern United States; Late Quaternary Warping. *Science*, 160: 1110–1112.

PINOT, J. P., 1965. Submerged shoreline and other evidences of a Quaternary sea level at 37 meters below the present one on the southern coast of Brittany (France). *Abstracts, Intern. Congr. Quaternary, 7th, Boulder, Colo.*, p.380.

Palaeogeography, Palaeoclimatol., Palaeoecol., 9(1971) 153–181

PINOT, J. P., 1966. Découverte d'un rivage submergé, et d'autre preuves d'un niveau marin quaternaire à 37 m sous l'actuel, sur la côte sud de la Bretagne. *Quaternaria*, 8: 225–230.

PINOT, J. P., 1968. Littoraux wurmiens submergés à l'ouest de Belle-Ile. *Bull. Assoc. Franc. Quatern.*, 5: 197–216.

SCHOLL, D. W., 1964. Recent sedimentary record in mangrove swamps and rise in sea level over the southwestern coast of Flórida, Part 1. *Marine Geol.*, 1: 344–366.

SCHOLL, D. W. and STUIVER, M., 1967. Recent submergence of Southern Florida: a comparison with adjacent coasts and other eustatic data. *Geol. Soc. Am. Bull.*, 78(1): 437–454.

SCHOLL, D. W., CRAIGHEAD, F. C. and STUIVER, M., 1969. Florida submergence curve revised; its relation to coastal sedimentation rates. *Science*, 163: 562–564.

SHEPARD, F. P., 1960. Rise of sea level along northwest Gulf of Mexico. In: F. P. SHEPARD et al. (Editors), *Recent Sediments, Northwest Gulf of Mexico*. Am. Assoc. Petrol. Geologists, Tulsa, Okla., pp.338–344.

SHEPARD, F. P., 1963a. *Submarine Geology*. Harper and Row, New York, N.Y., 2nd ed., pp.1–557.

SHEPARD, F. P., 1963b. Thirty-five thousand years of sea level. In: T. CLEMENTS (Editor), *Essays in Marine Geology in Honor of K. O. Emery*, Univ. S. Calif. Press., Los Angeles, Calif., pp.1–10.

SUGGATE, R. P., 1968. Post-glacial sea-level rise in the Christchurch Metropolitan area, New Zealand. *Geol. en Mijnbouw*, 47:291–297.

TERS, M. and PINOT, J. P., 1969. Livret-guide de l'excursion A 10; Littoral Atlantique. *Intern. Congr. Quaternary, 8th, Paris, 1969, Publ.*, pp.1–110.

WAALEWIJN, A., 1966. Investigation into crustal movements in The Netherlands. *Ann. Acad. Sci. Fennicae, A. III.*, 90: 401–412.

VEACH, A. C. and SMITH, P. A., 1939. Atlantic submarine valleys of the United States and the Congo submarine valley. *Geol. Soc. Am. Spec. Paper*, 7: 1–101.

VAN VEEN, J., 1964. Tide-gauges, subsidence-gauges and floodstones in The Netherlands. *Geol. en Mijnbouw*, 16: 214–219.

VINOGRADOV, A. P., DEVITS, A. L., DODKINA, E. I. and MARKOVA, N. G., 1966. Radiocarbon dating in the Vernadsky Institute, I–IV. *Radiocarbon*, 8: 292–323.

ZENKOVICH, V. P., 1969. Nature of the U.S.S.R. shelves and coasts. Lecture (multigraphed), *Symposium on the Evolution of Shorelines and Continental Shelves during the Quaternary, 1969*, UNESCO, Paris (in press).

Editor's Comments on Paper 22

Newman, Fairbridge, and March: *Marginal Subsidence of Glaciated Areas: United States, Baltic and North Seas*

Vertical movements caused by glacio-isostatic adjustments have been primarily studied within the borders of the former ice caps. However, as noted by Daly (Paper 7), the ice load should have caused a "forebulge" to develop in front of the ice cap that would collapse as the ice load was removed. Kupsch (Paper 8) concluded that direct evidence for a forebulge is generally lacking. However, Newman, Fairbridge, and March, in this paper, review the arguments for and against the peripheral bulge but primarily cite evidence in favor.

The negative arguments against the bulge are not overwhelmingly strong. The arguments for the existence of the peripheral bulge come mainly from the eastern coast of North America, although it should be noted that Bloom (1967, *Bull. Geol. Soc. Amer.*) used very much the same data to argue for the importance of the late-glacial Holocene water load on the subsidence of the continental shelf. One possible negative argument against both the water load and peripheral bulge concepts is that evidence for subsidence is absent, or very recent in age, at sites in comparable proximity to the ice margin along the northern and northeastern margin of the Laurentide Ice Cap. Nevertheless, Figs. 3 and 4 of Newman et al. present important data and suggest an inward (to the Laurentide ice center) migration of the forebulge wavecrest at 100 m yr^{-1}.

Evidence is presented for the existence and dynamics of the peripheral forebulge. It remains to be seen if evidence in the Great Lakes and other areas can be used to support the model; if not, other physical explanations of the curves of Fig. 3 will have to be proposed.

Walter S. Newman is Associate Professor of the Department of Earth and Environmental Sciences, Queen's College of the City University of New York. He has done considerable work on changes of sea level in southern New England. Rhodes W. Fairbridge is Professor of Geology at Columbia University and has published extensively in many fields, although he has long been interested in problems of glacio-eustatic and glacio-isostatic recovery (see Fairbridge, 1961).

Reprinted from *Étude sur le quaternaire dans le monde*, 795–801 (1971)

22

Marginal subsidence of glaciated areas : United States, Baltic and North Seas

by Walter S. NEWMAN
Queens College
City University of New York, Flushing, N.Y. 11367
Rhodes W. FAIRBRIDGE
Columbia University, New York, N.Y. 10027
and Stanley MARCH
Service Bureau Corporation
1350 Avenue of the Americas, New York, N.Y. 10019

INTRODUCTION

For more than a decade, we have been puzzled by anomalies in the Holocene « eustatic » sea level curves. We have discussed this situation repeatedly (NEWMAN and FAIRBRIDGE, 1960 a, 1960 b, 1962 a, 1962 b, 1969 ; FAIRBRIDGE and NEWMAN, 1965, 1968 ; NEWMAN, 1968 a, 1968 b ; NEWMAN and RUSNAK, 1965 ; NEWMAN and MUNSART, 1968 ; and NEWMAN and MARCH, 1968) and find that since the submergence curves of various authors diverge with increasing age, some form of Holocene crustal warping must be affecting the world's coasts. On the other hand, when a correction for tectonic warping is applied, many of the curves are found to be rather consistent. MÖRNER (1969 a, b) has demonstrated a curious divergence that only affects the early Holocene.

Study of the available data both for the northeast coast of North America and northern Europe indicates that portions of these areas are at present subsiding in zones roughly peripheral to those areas which were ice-covered during the last glaciation. The data further suggest that the boundary between rising and sinking areas is gradually migrating back toward the areas of maximum glacial depression. We have so far concluded that these crustal distortions represent the slow collapse of a former peripheral bulge (such as proposed by DALY) that developed somewhat beyond the limits of the last continental ice sheets as a result of the displacement of mantle material away from the ice loaded area.

Some geophysicists argue that such a bulge theoretically should exist because the assumed viscosity of mantle is such that the material displaced could not have been uniformly spread around the earth within the time interval available, although the assumptions involved are not easily justified. Nevertheless, on the basis of both our own and newly published data, there is still a certain persuasiveness in the peripheral bulge concept.

THE ARGUMENTS

THE NEGATIVE VIEW

The major argument against the peripheral bulge (or forebulge) concept in North America appears to be the « Algonquin hinge line » of LEVERETT and TAYLOR (1915) across Lake Michigan which is the zero isobase for the Algonquin beach at an altitude of 184 meters. According to them, south of this line, the Algonquin beach is horizontal and accordingly there is no field evidence to support the argument of peripheral subsidence. However, HOUGH (1958, p. 216) found that the LEVERETT and TAYLOR conclusion is not supported by field observations and that « ... all beaches above the present lake shore have been removed by erosion in all of the critical stretches of shore where the tilted beaches presumably come to horizontality. This absence of the Algonquin beach in the critical areas on both the eastern and western shores of both the Huron and Michigan basins becomes apparent only by a perusal of the detailed descriptions of the beach... » Moreover, FRYE (1963) observes that the anomalous bedrock thalweg of the « Ancient Mississippi Valley » declines less than 15 m in altitude through a distance of 500 km south of the border of northern Illinois, and may be explained by the collapse of a « forebulge » caused by glacial loading. FRYE points out that according to one set of assumptions a peripheral bulge would result that would be 1 200 km wide and 43 m high during the last Wisconsin (Würm) glacial maximum.

McGINNIS (1968) has also recently reviewed the peripheral bulge situation and concludes that theoretically a « low » upward should develop along the margin of an ice sheet but considers that it would be difficult to distinguish from other geologic phenomena.

A POSITIVE VIEW

To test the peripheral bulge hypothesis we have calculated the width and height of such a bulge using various assumptions. We used the data computed by DONN et al. (1962) who found that at its maximum extent the late Wisconsin Laurentide ice sheet covered an area of some 12.74×10^6 km² and attained a volume of some 25.48×10^6 km³. We made the following simplifying assumptions : (a) the ice sheet occupied a circular area and the peripherally affected area was ring shaped (Fig. 1) ; (b) the

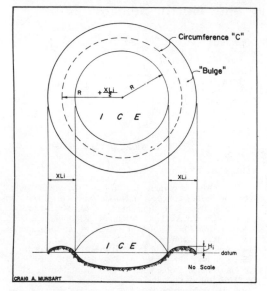

Fig. 1. — Sketch illustrating relationships between Laurentide ice sheet crustal loading and peripheral or marginal bulge.

ratio of ice volume to mantle volume displaced by glacial loading was 4:1 rather than the usually assumed 3:1, the difference assumed to have been taken up by a combination of phase changes and elastic compression ; (c) the bulge in cross-section assumed the shape of an ellipse bisected parallel to its major axis ; and (d) there is a direct relationship between glacial loading and the development of a marginal or peripheral bulge. Using the following formula :

$$H_1 = \frac{V}{\pi^2/2 (R + \frac{XLi}{2}) \, XLi}$$

where H_1 = height of « bulge »,
  V = volume of displaced material
    (= 0.25 volume of ice),
  R = radius of ice sheet,
  XLi = width of « bulge ».

We computed the height of the warp (Fig. 1) by varying the bulge width, at 100 km intervals, from 100 to 3 000 km while also decreasing, in ten per cent steps, both the area and the volume from 100 to 10 per cent of their respective assumed maxima. This was done in order to simulate the waning of the Laurentide Ice Sheet. Since, according to FAR-

RAND (1964), the ice volume probably decreased faster than the ice area owing to thinning of the ice sheets, we held percentages of ice volumes less than or equal to corresponding ice areas.

The radiocarbon dates of deglaciation utilized in a previous study (NEWMAN and MARCH, 1968) ranged from about 7 000 to 15 000 years B.P. whereas the late Wisconsin glacial maximum was believed to have been between 15 000 and 18 000 years ago. We have therefore no data for the height of the hypothetical « bulge » at the glacial maximum. Even if the Laurentide ice sheet had achieved isostatic compensation, fully 50 per cent of this ice had disappeared by 13 000 years B.P. (FARRAND, 1964). The NEWMAN and MARCH median radiocarbon date is 10 600 years B.P. (Table 1). The published data nevertheless indicate a maximum « bulge » height of 100 meters. We can only speculate as to width of this « bulge » (FAIRBRIDGE and NEWMAN, 1968), a series of assumptions up to 3 000 km width being considered.

We plotted a series of curves (Fig. 2) relating ice volume, ice area, « bulge » height and « bulge » width.

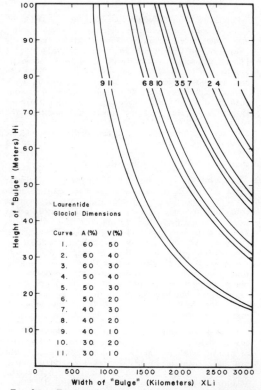

Laurentide
Glacial Dimensions

Curve	A (%)	V (%)
1.	60	50
2.	60	40
3.	60	30
4.	50	40
5.	50	30
6.	50	20
7.	40	30
8.	40	20
9.	40	10
10.	30	20
11.	30	10

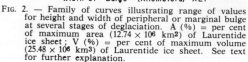

Fig. 2. — Family of curves illustrating range of values for height and width of peripheral or marginal bulge at several stages of deglaciation. A (%) = per cent of maximum area (12.74×10^6 km²) of Laurentide ice sheet ; V (%) = per cent of maximum volume (25.48×10^6 km³) of Laurentide ice sheet. See text for further explanation.

RADIOCARBON YEARS B.P. (x10³)

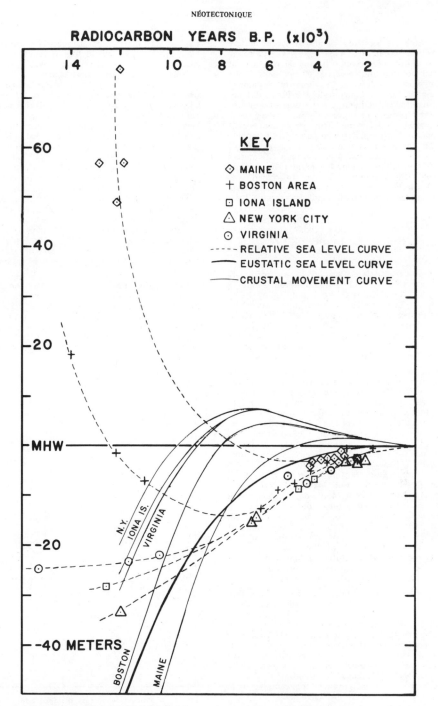

FIG. 3. — Apparent sea level curves, the SHEPARD and CURRAY (1967) eustatic sea level curve, and crustal movement curves for five stations along the north-eastern Atlantic Coast of the United States. Note displacement of wavecrests with time. See text for further explanation.

Only data giving a « bulge » height of 100 m or less were considered realistic. This « bulge » represents, at most, a continental glacier having 50 per cent of its original volume and covering 60 per cent of its maximum area. Following this line of reasoning, we find eigher (a) the Laurentide Ice Sheet did not achieve isostatic adjustment, or (b) our data record only the later portion of bulge collapse ; indeed, both explanations are conceivable. Our test indicates that the peripheral bulge model is quantitatively tenable within the following contraints : in its maximum expression, the bulge crest was less than 100 meters high while its half wave-length was almost certainly more than 1 000 kilometers. After this paper was completed another interesting treatment of this problem was added by BROTCHIE and SILVESTER (1969).

THE FIELD EVIDENCE

Using radiocarbon dated sea levels, we previously proposed the existence of low amplitude and long wave-length crustal waves migrating progressively inward toward the center of the former Laurentide Ice Sheet (FAIRBRIDGE and NEWMAN, 1965, 1968). With additional data that has become available during the past few years, we have now reconstructed our presentation which appears to demonstrate anew the reality of these assumed crustal waves.

The first step in our construction (Fig. 3) was to plot radiocarbon dated sea levels for these localities or areas which have available data going back in time for ten millennia or more. Such sites prove to be regrettably rare. We used data from five areas along the northeast coast of the United States : Cape Charles, Virginia (HARRISON et al., 1965 ; NEWMAN and MUNSART, 1968) ; New York City (NEWMAN, 1966) ; Iona Island in the Hudson River Estuary, some 45 km north of New York City (NEWMAN et al., 1969) ; the Boston area (McINTIRE and MORGAN, 1964 ; KAYE and BARGHOORN, 1964) ; and Maine (BLOOM, 1963 ; BORNS, 1963). Note that our Maine data include much of coastal Maine : a compromise adopted in order to assemble sufficient dates to be useful. From all these data, we have drawn five relative sea level curves (Fig. 3) : these curves diverge dramatically as we go back in time, much of the deleveling being caused by postglacial isostatic crustal rebound in Maine and eastern Massachusetts. Note the absence of data in the interval 7 000-10 000 years B.P.

Our second step was to try and construct a universal eustatic sea level curve. If Holocene deformation has indeed affected the North Atlantic coast, it should be possible to measure the extent, sign and rate of crustal warping if a stable coastal datum could be recognized. However, this proves to be a matter of some controversy. Some Holocene sea level investigators assume they have a completely stable world-wide tide gauge in their area. Should data from other areas conflict, they imply, it is because these other coasts are unstable.

We do not use the FAIRBRIDGE (1961) curve for two reasons : the curve is now a decade old and there

has been copious research published during the past ten years, and we would like to avoid the dangers of bias and circular reasoning. We have therefore plotted the SHEPARD and CURRAY (1967) eustatic curve because it is simple, uncomplicated, widely (if uncritically) accepted. Differences in detail from the FAIRBRIDGE curve of 1961 and the new MÖRNER curve of 1969 are rather minor, and these do not seriously affect our argument. SHEPARD and CURRAY constructed their curve back to 7 500 years B.P., the data prior to that time having a rather wide scatter. By inspection of other data and using their point plots, we extrapolated their curve back to 12 000 years B.P. Again, the exact detail of the eustatic curve is not crucial to our argument while a simple eustatic curve eases the computation problem. Inserting the eustatic sea level curve, accordingly, we discover that all northwestern Atlantic coastal data for the past 7 000 years plot below the eustatic curve while all older data plot above it. Qualitatively, this suggests that all stations must have been initially upwarped and have subsequently subsided.

Graphically, the difference between the relative sea level curve and the eustatic curve results in the generation of crustal movement (or deleveling) curves. The latter curves (indicated on Fig. 3) depict the evolution of points now on the coast at Mean High Water through the past 12 000 radiocarbon years. Similar crustal movement curves have pre-

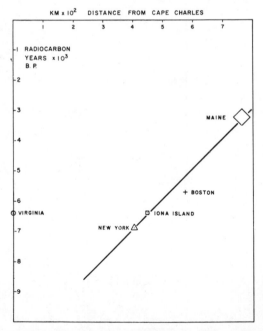

FIG. 4. — Plot of wavecrest distance from Cape Charles (Virginia) against time. Apparent velocity of crestal displacement of four northern stations is about 100 m/year toward north.

viously been constructed for Cape Charles, Virginia area (HARRISON et al., 1965), the Boston area (KAYE and BARGHOORN, 1964) and BLOOM (1963).

The five crustal movement curves plotted on Figure 3 illustrate a number of salient points. First, the curves decay nearly exponentially with time, a characteristic of most postglacial isostatic rebound curves. Secondly, the crustal movement curves are nearly parallel to one another suggesting similar mechanisms of crustal warping. Finally, and most important, the curves and curve crests are displaced from each other in a manner suggesting that the wave crest has migrated over a period of time.

The localities and areas involved are shown on Figure 4 while the distance (extrapolated so as to be normal to ice margin positions and presumably to the isobases) from Cape Charles is plotted against dates of maximum crestal height in Figure 5. Although we lack data from the critical interval 7 000-10 000 years B.P., the Figure 4 plot reveals several interesting relationships. The Virginia data have

no obvious relationship with the other four stations suggesting that either the older data are spurious, or represent still another crest distinct from that of the other four stations, or manifest crustal warping not associated with glacial isostasy. The Maine plot represents a large region and is plotted as an area rather than a point. The remaining three stations plot very nearly on a straight line with a displacement of the crest in the order of 100 m/year. Figure 3 indicates an inward migrating peripheral bulge having a height of less than 20 m while Figure 5 suggests a half wave-length of at least 350 km.

DISCUSSION

The model of an inward migrating and decaying peripheral bulge presented here is identical to that initially proposed and subsequently rejected by R. A. DALY (1934, see especially his Figure 69, p. 121). The

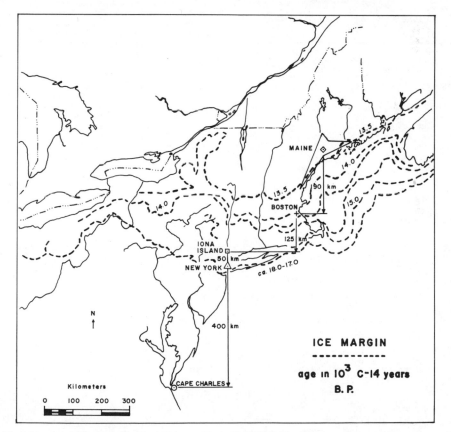

FIG. 5. — Location map of stations discussed in text and their relationship with Late Wisconsin ice margin positions plotted by PREST (1969).

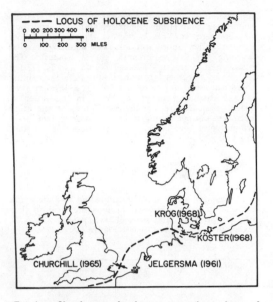

FIG. 6. — Sketch map showing an approximate locus of Holocene (postglacial) subsidence in northwestern Europe. Principal reference sources to the different areas are indicated.

family of crustal movement curves depicted in Figure 3 finds the bulge crest less than 20 meters high in the interval 2 500-7 000 radiocarbon years B.P. while the half wavelengths plotted on Figures 4 and 5 indicate a minimum of 350 km (i.e. a width of at least 700 km). Plots of these wavelengths and

crestal heights on Figure 2 fall in a portion of the field that would indicate that less than 10 per cent of maximum possible isostatic adjustment remained to be accomplished within the 2 500-7 000 radiocarbon years B.P. interval. This finding is in close agreement with the ANDREWS (1968) analysis which suggests that 90 per cent of uplift is accomplished within the 6 000 years subsequent to deglaciation.

We believe that the arguments presented here support the peripheral bulge hypothesis and present its dimensions and dynamics for North America in a rather general, if approximate, manner. With respect to the Scandinavian ice sheets, a review of the literature suggests that coastal areas near the periphery of the former ice also suffered differential subsidence during the Holocene. The latter conclusion seems supported by the works of CHURCHILL (1965) in England, JELGERSMA (1961) in the Netherlands, KROG (1968) in Denmark, and DECHEND and GRONWALD (1961), KÖSTER (1968) in Germany. We suggest that a careful sifting or northern European data would disclose a phenomena similar to the DALY model. In this way a belt of late Holocene subsidence appears to run from southern England, across north Germany and southern Denmark to the Baltic. The level of the mid-Holocene Littorina beach in the southeastern Baltic today stands at + 6 m (BACZYK, 1966), which might suggest perhaps a second crustal wave (cf. our Virginia data), but more research is needed. An interesting discovery was made in the western English Channel by BOILLOT (1964), which was not glaciated but should have been in the peripheral bulge area ; Eocene and Cretaceous limestones have been deeply affected by karst solution (essentially a subaerial process) to at least — 190 m ; if maximal eustatic lowering were — 130 m, this would call for a marginal bulge of 60 m only 200 km south of the maximal ice advance (Devonshire and Scilly Isles).

REFERENCES.

ANDREWS J. T. (1968). — Postglacial rebound in Arctic Canada : similarity and prediction of uplift curves. *Can. J. Earth Sci.*, vol. 5, pp. 39-47.

BACZYK J. (1966). — Formes littorales d'âge Atlantique sur les côtes du Golfe de Gdansk. *Rev. Géomorph. Dyn.*, vol. 16, pp. 162-172.

BLOOM A. L. (1963). — Late Pleistocene fluctuations of sea level and postglacial crustal rebound in coastal Maine. *Am. J. Sci.*, vol. 261, pp. 862-879.

BOILLOT G. (1964). — Géologie de la Manche occidentale. Fonds rocheux, dépôts quaternaires, sédiments actuels. *Ann. Inst. Océanogr.*, vol. 42, 220 p. (English summary, pp. 189-196).

BORNS H. W., Jr. (1963). — Preliminary report on the age and distribution of late Pleistocene ice in north central Maine. *Am. J. Sci.*, vol. 261, pp. 738-740.

BROTCHIE J. F., SILVESTER R. (1969). — On crustal flexure. *J. Geophys. Res.*, vol. 74, n° 22, pp. 5240-5252.

CHURCHILL D. M. (1965). — The displacement of deposits formed at sea-level 6,500 years ago in southern Britain. *Quaternaria*, vol. 7, pp. 239-249.

DALY R. A. (1934). — The changing world of the Ice Age. Yale University Press, 271 p.

DECHEND W., GRONWALD W. (1961). — Krustenbewegungen und Meeresspiegelschwankungen im Küstenbereich der südlichen Nordsee. *Geol. Jb.*, vol. 79, pp. 23-60.

DONN W. L., FARRAND W. R., EWING M. (1962). — Pleistocene ice volumes and see level lowering. *J. Geol.*, vol. 70, pp. 206-214.

FAIRBRIDGE R. W. (1961). — Eustatic changes in sea level in (AHRENS L. H. et al., eds.). « Physics and Chemistry of the Earth », vol. 4, pp. 99-185.

FAIRBRIDGE R. W., NEWMAN W. S. (1965). — Sea level and the Holocene boundary in the eastern United States. Report of VIth Intern. Congress on the Quaternary, vol. 1, pp. 398-418.

FAIRBRIDGE R. W., NEWMAN W. S. (1968). — Postglacial crustal subsidence of the New York area. *Zeit. für Geomorph.*, vol. 3, pp. 296-317.

FARRAND W. R. (1964). — The deglacial hemicycle. *Geol. Rund.*, vol. 54, pp. 385-398.

FRYE J. C. (1963). — Problems of interpreting the bedrock surface of Illinois. *Ill. Acad. Sci. Trans.*, vol. 56, pp. 3-11.

HARRISON W., MALLOY R. J., RUSNAK G. A., TERASMAE J. (1965). — Possible Late Pleistocene uplift, Chesapeake Bay entrance. *Jour. Geol.*, vol. 73, pp. 201-229.

HOUGH J. L. (1958). — Geology of the Great Lakes. University of Illinois Press, 313 p.

JELGERSMA S. (1961). — Holocene sea level changes in the Netherlands. *Meded. van de Geol. Stich.*, ser. C, vol. VI, n° 7.

KAYE C. A., BARGHOORN E. (1964). — Late Quaternary sea-level change and crustal rise at Boston, Massachusetts, with notes on the autocompaction of peat. *Geol. Soc. Amer. Bull.*, vol. 75, pp. 63-80.

KÖSTER R. (1968). — Postglacial sea-level changes in the western Baltic region in relation to worldwide eustatic movements, in (MORRISON R. B., WRIGHT H. E., Jr., eds.), « Means of Correlation of Quaternarv Successions », University of Utah Press, pp. 407-419.

KROG H. (1968). — Late-glacial and postglacial shoreline displacement in Denmark, in (MORRISON R. B., WRIGHT H. E., Jr., eds.), « Means of Correlation of Quaternary Successions », University of Utah Press, pp. 421-435.

LEVERETT R., TAYLOR R. B. (1915). — The Pleistocene of Indiana and Michigan and history of the Great Lakes. *U.S. Geol. Surv. Mono.* 53, 529 p.

MCGINNIS L. D. (1968). — Glacial crustal bending. *Geol. Soc. Amer. Bull.*, vol. 79, pp. 769-776.

MCINTIRE W. G., MORGAN J. P. (1964). — Recent geomorphic history of Plum Island, Massachusetts, and adjacent coasts. Louisiana State University Studies, Coastal Studies Ser., 44 p.

MÖRNER N. A. (1969 a). — The late Quaternary history of the Kattegat Sea and the Swedish West Coast. *Sveriges Geolog. Unders.*, arsbok 63 (3), 487 p.

MÖRNER N. A. (1969 b). — Eustatic and climatic changes during the last 15 000 years. *Geol. en Mijnb.*, vol. 48, n° 4, pp. 389-399.

NEWMAN W. S. (1966). — Late Pleistocene environments of the western Long Island area. Ph. D. Dissertation, New York University.

NEWMAN W. S. (1968). — Late Quaternary sea levels along the North Atlantic coast of the United States. *Geol. Soc. Amer. Spec.*, Paper 101, pp. 444-445.

NEWMAN W. S. (1968). — Coastal Stabilitv, in (FAIRBRIDGE R. W., ed.), « Encyclopedia of Geomorphology », New York, Reinhold Book Corp., ppè 150-156.

NEWMAN W. S., FAIRBRIDGE R. W. (1960). — Glacial lakes in Long Island Sound. *Geol. Soc. Amer. Bull.*, vol. 71, p. 1936 (abs.).

NEWMAN W. S., FAIRBRIDGE R. W. (1960). — Active subsidence in the New York area. *Geol. Soc. Amer. Bull.*, vol. 71, pp. 2107-2108.

NEWMAN W. S., FAIRBRIDGE R. W. (1962). — Postglacial subsidence of coastal New England. *Geol. Soc. Amer.*, Spec. Paper 68, pp. 239-240.

NEWMAN W. S., FAIRBRIDGE R. W. (1962). — Postglacial sea level, coastal subsidence and littoral environments in the Metropolitan New York City area. First Natl. Coastal and Shallow Water Conf., Natl. Sci. Foundation and Office of Naval Research, pp. 188-190.

NEWMAN W. S., FAIRBRIDGE R. W. (1969). — Marginal subsidence of glaciated areas : United States, Baltic and North Seas. VIII Congress INQUA, Résumés des Communications, Paris, p. 312.

NEWMAN W. S., MARCH S. (1968). — The North American littoral : Late Quaternary warping. *Science*, vol. 160, pp. 1110-1112.

NEWMAN W. S., MUNSART C. A. (1968). — Holocene geology of the Wachapreague Lagoon, Eastern Shore Peninsula, Virginia. *Marine Geol.*, vol. 6, pp. 81-105.

NEWMAN W. S., RUSNAK G. A. (1965). — Holocene submergence of the eastern shore of Virginia. *Science*, vol. 148, pp. 1464-1466.

NEWMAN W. S., THURBER D. H., ZEISS H. S., ROKACH A., MUSICH L. (1969). — Late Quaternary geology of the Hudson River estuary : a preliminary report. *N.Y. Acad. Sci. Trans.*, vol. 31, pp. 548-570.

PREST V. K. (1969). — Retreat of Wisconsin and Recent ice in North America. *Geol. Surv. Canada*, Map 1257 A.

SHEPARD F. P., CURRAY J. R. (1967). — Carbon-14 determination of sea level changes in stable areas. *Progress in Oceanography*, vol. 4, pp. 283-291.

Editor's Comments on Paper 23

Okko: *The Relation Between Raised Shores and Present Land Uplift in Finland During the Past 8000 Years*

One of the major areas of interest in Quaternary studies of raised marine features is the relationship, or otherwise, of present geophysical measurements (such as leveled rates of crustal movement, gravity anomalies, etc.) and movements during the postglacial period of crustal recovery. Do the regions of highest postglacial recovery coincide with those which are currently rebounding fastest? The selection by Okko considers this question. The review of earlier Finnish work that precedes the main discussion is useful, and the shoreline model (Fig. 1) is a good graphic portrayal of the interactions between isostatic recovery and eustatic sea level variations.

The main interest of the paper focuses on Fig. 3, where the present elevation of the Litorina water plane is plotted against current rate of uplift. Okko then defines an uplift equation based on the present rate of uplift (equations 3 and 4) and uses the equation to develop a series of uplift curves (Figs. 5–11) and relative sea level changes. The relative sea level curve is developed by reference to a "shore-displacement curve" from a site near the zero isobase (see Fig. 4).

The papers by Bloom (Paper 18), Mörner (Paper 21), and Okko are all concerned with the unraveling of the isostatic/eustatic components in relative sea level curves. They employ different methods but all contribute to our understanding of Quaternary crustal and sea level movements.

Reprinted from *Ann. Acad. Sci. Fennicae*, Ser. A, 5–11, 13, 54–59 (1967)

23

ANNALES ACADEMIAE SCIENTIARUM FENNICAE

Series A

III. GEOLOGICA — GEOGRAPHICA

93

THE RELATION BETWEEN RAISED SHORES AND PRESENT LAND UPLIFT IN FINLAND DURING THE PAST 8000 YEARS

BY

MARJATTA OKKO

Geological Survey of Finland
Otaniemi

HELSINKI 1967
SUOMALAINEN TIEDEAKATEMIA

INTRODUCTION

The data on land uplift in Finland fall into two categories, namely, recent measurements and prehistoric data. The former category consists of computations based on measurements by oceanographic, hydrographic and geodetic methods. The latter includes data on former shore lines obtained through the study of late-glacial (Late Weichselian) and postglacial shore-line displacement by geological and archaeological methods. The historical records proper on land uplift are relatively few (see M. OKKO 1963, pp. 68, 75—76).

The absolute rates of land uplift have been calculated from tide-gauge data. The longest series of tide-gauge records begins with the year 1888. On the Finnish mainland two series of levellings of high precision have been carried out, in the years 1892—1910 and 1935—1955. The system of land uplift differences, obtained from the levellings, has been connected with the absolute land uplift values at the tide gauges. The most recent map showing the isobases of land uplift in Finland is that published by KÄÄ-RIÄINEN (1963, 1966a, 1966b).

Geological and archaeological studies on prehistoric shore lines and their displacement have revealed that the rate of land uplift has diminished towards its present-day values.

The opinions regarding the mode of prehistoric land uplift in Finland are rather divergent. RAMSAY (1926, p. 27) introduced the concept of zones of tightly spaced isobases of land uplift, which are expressed as flexures in systems of raised shore surfaces. These flexures are not stagnant but move from place to place in the course of time. They do not coincide with any major tectonic zones but run along narrow sets of isobases of land uplift (RAMSAY 1926, p. 53). SAURAMO (1939, 1944, 1955a, 1955b, 1958) developed the theory of land uplift with stagnant hinge-lines. The hinge-lines divide the region of land uplift into belts in which groups of raised shore surfaces are differently tilted. HYYPPÄ (1963, 1966), among others, opposes the hinge-line theory and is of the opinion that land uplift has proceeded in its present general style through the Late Quaternary time. This concept is based on the observation that the isobases representing the uplifted oldest shore system of the Litorina Sea show that the relation between the rates of land uplift in different places has been nearly constant since the development of the Litorina limit (Ramsay 1926, pp. 22, 27; Hyyppä 1963, p. 38). The concept of mosaic-like block movements has also been considered (Sauramo 1933, 1934; Härme 1963, 1966; Paarma 1963; Simola 1963).

Ramsay (1924, 1926) made a thorough theoretical study of the relations between land uplift and changes in sea level in Late Quaternary time. Despite this pioneer work, the correlation of geodetic and geophysical measurements to geologic data on shore-line displacement is still an almost unexplored field of study. Kääriäinen (1953, pp. 64—70) has employed the altitudes of the different Litorina shores to estimate the retardation of land uplift in Finland. The direction of the base line of local and regional equidistant shore-line diagrams has been studied or controlled by reference to present isobases of land uplift (see e.g., Donner 1964). Recently, R. Aario (1965b) endeavoured to study prehistoric land uplift by comparing the altitudes of two shore systems to the numerical values of present uplift.

The purpose of the paper at hand is to examine the applicability of the numerical values of the present rates of land uplift into the study of shore-line displacement.

GENERAL

The numerical values of the rates of vertical crustal movements, as shown on the map of land uplift in Finland (Kääriäinen 1963, 1966a, 1966b), indicate the rate at which the bench marks in the network of precision levelling are being lifted up in relation to the absolute uplift at the tide gauges. According to Kääriäinen's map, Finland is an area where the vertical crustal movements are positive in relation to the present sea. In such an area the shore line moves seawards, i.e., the land emerges from the sea.

Geodetically, the values of absolute land uplift show the rate at which the crust is drawing away from the Earth's centre or from a given niveau. If the sea level had remained constant, the present altitude y_s of a shore formed t years ago would indicate the amount of uplift during t years. If the sea level stood at a vertical distance h_s from the present sea level in the year t, the amount of uplift y in t years is $y = y_s - h_s$. In other words, the altitudes of the raised shore surfaces include both a crustal and a sea-level component. In any estimation of the amount of uplift during a given span of time by means of dated shore lines, a correction of the shore altitudes by h_s is necessary.

RAISED MARINE SHORE SURFACES IN RELATION TO THE VALUES OF PRESENT LAND UPLIFT

Theoretical model

Along the Finnish coast the shore-line displacement is a combination of land uplift and eustatic changes in the sea level. When the sea level changes insignificantly, land uplift causes a seaward displacement of the shore. A sinking sea level would add to the apparent rate of this displacement.

During a sea-level rise, the shore is stable where the rate of the sea-level rise and the contemporaneous rate of land uplift are equal. A coast of emergence exists in areas where the uplift exceeds the sea-level rise. A coast of submergence develops in areas where the rate of land uplift is algebraically lower than the rate of the sea-level rise. The height of the transgression increases steadily from the stable shore towards the zero isobase, where the total height of transgression equals the height of the sea-level rise.

Fig. 1 shows a simplified model of a sea-level rise along a coastal uplift region. The most important simplifications are a rapid, wave-like sea-level rise and the assumption that the rate of crustal uplift diminishes linearly. It is assumed that the sea level rises as shown in the upper left corner in the figure. The sea rises h meters during four periods, from h_0 to h_4 beginning in t_0 and ending in t_4, and remains stable during the fifth period t_4 to t_5. During the same five periods, the land is lifted up as shown by the uplift graph to the right of the sea-level graph in the figure. In the uplift graph the lines show the successive amount of vertical movement of points being lifted up at a different rate from the level h_0. The oldest uplifted level is situated highest in the graph.

In the model the sea level is stable prior to t_0; this is the third simplification devised to give a definite beginning date t_0 for the sea-level rise. The shore line corresponding to h_0 and t_0 is marked s_0. During the period $t_0 \rightarrow t_1$ (I) the sea rises from h_0 to h_1. Simultaneously, shore line s_0 is lifted up but the rising sea transgresses its lowest part until the point where the amount of sea-level rise and crustal uplift are equal. On the coast where s_0 emerges from the sea, its vertical distance from s_1, the shore line of h_1 and t_1, is h_1 less than the true vertical crustal uplift.

During the period $t_1 \rightarrow t_2$ (II) the sea level rises from h_1 to h_2, which exceeds the rate of the initial rise. Shore lines s_0 and s_1 are lifted up. The sea transgresses s_1 up to the point where the amount of the sea-level rise $h_2 - h_1$ equals the amount of crustal uplift. This marks the stable shore of the second period. The sea also transgresses shore line s_0, but the point of intersection of s_0 and s_2, the shore line of h_2 and t_2, is not as high up

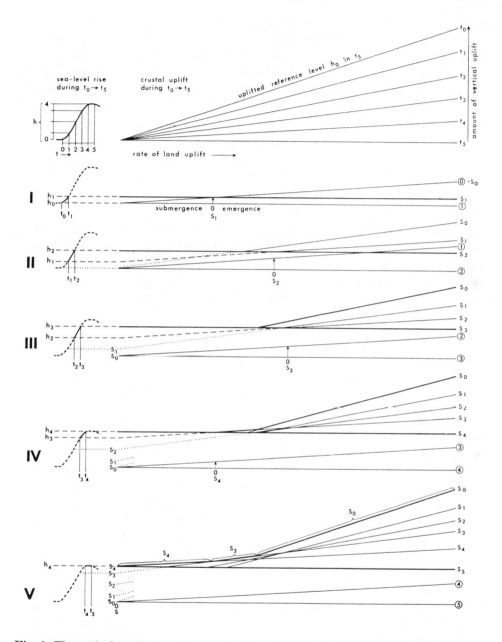

Fig. 1. Theoretical model of the relationship between isostatic land uplift and a sea-level rise. The resulting raised shore surface is metachronous. For explanation, see pp. 7—9.

on the uplift coast as is the stable shore. The point of intersection lies at the point where the rise in sea level during two periods $(h_2 - h_0)$ equals the amount of crustal uplift during the same two periods.

The sea-level rise proceeds at the same rate from h_2 to h_3 in the period $t_2 \rightarrow t_3$ (III). Owing to the retardation in the rate of uplift the stable shore moves somewhat towards the centre of uplift. The sea transgresses shore s_2 up to the stable shore but shore line s_1 begins to emerge here from the shore level s_3 of h_3 and t_3. The oldest shore line s_0 intersects s_3 at the point where the crustal uplift during $t_0 \rightarrow t_3$ equals the sea-level rise from h_0 to h_3.

During the end of the sea-level rise, from h_3 to h_4, in $t_3 \rightarrow t_4$ (IV), the stable shore moves to the point where the amount of crustal uplift during this period equals $h_4 - h_3$. Where the coast rises faster than the stable shore, the older shore lines are lifted up, whereby they preserve their relationships from t_3. Thus the point of intersection of the oldest shore line s_0 and the transgression limit s_3 emerges before the sea level reaches its maximum height, h_4. The transgression on this coast ends before the end of the sea-level rise.

The whole shore system formed during the sea-level rise is lifted up when the sea-level again becomes stable (V). The uppermost shore line is metachronous. Its steepest inclined part on the fastest rising coast dates from the beginning of the sea-level rise. It is s_0 in the model. The least inclined part on the slowest rising coast dates from the end of the sea-level rise, s_4 in the model. These two parts are linked up with a shore line that formed slightly before the sea-level rise slowed down and ended. It is s_3 in the model. The transgression limit extends farther from the zero isobase than does the point of intersection of the oldest and next to youngest shore lines.

In the discussion to follow the coast characterized by s_4 in the model will be called the transgression coast of the first order; the coast characterized by s_3 as the uppermost shore line is a transgression coast of the second order. The coast characterized by s_0 is the coast of emergence; its lowest part contains a transgression coast of the third order.

In his pioneer work on the relations between eustatic and crustal components, RAMSAY (1926) worked out a model for a shore limit of a sea-level rise along a coast of land uplift. In his model the metachronous shore limit has the same shape and age relationships as the model presented in the foregoing. The difference between the two models is in the treatment of the transgression limit proper. L. AARIO (1935 a, p. 128; 1935 b) subsequently recognized the transgression limit extending beyond the intersection of the initial shore line and the transgression limit.

In order to clarify the relationships between a sea-level rise and a crustal uplift, as proposed in the model, in terms of the present uplift in Fenno-

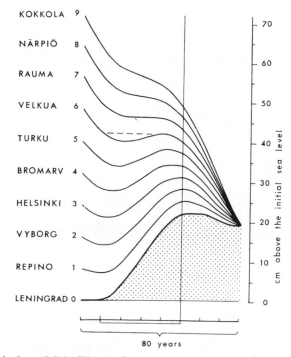

Fig. 2. Theoretical model in Fig. 1. drawn as a set of curves of shore displacement at different isobases of land uplift. The numbers of the curves show the rate of uplift in cm per 10 yr. The names of places refer to coastal sites corresponding to those rates.

The hypothetical sea-level rise causes a transgression of the initial shore at the zero isobase at the base of the figure. The transgression diminishes as the rate of uplift increases and it fades out at a point (7) that is lifted up faster than the point where the initial shore and the transgression limit occur at the same altitude (6).

scandia, nine curves of shore displacement for an arbitrary sea-level rise were calculated (Fig. 2). The sea level is assumed to rise 22 cm during four decades. During the first decade the sea level rises 4 cm, during the second and third decades 7 cm respectively, and 4 cm during the fourth decade. The sea level is stable during the fifth decade and sinks slowly during the next two decades. The curves of shore displacement show the situation at the end of the seventh decade after the beginning of the sea-level rise. The curves were calculated for each full centimeter of uplift during a decade; no retardation in the rate of uplift was considered. The coastal sites at each isobase are marked in the figure.

During the first decade, the sea would transgress the initial shore line until Bromarv (4 cm in 10 years), where the shore would remain stable. In areas rising faster than 4 cm per decade, the initial shore would emerge.

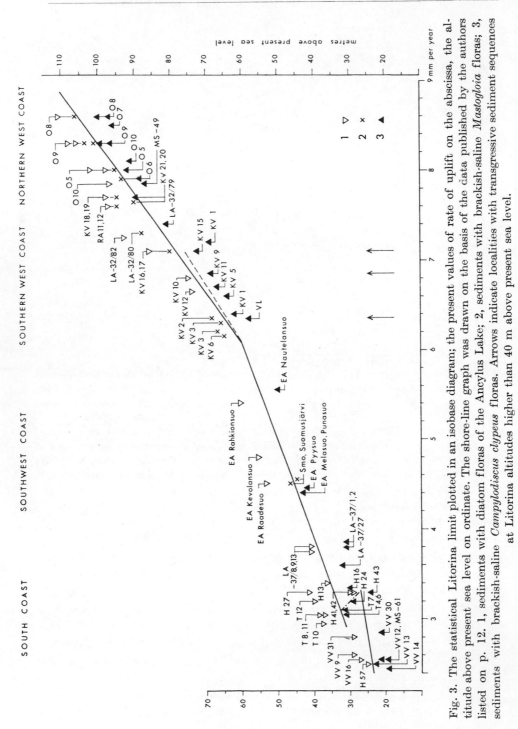

Fig. 3. The statistical Litorina limit plotted in an isobase diagram; the present values of rate of uplift on the abscissa, the altitude above present sea level on ordinate. The shore-line graph was drawn on the basis of the data published by the authors listed on p. 12. 1, sediments with diatom floras of the Ancylus Lake; 2, sediments with brackish-saline *Mastogloia* floras; 3, sediments with brackish-saline *Campylodiscus clypeus* floras. Arrows indicate localities with transgressive sediment sequences at Litorina altitudes higher than 40 m above present sea level.

Its vertical distance from the $+4$ cm shore-line would be 4 cm less than the amount of uplift during that decade. During the second and third decades, the stable shore would develop at Rauma (7 cm in 10 years). Meanwhile, the initial shore would emerge from Velkua (6 cm in 10 years) onwards. During the final phase, the fourth decade of sea-level rise, the shore would be stable back at Bromarv; and during the fifth decade, it would be back in Leningrad at the zero isobase. The uppermost shore line of the sea-level rise would be that of the initial stage between Kokkola (9 cm in 10 years) and Velkua, with the transgression coast of the third order extending from Velkua to Rauma. The transgression coast of the second order would extend from Velkua to Bromarv and that of the first order from Bromarv to Leningrad.

The model presented herein deals with a single sea-level rise. The height of a subsequent sea-level rise, measured from the initial sea level h_0, may reach a higher level than the preceding rise. The younger rise would cause transgression of the older shore line, whereby a less tilted part is added to the older transgression limit. The transgression caused by the younger rise fades out in an area that is being lifted up at a rate faster than that existing at the intersection of the older and younger shore surfaces.

Along the coast of land uplift, it is only at the absolute zero isobase that the full extent of a sea-level rise may be measured. There the sea-level rise equals the height of transgression of the initial shore. On the coast being lifted up there is, of course, one place where the amount of land uplift equals the amount of sea-level rise; in the numerical example it would be at the isobase 5.5 cm in 10 years. This critical point belongs to the transgression coast of the second order, but it is not characterized by any specific features on the shore line. If, however, the sea-level rise is fast and fades out quickly, the point of intersection of the initial shore and the transgression limit of the second order develops close to the critical point.

Provided that the land uplift has maintained its general course since a former sea-level rise that is known as to its magnitude and age, the retardation of the rate of land uplift since the beginning of that rise can be estimated by means of the shore altitudes on the coast of emergence. On the transgression coast the shore altitudes include a eustatic component. If this component is unknown, no reliable values for retardation are to be expected.

[*Editor's note:* The remainder of Okko's paper is a detailed evaluation of sea level curves from Finland employing archaeological and palynological data to provide age estimates for various portions of the curves. She concludes that apart from an area in the eastern part of the country, isostatic recovery has been regular.]

References

AARIO, L., 1932: Pflanzentopographische und palaogeographische Mooruntersuchungen in N-Satakunta. Fennia 55, No. 1. 179 p. — *Also* Communic. Inst. Forest. Fenn. 17, 1.

—»— 1935 a: Die postglazialen Niveauverschiebungen im mittleren Uusimaa mit Berücksichtigung ihrer Beziehungen zu steinzeitlichen Wohnplätzen. Ann. Acad. Scient. Fenn., Ser. A, XLIV, 1. 161 p.

—»— 1935 b: Ensimmäinen Litorina-transgressio ja Clypeusraja. Referat: Erste Litorinatransgression und Clypeus-Grenze. Terra, Vsk. 47, pp. 50—55. Helsinki.

—»— 1943: Über die Wald- und Klimaentwicklung an der lappländischen Eismeerküste in Petsamo. Ann. Bot. Soc. 'Vanamo' 19, No. 1. 155 p.

AARIO, RISTO, 1965 a: Development of ancient Lake Päijänne and the history of the surrounding forests. Ann. Acad. Scient. Fenn., Ser. A, III, 81. 191 p.

—»— 1965 b: Die rezente Landhebung als Grundlage eines Relationsdiagramms. Ann. Acad. Scient. Fenn., Ser. A, III, 85. 16 p.

—»— 1965 c: Die Fichtenverhäufigung im Lichte von C14-Bestimmungen und die Altersverhältnisse der finnischen Pollenzonen. C.R. Soc. Géol. Finlande 37, pp. 215—231. — *Also* Bull. Comm. géol. Finlande No. 218, pp. 215—231.

AARTOLAHTI, TOIVE, 1965: Oberflächenformen von Hochmooren und ihre Entwicklung in Südwest-Häme und Nord-Satakunta. Fennia 93, No. 1. 268 p.

—»— 1966: Über die Einwanderung und Verhäufigung der Fichte in Finnland. Ann. Bot. Fennici 3, pp. 368—379.

ALHONEN, PENTTI, 1965: C14-Datierung der vorgeschichtlichen Schlittenkufe aus Kullaa in Satakunta (Westfinnland). Suomen Museo 72, pp. 16—21.

ASKLUND, BROR, 1935: Gästrikländska fornstrandlinjer och nivåförändringsproblemen. Sveriges Geol. Unders., Ser. C, No. 391. 119 p.

ATLAS OF FINLAND, 1960. Published jointly by the Geographical Society of Finland and the Department of Geography, University of Helsinki.

AUER, VÄINÖ, 1924: Die postglaziale Geschichte des Vanajavesisees. Communic. Inst. Quaest. Forest. Finlandiae 8. 132 p. — *Also* Bull. Comm. géol. Finlande No. 69.

—»— 1928: Über die Einwanderung der Fichte in Finnland. Communic. Inst. Quaest. Forest. Finlandiae 13, No. 3. 24 p.

—»— 1958: The Pleistocene of Fuego-Patagonia, II. Ann. Acad. Scient. Fenn., Ser. A, III, 50. 239 p.

—»— 1963: Lateglacial and postglacial shoreline displacement in South America as established by tephra-chronology, compared with displacements of the Baltic shorelines. Fennia 89, No. 1, pp. 51—55.

Aurola, Erkki, 1938: Die postglaziale Entwicklung des südwestlichen Finnlands. C. R. Soc. Géol. Finlande 11, pp. 1—168. — *Also* Bull. Comm. géol. Finlande No. 121.

Backman, A. L., 1941: *Najas marina* in Finnland während der Postglazialzeit. Acta Bot. Fenn. 30. 38 p.

—»— 1943: *Ceratophyllum submersum* in Nordeuropa während der Postglazialzeit. Acta Bot. Fenn. 31. 38 p.

—»— 1948: *Najas flexilis* in Europa während der Quartärzeit. Acta Bot. Fenn. 43. 44 p.

Bendefy, L., 1965: Grundlegende Probleme der Erforschung der rezenten Erdkrustenbewegung. Gerlands Beitr. Geophysik, Bd. 74, pp. 484—495.

Berglund, Björn E., 1964: The post-glacial shore displacement in eastern Blekinge, south-eastern Sweden. Sveriges Geol. Unders., Ser. C, No. 599. 47 p. — *Also* Publ. Inst. Mineral., Paleont., Quaternary Geology, Univ. Lund, No. 129.

Brander, G. och Brenner, Thord, 1933: Fredriksbergsmossen. Referat: Das Fredriksberger Moor. Fennia 57, No. 5. 31 p.

Brandt, A., 1949: Über die Entwicklung der Moore im Küstengebiet von Süd-Pohjanmaa am Bottnischen Meerbusen. Ann. Bot. Soc. 'Vanamo' 23, No. 4. 134 p.

Curray, Joseph R., 1961: Late Quaternary sea level: a discussion. Geol. Soc. America Bull., Vol. 72, pp. 1707—1712.

Donner, J. J., 1964: On the late-glacial and post-glacial emergence of south-western Finland. Soc. Scient. Fenn., Comment. Phys.-Math. 30, 5. 47 p.

—»— 1965: Shore-line diagrams in Finnish Quaternary research. Baltica 2, pp. 11—20. Vilnius.

Edgren, Torsten, 1966: Jäkärlä-gruppen, en västfinsk kulturgrupp under yngre stenålder. Suomen Muinaismuistoyhd. Aikak. — Finska Fornminnesfören. Tidskr. 65. 159 p.

Fairbridge, Rhodes W., 1961: Eustatic changes in sea level. Pp. 99—185 *in* 'Physics and chemistry of the Earth', Vol. 4; ed. by L. H. Ahrens *et al.* Pergamon Press.

Florin, Sten, 1963: Bodenschwankungen in Schweden während des Spätquartärs. Baltica 1, pp. 233—264. Vilnius.

Fries, Magnus, 1963: Pollenanalytiskt bidrag till vegetations- och odlingshistoria på Åland. Referat: Pollenanalytischer Beitrag zur Vegetations- und Siedlungsgeschichte von Åland. Finskt Museum 68 (1961), pp. 5—20.

Härme, Maunu, 1963: On the shear zones and fault lines in Finnish Pre-Cambrian strata. Fennia 89, No. 1, pp. 29—31.

—»— 1966: On the block character of the Finnish Precambrian basement. Ann. Acad. Scient. Fenn., Ser. A, III, 90, pp. 133—134.

Hausen, H., 1964: Geologisk beskrivning över landskapet Åland. Skr. utg. av Ålands Kulturstiftelse IV. 196 p.

Hela, Ilmo, 1953: A study of land upheaval at the Finnish coast. Fennia 76, No. 5. 38 p. — *Also* Merentutkimuslaitoksen julkaisu No. 158.

Hyyppä, Esa, 1935: Kivikautinen asutus ja rannan siirtyminen Helsingin seudulla. Referat: Die steinzeitliche Besiedlung und Verschiebung des Ufers in der Gegend von Helsinki. Terra, Vsk. 47, pp. 31—49. Helsinki.

—»— 1937: Post-glacial changes of shore-line in South Finland. Bull. Comm. géol. Finlande No. 120. 225 p.

—»— 1950: Lapuan ja Pattijoen muinaissuksien geologinen ikäys. Referat: Geologische Datierung vorzeitlicher Schier aus Lapua und Pattijoki. Suomen Museo 57, pp. 24—40.

HYYPPÄ, ESA, 1956: En i Helsingfors funnen stenålderskanot. II. Geologisk datering. Referat: Ein in der Stadt Helsinki gefundenes steinzeitliches Kanoe (II). Finskt Museum 63, pp. 19—25, 29.

—»— 1963: On the Late-Quaternary history of the Baltic Sea. Fennia 89, No. 1' pp. 37—48.

—»— 1966: The Late-Quaternary land uplift in the Baltic sphere and the relation diagram of the raised and tilted shore levels. Ann. Acad. Scient. Fenn., Ser. A., III, 90, pp. 153—168.

JAKUBOVSKY, O., 1966: Vertical movements of the Earth's crust on the coasts of the Baltic Sea. Ann. Acad. Scient. Fenn., Ser. A., III, 90, pp. 479—488.

JELGERSMA, S., 1966: Sea-level changes during the last 10,000 years. Pp. 54—69 in 'World climate from 8000 to 0 B.C.', Proceedings of the International Symposium held at Imperial College, London, 18 and 19 April 1966. Royal Meteorological Society, London.

DE JONG, JAN, 1967: The Quaternary of the Netherlands. Pp. 301—426 in 'The Quaternary', Vol. 2; ed. by Kalervo Rankama. Interscience Publ.

KÄÄRIÄINEN, ERKKI, On the recent uplift of the Earth's crust in Finland. Fennia 77, No. 2. 106 p. — Also Veröff. Finn. Geod. Inst. No. 42.

—»— 1963: Land uplift in Finland computed by the aid of precise levellings. Fennia 89, No. 1, pp. 15—18.

—»— 1966 a: The Second Levelling of Finland in 1935—1955. Veröff. Finn. Geod. Inst. No. 61.

—»— 1966 b: Land uplift in Finland computed with the aid of precise levellings. Ann. Acad. Scient. Fenn., Ser. A, III, 90, pp. 187—189.

KENNEY, T. C., 1964: Sea-level movements and the geologic histories of the post-glacial marine soils at Boston, Nicolet, Ottawa and Oslo. Géotechnique, T. 14, pp. 203—230. — Also Norwegian Geotech. Inst., Publ. No. 62.

KING, CUCHLAINE A. M. and HIRST, RACHEL A., 1964: The boulder fields of the Åland Islands. Fennia 89, No. 2. 41 p.

KIVIKOSKI, ELLA, 1961: Suomen esihistoria. 310 p. Werner Söderström Oy.

KROG, HARALD, 1960: Post-glacial submergence of the Great Belt dated by pollen-analysis and radiocarbon. Internat. Geol. Congress, XXI Session, Norden, 1960, Pt. IV, pp. 127—133.

KUKKAMÄKI, T. J., 1939: Über zwei dem Präzisionsnivellement sich anschliessenden Fragen. Veröff. Finn. Geod. Inst. No. 26, pp. 119—125.

LAPPALAINEN, VEIKKO, 1965: Geologiska och geotekniska undersökningar vid Heinoo banuträtning, SW-Finland. Geologi, Vsk. 17, pp. 135—136. Helsinki.

LISITZIN, EUGENIE, 1964: Contribution to the knowledge of land uplift along the Finnish coast. Fennia 89, No. 4. 22 p.

—»— 1966: Land uplift in Finland as a sea level problem. Ann. Acad. Scient. Fenn., Ser. A, III, 90, pp. 237—239.

LUHO, VILLE, 1965: Helsinginpitäjän esihistoria. Pp. 7—92 in Helsinginpitäjä I. Helsinki.

LUKKALA, O. J., 1933: Tapahtuuko nykyisin metsämaan soistumista. Referat: Vollzieht sich gegenwärtig Versumpfung von Waldböden. Communic. Inst. Forest. Fenniae 19, No. 1. 127 p.

—»— 1934: Lounais-Suomen metsien puulajihistoriasta. Referat: Über die Holzgeschichte der SW-finnischen Wälder. Acta Forest. Fenn. 40, No. 18. 17 p.

LUNDQVIST, G., 1962: Geological radiocarbon datings from the Stockholm station. Sveriges Geol. Unders., Ser. C, No. 589. 23 p.

LUNDQVIST, G. 1963: De kvartära bildningarna. Pp. 400—647 in 'Sveriges geologi', 4 uppl., by Nils H. Magnusson, G. Lundqvist, Gerhard Regnéll. Svenska Bokförlaget.

MATTHEWS, BARRY, 1966: Radiocarbon dated postglacial land uplift in northern Ungawa, Canada. Nature, Vol. 211, pp. 1164—1166.

MEINANDER, C. F., 1954: Die Kiukaiskultur. Suomen Muinaismuistoyhd. Aikak. — Finska Fornminnesfören. Tidskr. 53. 192 p.

—»— 1957: Kolsvidja. Suomen Muinaismuistoyhd. Aikak. — Finska Fornminnesfören. Tidskr. 58, pp. 185—213.

MÖLDER, K. and SALMI, MARTTI, 1955: Explanation to the map of superficial deposits, sheet B 3, Vaasa. General Geological Map of Finland, 1:400 000.

NILSSON, TAGE, 1964: Standardpollendiagramme und C^{14}-Datierungen aus dem Agerödsmosse im mittleren Schonen. Lunds Univ. Årsskr. N.F., Avd. 2, Bd. 59, No. 1. 52 p. — Also Publ. Inst. Mineral., Palaeont., Quaternary Geol., Univ. Lund, No. 124.

OKKO, MARJATTA (editor), 1963: Bibliography of recent crustal movements; Finland and Finnish studies abroad. Fennia 89, No. 1, pp. 61—89.

OKKO, VEIKKO, 1949: G. Brander's data of the Littorina shore-line in North and Middle Ostrobothnia. C. R. Soc. Géol. Finlande 22, pp. 117—127. — Also Bull. Comm. géol. Finlande No. 144, pp. 117—127.

—»— 1964: Maaperä. Pp. 239—332 in 'Suomen geologia'; ed. by Kalervo Ránkama. Kirjayhtymä.

PAARMA, HEIKKI, 1963: On the tectonic structure of the Finnish basement, especially in the light of geophysical maps. Fennia 89, No. 1, pp. 33—36.

PALOMÄKI, MAURI, 1963: Über den Einfluss der Landhebung als ökologischer Faktor in der Flora flacher Inseln. Fennia 88, No. 2. 75 p.

RAMSAY, WILHELM, 1924: On relations between crustal movements and variations of sea-level during Late-Quaternary time, especially in Fennoscandia. Fennia 44, No. 5. 39 p. — Also Bull. Comm. géol. Finlande No. 66.

—»— 1926: Nivåförändringar och stenåldersbosättning i det baltiska området. Referat: Niveauverschiebungen im baltischen Gebiete. Fennia 47, No. 4. 68 p.

SALMI, MARTTI, 1944: Ein Seehundfund aus Ruukki und die Salzwasserkonzentrationsschwankungen des Wassers in dieser Gegend auf Grund fossilen Diatomeen-Floren. C. R. Soc. Géol. Finlande 16, pp. 165—187. — Also Bull. Comm. géol. Finlande No. 132, pp. 165—187.

—»— 1949: Die Litorinagrenze in der Umgebung von Alajärvi in Süd-Ostbottnien. C. R. Soc. Géol. Finlande 22, pp. 31—40. — Also Bull. Comm. géol. Finlande No. 144, pp. 31—40.

—»— 1955: Geologische Darlegung des Bronzschwertfundes von Kiukainen. Suomen Museo 63, pp. 77—83.

—»— 1961: Two Littorina transgressions in Virolahti, southeastern Finland. C. R. Soc. Géol. Finlande 33, pp. 417—436. — Also Bull. Comm. géol. Finlande No. 196, pp. 417—436.

—»— 1963: Drei subfossile Sattelrobben aus Ostbottnien; Geologische Datierung der Funde und einige chronologische Beobachtungen. Arch. Soc. 'Vanamo' 18:2, pp. 82—95.

SAURAMO, MATTI, 1933: Yoldiameri entisten ja nykyisten tutkimusten valossa. Summary: The Yoldia Sea in the light of earlier and modern research. Terra, Vsk. 45, pp. 1—14. Helsinki.

—»— 1934: Zur spätquartären Geschichte der Ostsee; Vorläufige Mitteilung. C. R.

Soc. Géol. Finlande 8, pp. 28—87. — *Also* Bull. Comm. géol. Finlande No. 104, pp. 28—87.

SAURAMO, MATTI, 1939: The mode of land upheaval in Fennoscandia during late-Quaternary time. C. R. Soc. Géol. Finlande 13, pp. 39—63. — *Also* Bull. Comm. géol. Finlande No. 115, pp. 39—63, *and* Fennia 66, No. 2.

—»— 1941: Die Geschichte der Wälder Finnlands. Geol. Rundschau, Bd. 32, pp. 579—594.

—»— 1944: Landhöjningens mekanism. Geol. Fören. Stockholm Förh., Bd. 66, pp. 536—550.

—»— 1949: Das dritte Scharnier der fennoskandischen Landhebung. Soc. Scient. Fenn., Årsb. — Vuosik. 27, B, 4. 26 p.

—»— 1955 a: Land uplift with hinge-lines in Fennoscandia. Ann. Acad. Scient. Fenn., Ser. A, III, 44. 25 p.

—»— 1955 b: On the nature of the Quaternary crustal upwarping in Fennoscandia. Acta geogr. 14, pp. 334—348. Helsinki.

—»— 1958: Die Geschichte der Ostsee. Ann. Acad. Scient. Fenn., Ser. A, III, 51. 522 p.

SCHWARTZ, MAURICE, 1965: Laboratory study of sea-level rise as a cause of shore erosion. J. Geol., Vol. 73, pp. 528—534.

—»— 1967: The Bruun theory of sea-level rise as a cause of shore erosion. J. Geol., Vol. 75, pp. 76—92.

SHEPARD, FRANCIS P., 1961: Sea level rise during the past 20,000 years. Z. f. Geomorph., Suppl. 3, pp. 30—35.

SIMOLA, LIISA KAARINA, 1963: Über die postglazialen Verhältnisse von Vanajavesi, Leteensuo und Lehijärvi sowie die Entwicklung ihrer Flora. Ann. Acad. Scient. Fenn., Ser. A, III, 70. 64 p.

STENIJ, S. E. and HELA, ILMO, 1947: Suomen merenrannikoiden vedenkorkeuksien lukuisuudet. Summary: Frequency of the water heights on the Finnish coasts. Merentutkimuslaitoksen julkaisu No. 138. 21 p.

TANNER, V., 1930: Studier över kvartärsystemet i Fennoskandias nordliga delar, IV. Résumé: Études sur le systéme quaternaire dans les parties septentrionales de la Fennoscandie, IV. Bull. Comm. géol. Finlande No. 88. 589 p. — *Also* Fennia 53, No. 1.

TYNNI, RISTO, 1966: Über spät- und postglaziale Uferverschiebung in der Gegend von Askola, Südfinnland. Bull. Comm. géol. Finlande No. 223. 97 p.

VALOVIRTA, VEIKKO, 1965: Zur spätquartären Entwicklung Südost-Finnlands. Bull. Comm. géol. Finlande No. 220. 101 p.

VIRKKALA, K., 1950: Kuusen yleistymisen ajankohta Länsi-Suomessa. Summary: The date of the beginning of the general spread of the spruce in West Finland. Terra, Vsk. 62, pp. 35—41. Helsinki.

—»— 1953: Altitude of the Littorina limit in Askola, southern Finland. C. R. Soc. Géol. Finlande 26, pp. 59—72. — *Also* Bull. Comm. géol. Finlande No. 159, pp. 59—72.

—»— 1959: Über die spätquartäre Entwicklung in Satakunta, W-Finnland. Bull. Comm. géol. Finlande No. 183. 56 p.

—»— 1966: Radiocarbon ages of the Råbacka bog, southern Finland. C. R. Soc. Géol. Finlande 38, pp. 237—240. — *Also* Bull. Comm. géol. Finlande No. 222, pp. 237—240.

WAALEWIJN, A., 1966: Investigations into crustal movements in the Netherlands. Ann. Acad. Scient. Fenn., Ser. A, III, 90, pp. 401—412.

WITTING, ROLF, 1918: Hafsytan, geoidytan och landhöjningen utmed Baltiska hafvet och vid Nordsjön. Referat: Die Meeresoberfläche, die Geoidfläche und die Landhebung den Baltischen Meere entlang und an der Nordsee. Fennia 39, No. 5. 346 p.

YLINEN, MAUNO, 1966: Subatlanttisesta transgressiosta Kiimingissä, Oulujoen pohjoispuolella. Summary: On a Subatlantic transgression near Oulu NE of Gulf of Bothnia. Geologi, Vsk. 19, pp. 6—9. Helsinki.

Printed October 1967

Editor's Comments on Paper 24

Einarsson: *Late- and Post-glacial Rise in Iceland and Sub-crustal Viscosity*

Few of the papers in this section have specifically mentioned the *causes* and *mechanisms* of the glacio-isostatic recovery that the authors studied. The studies were conducted from the interests of the Quaternary geologist/geomorphologist and written for this audience, not for geophysicists interested in problems of geodynamics. Fillon's paper partly bridges this communication gap, as does Einarsson's paper reprinted here.

Iceland was, of course, glaciated during the last glaciation and rebounded very rapidly upon removal of the ice load (Einarsson initially considers the effect of the isostatic movements of the ocean floor (Fig. 1) and tentatively concludes that the ocean floor will not respond to changes in water depth of the order of 100 m. His analysis of the Fennoscandian shorelines after the *Yoldia* Sea (ca. 9500 BP) indicated that there was no systematic "wave-motion" in toward the center of glaciation, a result that contradicts the conclusion of Newman et al. from North America (Paper 22). Einarsson proceeds to define an equation for uplift based on the unloading curve for the ice cap, and relative sea level movements are then predicted based on Fairbridge's eustatic sea level curve. Three values of the constant k are selected, where k is a function of viscosity and the diameter of the load. The concept of the proportionality of uplift (i.e., the shoreline relation diagram or the rate of uplift/amount of uplift diagram of Okko) is then used to determine the deleveling curves at other sites in Iceland.

Finally, Einarsson computes the relative difference in crustal viscosity between Iceland and Fennoscandia, finding them to be in the ratio 1:100.

Trausti S. Einarsson was born in 1907 in Reykjavik, Iceland, and is Professor of Mechanics and Physical Geology and Geophysics at the University of Iceland. He has published about 90 scientific papers on a variety of problems in geology and geophysics.

Reprinted from *Jökull*, **III**, 157–166 (1966)

24

TRAUSTI EINARSSON[1]:

Late- and Post-glacial Rise in Iceland and Sub-crustal Viscosity

It has been known for some time that the rise of Iceland at the end of the Last Ice Age was relatively rapid, when compared with the rise of Scandinavia, being largely completed in a few thousand years. On this basis I have made the rough estimate that subcrustal viscosity for Iceland is one order of magnitude less than that for Scandinavia (Einarsson 1953). As the value found for Iceland is probably representative for a considerable part of the Middle Atlantic Ridge, emphasis should be laid on obtaining this value with more certainty. A new approach has been made possible through a number of new data. The procedure in the present paper is aimed at finding a relative value, with Scandinavia as a standard. At the same time a general picture of the rise in Iceland is obtained.

The relative value of viscosity now found

for Iceland is a little less than one tenth, i.e. close to the earlier estimate.

In a first section we consider generally the changes of sea-level in unglaciated areas. Then we summarize relevant facts about Scandinavia, and finally discuss the Icelandic material and compare it with the Scandinavian one. Remarks on glacial sea-level in the Faeroe Islands are added.

1. UNGLACIATED AREAS.

The rise of late- and post-glacial sea-level relative to coasts far away from the glaciated regions has been studied by many authors. We shall use here data given by Fairbridge (1960, 1961), curve S_1 in Fig. 2, and for comparison data by Godwin et. al. (1958), curve S_2 in the same figure.

In connection with the application of such curves, it is necessary to consider the possible isostatic movements of the ocean floor and

1) Faculty of Engineering, University of Iceland, Reykjavík.

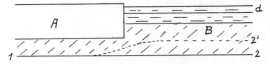

Fig. 1. Explanation of isostatic and eustatic changes. See text.

1. mynd. Til skýringar á lyftingu sjávarbotns á ísöld.

coastal areas. A and B in Fig. 1 are a section of a continental and oceanic block, respectively. By isostatic equilibrium before glaciation, *1—2* is a horizontal line of equal pressure. During glaciation an oceanic layer of thickness d, assumed here to be 100 m, is removed and deposited as ice in special areas outside the block A. As a result, pressure becomes inequal along *1—2* and we may consider four types of reaction to this. 1) Sub-crustal material is so fluid that it yields immediately and fully to the change in pressure. The ocean floor is then generally raised by $100/3.4 = 29.4 \backsim 30$ m, line *1—2* changes to *1—2¹*. The coastal strip is raised by half this value if the sea was at least 100 m deep at the original coast; if the shelf dips ½ degree, a correction of only a fraction of a metre must be applied, and thus can be dropped.

The real vertical movements, relative to the pre-glacial state, depend on the relative area of oceans and continents and on the downwarping in glaciated areas but this is of no avail here; only the relative movements of ocean floor and of continental margin, as given above, is of interest. We thus find that the removal of an ocean layer of 100 m thickness gives a measurable drop of sea-level by 85 m, relative to the continental coast, or a corresponding relative uplift of the ocean floor by 15 m.

2) Sub-crustal material yields with a time lag. 3) Sub-crustal material is plastic but the yield value is larger than about 10 kg/cm² (the pressure difference corresponding to 100 m of water). The ocean bottom will then remain stationary when layer d is removed, and the measurable drop of sea-level is 100 m. 4)The yield value is lower than in case 3, and there is only partial response to the pressure difference.

Our main concern is the rise of sea-level in late- and post-glacial time. This depends on

the two factors: sinking of the ocean floor relative to the continent A, and the water mass coming back to the ocean from the shrinking glaciers. As a measure of the latter let us tentatively take the shrinkage of the Scandinavian glacier.

The size of the latter is inferred for three epochs: 8800 B.P. (0.21 10⁶ km²), 10.000 B.P. (1.06 10⁶ km²), and 18.000 B.P. (2.5 10⁶ km²), from *Sveriges geologi* (Magnusson et. al. 1963, p. 462), and a *smooth* curve, G_a in Fig. 2, constructed. If the volume of the glacier were proportional to the area, the curve A in the same figure shows the expected rise of sea-level for the case of a fix ocean floor (case 3). If the more realistic assumption is made that the thickness of the glacier was at each time proportional to the diameter, the curve for the volume is given by G_v ($G_v \propto G_a^{3/2}$).

The corresponding rise of sea-level for case 3 is shown by the curve V_3. For case 1 the ocean floor sinks relative to the coastal area along the curve B_1, and relative sea-level then rises along the curve V_1. B_2 and B_4 are examples of cases 2 and 4, respectively. V_4 would lie between V_1 and V_3, and V_2 would lie close to V_1 except at the right end.

The curve S_1 (the ordinate of the minimum near 18.000 B.P. is taken as unity) lies mostly between V_1 and V_3 and it seems on the whole hardly possible to distinguish between the cases 1—4 by use of S_1 on one side and such a general picture of glacier shrinkage as we have used on the other. Sub-crustal material is probably plastic and a yield value of about 10 kg/cm² is mostly assumed (for the Icelandic area I have found that this is an upper limit). This would very likely be equivalent to case 3, i.e. a fix ocean floor, and this case will be assumed in the following. In the interval 11.000 —9.000 B.P. this appears to be a good approximation. We shall also take G_v as a fair approximation to the relativ decrease of glaciers in Iceland.

2. THE GLACIATED SCANDINAVIAN AREA.

When the glaciers of the Ice Age shrank, the glaciated areas began to rise isostatically. Concerning the mode of rise, it has been suggested in connection with the North American region that the marinal areas began to rise earlier than the central ones, in other words, that there was a kind of wave advancing towards the centre, following the shrinking ice margin. Such a process is likely for a very large area; we shall inquire whether it took place in the Scandinavian region.

Fig. 3 shows ancient shorelines and the present velocity of rise in Scandinavia,based on *Sveriges geologi* (p. 525, 528, 529, and 537). All sections are taken along the line Gedynia—Moskenesöy (southern Lofoten) which goes through the centre of uplift for the Littorina level. Line 1 is the level of the Yoldia sea (about 9500 B.P.), line 2 is the Ancylus lake (about 8500 B.P.), line 3 is the Littorina sea (about 7000 B.P.), and line 4 shows the velocity of present uplift, the scale being chosen so as to facilitate comparison with the Littorina level.

At the time of the Yoldia level here used, absolute sea-level was about 20 m below the present stage, according to curve S_1 in Fig. 2, and for the following comparison we lift the Yoldia level by 20 m to get line 1'. At the time of the Ancylus level the sea was 10—20 m

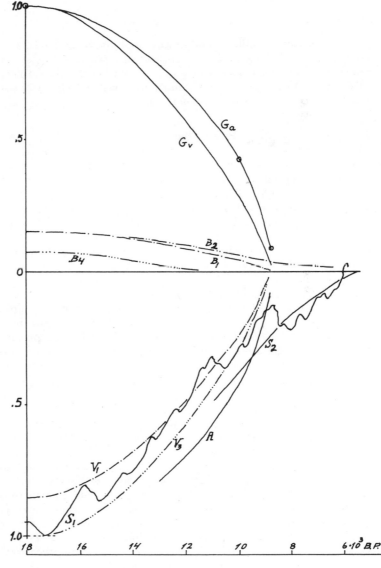

Fig. 2. Decrease of the Scandinavian ice shield at the end of the Ice Age; G_a: by area; G_v: by volume. B_1, B_2 and B_4 are different alternatives of isostatic depression of the general ocean floor. S_1 rise of sea-level relative to unglaciated areas after Fairbridge; S_2 same after Godwin et al. V_1, V_3, A are calculated curves for the rise of sea-level. Time scale at the bottom of figure.

2. mynd. Dvínun jökla í Skandinavíu í lok ísaldar: G_a og G_v. Hugsanlegt sig sjávarbotns: B_1, B_2 og B_4. Mælt ris sjávar utan jöklasvæða: S_1 og S_2, og reiknað: V_1, V_3 og A. Neðst: Tími í þúsundum ára fyrir nútímann.

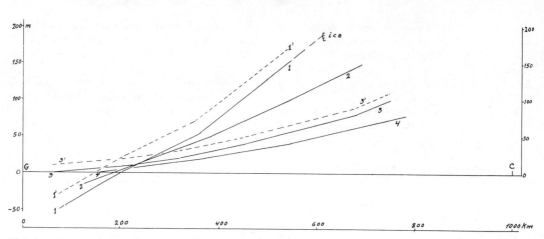

Fig. 3. Ancient shorelines and present uplift of Scandinavia. G = Gedynia (ice margin about 14000 B.P.), C = centre of uplift of Littorina level. 1. Yoldia sea, about 9500 B.P. 1'. Same corrected for stand of absolute sea-level. 2. Ancylus lake, about 8500 B.P. 3. Littorina sea, about 7000 B.P. 3'. Same corrected for stand of sea-level. 4. Velocity of present uplift in arbitrary units. *3. mynd. Gamlar strandlínur í Skandinavíu (1—3) og lyftingarhraði nú á tímum (4).*

below the present stage. On the other hand the lake was probably 10—20 m above sea-level of that time, but an accurate figure is not available; so we assume that the two corrections compensate each other. For the Littorina level we apply a correction of 10 m to get line 3'.

Comparing curves 1', 2, 3' and 4, we must conclude that after Yoldia time there was no systematic movement of a hingeline or a „wave-motion" towards the centre of glaciation. All the levels cut each other in very much the same area, indicating synchronous movement in the whole affected area. See also Hyyppä (1963).

It seems at first strange that the levels meet in a zone about 200—300 km inside the margin of the glacier at its maximum stage, see Fig. 4. This may be understood with reference to Fig. 5. Suppose a thick glacier reaching to the stage 1. It will create a bowl below itself and a corresponding low rise outside. Suppose now that the glacier extends with a *thin* marginal sheet to stage 2, then the weight of this sheet may have very little isostatic effect. It may perhaps lower the rise somewhat, but there need not be created a new and wider bowl. Furthermore, just as it is questionable whether a 100 m ocean layer has an isostatic effect, the same applies to an ice-sheet of 110 m thickness.

The hingeline of the movement, or axis of rotation of the levels, then seems not to have advanced gradually towards the centre, and lines 3 and 4 in Fig. 3 demonstrate very clearly that the present uplift is nearly identical in character with the average picture during the last 7000 years. Only in the first 1000 years after the ice of the Yoldia time disappeared, it seems (cf lines 1 and 2 in Fig. 3) that there was a temporary rapid uplift of the newly deglaciated area.

For the earliest time the synchronous character of the uplift of the Scandinavian area may in part be due to the fact that retreat of the margin and thinning of the glacier were synchronous but at any rate after about Ancylus time this character is a purely hydrodynamical feature on which studies of viscosity may be based.

3. ICELAND.

Iceland is roughly of a slightly oval shape, with an average diameter of about 450 km. Raised beaches are found all around the coast except at the mouth of Eyjafjördur on the middle northern coast, where the terraces along the fjord come just down to sea-level (Einarsson 1959). The age of the highest beaches is about 11.000 years (see below) at which time the sea was about 30 m below the present level.

With the addition of these 30 m the level of the maximum raised beaches would reach present sea-level some 80 km outside the coast at Eyjafjord, and some 100 km outside the middle southern coast. Thus the observable ancient beaches, extending 50—60 km into the country, lie well inside the region corresponding to the hingeline zone in Scandinavia. Considering further the smallness of the country compared

Fig. 4. Ice margins and Littorina zero. 1. Ice margin about 27000 B.P. 2. margin about 14000 B.P. 3. Littorina zero = intersection of L-level and present sea-level. After *Sveriges geologi,* p. 462 and 529.

4. mynd. Tvær stöður ísjaðarsins (1, 2) og skurðlína sjávar og Littorina flatarins.

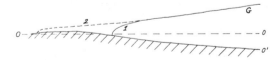

Fig. 5. Schematic section through southern margin of the Scandinavian glacier to illustrate position of hingeline.

5. mynd. Skýringarmynd af suðurjaðri Skandinavíujökuls til að gera grein fyrir legu hverfiáss strandlínuflatanna.

with Scandinavia, we are justified to assume a fully synchronous rise of the country, unless some special reasons for the countrary were found.

We also remark that the coast of Iceland would fully correspond to a straight continental coast in case the country had been unglaciated — the correction due to the curvature of the coast amounting to only about $\frac{1}{2}$ m — so during glaciation elsewhere the relative sea-level would in that case have dropped between 85 and 100 m. But as the oldest shorelines are only about 11.000 years old the distinction between a yielding and an unyielding ocean floor becomes unimportant.

The rapidity of post-glacial rise in Iceland, already referred to, is borne out by two main facts, using the latest material: 1) sea-level at Stokkseyri on the south coast was as low as now or lower 8000 B.P. This age rests on two C^{14}-ages of peat: 8065 ± 400 yr. and 8170 ± 300 yr. (Kjartansson 1964). 2) sea-level at Reykjavík (Seltjörn) was practically as low as now 9000 B.P. The two C^{14}-ages of peat give 9030 ± 280 yr. and 8780 ± 150 yr. (Thorarinsson 1964). Full rise in observable parts of the country appears to have been completed by 6000 B.P. together with that of absolute sea-level.

These facts, combined with the age and height of the raised beaches, enable us to construct curves for the rise of various localities. The age of about 11.000 years for the maximum beaches is based on C^{14}-ages of sea shells from localities in Southern Iceland as shown in Table I (Thorl. Einarsson 1964).

TABLE I.

Ages of sea shells from three localities.

Locality	Height of locality, m	Age, Yr. B.P.
Reykjavík, Airfield	13	9940 ± 260
,, ,, ...	13	10450 ± 160
,, ,, ...	13	10230 ± 190
,, ,, ...	13	10310 ± 260
Spóastadir	55	9930 ± 190
Hellisholtslækur	75	9580 ± 140
,, ...	75	9800 ± 150

Four different constructions of the curve of rise for Reykjavík are shown in Fig. 6. The

JÖKULL 161

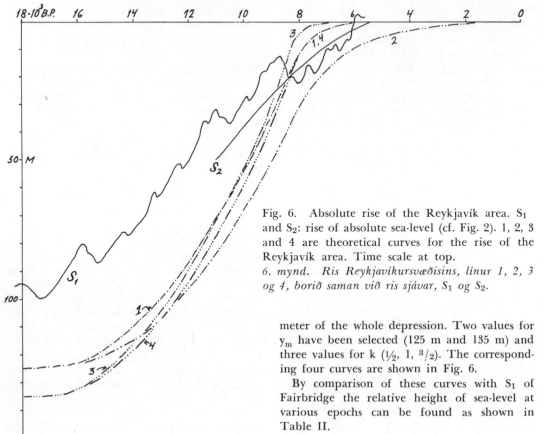

Fig. 6. Absolute rise of the Reykjavík area. S_1 and S_2: rise of absolute sea-level (cf. Fig. 2). 1, 2, 3 and 4 are theoretical curves for the rise of the Reykjavík area. Time scale at top.
6. mynd. Ris Reykjavíkursvæðisins, línur 1, 2, 3 og 4, borið saman við ris sjávar, S_1 og S_2.

method of constructing the curves is the following. For the decrease of glacier load we use the curve G_v of Fig. 2. The ordinate of the curve for absolute rise of the locality be y (positive downward), its maximum value y_m at 18.000 B.P. is taken to correspond to equilibrium under full load, ordinate of $G_v = 1$. The rapidity of rise of the locality is then $dy/dt = -k (y - y_m \cdot G_v)$, where k is a constant for the country, depending on viscosity *and* the dia-

meter of the whole depression. Two values for y_m have been selected (125 m and 135 m) and three values for k ($1/2$, 1, $3/2$). The corresponding four curves are shown in Fig. 6.

By comparison of these curves with S_1 of Fairbridge the relative height of sea-level at various epochs can be found as shown in Table II.

As criteria for the selection of the fitting curve we have the demands that the highest sea-level must come near 43 m, the maximum height of the raised beach at the locality, and that the curve must intersect S_1 near the time 9000 B.P. The intersection for curve 3 is 8500 B.P. and for curve 4 it is 8300 B.P.

It is clear that y_m must be considerably larger than 100 m, otherwise the curve would have to be flat in the beginning and later rise rather suddenly. But this is not possible if the early and late parts shall be consistent, i.e. if k is

TABLE II.

Relative heights of sea-level at Reykjavik at various time.

y_m	k	T =	17.3	15.8	13.3	12.0	11.0	10.0	9.0	8.0	7.5	7.0	6.5	6.0×10^3 B.P.
125	1	Curve 1	25 m	42	41	37	39	24	19	− 11	− 13	− 7	− 10	0
125	$1/2$	— 2	25	43	48	46	52	38	35	8	4	7	2	10
135	$3/2$	— 3	35	51	46	39	40	23	15	− 19	− 18	− 9	− 11	0
135	1	— 4	35	52	49	44	45	29	22	− 10	− 13	− 7	− 10	0

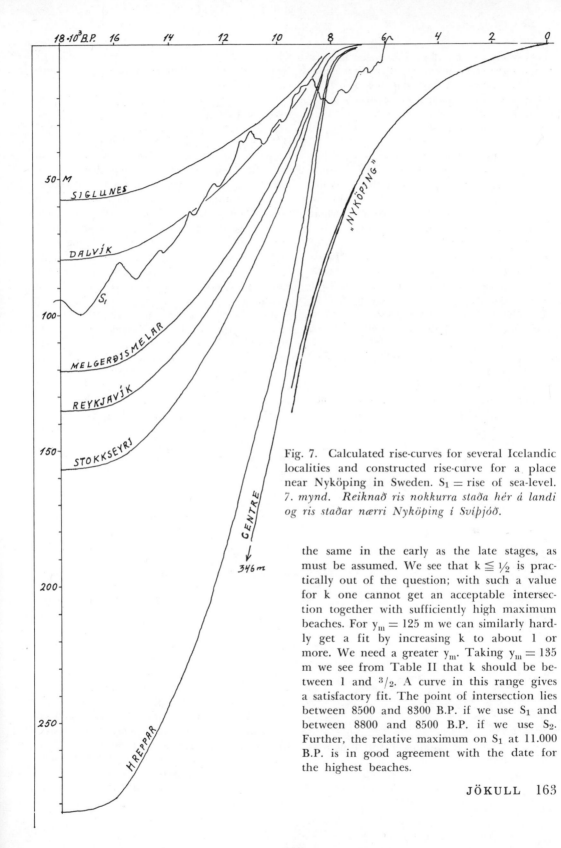

Fig. 7. Calculated rise-curves for several Icelandic localities and constructed rise-curve for a place near Nyköping in Sweden. S_1 = rise of sea-level. *7. mynd. Reiknað ris nokkurra staða hér á landi og ris staðar nærri Nyköping í Svíþjóð.*

the same in the early as the late stages, as must be assumed. We see that $k \leqq \frac{1}{2}$ is practically out of the question; with such a value for k one cannot get an acceptable intersection together with sufficiently high maximum beaches. For $y_m = 125$ m we can similarly hardly get a fit by increasing k to about 1 or more. We need a greater y_m. Taking $y_m = 135$ m we see from Table II that k should be between 1 and $^3/_2$. A curve in this range gives a satisfactory fit. The point of intersection lies between 8500 and 8300 B.P. if we use S_1 and between 8800 and 8500 B.P. if we use S_2. Further, the relative maximum on S_1 at 11.000 B.P. is in good agreement with the date for the highest beaches.

JÖKULL 163

413

Taking either curve 3 or 4 to correspond in the main to reality, it emerges that after about 8800 B.P. sea-level at Reykjavík fell below the present level and remained so for nearly 3 millennia. This result is retained if S_2 is used instead of S_1 but the drop in sea-level is less in that case.

On account of the synchronous rise of the country (which must at least be true for Southern Iceland) we can find the rise-curves for other localities by multiplying the Reykjavík-curve by a locality factor. For Stokkseyri, where the maximum raised beach is assumed to be 55 m (by comparison with Hjalli in Ölfus), the factor is $(32 + 55)/(32 + 43) = 1.16$, the figure 32 being the ordinate of the S_1 curve at 11.000 B.P. The curve for Stokkseyri lies insignificantly below the Reykjavík-curve after 9000 B.P., and sea-level must have dropped in the following time by nearly the same amount. Using curve 3 for Reykjavík, I have drawn, Fig. 7, the corresponding curves for Stokkseyri, Hreppar (125 m beach), Melgerdismelar, Eyjafjord (35 m), Dalvík, Eyjafjord (12 m), Siglunes, mouth of Eyjafjord (0 m), and for the assumed centre of the depression where the (unobservable) beach height is taken as 160 m. The corresponding maximum depression at this centre was 346 m.

In this way theoretical data. may be obtained for various localities and tested by local findings. We see for instance that at about 18.000 B.P. sea-level should have fallen about 40 m below the present one at the coast near Siglunes, and a rim of low land would have been added that may in part have been unglaciated. Grímsey must have been much larger than now, cf Fig. 8, and at least nearly connected with the mainland. But local data for testing are scarse and on the whole it would carry us too far to enter upon such discussions here. For the time being we shall restrict us to the material already given and the rise-curves built on this as shown above.

We now proceed to obtain the relative viscosity. In the theory of Haskell for the rise of Scandinavia (after Scheidegger 1958) is considered an uncompensated depression of depth d and radius R (R being the distance from the centre to a point where the depth has decreased to a certain small part of the maximum depth d). Viscosity of the underlying material is η. Then the velocity of rise at the centre of the depression is $V_0 = q \cdot d \cdot R/\eta$, where q is a constant wich we can take as unity. This depression and its progressive shallowing correspond to the stage after the ice load has disappeared.

Comparing two such depressions of radii R_1 and R_2 and depths d_1 and d_2 respectively, the central velocities of rise are $V_1 = d_1 R_1/\eta_1$ and $V_2 = d_2 R_2/\eta_2$. If the form were the same, i. e. $(d/R)_1 = (d/R)_2$, and the viscosity the same in both cases then we find that the ratio of the velocities is the same as the ratio of the squares of the radii. If we apply this to our case and take for Scandinavia $R = 900$ km and for Iceland $R = 240$ km, we find that Scandinavia would rise with a 14 times greater velocity than Iceland, and a filling up of the depressions would take 3.75 times longer time in Iceland than in Scandinavia. As the actual times for filling up are in reality about reversed, the main reason must be much less viscosity for Iceland than for Scandinavia.

For a final comparison of the two depressions we need not use the centres, as the synchronous movement enables us to use any point in the

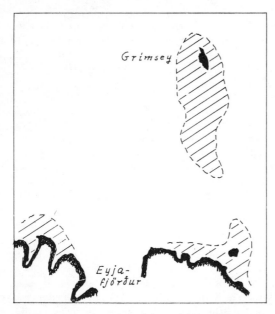

Fig. 8. Approximate position of strandline near Grímsey 18000 B.P.

8. mynd. Áætluð lega strandlinu kringum Grímsey 18000 árum fyrir nútimann.

depressions. For Iceland we take Stokkseyri and for Scandinavia a point with the same relative distance from the centre, on the section Gedynia-Moskenesöy. This point is near Nyköping south of Stockholm, we shall call it „Nyköping". We construct a curve of rise for this point, Fig. 7, using Fig. 3. There are two alternatives, taking absolute sea-level from S_1 or from S_2. The rise at „Nyköping" at a convenient time, 9000 B.P., is found to be 40 m and 45 m per 1000 years respectively. This gives the viscosities $(127 + a) \cdot 900/40$ and $(137 + a) \cdot 900/45$ in arbitrary units, where a is the present amount of downwarping. To give consistency with the present uplift, about 3 m/1000 yr, a must be about 10 m, and we get the viscosities 3080 and 2940. For Stokkseyri we get $30 \cdot 240/29 = 248$. The ratio of the viscosities is then 12.4 and 11.9. The above value for Stokkseyri is independent of the time chosen, and this would also be the case for Nyköping if the curve had been constructed on the same principle. But it has simply been drawn smoothly through 3 points and it gives a slightly less viscosity for the latest time than for 9000 B.P.

We conclude that it must be a good approximation to say that viscosity for the Iceland region is one order of magnitude less than for Scandinavia, as earlier estimated.

It would be tempting to assume that the low viscosity for Iceland is due to its position on the mid-Atlantic Ridge. But it has been pointed out (Broecker 1966) that the uplift of Scandinavia may lead to an abnormally high calculated viscosity, and the value for Iceland is then closer to what might be a normal value.

If the great mobility for Iceland were due to a special thin fluid layer, then the models for Iceland and Scandinavia would be quite different and the above comparison would be invalid. But it is clear that in this case the viscosity of the thin layer would have to be less than the value we found.

Another centre of Pleistocene glaciation are the Faeroe Islands. The radius of the glacier can be taken as 50 km. If the maximum thickness was 500 m, and the thickness followed the formula $H = 500 \, (1 - (x/50)^2)$, where x is the distance from the centre in km, then the centre would be depressed 24 m while it would rise isostatically 22 m on account of lowering of the surrounding sea, thus giving a net sinking of only 2 m (Found by method given in Einarsson 1953). Under these circumstances raised beaches would not occur, which is in keeping with observations. (Rasmussen personal communication.) This small glacial centre, therefore, does not lend itself to viscosity studies, which shows the limit to the application of the above considerations.

Finally, it may be remarked that when general sea-level dropped by 85—100 m during the last glaciation it dropped by that amount at the Faero Islands. A quite considerable area was thereby laid dry and unless it was entirely covered by the glacier it presented a possible refugium for plants and lower animals.

ÁGRIP

RIS ÍSLANDS Í LOK ÍSALDAR OG SEIGJA DJÚPLAGA UNDIR JARÐ-SKORPUNNI.

Eftir Trausta Einarsson.

Ris Íslands eftir að jökulfarginu létti af því var mun hraðara en hliðstætt ris Norðurlanda. Á grundvelli þess áætlaði ég 1953, að seigja djúplaga væri hér eitthvað um $1/10$ af gildi hennar undir Norðurlöndum og stæði þetta lága gildi væntanlega í sambandi við legu Íslands á Mið-Atlantshafshryggnum. Síðan hafa verið gerðar ýmsar aldursmælingar, sem snerta ris Íslands, og er nú hægt að gera málinu fyllri skil.

Fyrst er athugað, hvort hinn almenni sjávarbotn muni hafa sigið, er 100 m þykkt sjávarlag bættist á hann við bráðnun jökla í lok ísaldar. Niðurstaðan er óviss, en sýnt, að viðunandi nákvæmni fæst, ef gert er ráð fyrir óhreyfðum sjávarbotni.

Rýrnun jökla á Norðurlöndum undir lok ísaldar er borin saman við mælingar á hækkun almenns sjávarborðs og reynist í stórum dráttum í góðu samræmi við þær. Er hún því einnig notuð sem mælikvarði á rýrnun jökla hér.

Strandlínur frá ýmsum tímum eru þekktar við Eystrasalt, og sýna þær, svo og nútíma ris svæðisins, að sigdældin hefur grynnkað sem heild án þess að skreppa saman, og rishraði á hverjum stað hefur verið í réttu hlutfalli við dýpið. Sömu reglu má ganga út frá hér, en af henni leiðir, að rislínu er hægt að teikna fyrir sérhvern stað á landinu, þar sem forn sjávarmörk eru þekkt, ef búið er að teikna rislínu fyrir einhvern einn stað, og er Reykjavík valin sem slíkur grundvallarstaður. Sem dæmi um niðurstöður má nefna, að fyrir um 18000 árum stóð sjór um 40 m neðar en í dag við mynni Eyjafjarðar, og á Grímseyjarsundi féll hann þá yfir 50 m. Grímsey var þá miklu stærri en nú, en óvíst, hvort hún hefur verið landföst. Mun þarna hafa verið íslaust svæði, eins og beinar athuganir benda til.

Nú eru valdir tveir staðir, sem hafa sömu hlutfallsfjarlægð frá miðju sigdældar, Stokkseyri og staður nærri Nyköping í Svíþjóð, og bornar saman rislínur þeirra. Með hliðsjón af reikningi Haskells á seigju undir Norðurlöndum fæst þá, að seigjan þar er um 12 sinnum meiri en hér.

Því hefur nýlega verið haldið fram, að seigjan, sem ris Norðurlanda bendir til, sé óeðlilega mikil og venjulegt gildi muni um $1/10$ þess.

Sé þetta rétt, er seigjan undir Íslandi ekkert óeðlilega lág. En mælingar á seigju djúplaga eru enn fáar og því of snemmt að fullyrða nokkuð um þetta atriði.

Loks er litið á Færeyjar. Hefur sig þeirra undan jökli sem næst vegið upp á móti lyftingu, sem stafaði af því að sjór hvarf af grunnsvæðinu í kring. Skýrir þetta þá staðreynd, að fornir marbakkar eru óþekktir í Færeyjum. Í heild féll sjávarmál um 85—100 m við Færeyjar á síðustu ísöld.

REFERENCES - HEIMILDARRIT

Broecker, W. S. 1966. Glacial Rebound and the Deformation of the Shorelines of Proglacial Lakes. J. G. R. 71, 4777—83.

Einarsson, Thorl. 1964. Náttúrufræðingurinn, 34, p. 127.

Einarsson, Trausti. 1953. Depression of the Earth's Crust under Glacier Load. Various Aspects. Jökull, 3, p. 2—5.

— 1959. Studies of the Pleistocene in Eyjafjörður, Middle Northern Iclend. Vís. Ísl. No. 33.

Fairbridge, R. W. 1960. The Changing Level of the Sea. Scientific American, May 1960. p. 70—79.

— 1961. Eustatic Changes in Sea Level; in Physics and Chemistry of the Earth. Vol. 4, p. 99—185. Pergamon Press.

Godwin, H., Suggate, R. P., and Willis, E. H. 1958. Radiocarbon dating of the eustatic rise in Ocean Level. Nature, 181, p. 1518—19.

Hyyppä, E. 1963. On the late-Quaternary history of the Baltic Sea. Fennia, 89, No. 1, 37—50.

Kjartansson, G. 1964. Náttúrufræðingurnn 34, p, 101.

Magnusson, N. H., Lundquist, G., Regnéll, G. 1963. Sveriges geologi, 4th ed. Stockholm.

Thorarinsson, S. 1964. Náttúrufræðingurinn, 34, p. 125.

Scheidegger, A. S. 1958. Principles of Geodynamics, p. 266. Springer.

(Manuscript received February 1967.)

Geophysical Research
in Glacial Isostasy

III

Editor's Comments on Paper 25

McConnell: *Viscosity of the Mantle from Relaxation Time Spectra of Isostatic Adjustment*

Many of the preceding papers have been concerned with the observation and recording of changes in level and deformation of marine or lake beaches in areas affected by glacial loading. These data provide a unique source of information to the geophysicist concerned with the strength and response of the crust to loads of large size and long duration.

McConnell derives the relaxation time spectra by the decomposition of Sauramo's shoreline diagram (Fig. 2). This approach required that deformation did not extend out farther than 800 km from the center of uplift, and that a circular approximation to the actual shape (Fig. 1) of the loaded area was adequate. Some of the results of the analysis are probably dependent upon the basic geological data, and Sauramo's diagram is far from universally accepted by Fennoscandian field workers. Figure 4 shows the transformed shoreline elevations and indicates that two peaks are present at wave numbers 3.5 and 12×10^{-8} cm^{-1} (wave number where a is the radius of the earth and n is the degree of surface harmonics). The relaxation time plot of Fig. 5 indicates nearly constant relaxation time for most wave numbers with a tendency to more rapid recovery at the shorter wavelengths.

McConnell used the relaxation time spectra to suggest that the postglacial rebound could be modeled by assuming an elastic layer 120 km thick, a viscosity decrease within the asthenosphere, and then an increase in viscosity at depths greater than 1200 km. Only limited further rebound (20 m) is predicted, despite current high rates of postglacial uplift near the center of the ice cap (~ 1 m/100 yr).

Robert K. McConnell, Jr., received his B.S. from Princeton University and his Ph.D. in geophysics from the University of Toronto. He is currently President of Earth Sciences Research, Inc., Cambridge, Mass. He has worked extensively in the area of geophysics and published over 25 papers. He has been involved in university teaching at M.I.T. and the University of Rhode Island.

Copyright © 1968 by the American Geophysical Union

Reprinted from *J. Geophys. Res.*, **73**, 7089–7105 (1968)

Viscosity of the Mantle from Relaxation Time Spectra of Isostatic Adjustment

ROBERT K. MCCONNELL, JR.

Arthur D. Little, Inc., Cambridge, Massachusetts 02140

25

After allowing for the effects of sea level fluctuations, harmonic analysis of the present level of former shoreline features in southeast Fennoscandia indicates that shorter wavelength departures from equilibrium relax more rapidly than longer ones. These observations together with indicated recent secular decreases in the length of the day would seem to confirm that flow in the upper-mantle low-viscosity channel as a result of the Pleistocene glacial loading is the most likely source of the earth's nonhydrostatic bulge.

INTRODUCTION

Recently a great deal of attention has been paid to the derivation of theoretical relaxation times of isostatic adjustment in layered earth models [*McConnell*, 1965; *Takeuchi and Hasegawa*, 1965; *Anderson and O'Connell*, 1966]. That such adjustment follows melting of the Pleistocene ice sheets has been confirmed in Fennoscandia, North America, Britain, and Antarctica (see, for example, *Gutenberg* [1941], *Flint* [1957], and *Wollard* [1962]), and following the disappearance of Lake Bonneville [*Crittendon*, 1963]. More recently the possibility that the nonhydrostatic portion of the earth's equatorial bulge may also result from glacial loading has been the subject of considerable controversy [*Wang*, 1966; *McKenzie*, 1966; *McConnell*, 1968]. In this paper we shall make use of some published data on postglacial upwarping of Fennoscandia, gravity anomalies, and recent changes in the length of the day in an attempt to devise a consistent relaxation time spectrum for the earth when deformed by surface loads and to then interpret this spectrum in terms of a viscosity gradient in the mantle.

DATA FROM FENNOSCANDIA

The continuous upwarping of Fennoscandia since the early stages of deglaciation of the Baltic region over 10,000 years ago is supported by evidence from tide gages, precise leveling, and paleoshorelines. Several workers, including *Bergsten* [1930], *Gutenberg* [1941], and *Sirén* [1951], have analyzed data from gages placed along sea and lake shores. They have shown that vertical land movement is now taking place over much of the deformed area with rates of up-warping as great as 1 cm per year occurring in the upper parts of the Gulf of Bothnia. The evidence from the tide gages is confirmed by a comparison of results of precise leveling over Finland [*Kääriäinen*, 1949], which show similar rates of upwarping.

The most helpful data for the determination of relaxation time spectra at short wavelengths are provided by old shoreline features which are now warped and elevated far above the present sea level. Since the ages of these shorelines that cover much of the glaciated area can be determined by pollen studies, radiocarbon methods, and varve counting, they provide a relatively complete history of the upwarping since the retreat of the last ice sheet. A typical isobase map that shows contours of the present elevations of former shorelines is given in Figure 1. This map and all other maps constructed by *Sauramo* [1955, 1958] and other workers indicate that the most intense upwarping in the past has occurred in the Ångermanland region of Sweden where the greatest thickness of ice is believed to have been.

A reliable determination of the relaxation time spectrum of a region requires that a complete history of isostatic adjustment be known out to the limits of the area being deformed. In Fennoscandia the record of the beaches is both incomplete and obscured by the effects of sea level fluctuations unrelated to isostatic adjustment. The difficulties are well illustrated by Figure 2, which shows *Sauramo*'s [1958] interpretation of the present levels of shoreline features of various ages that fall on or near

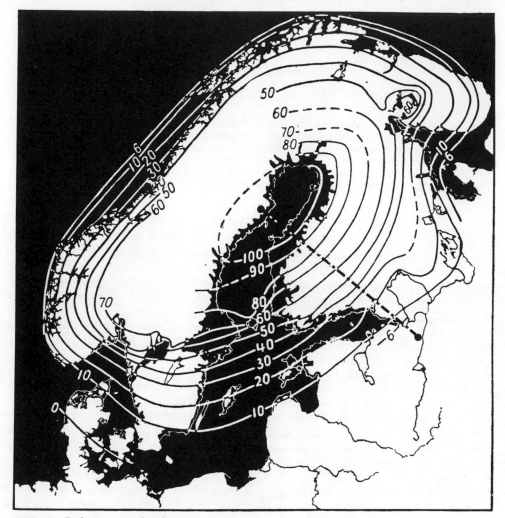

Fig. 1. Isobase map showing present elevations in meters of features which were at sea level during the middle stages of the Litorina Sea (about 3000 B.C.), after *Sauramo* [1958]. Dot-dash line indicates position of profiles analyzed.

the line shown in Figure 1. To make this interpretation, Sauramo found it necessary to include data from a large area surrounding the indicated line. However, in an attempt to clarify the significant parts of the motion, to provide the maximum consistency between different levels, and to eliminate random errors and small local effects, he did not always project the data from offset stations strictly at right angles. This method of diagram construction, of course, may introduce additional uncertainty into the data and for this reason it has been criticized by *Niskanen* [1939].

Reduction of Shoreline Data to Show Isostatic Adjustment

Even if Sauramo's data as presented in Figure 2 were exact, two additional reduction steps involving assumptions about the form of the uplift and the effects of sea level fluctuations would be necessary before attempting to derive the relaxation time spectrum. To illustrate these steps and the subsequent spectral analysis, six sets of shoreline features of different ages were selected (see Figure 2).

For several of these selected beach levels, Sauramo did not have data for the periphery

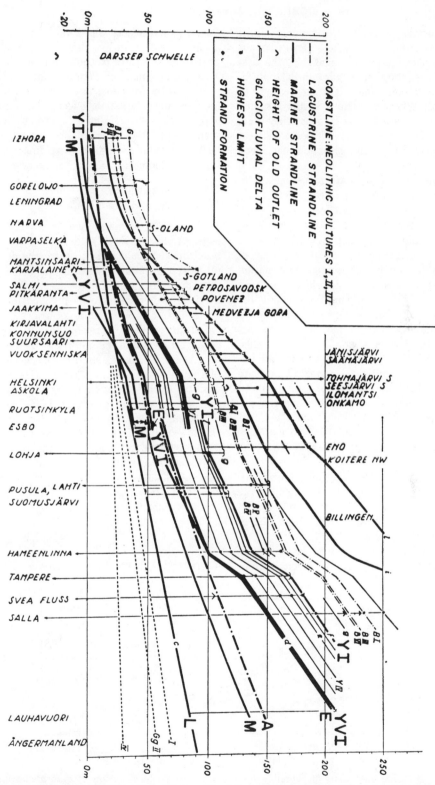

Fig. 2. Diagram showing the relationship between the most important strand lines in southeastern Fennoscandia. The larger letters indicate the levels used in analysis: YI, Yoldia; YVI, Yoldia VI; E, Echineis; A, Ancylus; M, Mastogloia; and L, Litorina. The unreduced shoreline data is from *Sauramo* [1958].

of the deformation or for the assumed uplift center in Ångermanland. Where elevations out to the periphery were not available, the first data reduction step involved interpolation of the missing values from data on beaches of similar ages. Where elevations at the assumed center of deformation were missing, they were assumed to be the same as the points nearest to the center where the elevation had been given.

The second step in the data reduction was to make corrections for sea level fluctuations, leaving only the bona fide effects of warping. The best measure of the sea level changes is, of course, the present elevation of each shoreline where no deformation is known to have occurred. Unfortunately, our present information does not allow this to be determined without introducing additional assumptions about the form of the uplift. One reasonable assumption is that any shoreline near the periphery of the deformed area that has not undergone much tilting has not been involved in isostatic adjustment. If this assumption is correct, a measure of the uplift between the time of formation and the present is obtained by subtracting the elevation of the untilted portion from the elevations of all other portions of the same shoreline.

After extrapolating any missing data, the elevations of the six selected shorelines were recorded at 36-km intervals from Ångermanland to beyond Leningrad and corrected for the sea level changes. The corrected levels, which now presumably show only the effects of isostatic adjustment, are shown in Figure 3.

Examination of these corrected levels clearly shows that soon after the ice melted from Ångermanland there was a rapid upwarping of a small central region, leaving only a broad depression to return to equilibrium more slowly. This phenomenon, which was pointed out many years ago by Sauramo, has usually been attributed to 'delayed elastic effects.' In a previous paper [*McConnell*, 1965] it was shown that such rapid upwarping of shorter wavelength disturbances may also be explained by low-viscosity channels in the upper mantle.

Transformed Beach Levels and Departures from Equilibrium

Although the large number of corrections and assumptions necessary to arrive at the reduced beach levels would seem to leave the derivation of relaxation time spectra from the available data open to considerable uncertainty, what is available would seem to be justified for lack of better data computation of spectra.

Let us consider a region that is freely returning to equilibrium after the melting of an ice sheet. We may describe the vertical displacement of the surface from the equilibrium position by the function $w_z(x_i, t)$, where x_i is the set of coordinates of the point where the displacement was measured and t is the time of the observation. If the displacement is circularly symmetrical about an axis so that the vertical displacement may be written as $w_z(r, t)$, where r is the distance from the axis, we may define $Z(u, t)$ to be the zero-order Hankel transform of the displacement

$$Z(u, t) = \int_0^\infty w_z(r, t) J_0(ur) r \, dr \qquad (1)$$

where $J_0(ur)$ is the Bessell functon of the first

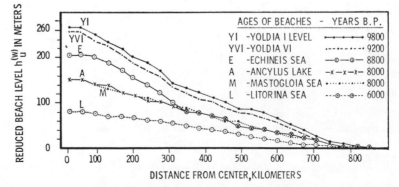

Fig. 3. Relative elevations of shoreline features in southeastern section of deformed region (after *Sauramo* [1958]) after adjustments for changing datum have been made.

kind of order zero and u is a wave number. It has been shown [*McConnell*, 1965] that for simple Newtonian flow, a mantle of constant density but variable viscosity Z will be of the form

$$Z(u, t) = Z(u, t_0) \exp [(t_0 - t)/\tau(u)] \quad (2)$$

In this expression the relaxation time spectrum $\tau(u)$ will depend on the variation of viscosity as a function of depth.

It, instead of a circularly symmetric deformation, we consider one in which the vertical displacement from equilibrium $w_z(x, t)$ is independent of the y coordinate, $Z(u, t)$ is more appropriately defined as the one-dimensional Fourier transform

$$Z(u, t) = \int_0^\infty w_z(x, t) \exp [iux] \, dx \quad (3)$$

In the previous paper it was shown that, when $Z(u, t)$ is defined in this manner, it will also have the form (2), and for the same variation of rheological properties with depth it will have a relaxation time spectrum $\tau(u)$ that is identical to that for the circularly symmertical conditions. Although the relaxation time spectrum is based on the vertical departure from equilibrium w_z whereas the raw data yield only the present displacements of old shorelines from sea level, the use of the Fourier transform allows each of these to be obtained from the other without difficulty. Consider a shoreline formed at a time t_n with respect to the present time t_0 whose present height at a point with coordinates x_i is now $h(x_i, t_n)$.

The displacement $w_z(x_i, t_n)$ from equilibrium at time t_n is given by

$$h(x_i, t_n) = w_z(x_i, t_n) - w_z(x_i, t_0) \quad (4)$$

Making use of the linear property of Fourier transforms

$$H(u, t_n) = Z(u, t_n) - Z(u, t_0) \quad (5)$$

where $H(u, t_n)$ and $Z(u, t_n)$ are the transforms (either Fourier or Hankel) of $h(x_i, t_n)$ and $w_z(x_i, t_n)$, respectively. Substituting (2) into (5) yields

$$Z(u, t_0)(\exp [(t_0 - t_n)/\tau(u)] - 1) \doteq H(u, t_n) \quad (6)$$

Thus, having determined the function $H(u, t_n)$

by transformation of the observed beach levels, it is possible to determine the relaxation times $\tau(u)$ and amplitude $Z(u, t_0)$ that best satisfy (6) for each value of u. The most obvious advantage of this technique is that, since it utilizes only data obtained after the melting of the ice sheets, no assumptions about the detailed form of the glacial retreat are required.

Because of the roughly circular form of the Fennoscandia deformation, it would seem to be most appropriate to carry out the harmonic analysis by using the Hankel transform. Figure 4 shows the function $uH_n(u)$ obtained by numerical integration of (1) for each of the corrected shoreline elevations from Figure 3 assuming that there is no warping beyond 800 km from the center of the deformed region.

The central depression pointed out with reference to Figure 3 appears again in the transformed profile as a secondary peak in the neighborhood of $u = 12 \times 10^{-8}$ cm^{-1} which disappears far more rapidly than the main peak near 3.5×10^{-8} cm^{-1}.

Exponentially decaying functions of the form (6) were fitted, by using a least-squares criterion, to various combinations of the most recent 3, 4, 5, and 6 beaches. Although the constraint was imposed that the fitting function must predict that present shoreline features are now at sea level, no attempt was made to force the functions to yield the present rates of uplift based on the data of other workers. However, as discussed below, the discrepancies between the predicted rates and the observed ones may be used as an indication of where this approach may yield errors. In considering the reliability of this method, one must also constantly keep in mind the many uncertainties in the data, especially those resulting from the possible assumption of incorrect dates for the shorelines.

The relaxation times resulting from the curve fitting are shown as points in Figure 5. Before drawing curves of the function $\tau(u)$ through the points, the ones that seemed most likely to be reliable were selected by assigning a semiquantitative reliability grade reflecting how well the observations approach a true exponential decline. Relaxation times that yielded a rms deviation between the observed and the theoretical uplift of less than 10% of the present transformed height of the oldest shoreline were considered to be good. If the deviation was

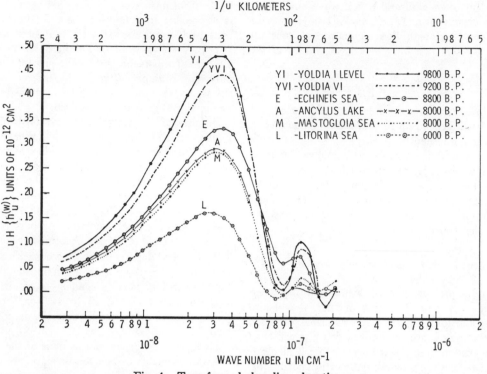

Fig. 4. Transformed shoreline elevations.

between 10 and 20%, then the value was considered to be fair; if greater than 20%, poor. The most consistent relaxation times are derived from data near the prominent spectral peaks around 3.5×10^{-8} cm^{-1} and 12×10^{-8} cm^{-1}. Few reliable values are available for wave numbers between 8×10^{-8} and 10×10^{-8} cm^{-1} or greater than 14×10^{-8} cm^{-1} because of minima in the shoreline spectra. As far as possible, the curves shown in Figure 5 were drawn smoothly through the point to which a good grade had been assigned, whereas poor points and points were no meaningful relaxation time could be assigned were usually ignored.

Of the four curves shown, only the one representing the most recent three levels (L, M, A) shows any significant trend toward longer times at large wave numbers for the main peak. However, this spectrum also yields fair values of very short relaxation times near the secondary peak. Originally this curve was considered unreliable, since two of the beaches used are of approximately the same age; however, it does suggest that the most recent deformation is more consistent with longer relaxation times at

wave numbers around 6×10^{-8} cm than our interpretation using older beaches. Probably the most representative relaxation spectrum is the one derived from the combination L, M, A, E, and YVI; it averages over most of the available data while ignoring the shoreline features formed at a time (YI) when there may have still been enough ice in Ångermanland to prevent free return to equilibrium.

Two significant features of all the relaxation time curves are apparent. Most important is the decrease in relaxation time as the higher wave numbers are encountered; this would appear to imply a low-viscosity layer and/or an elastic layer near the surface [McConnell, 1965]. The second is the approach of the curves to a constant value at smaller wave numbers; this trend results from the assumption that the end point of the profile and all points beyond it have suffered no important deformation since the ice left Ångermanland. If the assumption is correct, a continued viscosity increase with depth is implied; if it is not correct, these portions of the curve at smaller values of u must be disregarded. To resolve this question, we turn to the data on present gravity anomalies and deformation rates.

Evidence from Tilting and Gravity Anomalies

The preceding analysis of the relaxation time spectra from shoreline data would seem to indicate that little isostatic adjustment remains to be accomplished at any wavelength. On the other hand, calculations by *Niskanen* [1939] and *Kääriäinen* [1953] indicate that up to 200 meters of uplift remain before equilibrium is restored and that adjustment is proceeding at a rate of over 9 mm per year.

Figure 6 compares the present rates of upwarping $\dot{w}_z$ over Fennoscandia as interpreted by Kääriäinen with the remaining uncompensated deformation w_z as estimated by Niskanen. The results shown were scaled from maps by *Heiskanen and Vening Meinesz* [1958]. Although the data do not extend out to the limits of tilting, and the interpretation of the gravity

anomalies is not very precise, if one carries out the transformation by using what is available, the diagram of the transformed profiles of w_z and $\dot{w}_z$ would show peaks in the neighborhood of $u = 3.6 \times 10^{-8}$ cm^{-1} and $u = 3.0 \times 10^{-8}$ cm^{-1}, respectively (Figure 7). These peaks are at about the same place as those interpreted for the old strand lines. The maximum amplitude of the transformed upwarping is $u\dot{Z} = 0.20 \times 10^{-16}$ cm^2/yr or 0.632×10^{-24} cm^2/sec and of the estimated displacement is $uZ = 0.36 \times 10^{-12}$ cm^2, yielding a value for the relaxation time $\tau = Z/\dot{Z} = 5 \times 10^{11}$ sec. It is interesting to note that, when the relaxation times obtained by *Haskell* [1937] and *Heiskanen and Vening Meinesz* [1958] (who used present uplift and strand line data for their calculations) are compared with this value, they fall approximately on a line joining it to the very short wavelength shoreline data (Figure 8). Surprisingly enough, the values derived by Crittendon (for Lake

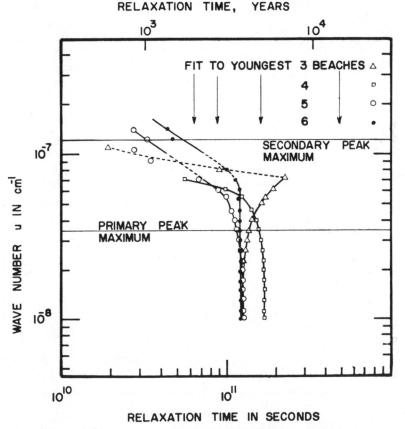

Fig. 5. Relaxation time plot for reduced shorelines of Figure 4.

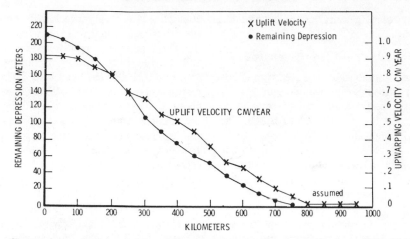

Fig. 6. Comparison of present rate of upwarping over Fennoscandia from Kääriäinen (×) with remaining uncompensated deformation (•) as estimated by Niskanen.

Bonneville) predict longer relaxation times at shorter wavelengths than the Fennoscandian values. To determine whether this difference results from real differences between the crust and upper mantle properties will require further analysis.

EARTH'S EQUATORIAL BULGE

To extend the relaxation time spectrum to longer wavelengths, it becomes appropriate to re-examine the possibility that the nonhydrostatic portion of the earth's gravitational field may be due to Pleistocene glaciation. For our purposes it is sufficient to consider the gravita-

tional potential and moment of inertia of a surface distribution of mass $a_2 P_2 (\cos \theta)$, where P_2 is the familiar second-order Legendre polynomial and θ is the (geocentric) colatitude. A negative value of a_2 would correspond to the transport of mantle material from the poles to the equator, as would be expected to have resulted from loading by an ice sheet. A correlation between wave number u appearing in the Hankel transform, radius R, and polynomial order n can be made by making the substitution

$$[n(n + 1)]^{1/2}/R = u \qquad (7)$$

Taking R to be 6.37×10^8 cm, the mean radius

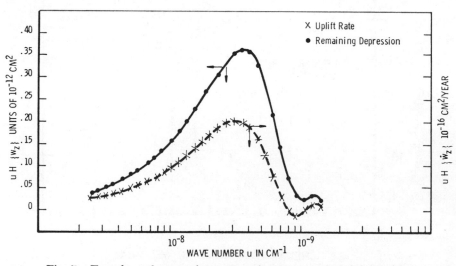

Fig. 7. Transformed upwarping rates and uncompensated deformation.

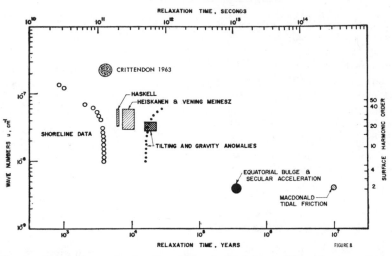

Fig. 8. Comparison of relaxation times derived from different assumptions.

of an equivalent spherical earth, the P_2^0 term has an equivalent wave number u of 2.24×10^{-9} cm^{-1}. If the earth did not deform under the load, the increased gravitational potential U_2' due to this surface distribution of mass would be

$$U_2' = \frac{4\pi}{5} G \frac{R^4}{r^3} a_2 P_2(\cos \theta) \qquad (8)$$

where G is the gravitational constant and r is the distance from the center of the earth, which we may write as

$$U_2' = -\frac{GM^2}{r} \left[J_2' \left(\frac{R_e}{r} \right)^2 P_2(\cos \theta) \right] \qquad (9)$$

where M is the mass of the earth 5.98×10^{27} grams, J_2' is the difference between the equilibrium value of J_2 and the observed value, and R_e is the equatorial radius of the unloaded earth. Now, as $R_e = 6.378R/6.371 = 1.001\ R$, we need make no distinction between R_e and R and can rearrange (9) to be

$$a_2 = -\frac{1.25}{\pi} \frac{M}{R^2} J_2' = -5.87 \times 10^9 J_2' \quad (10)$$

Following *McKenzie* [1966], we may note the difference between the theoretical value J_2 calculated by *Jeffreys* [1963] for the hydrostatic coefficients of the gravitational field of -1072.1×10^{-6} and the value observed by *Kaula* [1966] of -1082.7×10^{-6}. From (10) the unexplained anomaly of $J_2' - 1.06 \times 10^{-5}$

would correspond to an excess mass coefficient of

$$a_2 = -6.22 \times 10^4 \quad \text{g/cm}^2 \qquad (11)$$

The moment of inertia I' of such a mass distribution may be calculated by integrating over the surface of the sphere

$$I' = a_2 \int_S (R \sin \theta)^2 P_2(\cos \theta)\ dS$$

$$= -\tfrac{8}{15}\pi R^4 a_2 \qquad (12)$$

$$= -2.68 \times 10^{35} a_2$$

And substituting for a_2 from (10) gives

$$I' = 1.67 \times 10^{40} \quad \text{g/cm}^2 \qquad (13)$$

Now neglecting the effects of tidal deceleration by the moon, for time scales of the order of 10^6 years, we may write for the earth

$$\frac{d}{dt}(I\omega) = 0 \qquad (14)$$

If the nonhydrostatic portion of the bulge relaxes exponentially such that

$$a_2(t) = a_2(t_0) \exp\left[(t_0 - t)/\tau\right] \qquad (15)$$

$$\tau = \frac{I'}{I} \frac{\omega}{\dot{\omega}} \qquad (16)$$

Following McKenzie and choosing $I = 0.33$ MR^2, (16) becomes

$$\tau = 2.1 \times 10^{-5} \omega/\dot{\omega} \qquad (17)$$

Munk and MacDonald [1960] have summarized the evidence for historical changes in the length of the day, calling attention to an unexplained nontidal increase in ω by 1.0 to 1.4 parts in 10^7 during the last 2000 years based on interpretations by *Fotheringham* [1920] of ancient eclipses. This would correspond to values of $\omega/\dot\omega$ of 6.3×10^{17} and 4.5×10^{17} sec, respectively. Substituting these values into (17) yields values for the relaxation times τ of 1.3×10^{13} and 0.95×10^{13} sec, more than an order of magnitude less than the values obtained by *Munk and MacDonald* [1960] and *MacDonald* [1963], assuming that the equatorial bulge is a result of tidal deceleration. (Figure 8).

VISCOSITY OF THE UPPER MANTLE

Any attempt to estimate the flow properties of the mantle from the relaxation time spectra is limited by uncertainties in the rheological model and in the data. The use of Newtonian viscosity as a model for mantle flow has been criticized by *MacDonald* [1966] and *Orowan* [1965] and supported by *McKenzie* [1966] and *Gordon* [1965]. In the following discussion we will make no attempt to justify such a model on theoretical grounds, but shall try to show that the observed relaxation time spectra are consistent with a Newtonian type flow in the upper mantle.

A problem arises from the apparent necessity of choosing one of the following three possible relaxation times at long wavelengths:

1. A value of about 1.2×10^{11} sec indicated by the beach level data alone on the assumption of no tilting beyond Leningrad.

2. A relaxation time of about 1×10^{13} sec assuming that historical changes in the length of the day result from collapse of the equatorial bulge.

3. The value of 3×10^{14} sec suggested by *Munk and MacDonald* [1960] and *MacDonald* [1963] assuming that the equatorial bulge results from tidal deceleration.

All the different values together with the shoreline data are shown in Figure 8.

We shall first make use of the beach level data, especially the values obtained for short wavelengths (large wave number) to estimate the earth's near-surface properties. We shall then compare the viscosity distributions required by each of the different long wavelength relaxation times.

Viscosity Distributions Based on Shoreline Features Alone

Although there is some disagreement among the different experimental relaxation time spectra calculated by using different combinations

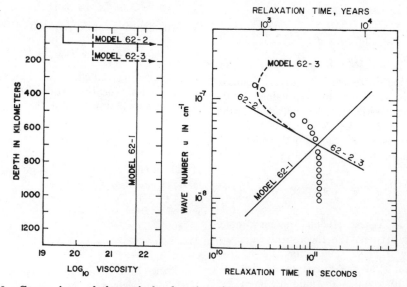

Fig. 9. Comparison of theoretical relaxation times as a function of wave number for a viscous half-space and a single low viscosity layer overlying a rigid half-space with times derived from Fennoscandian deformations. Open circles: shoreline data.

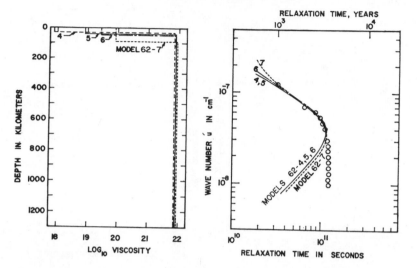

Fig. 10. Relaxation spectra for viscous half-space with overlying low viscosity layer compared with Fennoscandian shoreline data. Open circles: shoreline data.

of beaches, they would seem to have enough features in common to justify the selection of a single curve as being representative of the entire group, and for a number of reasons the spectrum derived from the five most recent of the six shoreline profiles shown in Figure 3 has been used.

In making the comparison between any spectrum interpreted from the shoreline data and the various dimensionless theoretical spectra when the two have little in common, the constraint that the relaxation time corresponding to the maximum amplitude of the function $uH(u)$ must coincide with the theoretical curve results in consistent interpretations. For the shorelines analyzed in the first part of this paper, this maximum occurred at about 3.5×10^{-8} cm^{-1} (Figure 8).

For the calculation of rheological parameters, a single density value of 3.5 g cm^{-3}, which is approximately that of the mantle material at a depth of 100 km, was chosen. As the gravitational acceleration in most of the upper mantle is nearly constant, a single value of 983 cm sec^{-2}, a rough average for the upper mantle, was selected. Any errors in apparent viscosity arising from the choice of these two parameters should be small and should not significantly affect the relative viscosity changes with depth. A selection of models that fit the representative curve of the beach level data alone are

shown in Figures 9 through 12 and a discussion of each follows.

Model 62-1. As a first approximation, the experimental results were compared with the theoretical curve for a homogeneous viscous half-space (Figure 9), the model used by all previous investigators. The inadequacy of the fit is apparent. However, if the criterion of matching at the maximum amplitude is used, the model yields a viscosity of 5.7×10^{21} poises, which is about half the value obtained by *Haskell* [1937] and *Heiskanen and Vening Meinesz* [1958]. The difference between their values and ours results both from the assumption that no deformation extends beyond Leningrad (and hence that the approach to a constant relaxation time at long wavelengths is correct) and from the neglect of data on present rates of upwarping.

The most important discrepancy between the calculated and predicted spectra is in the shape of the curve; the observed long wavelength relaxation times are much less than the theory would predict. One is thus led directly to the conclusion that the homogeneous fluid is not a very realistic representation of the actual mantle and its behavior.

Models 62-2 and 62-3. One way to attempt to improve the theoretical curve is to consider what would happen if the flow were to take place entirely within a relatively thin hori-

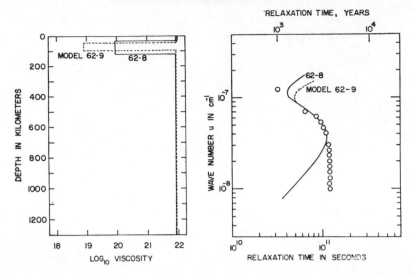

Fig. 11. Relaxation spectra for a viscous half-space underlying two equal-thickness low-viscosity layers compared with Fennoscandian shoreline data. Open circles: shoreline data.

zontal layer underlain by a rigid boundary and overlain by an elastic layer which is thin enough that its effects (Model 62-2) are negligible. With this model the fit at short wavelengths is improved, but at the long wavelengths the observed recovery times are much too short (Figure 9), and it seems impossible to cause both the 'maximum' amplitude and any other parts of the observational data to fit the theoretical curve well. Although there are many combinations of depth and viscosity in the single-layer model that would provide as good a fit as that shown for Model 62-2 in Figure 9, they all seem to be related by the approximate expression $\eta \simeq (3.8 \times 10^{-2})h^3$, $h < 2 \times 10^7$, where h is the thickness of the flowing layer in centimeters.

If we arbitrarily choose the thickness of the flowing layer to be 100 km, then the appropriate viscosity is 3.8×10^{19} poises. If the thickness is increased to 200 km (Model 62-3), a viscosity of 2.9×10^{20} poises would be required.

It is important to note that the slope of this curve for the soft layer over the viscous half-space provides an upper limit to the rate at which the relaxation time can increase with increasing wavelength for any layered viscous halfspace. Although this model in its present form is not consistent with the shoreline data, it is shown elsewhere [*McConnell*, 1968] that it may be used as evidence against tidal fric-

tion as a possible cause of the earth's nonhydrostatic bulge.

By removing the restriction that the lower material is rigid (Models 62-4, 62-5, 62-6, and 62-7, Figure 10), it is easy to improve the quality of the fit in the neighborhood of the bend in the curve. The theoretical times at the long-wavelength, low wave number end now become subject to the same objections as those of Model 62-1.

If the same criteria are used to match each of the four curves, they form a series in which each model appears to represent the experimental data equally well. A comparison of the four interpretations shows that the viscosity of the lowest layer is not very sensitive to the model chosen and that all yield viscosity estimates between 7.8×10^{21} and 8.5×10^{21} poises. It appears then that there is such a sufficiently good separation between the effects of the upper and lower layers that any estimates for the viscosity of the lower layers (based on long wavelength deformations) are unlikely to be modified to any great extent by the introduction of new short wavelength data. On the other hand, the depth and viscosity of the upper layer are not uniquely determined at all, but rather they are approximately related by the expression $\eta \simeq 0.13h^3$, $h < 10^7$ cm. Although all these models yield large wave number relaxation times that are consistent with the shoreline data, they require a low-viscosity layer extending to the surface. On the other hand, the geological, heat flow, and seismic evidence all suggest that in stable continental areas the closest approach to melting occurs somewhere in the upper mantle. Thus it is appropriate to see whether we can construct a consistent model by using only viscous layers with the lowest viscosity somewhere below the surface.

Models 62-8 and 62-9. Models 62-8 and 62-9 (Figure 11) represent attempts to fit the shoreline data with a layer of low viscosity sandwiched between a half-space and a capping layer, each with much higher viscosity. Model 62-8 assumes that the depth to the bottom of the second layer is four times the depth to the bottom of the first layer and that the viscosities are in the ratios 100 : 1 : 100. The interpretation suggests a low viscosity layer in the upper mantle, but its lower limit at 118 km would seem to be too near the surface to correspond to any known seismological boundary appropriate to a stable continental crust. Still, this is some improvement over the single layer models.

Besides the poor agreement at long wavelengths that has been apparent in all the earlier examples, the most important difficulty is that all the rheological boundaries have been much too shallow to agree with the known structure. Model 62-9 represents an attempt to improve

the fit by adding a thicker viscous layer above the soft material. Although the attempt does not result in much improvement over Model 62-8, several interesting effects do appear. The lowest viscosity, 8.6–10^{21} poises, agrees well with all the previous one- and two-layer models. The second layer has a thickness of 40 km and a viscosity of 8.6 $\times$ 10^{18} poises. These values are almost identical with those for the upper layer of Model 62-5 (which was a single-layer example), thus indicating that the portion of the curve being fitted near the bend depends mainly on the viscosity of the underlying half-space and the thickness and viscosity of the layer immediately above it. The position of the large wave-number branch would seem to be determined almost entirely by the properties of the upper layer.

Since it is unrealistic to assume that the uppermost part of the crust is less viscous than the mantle proper, some other way must be found to increase the depth at which the flow is taking place.

Models 62-10 and 62-11 (Figure 12). It should now be obvious that there is little hope of obtaining a satisfactory fit to the shoreline spectra by constructing a model that consists entirely of viscous layers whose boundaries correspond to seismological discontinuities.

The one effect that has been neglected to this point is the ability of the material to deform elastically. It has been shown [*McConnell,*

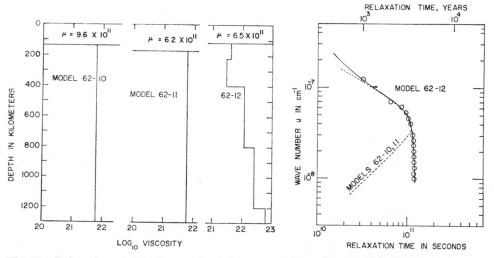

Fig. 12. Relaxation spectra for an elastic layer overlying a layered viscous half-space compared with Fennoscandian shoreline data. Open circles: shoreline data.

1963, 1965] that an elastic layer of appropriate dimensions will decrease the relaxation time at short wavelengths in much the same way that a low-viscosity layer will.

It is an interesting characteristic of this elastic layer that the thickness over which it is effective in decreasing the relaxation time with increasing wave number is closely centered around the value of $uh = 1$ and requires that $\tilde{\mu} \equiv \mu/pgh > 10$. Thus, even the crudest comparison of the observations with the theoretical curves [*McConnell*, 1965] indicates that the effective thickness h for the elastic layer must be about 10^7 cm and thus μ somewhat greater than 3×10^{11} dynes/cm², i.e., comparable with crustal rocks.

Models 62-10 and 62-11 represent attempts to interpret the short wavelength portions of the data in terms of a single elastic layer with Poisson's ratio 0.25 overlying a viscous half-space. Both curves show good agreement over the upper portions but have the usual failing that the predicted relaxation times are too short at the long wavelengths.

Model 62-10 consists of an elastic layer 140 km thick with a rigidity of 9.6×10^{11} dynes cm⁻² overlying material with a viscosity of 6.0×10^{21} poises.

Model 62-11 is similar, but the elastic layer is 175 km thick and its rigidity 6.2×10^{11} dynes cm⁻². There is no change in the viscosity.

Since the average rigidity of the crust and upper mantle determined seismically is of the order of 6.7×10^{11} dynes cm⁻², the similarity of the values calculated from the relaxation time spectra and those determined seismologically is striking. This observation suggests that if the decrease of relaxation time at large wave numbers is caused only by the elastic properties of the crust and upper mantle under Fennoscandia, these regions must have been able to maintain elastic stresses of the order of 10^8 dynes cm⁻² without creep for the entire duration of the last glaciation. The question of whether it is the elastic effect alone that is responsible for this characteristic of the spectrum will be discussed in more detail.

Model 62-12. By drawing on the experience gained from the earlier attempts at curve matching, a model was constructed whose boundaries coincide roughly with the depth at which seismic velocities are known to change and whose

relaxation times fit the shoreline data over the entire range of wavelengths. Model 62-12, which fulfills these specifications, has the following features:

1. An elastic layer 120 km thick and with a rigidity of 6.5×10^{11} dynes cm⁻²;
2. Viscosities that decrease from 4.1×10^{21} poises just below the elastic layer to 2.7×10^{21} poises between 220 and 400 km and then rapidly increase to more than 68.5×10^{21} poises below 1200 km.

It is of considerable interest that this model (except for the assumption of rigidity) has its lowest viscosity in the neighborhood of, but somewhat below, the low shear velocity zone of the upper mantle. The discrepancy might be due to observational error, to effects of a phase change in the lower region, or to a lack of uniqueness in the model. Accordingly, an attempt was made to test the last possibility by altering the model in such a way that the two zones would coincide. Although the attempt was not successful, it will be shown below that a similar discrepancy does not appear for the relaxation time spectra using the long wavelength data from the equatorial bulge.

Viscosity Distribution Assuming Equatorial Bulge Caused by Pleistocene Glaciation

In a companion paper [*McConnell*, 1968], it was concluded that although a relaxation time of 10^7 years (3×10^{14} sec) for P_2^0 deformations as interpreted by *Munk and MacDonald* [1960] and *MacDonald* [1963] was not consistent with relaxation times from Fennoscandia, there was no such problem with times between 3×10^{11} and 3×10^{13} sec corresponding to flattening of the earth by Pleistocene ice sheets. In that paper we discussed a number of models that might provide relaxation times in the appropriate range and concluded that all required a much steeper viscosity gradient below 900 km than that suggested by Model 62-12. We now consider what models might be most consistent with a relaxation time of $\simeq 10^{13}$ sec, that is indicated by the data on recent secular acceleration of the earth.

When we try to predict the long wavelength relaxation times to fit the data from the equatorial bulge, it is clear that the planar layered half-space model is not entirely appropriate.

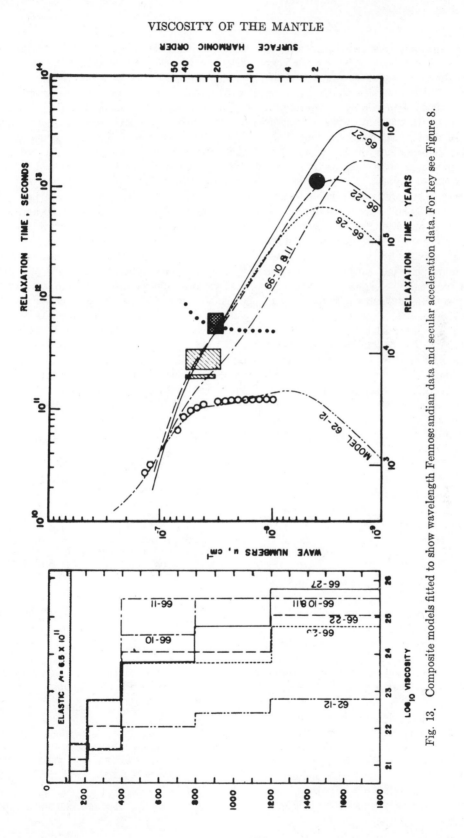

Fig. 13. Composite models fitted to show wavelength Fennoscandian data and secular acceleration data. For key see Figure 8.

However, it can easily be shown that relaxation times for a spherical earth will not differ very much from those estimated by using the half-space values and (7). For example, by using *McKenzie's* [1966] expression for the uniform spherical earth with constant values of η/pg as a function of depth, the relaxation time of the second-order deformations would be only 1.83 times that predicted from this 'earth stretching' approximation. Since this is comparable with the errors in the relaxation time estimates at long wavelengths and since we are interested only in order-of-magnitude effects, neglect of the earth's curvature would seem to be justified. If, as is indicated by the calculations, most of the flow takes place in a narrow upper mantle layer, then the neglect of sphericity should introduce even less difficulty [*Anderson and O'Connell*, 1966].

Figure 13 compares Model 62-12 with five models (66-10, 66-11, 66-22, 66-26, and 66-27) described in *McConnell* [1968] that fit short wavelength data from Fennoscandia and predict long wavelength relaxation times of the order of 10^{13} sec. It is clear that all the latter models predict a much steeper viscosity gradient in the upper mantle than the original interpretation based on Fennoscandian beach levels alone. This result suggests that the assumption of no deformation beyond the Leningrad region was not justified, and hence the long wavelength portion of the spectra derived from the Fennoscandian data must be disregarded.

CONCLUSIONS

Individually, the relaxation times would not be sufficiently reliable to warrant any detailed interpretation, but when taken together, the calculations from the Fennoscandian beaches, gravity anomalies, and present Fennoscandian upwarping would seem to indicate a consistent pattern of a very rapid increase in relaxation times at long wavelengths under surface loads.

The most reasonable interpretation for the very long wavelength data indicates that most of the observed nonhydrostatic component of the equatorial bulge and the historical increase in the length of the day also may be explained by isostatic adjustment after the melting of Pleistocene glaciers. When the relaxation time for the P_2^0 deformation of about 10^{13} sec is com-

pared with the short wavelength Fennoscandian data, all the points would seem to be very close to the limiting curve for a low-viscosity upper mantle over a high-viscosity lower mantle.

When interpreted in terms of elastic and viscous models, the data would suggest a rigid upper mantle and crust down to about 120 km that can withstand stresses of the order of 3×10^8 dynes/cm^2 over periods of 10^5 years or more. Below this, material deforms by creep with viscosities of the order of 10^{21} poises in a zone between 100 and 300 km deep and with viscosities as high as 10^{25} poises below 800 km.

The strong-viscosity contrast between the upper and lower mantles would seem to have a number of implications. In the first place, the high-viscosity lower mantle seems to make thermal convection in this region highly unlikely. Secondly, the low-viscosity layer would seem to be so well developed that it will be not only the locus of most of the material movement but also will be such an effective decoupling zone that one must look for the immediate causes of such large scale crustal motions as continental drift in the upper mantle and crust rather than in the more traditional mantle-wide convection currents.

Acknowledgments. I wish to thank Dr. Dan P. McKenzie for calling my attention to the unexplained apparent *decrease* in the length of the day upon which the relaxation time calculation for the P_2^0 component is based and Professors F. S. Grant, C. Barnes, R. M. Farquhar, J. T. Wilson, and D. York of the University of Toronto for their advice and encouragement. The Finnish Academy of Science and Letters kindly gave their permission to reproduce Figures 1 and 2 from *Annales Academiae Scientiarum Fennicae.*

REFERENCES

Anderson, D. L., and R. O'Connell, Viscosity of the earth. *Geophys. J., 14,* 287, 1967.
Bergsten, F., Changes of level on the coasts of Sweden, *Geograf. Ann.,* 21–55, 1930.
Crittendon, M. D., Jr., Effective viscosity of the earth derived from isostatic loading of Pleistocene Lake Bonneville, *J. Geophys. Res., 68,* 5517–5530, 1963.
Flint, R. F., *Glacial and Pleistocene Geology,* John Wiley, New York, 1957.
Fotheringham. J., Secular accelerations of sun and moon as determined from ancient lunar and solar eclipses, occultations, and equinox observations, *Monthly Notices, Roy. Astron. Soc., 80,* 578, 1920.
Gordon, R. B., Diffusion creep in the earth's mantle, *J. Geophys. Res., 70,* 2413–2418, 1965.

Gutenberg, B., Changes in sea level, postglacial uplift, and mobility of the earth's interior, *Bull. Geol. Soc. Am., 52*, 721-772, 1941.

Haskell, N. A., The viscosity of the asthenosphere, *Am. J. Sci., 33*, 22-28, 1937.

Heiskanen, W. A., and F. A. Vening Meinesz, *The Earth and Its Gravity Field*, McGraw-Hill, New York, 1958.

Jeffreys, H., On the hydrostatic theory of the figure of the earth, *Geophys. J., 8*, 196-202, 1963.

Kääriäinen, E., Beitrage zur Landhebung in Finnland, *Veroeff Finn. Geod. Inst., 36*, 91-94, 1949.

Kääriäinen, E., On the recent uplift of the earth's crust in Finland, *Publ.-Finn Geod. Inst., 42*, 1953.

Kaula, W. M., Tesseral harmonics of the earth's gravitational field from camera tracking of satellites, *J. Geophys. Res., 71*, 4377-4388, 1966.

MacDonald, G. J. F., The deep structure of oceans and continents, *Rev. Geophys., 1*, 587-665, 1963.

MacDonald, G. J. F., The figure and long term mechanical properties of the earth, in *Advances in Earth Science*, edited by P. M. Hurley, M.I.T. Press, Cambridge, 1966.

McConnell, R. K., Jr., The viscoelastic response of a layered earth to the removal of the Fennoscandian ice sheet, Ph.D. thesis, University of Toronto, 1963.

McConnell, R. K., Jr., Isostatic adjustment in a layered earth, *J. Geophys. Res., 70*, 5171-5188, 1965.

McConnell, R. K., Jr., Viscosity of the earth's mantle, in *Proc. Conf. History Earth's Crust, 1966*, edited by R. A. Phinney, Princeton University Press, Princeton, N. J., in press, 1968.

McKenzie, D. P., The viscosity of the lower mantle, *J. Geophys. Res., 71*, 3995-4010, 1966.

Munk, W. H., and G. J. F. MacDonald, *The Rotation of the Earth*, Cambridge University Press, New York, 1960.

Niskanen, E., On the upheaval of land in Fennoscandia, *Ann. Acad. Sci. Fennicae, A, 53*(10), 1-30, 1939.

Orowan, E., Convection in a non-Newtonian mantle continental drift, and mountain building, *Phil. Trans., Roy. Soc. London, A, 258*, 284-313, 1965.

Sauramo, Matti, Land uplift with hinge-lines in Fennoscandia, *Ann. Acad. Sci. Fennicae, A*(3), 44, 1955.

Sauramo, Matti, Die Geschichte der Ostsee, *Ann. Acad. Sci. Fennicae, A*, 1958.

Sirén, A., On computing the land uplift from the lake water level records in Finland, *Hydrograf. Tomiston Tiedonantoja 14*, 1-182, 1951.

Takeuchi, H., and Y. Hasegawa, Viscosity distribution within the earth, *Geophys. J., 9*, 503-508, 1965.

Wang, C. Y., Earth's zonal deformations, *J. Geophys. Res., 71*, 1713-1720, 1966.

Woolard, G. P., The land of the Antarctic, *Sci. Am. 207*(3), 151-166, 1962.

(Received September 1, 1967; revised June 10, 1968.)

Editor's Comments on Paper 26

Broecker: *Glacial Rebound and the Deformation of the Shorelines of Proglacial Lakes*

The final three papers are concerned with developing geophysical models of crustal response to glacial unloading. Interestingly, the authors in Papers 26, 27, and 28 all use as an index of the "fit" of their models to reality, the well-developed and well-studied shoreline of Glacial Lake Algonquin. The fact that all three papers result in close approximations between their models and the observed shorelines indicates some of the conceptual problems in model building, namely that the ability to obtain a close fit from the field data by a specific earth model need not necessarily mean that the model is correct. All three papers (26, 27, 28) have been used in other areas with reasonable results.

Broecker makes four major assumptions, and specifies them: the rate of isostatic recovery is proportional to the size of the isostatic deflection; that the strength of the crust is limited so that very little of the ice load is. supported by the elastic strength of the lithosphere; and that the ice during retreat was dynamic and that isostatic equilibrium had been achieved prior to the retreat phase. The amount of residual downwarp during ice retreat is thus a function of the rate of ice retreat, the half-life of the crustal relaxation, ice thickness, mantle density, and the shape of the ice cap. Broecker uses a $t_{1/2}$ of 700 years, based on the Greenland uplift curve. One intriguing result of this analysis is that Broecker (p. 4781) suggests that the amount of uplift and gradient of the shorelines (or half-distance, $s_{1/2}$) is a function of the rate of glacial retreat. Thus, as $s_{1/2}$ can be calculated, the rate of retreat can be estimated. Broecker's model differs from McConnell's preferred model in that a very thin elastic crust (ca. 20 km?) is indicated rather than one of 120 km.

Wallace S. Broecker is Professor of Geochemistry at Lamont-Doherty Geological Observatory of Columbia University. He has published a great many papers in the areas of geophysics and geochemistry.

Reprinted from *J. Geophys. Res.*, **71**, 4777–4783 (1966)

Glacial Rebound and the Deformation of the Shorelines of Proglacial Lakes[1]

WALLACE S. BROECKER

Lamont Geological Observatory, Columbia University
Palisades, New York

A simple isostatic model explaining the pattern of deformation of the shorelines of proglacial lakes has been developed. The rate of glacial retreat before the formation of the shoreline can be derived from the curvature of its uplifted portion. The rate calculated in this way for the retreat preceding the formation of Lake Algonquin is 120 km/10^3 yr, a value not in conflict with the radiocarbon chronology for this interval. The agreement between the uplift predicted at the iceward extreme of the shoreline (260 meters) and the actual maximum uplift (250 $\pm$ 50 meters) provides an independent check on the validity of the model. If the model proves to be correct, the implications are as follows. (1) The continental ice sheets had shapes and total thicknesses during their retreat phases not dissimilar to those observed for present-day ice masses on Greenland and Antarctica, i.e., dynamic equilibrium was maintained; (2) rebound at the edge of large continental ice sheets is a simple isostatic process occurring with the Washburn-Stuiver time constant of about 700 years; and (3) the strength of the crust is sufficiently small to prevent the lateral influence of a continental ice sheet from extending more than a few tens of kilometers beyond its margins.

INTRODUCTION

Glacial lake shorelines show a characteristic pattern of deformation. Their iceward ends have been uplifted by amounts ranging up to several hundred meters. The magnitude of the uplift for any given shoreline decreases away from the former ice front, the shoreline becoming nearly horizontal within no more than a few hundred kilometers (see, for example, *Leverett and Taylor* [1915]). The shoreline of Lake Algonquin, which occupied the basins of present-day Lakes Michigan and Huron for a brief period during the retreat of the late Wisconsin ice sheet, provides a classic example of this phenomenon. The boundaries of this lake and its relationship to the ice front as given by *Hough* [1958], are shown in Figure 1. As shown in Figure 2, the elevation, h, of the deformed Algonquin shoreline as a function of distance, s, from the ice front can be approximated by the relationship

$$h = h_0 + z_D e^{-0.693 s / s_{1/2}}$$

where h_0 is the elevation of the horizontal southern part, z_D the extent of uplift at the

---

[1] Lamont Geological Observatory Contribution 943.

ice margin, and $s_{1/2}$ the distance from the ice margin at which the uplift is $\tfrac{1}{2} z_D$.

THE MODEL

Although shoreline upwarp is generally attributed to isostatic rebound resulting from the removal of the ice load, there is no broad agreement regarding details of this process [*Flint*, 1957]. In this paper a simple model is presented which quantitatively explains the deformation and relates it to the rate of glacial retreat. Four major assumptions are made: (1) the rate of isostatic recovery is proportional to the magnitude of the isostatic anomaly, (2) the strength of the crust is sufficiently small for isostatic anomalies to have negligible lateral influence, (3) the profile of the ice sheet during periods of retreat was the same as that observed for existing ice sheets, and (4) before the ice retreated, isostatic equilibrium was achieved under the ice. As shown in the appendix, these assumptions lead to the following predictions regarding the amount of residual land depression in front of a retreating glacier:

1. The magnitude of the residual downwarp, z_D, at the time of deglaciation will depend on the retreat rate, c, of the ice and the shape of the ice (i.e., the interior thickness, a, and the

4777

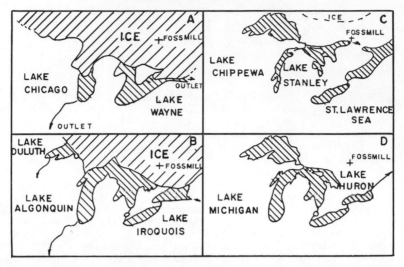

Fig. 1. Configuration [after *Hough,* 1958] of the boundaries of proglacial lakes at the Valder ice maximum (*A*), at the time of the main Lake Alogonquin stage (*B*), and after the retreat beyond the Fossmill outlet (*C*). The present-day configuration of the Great Lakes is given for comparison (*D*).

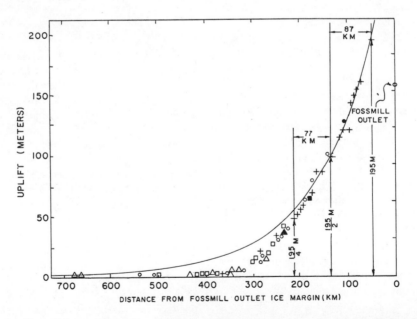

Fig. 2. Composite profile of the shoreline of Lake Algonquin as measured on the east side of Lake Huron [*Chapman,* 1954, designated by +], on the west side of Lake Huron [*Leverett and Taylor,* 1915, designated by □], on the east side of Lake Michigan [*Leverett and Taylor,* 1915, designated by ○], and on the west side of Lake Michigan [*Leverett and Taylor,* 1915, designated by △]. The distances are all referenced to Chapman's profile using the Fossmill outlet (and hence the northernmost possible position of the ice at Algonquin time) as a reference point. As the ice front positions for the other profiles are not well known, the highest point on each (filled symbol) has been fitted to the Chapman curve (the curve defined by +s). The solid curve is the theoretical equation for $s_{1/2}$ equal 84 km.

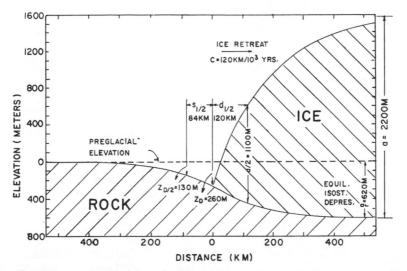

Fig. 3. Profile of the depressed land surface under and adjacent to an ice sheet retreating at the rate of 120 km per thousand years. The parameters shown are defined in the text. The vertical exaggeration is 300-fold.

distance, $d_{1/2}$, from the margin at which the ice thickness is $a/2$). The relationship is

$$z_D = \frac{a}{r} \frac{ct_{1/2}}{d_{1/2} + ct_{1/2}}$$

where $t_{1/2}$ is the time constant for rebound (and hence the time required for one-half the uncompensated land depression to disappear) and r the ratio of the density of the displaced mantle rock to that of glacial ice.

2. The profile of downwarp away from the ice front will be

$$z = z_D e^{-0.693s/s_{1/2}}$$

where $s_{1/2}$ is the distance the ice front retreats in one half-response time (hence $ct_{1/2}$). Figure 3 is a graphical presentation of these relationships.

Within a few thousand years after formation, rebound will have been completed and the originally horizontal shoreline will have assumed the profile

$$h = h_0 + z_D e^{-0.693s/s_{1/2}}$$

Hence the model generates a shoreline with the observed form. Details of these derivations are given in the appendix.

APPLICATION TO LAKE ALGONQUIN

Two independent checks on the validity of the model are available in the case of the Algonquin shoreline. First, the predicted maximum uplift, z_D, can be compared with that observed, and, second, the computed retreat rate can be compared with that based on radiocarbon data. In order to do this, we must assign values to $t_{1/2}$, $d_{1/2}$, a, and r. *Washburn and Stuiver* [1962] have shown by radiocarbon dating of a sequence of raised beaches that the half-response time for rebound in eastern Greenland was about 700 years. (see Figure 4). Profiles of the Antarctic and Greenland ice sheets yield values averaging 120 km for $d_{1/2}$. The measured thicknesses of the antarctic and Greenland ice masses (and hence a) average 2200 m [*Holtscherer and Robin*, 1954; *Woollard*, 1958; *Crary*, 1960; *Novikov*, 1960]. Since the earth's upper mantle has a density 3.5 g/cm³, r must be close to 3.5. Using the value of 84 km measured for the Algonquin shoreline for $s_{1/2}$ (see Figure 2), we predict z_D and c to be 260 m and 120 km per 10^3 years, respectively.

As shown in Figure 2 the maximum observed uplift for the Algonquin shoreline is 195 m. Geologic evidence (see Figure 1) suggests that the ice front could have been no more than 50 km north of this point. Extrapolation over this distance based on the best-fit equation for the profile yields an upper limit of 300 m for z_D. The best estimate of the actual z_D is then

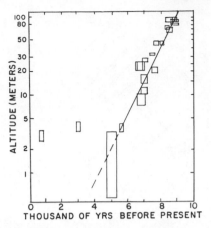

Fig. 4. Radiocarbon age of marine deposits in the northeast of Greenland as a function of elevation (corrected for eustatic rise of sea level) [*Washburn and Stuiver*, 1962].

250 ± 50 m. This compares favorably with the prediction of 260 m.

The glacial retreat preceding the formation of Lake Algonquin began shortly after the culmination of the Valders ice advance (Figure 1). The Two Creeks forest destroyed by this advance grew 11,800 years ago [*Broecker and Farrand*, 1963]. Although the age of Lake Algonquin itself is as yet uncertain, the best evidence suggests that the Fossmill outlet opened before 9700 years ago—probably between 10,000 and 11,000 years ago [*Terasmae and Hughes*, 1960]. Hence a period of 1200 ± 600 years was available for the ice retreat from the Valders maximum to its position north of the Great Lakes during Lake Algonquin time. Although the lobate form of ice makes establishment of the retreat distance difficult, the range 240 ± 100 km should include the actual distance. The retreat rate corresponding to these estimates is 200 ± 130 km/10^3 years. Hence, within these broad limits, the rate of 120 km/10^3 years predicted by the model is satisfactory.

SENSITIVITY TO THE ASSUMPTIONS

The sensitivity of these results to the four basic assumptions of the model must be considered. The results of *Washburn and Stuiver* [1962] clearly establish the validity of the first assumption for one area in Greenland, but the fact that Hudson Bay and central Scandinavia

are still rising suggests either that the assumption is not everywhere applicable or that the time constant is considerably longer in some places than others. The major criticism of this assumption would then be that the 700-year value for $t_{1/2}$ is too small.

If the lateral strength of the crust is not negligible, the $s_{1/2}$ distance measured for a given shoreline is not a measure of $ct_{1/2}$. In other words, part (or even all) of the depression away from the ice front could reflect the lateral influence of the residual ice load. Thus, if the second assumption is not valid, $ct_{1/2}$ must be less than the $s_{1/2}$ measured for a given shoreline.

The assumption is often made that during periods of retreat the continental ice sheets stagnated and lost their equilibrium profiles. If so, the ice thickness, a, selected here is too large and that of $d_{1/2}$ too small.

If isostatic adjustment were not achieved under the fully expanded ice sheets, the maximum land depression would be less than $a/3.5$.

The most likely errors in the assumptions would be in such a direction as to make $t_{1/2}$ too small, $ct_{1/2}$ too large, $d_{1/2}$ too small, and a too large. The effect on z_D and c of adjustments in accord with these potential biases are summarized in Table 1. In all cases any changes in z_D or c are toward lower values. As reduction of z_D by more than 30% or of c by more than a factor of 2 would violate observation, for the

TABLE 1. Sensitivity of the Model to the Choice of Basic Parameters

Assigned Parameter Values				Computed Values*	
$t_{1/2}$, yr	a, m	$d_{1/2}$, km	$ct_{1/2}/s_{1/2}$	z_D, m	c, km/10^3 yr
700	2200	120	1	260	120
1400	2200	120	1	260	60
350	2200	120	1	260	240
700	4400	120	1	520	120
700	1100	120	1	130	120
700	2200	240	1	165	120
700	2200	60	1	360	120
700	2200	120	0.5	160	60

* Observed value for z_D is 250 ± 50 m, and for c it is 200 ± 130 km/10^3 yr.

northern part of Lake Algonquin $t_{1/2}$ could have been no greater than 1400 years, the ratio $ct_{1/2}/s_{1/2}$ no less than 0.6, $d_{1/2}$ no more than 200 km, and a no less than 1500 m.

Implications to Pleistocene Geology

If this model is valid, the profiles of strandlines taken perpendicular to the ice front should be a family of curves similar to those shown in Figure 5. The shape should reflect primarily the rate of ice retreat. Rapid retreat (>100 km/10^3 yr) leads to large total uplifts (hundreds of meters) and to large values of $s_{1/2}$ (>60 km); slow retreat (<30 km/10^3 yr) leads to small total uplifts (<100 m) and to small values of $s_{1/2}$ (<25 km). Variations in the retreat rate would cause perturbations from the theoretical curve. For example, the deviation of the profile of the Algonquin shoreline from the theoretical curve suggests that the retreat rate was more rapid during the later than the earlier stages of retreat. If these predictions are correct, this system should be useful to the field geologist in his attempt to understand the pattern of retreat of the ice sheets at the close of the last glacial period.

Geophysical Implications

If the Washburn-Stuiver time constant is universally applicable, mantle viscosities must be lower than those determined by *Daly* [1934] and *Haskell* [1935] for the Scandinavian data. The present-day uplift of Scandinavia and the Hudson Bay region [see, for example, *Farrand*, 1962] may not be isostatic in origin but rather

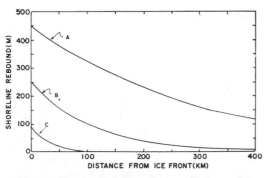

Fig. 5. Theoretical shoreline profiles as a function of the rate of glacial retreat. Curve A represents a rate of retreat of 300 km/10^3 yr; B, 100 km/10^3 yr; C, 30 km/10^3 yr.

may reflect some other process such as phase changes in the upper mantle [see *Broecker*, 1962]. The absence of any warping of the southern portions of the Algonquin shoreline suggests that the peripheral bulges often invoked by geologists [see *Flint*, 1957; *Daly*, 1934] are not universally present. The thickness of the layer in the mantle undergoing lateral transport must be sufficiently large to make depletions adjacent to the ice masses during periods of retreat insignificant. The model also places important limits on the strength of the crust.

Implications Regarding Theories of Changing Climates

Emiliani and Geiss [1957] proposed a theory of glaciation which requires the ratio of surface area to volume for ice sheets to increase considerably during periods of major ice retreat. The model presented here indicates that no such change took place. The ice must have retained its equilibrium shape.

Appendix
Derivation of Equation Predicting Shape of Shorelines Warped by Glacial Rebound

If z is the depression of the land surface, y the thickness of the overlying ice, and r the ratio density of the isostatically displaced rock to that of glacial ice,

$$\frac{dz}{dt} = -\frac{0.693}{t_{1/2}}\left(z - \frac{y}{r}\right)$$

where $z - y/r$ is the isostatically uncompensated depression of the land surface and $t_{1/2}$ the half-time for response.

The approximate shape of the ice is given by

$$y = a\left[1 - \exp\left(-\frac{0.693}{d_{1/2}}x\right)\right]$$

where x is the distance from the ice front, a the average ice thickness for the interior of the mass, and $d_{1/2}$ the distance from the ice front where half this interior thickness is found.

If the ice retreats at a uniform rate, c,

$$x = x_0 - ct$$

where x_0 is the initial distance from the ice front and t the time elapsed since the retreat began.

Hence

$$\frac{dz}{dt} = \frac{0.693}{t_{1/2}}$$

$$\cdot \left\{ z - a \left[1 - \exp\left(-\frac{0.693}{d_{1/2}} (x_0 - ct) \right) \right] \right\}$$

The solution for the initial condition that isostatic equilibrium existed at $t = 0$ (and hence $z = y/r$ for all x) is

$$z = \frac{a}{r} \left\{ \left(1 - \frac{d_{1/2}}{d_{1/2} + ct_{1/2}} \right) \right.$$

$$\cdot \exp\left[-\frac{0.693}{d_{1/2}} (x_0 - ct) \right]$$

$$- \left(1 - \frac{d_{1/2}}{d_{1/2} + ct_{1/2}} \right)$$

$$\left. \cdot \exp\left[-0.693\left(\frac{x_0}{d_{1/2}} + \frac{t}{t_{1/2}} \right) \right] \right\}$$

at deglaciation (and hence when $x_0 = ct$) the residual land depression will be

$$z_D = \frac{a}{r} \left(\frac{ct_{1/2}}{d_{1/2} + ct_{1/2}} \right)$$

$$\cdot \left\{ 1 - \exp\left[-0.693\left(\frac{x_0}{d_{1/2}} + \frac{x_0}{ct_{1/2}} \right) \right] \right\}$$

For $x_0 > 4d_{1/2}$ or $x_0 > 4ct_{1/2}$,

$$z_D = \frac{a}{r} \left(\frac{ct_{1/2}}{d_{1/2} + ct_{1/2}} \right)$$

where a/r is the equilibrium isostatic depression under the interior of the ice mass and $t_{1/2}$ and $d_{1/2}/c$ are, respectively, the time constants for compensation and for mass removal.

The profile of land depression away from the ice front will be

$$z = z_D \exp\left(-\frac{0.693s}{ct_{1/2}} \right)$$

where s is distance measured from the ice front. As $ct_{1/2}$ is the distance, $s_{1/2}$, the ice moves in the half-time for response,

$$z = z_D \exp\left(-\frac{0.693s}{s_{1/2}} \right)$$

Since z is the magnitude of the rebound of the shoreline, its profile after isostatic equilibrium has been reestablished will be

$$h = h_0 + z_D \exp -\frac{0.693s}{s_{1/2}}$$

where h is the elevation of the shoreline as a function of distance, s, from the ice margin at the time the shoreline formed and h_0 is the elevation of the shoreline at the time of formation (and hence of the present-day shoreline elevation at distances from the ice margin great enough for rebound to be complete before the shoreline was cut (and hence for $s > 4s_{1/2}$)).

Since

$$z_D = \frac{a}{r} \frac{ct_{1/2}}{d_{1/2} + ct_{1/2}}$$

and

$$s_{1/2} = ct_{1/2}$$

the two parameters, z_D and $s_{1/2}$, defining the warp of the shoreline are related as follows:

$$z_D = \frac{a}{r} \frac{s_{1/2}}{d_{1/2} + s_{1/2}}$$

The recessional velocity, c, of the ice is given by

$$c = \frac{s_{1/2}}{t_{1/2}}$$

Assigning values of 700 years to the rebound half-response time, $t_{1/2}$ (based on C^{14} dating of the Greenland uplift), of 2.2 km to the average interior ice thickness, a, (based on present-day Greenland and antarctic ice sheets), and of 120 km to distance from the ice margin to that point where it achieves one-half its interior thickness, $d_{1/2}$ (based on the present-day antarctic ice sheet), the following relationships are obtained:

$$z_D = 0.7 \frac{s_{1/2}}{120 + s_{1/2}} \text{ km}$$

and

$$c = \frac{s_{1/2}}{700} \text{ km/yr}$$

Acknowledgments. Discussions with Johannes Geiss, Barclay Kamb, Arthur Bloom, and G. J. Wasserburg helped to crystallize some of the ideas presented here.

Support of the Sloan Foundation is also gratefully acknowledged.

References

Broecker, W. S., The contribution of pressure-induced phase changes to glacial rebound, *J. Geophys. Res., 67,* 4837–4842, 1962.

Broecker, W. S., and W. R. Farrand, Radiocarbon age of the Two Creeks forest bed, Wisconsin, *Bull. Geol. Soc. Am., 74,* 795–801, 1963.

Chapman, L. G., An outlet of Lake Algonquin at Fossmill, Ontario, *Proc. Geol. Assoc. Can., 6,* 61–68, 1954.

Crary, A., Status of United States scientific programs in the Antarctic, *IGY Bull. 39, Trans. Am. Geophys. Union, 41,* 521–532, 1960.

Daly, R. A., *The Changing World of the Ice Age,* Yale University Press, 1934.

Emiliani, C., and J. Geiss, On glaciations and their causes, *Geol. Rundschau, 46,* 576–601, 1957.

Farrand, W. R., Postglacial uplift in North America, *Am. J. Sci., 260,* 181–199, 1962.

Flint, R. F., *Glacial and Pleistocene Geology,* John Wiley & Sons, New York, 1957.

Haskell, N. A., The motion of a viscous fluid under a surface load, *Physics, 6,* 265–269, 1935.

Holtscherer, J., and G. de Q. Robin, Depth of polar ice caps, *Geograph. J., 120,* 193–202, 1954.

Hough, J. L., *Geology of the Great Lakes,* University of Illinois Press, 1958.

Leverett, F., and F. B. Taylor, The Pleistocene of Indiana and Michigan and history of the Great Lakes, *U. S. Geol. Surv. Monograph 53,* 1915.

Novikov, V., The study of the Antarctic is continuing, *Priroda,* no. 8, 43–52, 1960.

Terasmae, J., and O. L. Hughes, Glacial retreat in the North Bay area, Ontario, *Science, 131,* 1444–1446, 1960.

Washburn, A. L., and M. Stuiver, Radiocarbon-dated postglacial deleveling in north-east Greenland and its implications, *Arctic, 15,* 66–73, 1962.

Woollard, G., Preliminary report on thickness of ice in Antarctica, *IGY Bull. 13, Trans. Am. Geophys. Union, 39,* 772–778, 1958.

(Manuscript received January 12, 1966; revised June 27, 1966.)

Editor's Comments on Paper 27

Brotchie and Silvester: *On Crustal Flexure*

Brotchie and Silvester model the earth as a thin elastic shell over an enclosed viscous liquid. The flexural stiffness of the crust is included in the equations. Two primary models are introduced: the first is for a long-term steady-state load when the viscosity of the liquid is of no concern, and the second is for a time-dependent load (ice cap) where the viscosity of the mantle is important.

The load of an ice cap is considered as a series of superimposed disks of thickness h. The "radius of relative stiffness," l, is a measure of the strength of the elastic shell and is a function of the thickness of the shell and the values for the elastic modulus and Poisson's ratio. It is thus a measure of how far local loads will affect the deformation of the shell. The effect of l at the margin of an ice cap is to cause a deflection at the margin and to cause a downwarp away from the edge of the load which decreases in height and merges into a peripheral forebulge. The effect of the parameter l is to increase the deflection at the margin as l increases. If the amount of depression at the ice margin can be determined by field work, then this should allow the length of the flexural parameter to be determined.

The static model is then used as a basis for a dynamic model that has as variables the rate of ice retreat and the viscous constant. Rate of rebound is considered to be driven by the magnitude of the unequilibrated deflection, w, between successive time steps. Figure 8 is the result of one dynamic model; the fit with the Lake Algonquin shoreline is reasonable, although the predicted rebound at the center of the ice sheet is 100 m greater than observed. Note (Table 3) that the model only predicts that 18 m of rebound remains at the center of the ice sheet. The computations of Brotchie and Silvester have been criticized because they assume an elastic shell only 32 km thick (see Paper 28).

John F. Brotchie is head of the Systems Research Section, Division of Building Research, C.S.I.R.O., Melbourne, Australia. He received his D.Eng. from Berkeley in 1961. He is the author of a number of papers on plate and shell mechanics. Richard Silvester is in the Department of Civil Engineering, University of Western Australia, where he has been since 1949. He has published primarily in the field of ocean wave phenomena and is author of *Coastal Engineering* (Elsevier).

Reprinted from *J. Geophys. Res.*, **74**, 5240–5252 (1969)

On Crustal Flexure

27

J. F. Brotchie

Division of Building Research, Commonwealth Scientific and Industrial Research Organization
Melbourne, Australia

R. Silvester

Asian Institute of Technology, Bangkok, Thailand

The deformation of the earth's crust under superposed loads is considered as a problem in structural mechanics. The crust is treated as a uniform, elastic, thin, spherical shell, and the mantle is treated as an enclosed viscous liquid. Equations relating loading and crustal response are developed. Loadings may be spatially concentrated, locally distributed, or global and may be applied as a steady-state or transient condition. Crustal displacements under reservoir loadings, glacial sheets, and sea-level rises are predicted and agree well with observations both in magnitude and in time. The Wisconsin glaciation in North America is treated in particular.

Introduction

The structure of the earth apparently varies continuously and discretely with distance from the surface [Gutenberg, 1959; King, 1962]. The discrete changes define a relatively thin crust enclosing a denser and far deeper mantle and even denser central core (Figure 1a). Crust thickness varies from 30–60 km on the continents down to 8 km or less beneath the oceans. Under rapid rates of loading, including seismic shock, the outer zone of the mantle behaves essentially as a solid, but, under the slower rates of loading from sedimentation and glaciation [Farrand, 1962], it has the response of a dense and viscous liquid. Gravitational evidence, rheology [King, 1962], and the theories and weight of evidence for mantle convection [Girdler, 1963] all support the concept of mantle flow.

In the present paper the crust is considered initially as an isotropically elastic, uniformly thin, spherical shell, and the mantle is considered as an enclosed viscous liquid. Displacements are assumed to be in the linear range. Gravity forces are assumed to act toward the center of the sphere. Loadings considered are concentrated (a volcanic cone or a reservoir), distributed over a limited area (ice, water, sediment, or lava), and global (a sea-level rise).

Two basic models are developed, one for

steady-state conditions in which viscosity of the liquid does not enter and one for time-dependent loads in which the viscosity of the mantle is considered.

The models differ from previous models in that both the flexural stiffness of the crust and its stiffness as a shell are included in the equations.

Steady-State Model

Theory

The effect of a local loading of this model is to produce a localized deformation of the shell and displacement of the liquid beneath. Forces are thereby induced which hold the load in equilibrium in this deformed state. In the region of the deformation these forces are primarily (1) a reaction of the liquid proportional to the normal displacement of the shell at each point, (2) moments and shears in the shell, and (3) in-plane forces acting at its middle surface.

The differential equation for deflection is obtained by considering these three primary equilibrating forces to be acting on an element of the shell. The resulting equation for small displacements in a thin, elastic, liquid-filled, uniform spherical shell [Brotchie, 1969] is

$$D\nabla^4 w + (ET/R^2)w + \gamma w = q \qquad (1)$$

in which w is the radial displacement of the shell under normal loading of intensity q, D is

5240

445

the flexural stiffness of the shell cross section $\equiv$ $[ET^3/12(1 - \nu^2)]$, T is the thickness of the shell, E is its modulus of elasticity, ν is Poisson's ratio for the shell material, R is the radius of its middle surface, γ is the density of the enclosed liquid, and ∇^4 is the biharmonic operator in the surface coordinates of the shell. Where deformations are confined to a shallow zone, ∇^4 may be defined in the coordinates of a plane tangent to an origin in this zone (Figure 1b), e.g.,

$$\nabla^4 \equiv \nabla^2 \cdot \nabla^2 \equiv \left(\frac{\partial^2}{\partial r^2} + \frac{1}{r}\frac{\partial}{\partial r} + \frac{\partial^2}{r^2 \partial \theta^2}\right)^2$$

$$r = R\phi$$

in which r and θ are plane polar coordinates and ϕ and θ are the spherical coordinates of the shell.

The three terms on the left side of (1) have the dimensions of pressure. The first represents the contribution to reaction provided by the stiffness of the shell cross section, acting as a plate. The second term is the effect of shell action, that is, of in-plane forces and of curvature of the shell. Here the shell is acting as a

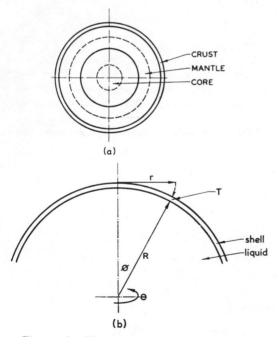

(a)

(b)

Fig. 1. (a) The internal structure of the earth and (b) simplified model proposed showing shell parameters and coordinates.

two-way arch. The third term is the contribution of the liquid enclosed, which in a radial gravity field provides an isostatic reaction of intensity proportional to radial displacement; that is, the reaction of the liquid at any point is γw. Together these three terms equilibrate the loading q.

In solving for w, (1) may be rewritten as

$$\nabla^4 w + (1/l^4)w = q/D \qquad (2)$$

in which $l^4 \equiv D/[(ET/R^2) + \gamma]$. The parameter l has the dimensions of length and is the 'radius of relative stiffness' of the composite model. It is a measure of stiffness of the model. The terms ET/R^2 and γ have the dimensions of density. The effects of curvature and of liquid reaction are thus analogous and complementary. For axisymmetrical loading about the origin of coordinates, the homogeneous solution [Brotchie et al., 1961] of (2) reduces to

$$w_0 = C_1 \text{ ber } x + C_2 \text{ bei } x + C_3$$
$$\text{ker } x + C_4 \text{ kei } x \qquad (3)$$

in which $x = r/l$ and ber, bei, ker, and kei are Bessel-Kelvin functions of zero order. The particular solution for uniform loading is

$$w_P = ql^4/D \qquad (4)$$

and the total deflection is

$$w = w_0 + w_P \qquad (5)$$

The corresponding moments, shears, and in-plane forces may be expressed as derivatives of deflection [Brotchie et al., 1961].

For a concentrated load of magnitude P (Figure 2), $C_1 = C_2 = C_3 = 0 = w_P$ and the equation for deflection reduces to

$$w = (Pl^2/2\pi D) \text{ kei } x \qquad (6)$$

Similarly for a 'hot spot' on the shell resulting in a localized change of curvature,

$$C_1 = C_2 = C_4 = 0 = w_P$$
$$w = C_3 \text{ ker } x \qquad (7)$$

These two loading conditions may be combined to describe other axisymmetrical loadings [Brotchie, 1969].

Applications

The crust parameters [Gutenberg, 1959; King, 1962] are taken as middle-surface radius

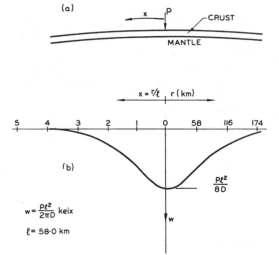

(a)

x → P
CRUST
MANTLE

x = r/ℓ   r (km)

5   4   3   2   1   0   58   116   174

(b)

$w = \dfrac{P\ell^2}{2\pi D}$ keix

$\ell = 58.0$ km

$\dfrac{P\ell^2}{8D}$

w

Fig. 2. Crustal deflection under a concentrated load at equilibrium, e.g. a volcanic cone or reservoir.

$R = 6350$ km, thickness $T = 37$ km, elastic modulus $E = 8.35 \times 10^{10}$ N m^{-2} and Poisson's ratio $\nu = 0.25$. The relative density of the mantle is assumed to be 3.37. Other relevant parameters derived from these are

$R/T = 171.622$.
$ET/R^2 = 76.62$ N m^{-3}.
$\gamma = 3370$ kg m^{-3} $(= 33048.4$ N m$^{-3})$.
$D = 3.7596 \times 10^{23}$ N m.
$l = 58.042$ km ≈ 58 km.

The magnitude of γ is seen to be of greater order than that of ET/R^2, and, if the ET/R^2 term is omitted from the expression for l, then

$l = 58.076$ km. Thus the effect of approximations in the curvature terms are negligible, and the term itself is not of great significance in the equations for l or w.

The sensitivity of the parameter l to changes in thickness and elastic modulus of the crust is indicated in Table 1. The effect of submergence is to reduce the effective density γ of the mantle to $\gamma - \gamma_w$, where γ_w is the density of water, and the effect of this on l is also shown. The last line in Table 1 gives the corresponding parameters assumed for the ocean crust.

Concentrated loading. If a volcanic cone is considered as a concentrated loading (Figure 2), displacements of the crust are given by (6), in which P is the total loading due to the cone. Maximum deflection occurs beneath the cone (at $x = 0$) and is $Pl^2/8D = P/(8l^2 \gamma')$, in which $\gamma' \equiv \gamma + ET/R^2$.

For a cone of height 10,000 meters, radius 40 km at the base, and mean relative density 2.4, the total load on the crust is 4.02×10^{16} kg. Deflection of the continental crust ($l = 58.0$ km) under this load is 442 meters at the center of the load. If the same concentrated load were applied to the oceanic crust ($l = 20$ km), the resultant deflection would be 5200 meters (neglecting load distribution and loss of weight of the cone with immersion or partial immersion). Downward deflections extend over a radius of approximately $4l$, that is, 230 km on the continent and 80 km for the ocean crust. Thus the volume of mantle displaced is (essentially) the same in each case, and its mass is (almost) equal to the load P. The effect of distribution of the load is shown later (Figure 6).

TABLE 1. Effect of Crustal Parameters on Model Stiffness, l

Crust Thickness T, km	Elastic Modulus E, N m^{-2} $\times 10^{10}$	Stiffness D, N m^2 $\times 10^{20}$	Relative Density of Mantle	Radius of Relative Stiffness l, km
37	8.35	3759.58	3.37	58.042
37	4.175	1879.79	3.37	48.822
37	8.35	3759.58	2.37*	63.367
50	8.35	9277.78	3.37	72.733
30	8.35	2004.00	3.37	49.600
8	8.35	38.00	3.37	18.412
8	8.35	38.00	2.37*	20.105

* Immersed value.

Other concentrated loadings such as reservoirs can be handled similarly. Lack of symmetry of the load is not significant when the loaded area is small. When the loaded area is very small, however, shear deformation and normal stress are significant and this case is considered later.

Uniform loading. The deflections of the crust under a circular ice sheet of radius A $(=al)$ and uniform depth h are given by (3), (4), and (5), in which the coefficient $q\,l^4/D$ reduces to $\gamma_{\text{ice}}h/\gamma'$, where $\gamma' = \gamma + ET/R^2$, γ_{ice} is the density of the ice, and γ is the density of the mantle. Within the loaded circle, deflection w_i is given by

$$w_i = \frac{\gamma_{\text{ice}}h}{\gamma'}\,(a\ \text{ker}'\ a\ \text{ber}\ x$$

$$- a\ \text{kei}'\ a\ \text{bei}\ x + 1) \qquad (8)$$

and, outside the ice sheet, deflection w_0 is

$$w_0 = \frac{\gamma_{\text{ice}}h}{\gamma'}\,(a\ \text{ber}'\ a\ \text{ker}\ x$$

$$- a\ \text{bei}'\ a\ \text{kei}\ x) \qquad (9)$$

Deflections are shown in Figure 3.

Uniform sediment or water-level rise may be handled similarly. The zone of significant deformation extends approximately $4l$ beyond the load. Isostatic displacement is approached at the center of the load only when the radius of load is greater than $3l$. For small radii deflection descreases almost linearly with area.

Variable loading. Deflections of the crust for an ice sheet of variable thickness are found by superposition using the uniform thickness solution. The variable thickness may be approximated by a stepped distribution (Figure 4). The sheet may then be considered to be composed of uniform layers of depth h and radius a_n; the step size h may be selected to give the degree of accuracy desired. Summation gives the total deflection.

A computer program for calculating and plotting deflections and stresses for various ice profiles has been developed. The relative density of the ice is taken as 0.9 and the maximum thickness at the ice center is taken as 3000 meters. The results for a parabolic profile are given in Table 2 and Figure 4. The corresponding stresses in the crust are shown in Table 2 and appear to be sufficient to cause surface cracking. A comparison of deflections for various

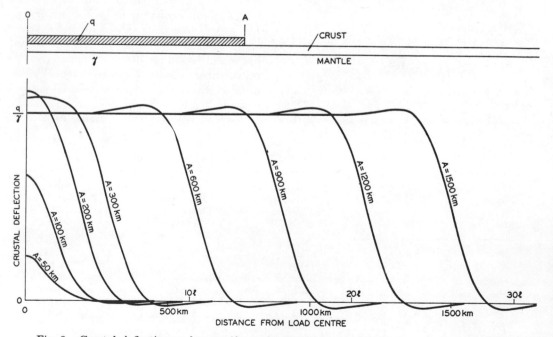

Fig. 3. Crustal deflection under a uniform circular load of radius A, e.g. a uniform ice sheet or water-level rise.

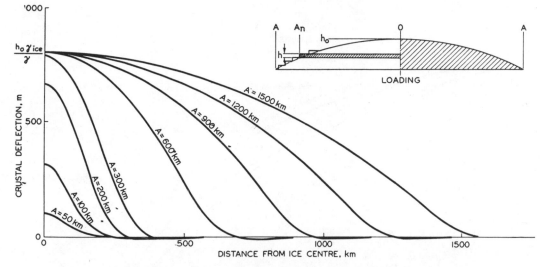

Fig. 4. Crustal deflection under an axisymmetric load of parabolic profile, e.g. an ice sheet of central depth 3000 meters and radius A ($l = 58.0$ km).

profiles of ice, including a theoretical equilibrium profile for a two-dimensional sheet [*Weertman*, 1961], is shown in Figure 5. For a parabolic ice profile of radius 1450 km the predicted downward deflection at the edge of the ice is 22.6 meters and downward displacement extends over 70 km beyond this edge. This is in approximate agreement with data [*Farrand*, 1962; *Broecker*, 1966; *Chapman*, 1954; *Leverett and Taylor*, 1915] obtained in North America from shorelines of glacial lakes formed during the Wisconsin glaciation. These lakes were formed in the zone of downward displacement at the edge of the ice, and the shorelines were uplifted when the ice retreated (Figures 7 and 8). Depth at the edge of the ice is sensitive to ice profile and the results for a parabola appear to fit the observations. This profile also appears to approximate the profiles of the existing ice caps on Greenland and Antarctica

[*Donn et al.*, 1962; *Holtzscherer and Robin*, 1954]. The maximum thickness assumed of 3000 meters is also based on these existing ice caps and on equilibrium [*Weertman*, 1961].

No evidence is available to confirm the predicted deflections beneath the ice. However, the land surface beneath the Greenland ice cap is concave downward [*Holtzscherer and Robin*, 1954], apparently approximating the predicted displacements of Figure 4. The center of the basin is of the order of 1000 meters below the edge and is below sea level. The land beneath the antarctic ice appears to be similarly displaced, but the deflected profile is complicated by the mountainous terrain.

Deflections of the oceanic crust under conically distributed loads of various radii are calculated similarly and are shown in Figure 6.

Unsymmetrical loads. Unsymmetrical loadings may also be handled by superposition.

TABLE 2. Deflections and Stresses in the Crust beneath an Ice Sheet of Parabolic Profile, Central Depth 3000 meters and Radius 1450 km ($l = 58$ km)

Distance from center, km	0	290	580	870	1160	1450	1500	1600	1700	1800
Deflection, meters	801.93	769.12	673.00	512.76	288.06	22.60	3.74	−3.86	−1.07	0.11
Flexural stress, radial, N m^{-2} × 10^{-3}	−1627	−1479	−1503	−1483	−1988	10,300	7883	1231	−402	−180
Flexural stress, tangential, N m^{-2} × 10^{-3}	−1627	−1495	−1500	−1495	−1620	1486	1326	263	−59	−33
Shear stress, N m^{-2} × 10^{-3}	0	16	9	8	−31	104	−684	−356	−14	26

449

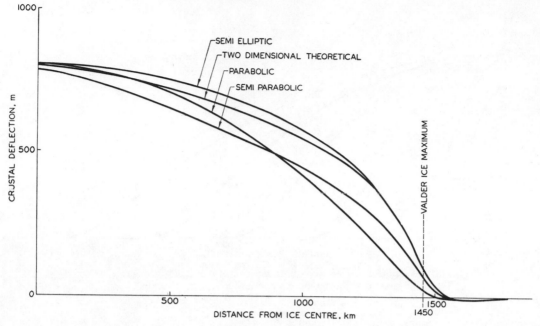

Fig. 5. Crustal deflections for axisymmetric loads of radius $A = 1450$ km, and different radial ice profiles.

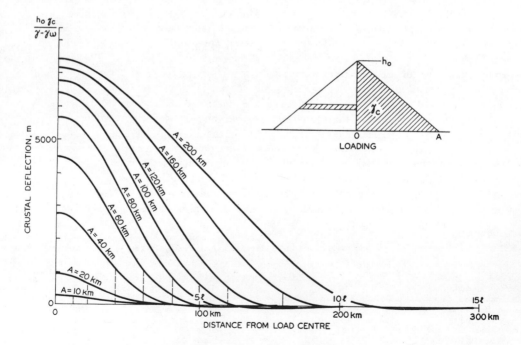

Fig. 6. Deflection of the ocean crust ($l = 20$ km) under a volcanic cone of height 10,000 meters, net density 2.0, and radius A.

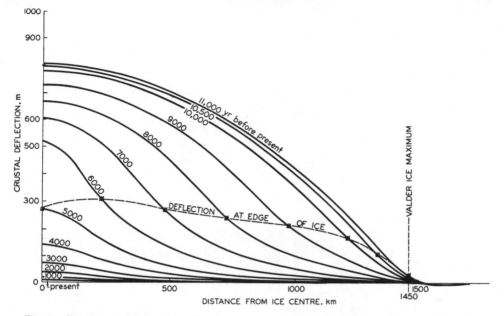

Fig. 7.   Predicted crustal rebound under constant ice retreat from the Valder ice maximum (11,000 B.P.). Initial ice radius is 1450 km, central depth 3000 meters, rate of retreat 0.242 km yr^{-1}. Time for half-rebound = 1000 years.

During the Pleistocene period, Lake Bonneville covered an irregular area of approximately 520 by 200 km and had a total mass of 1.05×10^{16} kg [*Crittenden*, 1963].

This loading may be approximated by two adjacent, uniform, axisymmetric lakes of radii 95 and 65 km and of depths 310 and 130 meters, respectively.

Deflections are given by (8) and (9) (see also Figure 3). Maximum deflection is 61 meters near the center of the larger load. The maximum measured uplift of the shoreline since removal of the water is 64 meters [*Crittenden*, 1963], and the observed profile approximates that predicted by the model. Closer agreement is obtained by more accurately approximating the distributed load.

Sedimentary basins may be handled in the same way. Displacement profiles may be plotted for each stratum, and where the basin is submerged, immersed densities are substituted.

Dynamic Model

Theory—Mantle Viscosity

In the steady-state model, the mantle was considered as a dense liquid, but viscosity did not apply. This model would handle static loads

or loading changes after a time interval sufficient for equilibrium to be restored. In the dynamic model another factor is included: the time-dependent effect of viscosity of the liquid.

The rate of displacement of the shell under mantle flow is assumed to be proportional to the unequilibrated pressure at the shell-mantle interface. This in effect is the assumption made in previous isostatic studies [*Gutenberg*, 1959; *Broecker*, 1966]. Thus, if a loading increment $\triangle q_0$ is applied to the shell, this unbalanced pressure is applied to the mantle, neglecting spread due to shell thickness, and the initial displacement rate is proportional to Δq_0, where Δq_0 is a continuous smoothly varying function of position [*Brotchie*, 1969],

$$dw/dt = - (C_1/\gamma') \, \Delta q_0 \qquad (10)$$

in which C_1 (and γ') is a constant. After time t, the interface is displaced a distance Δw, the unbalanced pressure reduces to Δq, and the rate of displacement reduces to

$$\frac{dw}{dt} = -C_1 \frac{\Delta q}{\gamma'} \approx -C_1(\Delta w_e - \Delta w) \qquad (11)$$

in which Δw_e is the steady-state deflection under the load Δq_0.

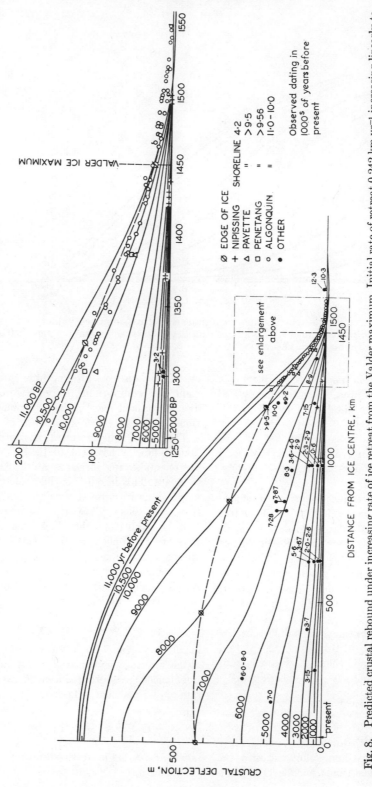

DISTANCE FROM ICE CENTRE, km

CRUSTAL DEFLECTION, m

Fig. 8. Predicted crustal rebound under increasing rate of ice retreat from the Valder maximum. Initial rate of retreat 0.242 km yr⁻¹ increasing linearly to complete deglaciation by 7000 B.P. Time for half-rebound = 1500 years. Observed displacements are also plotted, and carbon datings are given in thousands of years before present.

TABLE 3. Predicted Crustal Displacements and Displacement Rates in North America

Values are for displacements since the Valder ice maximum (11,000 years B.P.), for an initial ice radius of 1450 km, rebound constant $C_1 = 0.00045$ yr^{-1}, and deglaciation rate of initially 0.242 km yr^{-1} increasing linearly to complete deglaciation at 7000 B.P.

Time, years B.P.	Distance from Ice Center, km				
	0	483	966	1450	1930
Displacement, meters					
11,000	801.93	712.17	445.21	22.60	0.069
10,000	782.09	682.48	388.39	13.58	0.042
5,000	170.10	99.58	40.66	1.43	0.003
0	17.90	10.50	4.29	0.15	0.000
Displacement Rate, m/100 yr					
11.000	0	0	0	0	0
10,000	−2.46	−3.54	−6.94	−0.64	−0.003
5,000	−7.65	−4.48	−1.83	−0.063	−0.000
0	−0.807	−0.471	−0.192	−0.006	−0.000

Integration gives

$$\Delta w = \Delta w_e (1 - e^{-C_1 t}) \qquad (12)$$

and the rate of displacement at any time, from (11) and (12) is

$$dw/dt = -C_1 \, \Delta w_e e^{-C_1 t} \qquad (13)$$

Applications—Glacial Retreat

Consider the case of the axisymmetric uniform ice sheet of constant depth but with the edge radius decreasing at a constant velocity c.

The equilibrium deflection at any time t is given by (8) and (9), in which the ice radius a is

$$a = a_0 - (ct/l) \qquad (14)$$

and a_0 is the radius at time $t = 0$, e.g. the commencement of retreat.

The particular case of retreat of a parabolic ice sheet of initial radius 1450 km (900 miles), initial central depth of 3000 meters, constant rate of retreat $c = 0.242$ km yr^{-1} (0.150 miles yr^{-1}), and viscous constant of $C_1 = 0.00065$ yr^{-1} is plotted in Figure 7. The central depth is assumed to decrease linearly with the radius a (to 1500 meters at $a = 0$). Displacements are given at 1000-year intervals from the commencement of retreat at 11,000 years B.P. (before present). This is assumed to approximate the conditions during retreat from the Valder ice maximum [*Farrand*, 1962; *Broecker*,

1966; *Leverett and Taylor*, 1915], which resulted in removal of the last glacial ice sheet from North America. The case of a variable rate of retreat, initially 0.242 km yr^{-1} but increasing linearly to result in complete deglaciation by 7000 B.P., is also predicted (Figure 8, Table 3). Observed displacements based on shoreline uplifts in this region and on carbon dating of specimens from these shorelines are plotted in Figure 8. This latter case fits evidence on time of deglaciation, and, if $C_1 = 0.00045$, it also agrees well with observed rebounds and rebound rates [*Farrand*, 1962; *Broecker*, 1966; *Leverett and Taylor*, 1915; *Washburn and Stuiver*, 1962; *Barnett*, 1966], including present rebound rates obtained from tide gage records. Predicted present rebound rate near the ice center is 0.81 m/100 yr. Observed rebound rate from tide gage records is 0.73 m/100 yr at Hudson Bay [*Barnett*, 1966].

Predicted rebounds after removal of the ice are of the order of 300 meters or less, and this envelope (Figures 7 and 8) fits the maximum rebounds recorded. The observed points (Figure 8) above the predicted curve near the Valder maximum are in better agreement with the dynamic model if the earlier and larger loading due to the Wisconsin maximum glacial cycle is also included. Predicted rebound rates are greatest at the time of ice removal and decrease almost exponentially with time, in accordance with observations [*Farrand*, 1962].

The value of the rebound constant C_1 that best fits the observed data is sensitive to the rate of deglaciation assumed and lies in the range 0.00065 to 0.00045 yr^{-1}. The constant for Greenland appears to be larger [*Washburn and Stuiver*, 1962] ($C_1 = 0.001$), but the same effect would be produced with a lower value if the ice retreated and later advanced to its present position 8 km from the site observed. (If the ice retreated over 50 km, C_1 would lie in the range above.)

For the special case of constant ice retreat, and in the region where the deflection lag δw due to viscosity is of smaller order than the equilibrium deflection w_e, the approximation

$$dw/dt = \delta w/\delta t \qquad (15)$$

may be introduced.

The quantity δt is the time required to rebound δw. From (11), (12), and (15), we have

$$\delta t = 1/C_1 = (l/c)\, \delta a$$

in which δa is the movement of the ice front in time δt, which gives

$$\delta a = c/C_1 l$$

$$C_1 = c/l\delta a'$$

Thus the viscous lag is represented by a constant time lag $\delta t = 1/C_1$, which is independent of c, and an effective lag in ice radius δa, which is proportional to c. Under ice advance the sign of c and hence the signs of δw and δa are reversed.

REFINEMENTS OF MODELS

Several factors not included in the models presented but which may also be significant in the geoshell are (1) initial in-plane stress in the shell; (2) shear deformation of the shell; (3) creep deformation of the shell; (4) normal stress, and normal strain in the shell; (5) non-uniformity of the shell in thickness, elastic modulus, and curvature; (6) compressibility of the liquid enclosed; (7) global effects of localized loads, producing tension in the shell and compression in the liquid; (8) deformations in deeper zones of the shell where plane coordinates do not apply.

These various effects are considered in a refined model [Brotchie, 1969], two applications of which follow.

Concentrated loading. The loading on the crust resulting from the weight of water in Lake Mead [Westergaard and Adkins, 1934; Raphael, 1954], is a concentrated load of 3.6×10^{13} kg, applied over a relatively short period, commencing in 1935 with the completion of Hoover Dam.

From the initial steady-state model, maximum deflection beneath the load is $w_0 = Pl^2/8D = 0.40$ meter. If the refined model [Brotchie, 1969] including shear deformation is used and the effective radius of the loading at the middle surface is assumed to be 24 km, the equilibrium deflection is increased to $\sim 2.5\, w_0 = 1.0$ meter.

By using the dynamic model and assuming a uniformly distributed loading of radius 24 km at the mantle interface, the initial rate of this deflection at the center of loading from (10)

is $dw/dt = 0.004$ m yr^{-1}. The only initial deflection of the crust is that due to elastic normal strain. This value has been previously estimated [Westergaard and Adkins, 1934] at 0.25 meter for an elastic modulus E of 4.17×10^{10} N m^{-2}. The value for E assumed here is 8.35×10^{10}, which allows for an increase in E with depth [Gutenberg, 1959], giving an initial elastic deflection of 0.13 meter. The model deflections then are 0.13 meter initially, 0.14 meter in 3 years, 0.18 meter after 13 years, and over 1.0 meter eventually. Maximum measured deflections [Raphael, 1954] were 0.13 meter in 1940 and 0.18 meter in 1950, representing approximately 3 and 13 years of full load, respectively.

Global effects. The local effects of glaciation are described in the previous parts. Under ice advance the crust deforms downward (equations 8 and 9), displacing large volumes of mantle. Under ice retreat the mantle returns, resulting in crustal rebound.

During ice retreat the melted ice flows into the oceans, which causes a rise in sea level around the earth. This in effect is a distributed load equal in magnitude to the (melted) ice. It causes a deflection of the sea bed and a displacement of the mantle beneath. The mass of the mantle displaced is essentially that of the melted ice and that of the mantle previously displaced from beneath the ice. Thus, the global effect of deglaciation is to cause removal of the ice load and a simultaneous increase in hydrostatic load in the ocean bed; as the water flows from the ice caps to the oceans, the mantle is forced to flow in the opposite direction to maintain equilibrium (Figure 9). Because of the higher viscosity of the mantle, a time lag occurs before equilibrium is again achieved. The dynamic model again applies to this viscous lag. From the restricted case considered previously, the time lag in displacement of the ocean crust during constant rate of increase in water depth is $\delta t = 1/C_1$, i.e. 1400–2200 years.

Glaciation will result in in-plane forces in the crust due to localized deformation under the ice, removal of load from the ocean bed, and mantle flow.

The reduction in total ice volume throughout the world since the Wisconsin glaciation has been estimated by Donn et al. [1962] as be-

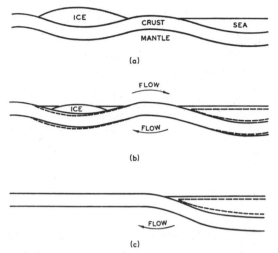

Fig. 9. Global effect of deglaciation showing (a) equilibrium displacement under ice and ocean, (b) flow of melting ice to ocean, lagging flow of mantle to compensate, and resulting crustal displacements, and (c) approaching equilibrium after deglaciation.

tween 4.2 and 4.96 × 10⁷ km³. The area of ocean is given as 3.62 × 10⁷ km² = 71% of the surface area of the globe. The total increase h in sea depth is then between 106 and 124 meters for an ice density of 0.9. The actual sea-level rise, allowing for equilibrium displacement of the sea bed under load, is approximately $h[(\gamma - \gamma_w)/\gamma] = 75$ to 87 meters.

On the continental shelves the displacement of the sea bed will decrease toward the shore and will vary with the terrain of the shelf and hence with the distribution of the increased load. Thus the strand lines formed during the Wisconsin glaciation will now lie at a depth between 75 and 106 meters (or 87 and 124 meters) below present sea level. (From two estimates [Donn et al., 1962] of ice volume during an earlier Pleistocene glaciation, a further strand line should occur at a depth within the range 97 to 138 meters (or 113 to 159 meters).)

Strand lines varying in depth within these limits have been observed [Donn et al., 1962; Fairbridge, 1961; Phipps, 1967] on various continental shelves.

The present ice volume, including the ice caps of Greenland and Antarctica, is similarly estimated [Donn et al., 1962] at between 1.60 and 2.01 × 10⁷ km³. Melting of these ice caps

would result in a further increase in sea depth of 69.3 to 87 meters. The corresponding rise in sea level would be 49 to 61 meters for equilibrium displacement of the crust. If the melting were to occur suddenly before displacement of the mantle, the rise would be almost 69.3 to 87 meters.

If the time lag in deflection is 1400 years and the sea depth was increasing at a rate of approximately 9.0 m/1000 yr, the lag in displacement would be $\delta w = 12.6$ meters. For a time lag of 2200 years, $\delta w = 19.8$ meters. This lag may have resulted in a fall in sea level after deglaciation ceased.

The stress in the crust resulting from an increase in sea depth of 106 meters is approximately

$$-\frac{0.71 h_w \gamma_w E}{\gamma R} = -2.96 \times 10^5 \text{ N m}^{-2}$$

CONCLUSIONS

Crustal stiffness, curvature, and mantle density combine to form the parameter l, the 'radius of relative stiffness' of the crust. Crustal deformation is expressed in terms of this parameter. The average continental value of l is about 58 km. The corresponding value for the ocean crust is 20 km, including the effect of immersion.

For concentrated loads such as volcanic cones and reservoirs, the zone of downward displacement extends approximately $4l$ from the load center, i.e. 230 km on the continent and 80 km beneath the sea.

For uniform loading the zone of downward displacement extends approximately 2 to $3l$ beyond the edge of the load. Isostatic equilibrium is obtained at the center only when the loaded radius is greater than $3l$.

The deformed zone as a whole is essentially in isostatic equilibrium under constant loads but areas within the zone are in flexural equilibrium only, allowing isostatic gravity anomalies, both positive and negative.

For concentrated loads, the effects of shear deformation and transverse strain are significant in the region of the load and are handled by the refined model.

Deflections beneath a concentrated load are of the order of 10 times greater for the oceanic crust than for the continental crust.

The steady-state model predicts the equilibrium deformation of the crust under long-term distributed loads.

The dynamic model predicts the response of the crust to changes in these loads with reasonable accuracy in both magnitude and time.

Predicted rates of rebound also agree with observation throughout the time range considered.

The values of the time constant that best fit the observed data are in the range 0.00065–0.00045. This gives a time for half-rebound $\log_e 2/C_1$ in the range 1000–1500 years.

Where the load is changing at a constant rate, e.g. sea-level rise, a time lag in rebound beneath the load of $\delta t = 1/C_1 = 1400$–2200 years is predicted.

A global effect of deglaciation is flow of the melted ice into the oceans, which causes a rise in sea level, and a compensating flow of the mantle from beneath the ocean to beneath the zone deformed by the ice, which lowers the ocean bottom. The flow of the mantle lags behind the flow of the water.

Various types of crustal deformation may be predicted. Under a distributed load of radius A, the zone of downward displacement has a radius greater than A. Under loading of other types than distributed (e.g., circumferential loads) or under combined bending and compression, the crust will tend to deform into ridges and troughs with period of the order of $2\pi l$. For concentrated loadings such as volcanic cones, troughs of width less than $4l$ may surround the cone. The width of the troughs decreases with increasing width of load. Creep in the crust due to flexure or compression will tend to deepen these troughs and reduce their width.

From the models presented points of maximum stress under a given loading and the magnitude of this stress may be predicted. If the strength of the crust is known, the points at which rupture or yielding may occur can be predicted. From the displacement under load, energy stored may be calculated and may prove useful in earthquake studies.

Stress concentrations will tend to occur at the edges of continents owing to sedimentation loads, sea-level rise, compression, change in thickness, and irregularity of the boundary.

Axisymmetric cases were considered in particular, but unsymmetrical solutions may also be obtained.

Displacements in sedimentary basins due to weight of sediment may be predicted and points of maximum and minimum displacement located, which will provide useful data for mineral exploration.

Where the crust is submerged, the effective relative density of the mantle is reduced from 3.37 to 2.37. The effect is to increase l from 58 to 63.4 km for the continental crust and from 18.4 to 20.1 km for the ocean crust.

The response of the mantle and the crust varies with time: The immediate response is that of an elastic, solid, layered sphere. The longer-term response is that of an elastic shell enclosing a viscous liquid. Over still longer periods the viscosity of the liquid may be neglected.

The models presented may be used to predict local and global responses to concentrated, locally distributed, and global loads, and to changes in these loads.

REFERENCES

Barnett, D. M., Re-examination and re-interpretation of tide gauge data for Churchill, Manitoba, *Can. J. Earth Sci., 3,* 77, 1966.

Broecker, W. S., Glacial rebound and the deformation of the shorelines of proglacial lakes, *J. Geophys. Res., 71,* 4777, 1966.

Brotchie, J. F., General elastic analysis of flat slabs and plates, *Proc. Amer. Concrete Inst., 56,* 127, 1959.

Brotchie, J. F., J. Penzien, and E. P. Popov, Analysis of stress concentrations in thin rotational shells of linear strain hardening material, *Res. Rep. 100-11, Inst. Eng. Res.,* University of California, Berkeley, Feb. 1961.

Brotchie, J. F., Liquid filled shell model for crustal flexure, *CSIRO Tech. Rep.,* in press, 1969.

Chapman, L. G., An outlet of Lake Algonquin at Fossmill, Ontario, *Proc. Geol. Ass. Can., 6,* 61, 1954.

Crittenden, M. D., Effective viscosity of earth derived from isostatic loading of Pleistocene Lake Bonneville, *J. Geophys. Res., 6,* 5517, 1963.

Donn, W. L., W. R. Farrand, and M. Ewing, Pleistocene ice volumes and sea level lowering, *J. Geol., 70,* 206, 1962.

Fairbridge, R. W., Eustatic changes in sea level, p. 99, in *Physics and Chemistry of the Earth,* vol. 4, Pergamon Press, London, 1961.

Farrand, W. R., Post-glacial uplift in North America, *Amer. J. Sci., 26,* 181, 1962.

Girdler, R. W., Rift valleys, continental drift, and convection in the earth's mantle, *Nature, 198,* 1037, 1963.

Gutenberg, B., *Physics of the Earth's Interior,* Academic Press, New York, 1959.

Holtzscherer, J. J., and G. de Q. Robin, Depth of polar ice caps, *J. Geogr., 120,* 193, 1954.

King, L. C., *The Morphology of the Earth,* Oliver and Boyd, London, 1962.

Leverett, F., and F. B. Taylor, The Pleistocene of Indiana and Michigan and history of the Great Lakes, *U.S. Geol. Surv. Monogr. 53,* 1915.

Phipps, C. V. G., The character and evolution of the Australian continental shelf, *Aust. Petrol. Explor. Ass. J.,* p. 44, 1967.

Raphael, J. M., Crustal disturbances in the Lake Mead area, *U.S. Bur. Reclam. Eng. Mon. 21,* 1954.

Washburn, A. L., and M. Stuiver, Radiocarbon dated post glacial relevelling in north east Greenland and its implications, *Arctic, 15,* 66, 1962.

Weertman, J., Equilibrium profile of ice caps, *J. Glaciol., 3,* 953, 1961.

Westergaard, H. M., and A. W. Adkins, Deformations of earth's surface due to weight of Boulder Reservoir, *U.S. Bur. Reclam. Tech. Mem. 422,* 1934.

(Received November 18, 1968;
revised June 9, 1969.)

Editor's Comments on Paper 28

Walcott: *Isostatic Response to Loading of the Crust in Canada*

The final paper in this volume is contained in a special issue of the *Canadian Journal of Earth Sciences* which contains papers on recent crustal movements in Canada, many specifically related to glacial unloading. Walcott's paper is an exciting contribution covering many aspects of interest.

The smoothed free air gravity anomaly map (Fig. 1) illustrates that large negative anomalies geographically correspond with the former central area of the Laurentide Ice Sheet, whereas positive anomalies fringe the margin of the former ice sheet. If these deviations from isostatic balance are the result of mass imbalances due to glacial unloading, they should correspond spatially with current positive and negative crustal movements (which they do as a first approximation). As negative anomalies in Hudson Bay amount to a suggested 250–450 m of remaining rebound, Walcott (Fig. 5) proposes that glacio-isostatic rebound may consist of two periods with relaxation times of 1000 and 50,000 years, respectively. Wallcott's estimate of remaining rebound is of course much larger than predicted in Paper 27.

Walcott proceeds to define the flexural parameter of the lithosphere, α, which is similar to the "radius of relative stiffness" in Paper 27. The effect of the glacial load on an elastic crust as opposed to a hydrostatic crust (i.e., Paper 26) is illustrated on Fig. 10. According to Walcott the depression at the ice edge is independent of the flexural parameter, and the gradient of the proglacial depression is a measure of σ/h_0 where h_0 is maximum ice thickness. The Lake Algonquin shoreline is then modeled by considering the profile of elastic warping with a flexural parameter of 180 km—the agreement (Fig. 11) is as good as those shown in Papers 26 and 27!

The increased interest of geophysicists in glacio-isostatic rebound is very welcome and argues well for the future. Quaternary field workers should be aware of the different assumptions of the various earth models and should attempt to find field data in support (or otherwise) of the various proposals. The dialogue between field and theoretical workers has not been extensive, although it appears to be improving, and should benefit both sides in studies of glacio-isostatic unloading processes.

Richard I. Walcott is a Research Scientist with the Earth Physics Branch, Department of Energy, Mines and Resources, Ottawa, of the Government of Canada. He is the Canadian representative on various international committees dealing with recent crustal movements. He has published several papers in the last few years on the effects of glacial loading/unloading and on relaxation processes in the earth.

Reprinted from *Can. J. Earth Sci.*, **7**, 716–727 (1970)

Isostatic response to loading of the crust in Canada[1,2]

R. I. WALCOTT

Gravity Division, Observatories Branch, Department of Energy, Mines and Resources,
Ottawa, Canada

Received May 28, 1969
Accepted for publication July 23, 1969

28

A smoothed free air anomaly map of Canada indicates that the central part of the region occupied by the Laurentide Ice Sheet is over-compensated. Due to the close association of the free air gravity, the apparent crustal warping, the time of deglaciation, and the congruence of the gravity anomalies and the Wisconsin Glaciation, it is concluded that the over-compensation is due to incomplete recovery of the lithosphere from the displacement caused by the Pleistocene ice loads. The amplitude of the anomalies, about −50 milligals, suggests that a substantial amount of uplift has yet to occur and that the relaxation time of crustal warping is of the order of 10 000 to 20 000 y.

The profile of the ground surface at the edge of a continental ice sheet on an elastic lithosphere is assessed using a value of the flexural parameter of the lithosphere calculated from gravity and deformation studies in the Interior Plains. The conclusions are: (*a*) a purely elastic forebulge is not likely to reach an amplitude of more than a few tens of meters; (*b*) the crust will be depressed for a considerable distance beyond the edge of the ice sheet; and (*c*) for large ice sheets crustal failure will probably occur in a preferential zone several hundred kilometers inside the maximum ice limit.

The Isostatic Response to Loading of the Crust in Canada

Two separate aspects of the isostatic response to the loads of the Laurentide Ice sheets are discussed in this paper. They are the gravity effect and the shape of the depression of the crust produced by the ice load. The first of these extends the work of Innes *et al.* (1968) of the Dominion Observatory on gravity anomalies of Hudson Bay. The second arose from a recent study of crustal bending beneath topographic loads in the Interior Plains. Results from that study have been applied to the investigation of bending of the crust at the margin of an ice sheet and have some relevance in the understanding of present vertical movements and past displacements of old lake levels peripheral to the Wisconsin ice sheet.

There is very little common ground between these two aspects and normally each would be discussed separately. However, this symposium affords an opportunity for the presentation of both so they are combined under the general title at the head of the paper.

Gravity Effect

Mean free air anomalies in West Antarctica (Bentley 1964) and Greenland (Hamilton *et al.* 1956) are close to zero although the existing ice sheets there have gravity effects exceeding 110 milligals. This indicates that in both areas isostatic adjustment of the earth to the load of the ice sheets has occurred and that if the ice were removed free air anomalies of about −100 milligals would be observed. In both Fennoscandia (Niskanen 1939) and Canada (Innes *et al.* 1968) free air anomalies become systematically more negative towards the center of the now disappeared Pleistocene ice sheets and this is interpreted as due to incomplete recovery to the Pleistocene isostatic disturbance. This gravity effect in Canada is examined in more detail below.

Although still incomplete, coverage of North America in gravity observations is now sufficient for the presentation of a small scale free air anomaly map of the area covered by the Wisconsin Glaciation (Fig. 1). In part of the area, notably northwest and immediately north of Hudson Bay, control is weak and some minor features of the gravity field may require revision as more data become available. However, major revision is unlikely. The input data

---

[1]Presented at the Symposium on Recent Crustal Movements, Ottawa, Canada, March 17–18, 1969.

[2]Contribution of the Dominion Observatory No. 298.

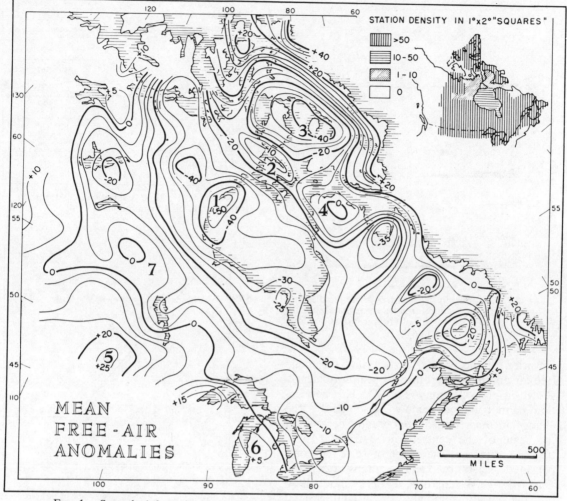

Fig. 1. Smoothed free air anomaly map of Canada and northern United States. Average free air anomalies in 1° × 2° 'squares' are smoothed twice by the method of averages and contoured. Number of observations for each average shown in the inset. 1) Eskimo Point; 2) Southampton Island; 3) Foxe Basin; 4) Ungava Peninsula; 5) Williston Basin; 6) Michigan Basin; 7) Flin Flon.

to Fig. 1 are the free air anomalies averaged over areas of 1° × 2° "squares". The inset shows the number of observations available in the calculation of each average. This data was then smoothed by hand using the method of averages to eliminate high frequency anomalies presumably related to local geology or elevation changes.

The major features of the map are: (*i*) elliptical symmetry of the gravity field with the major axis trending northwest; (*ii*) an anomaly low centred over Eskimo Point on the west coast of Hudson Bay with an amplitude of

about −50 milligals; (*iii*) a second low of about −40 milligals centered over the Foxe Basin; (*iv*) a ridge in the gravity field coinciding with the topographic ridge of Ungava Peninsula and Southampton Island across the northern part of Hudson Bay and separating the two gravity lows; (*v*) a positive outermost ring with the zero anomaly contour along the outer edge of the elliptical trough occupied by the St. Lawrence, Great Lakes, Lakes Winnipeg and Athabasca, Great Bear and Great Slave Lakes, and Melville Sound.

There is an obvious correlation between

3

Contribution No. 298

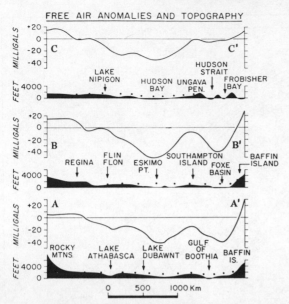

FIG. 2. Relationship between free air anomalies and topography. Cross sections trend approximately northeastward. Dots indicate uplift of ground surface required to produce zero free air anomaly.

elevation of the ground surface and the mean free air anomalies and this is even more clearly shown in cross sections (Fig. 2). This correlation cannot be adequately explained by the change in mass due to topography because of the scale of the central low. Jeffreys (1959, p. 183) discusses the effect of scale on compensated structures and demonstrates that where the wave length is, as here, about 2000 km the largest anomaly for a compensated structure is about 5 milligals. Yet the observed anomaly is more than 10 times this value and indicates broad over-compensation in the area. For shorter wave length anomalies a significant part may be due to the topography. Thus, using Jeffreys argument, about 10 milligals of the Southampton Island ridge anomaly may reflect topography. However, mean free air anomalies at sea level along the ridge are in the range −20 to +10 milligals whereas mean anomalies at the same elevation in Foxe Basin and Eskimo Point area are consistently less than −40 milligals. This difference is too great to be explained by the comparatively small elevation differences between the two areas.

The correlation of free air anomalies and topography suggests that both the gravity field and ground surface are depressed together;

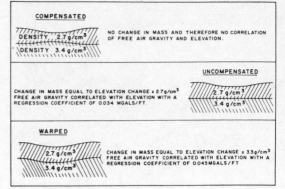

FIG. 3. Some possible theoretical relationships of topography and free air anomaly.

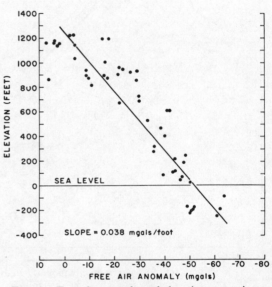

FIG. 4. Free air anomaly and elevation regression on the profile Flin Flon, Manitoba, to Eskimo Pt., District of Keewatin. Dots are the mean elevations and anomaly in 1° × 1° 'squares'. The high value of slope suggests crustal warping (c.f. Fig. 3).

that the central gravity lows are due to downwarping of the crust over Hudson Bay and the Foxe Basin. If the change in mass associated with the change in elevation is due to the displacement of heavy mantle rather than lighter crustal material, the slope of the regression should exceed 0.034 mgal/ft (~0.112 mgal/m) (Fig. 3). A value of 0.038 mgal/ft (~0.125 mgal/m) is found in the profile from Flin Flon to Eskimo Point, a distance of 1200 km (Fig. 4), and provides further circumstantial evidence of crustal downwarping.

4   Contribution No. 298

There are a number of reasons and observations which indicate that the over-compensation and associated crustal warping are closely related to continental glaciation.

(*i*) The correspondence in position and symmetry of the free air gravity anomalies with the area covered by the Wisconsin Glaciation. Both show elliptical symmetry and their major axes coincide.

(*ii*) The close correspondence between gravity anomaly and the time of deglaciation. The two gravity lows over Hudson Bay in the area of Keewatin and Foxe Basin lie in the close vicinity of the last remnants of the Wisconsin ice sheet. Deglaciation occurred about 1000 y earlier on Southampton Island which lies on the ridge between the two gravity lows (Craig 1968).

(*iii*) Studies of tilt of old marine surfaces indicate a correspondence of crustal uplift and the free air anomaly map. Specifically, isobases on these surfaces tend to be parallel to gravity contours as would be expected if uplift and gravity anomaly were both related to the rebound after deglaciation.

Remaining Uplift

The congruence of gravity anomalies with the Wisconsin Glaciation in general and with the Keewatin and Foxe Basin ice remnants in particular is an indication that a large part of the anomaly is probably due to the isostatic effect of the ice load. We cannot be sure of the amount because, at present, we have no way of knowing what the equilibrium value of gravity should be or what the contribution is to the anomaly from other sources; if anything we should expect the anomaly over the Hudson Bay Paleozoic basin to be positive rather than negative with respect to adjacent areas as in the similar Williston and Michigan basins (Fig. 1). On the basis that the striking correspondence in position and orientation of the free air anomaly with the glaciation is not accidental the free air anomaly due to the isostatic effect is estimated to be between -35 and -65 mgals. The amount of remaining uplift can be estimated on the assumption that mantle material of density 3.3 g/cm^3 will flow inward until equilibrium is attained and is between 250 and 450 m at Eskimo Point. An estimate of the remaining uplift in meters at any point may be made by multiplying the anomaly, in milligals, by 7.

Relaxation Time

A marine deposit at an elevation of 210 m in the vicinity of Eskimo Point has a radiocarbon age of about 7000 y (Lee 1959). Allowing for the eustatic change in sea level, the uplift which has occurred in 7000 y in that area is about 230 m, or perhaps a little more, depending on depth of deposition of the shells. That is, the estimated 250 to 450 m remaining uplift in the vicinity of Eskimo Point is between 52 and 66% of the total displacement 7000 y ago. Assuming the disturbance has a simple exponential decay the relaxation time from these figures is between 10 700 and 17 100 y.

This value of relaxation time conflicts with values calculated from uplift curves based on radiocarbon dates on old marine shore lines. These curves, also interpreted as simple exponential decay, give relaxation times almost an order of magnitude smaller than the above value, e.g. Andrews (1968) gives a relaxation time of 2500 y.

The conflict can be resolved if we abandon the hypothesis that the decay is a simple exponential one. The decay may conform to an expression such as:

$$w = 150e^{\frac{-t}{1000}} + 450e^{\frac{-t}{50\,000}}$$

where w is the remaining uplift at time t, involving the summation of two exponential terms which have relaxation times of 1000 and 50 000 y and amplitude constants of 150 and 450 m (Fig. 5). The curve for this expression

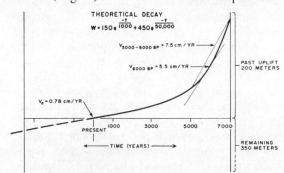

FIG. 5. Hypothetical uplift curve. The uplift is given by the addition of two exponential terms. The shape of the curve before the present is similar to observed uplift curves, and the remaining uplift is consistent with the amount inferred from gravity anomalies.

5

Contribution No. 298

agrees well with the gross features of the uplift curve, indicating a high rate of uplift 5000 to 7000 y ago, a low rate at present, and an uplift to the present of about 200 m. The curve also supports the major conclusion of the gravity study that the amount of remaining uplift in the geometric center of the isostatic disturbance is at least equal to, and probably more than, the amount of uplift that has already occurred over the last 7000 y.

Discussion

Some such expression as that given in the above equation is necessary to account for both the gravity and uplift data of the decay of the Laurentide depression. It may also be noted that detailed studies of the uplift itself such as those of Andrews (1968) require a more complicated mathematical model than simple exponential decay. This is implied for instance in Andrews' (1968) observation that the time constant, k, apparently decreases with time. However, while the form of the expression, involving two time constants, appears reasonable, reliable values of the constant and amplitude terms have yet to be evaluated and more rigorous work in this direction is required. Nevertheless, the existence of two time constants is of some interest in the light of work by Niskanen (1949), Ramberg (1967), and Orowan (1965).

Two time constants can arise in those earth models involving an elastic lithosphere overlying a fluid substratum. Niskanen (1949) argued that the stresses produced in the lithosphere by the downwarp under an ice load have decayed by the time of deglaciation and the equilibrium profile is the warped profile. Inward flowing mantle material has therefore to work against elastic stresses produced in the crust by uplift, and uplift is retarded.

Two time constants also appear explicitly in models involving the relaxation of two layers of different viscosity and thickness (Ramberg 1967, p. 196) and models of this sort may therefore lead to an explanation of the observations. Ramberg's example of a surface layer of thickness 100 km and viscosity 10^{22} poises overlying a layer of 10^{21} poises and similar thickness has a relaxation curve of the required form but the relative magnitude of the amplitude terms is reversed.

If the flow in the asthenosphere occurs by hot viscous creep of crystalline matter, then the apparent viscosity of the flow is likely to be extremely non-Newtonian, the creep rate rising as the fourth or fifth power of the stress (Orowan 1965). Such nonlinear flow will be expected to lead to a much more rapidly decreasing rate of uplift than simple exponential decay and hence explain curves of the above type.

These three examples are illustrative of the type of studies information on the dynamics of post-glacial uplift can be expected to throw light on and emphasize the importance data of the sort given at this symposium have in geophysical theory.

Shape of the Depression

The equilibrium shape of the depression produced by the load of a large ice sheet will not differ much from that of local isostatic compensation as described by Weertman (1961) except near the ice edge, and there, because of strength of the lithosphere, large divergences are possible. The stresses caused in the lithosphere by the load will produce a depression far beyond the limits of the load and, as suggested by McGinnis (1968), elastic upward bending of the crust may occur to form a forebulge in front of the ice sheet. In order to calculate the extent of the depression and the amount of uplift some measure of the flexural parameter of the lithosphere is needed.

Flexural Parameter of the Lithosphere

Topography produces differential loads over the earth's crust and by studying the nature of the deformation or the distribution of the compensation produced by these loads the isostatic response of the crust to loading can be assessed. Interpretation, however, is only effective for specific models of the earth. In the model chosen here rocks at a depth of no more than 200 or 300 km have no long term strength, and the stresses due to the topographic loading are entirely taken up within the strong superincumbent shell, that is within the earth's lithosphere. The analysis of stress and deformation follows the similar geophysical approaches of Vening–Meinesz (1931) and Gunn (1943a), in which the lithosphere is treated as an elastic sheet overlying a fluid substratum.

Contribution No. 298

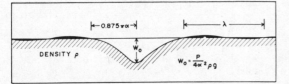

FIG. 6. Flexure of floating elastic sheet due to point load P. The flexural parameter α defines the wavelength, and, in part, the amplitude. Wavelength $\lambda = \pi \cdot \alpha$.

The fundamental lithospheric parameter which governs the wavelength and, in part, the amplitude of bending is the flexural parameter α (Fig. 6).

$$\alpha = \left(\frac{ET^3}{3(1 - \sigma^2) \cdot \rho g} \right)^{\frac{1}{4}}$$

where E = Young's Modulus
T = Sheet thickness
σ = Poisson's ratio
ρ = Density of the substratum
g = Acceleration of gravity.

The method used in determining the flexural parameter is straightforward. The deformation produced by known loads is computed for different values of the flexural parameter, and compared with the observed deformation.

Caribou Mountains Load

In northern Alberta there are a number of large, flat-topped hills, topographic residuals from late Tertiary erosion, which produce differential loading of the earth's crust. One of the more isolated and certainly the most symmetrical of these is the Caribou Mountains residual (Fig. 7). It is almost circular in shape with a diameter of about 150 km, a mean elevation of about 3100 ft (~940 m) and sides which fall steeply to the surrounding plain which is at an elevation of about 1000 ft (~304 m). The mean density of the rocks in the area as determined from nearby Bouguer density profiling is 2.3 g/cm³. The Caribou Mountains therefore constitute a roughly circular load of about 150 bars or 1.5×10^8 dynes/cm² over a radius of 75 km.

A large number of oil wells have been drilled in the area, and in the western half are sufficiently closely spaced to provide good control on the depths of underlying strata. One formation—the Muskeg—lying about 1000 ft (~304 m) above the basement was chosen as

a marker to study the deformation due to loading of the Caribou residual.

Structural contours on the horizon at the top of the Muskeg Formation (Fig. 7) are sub-parallel and indicate a dip to the southwest of about 4.4 m/km or about 0.44%. There is no noticeable deflection of the contours in response to the Caribou residual.

The theoretical deformation produced by the actual topographic loading in the vicinity of the Caribou Mountains was calculated for different values of the flexural parameter of the lithosphere, α, and plotted as structural contours (Fig. 8). The calculation was made in the following way: the contribution to the vertical displacement at the center of concentric rings was calculated by numerical solution of the integral given by Hertz (1884) in

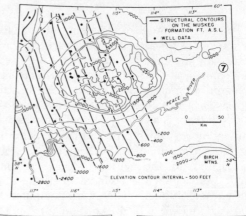

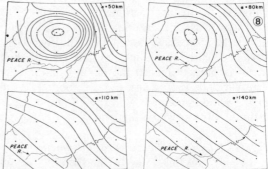

FIG. 7. Deformation of a surface underlying the Caribou Mountains, Alberta. Elevation contours show distribution of the load due to the topography. Structural contours on the Muskeg Formation show no local disturbance due to the load.

FIG. 8. Theoretical structural contours due to the Caribou Mountains load calculated for different flexural parameters (α). Contour interval is 200 ft (~61 m). The minimum flexural parameter to produce observed deformation (Fig. 7) is between 110 and 140 km.

his polar coordinate solution of the differential equation of a floating elastic sheet. The contribution for the rings were the weights by which actual mean elevations, determined for the Hayford Zones A-15, were multiplied: the deflections were computed on the grid shown by the black dots in Fig. 8 and contoured at 200 ft (~61 m) intervals.

For values of α of 50 and 80 km the disturbance of surfaces underlying the load should be quite obvious. For α of 110 km the disturbance is still noticeable as a deflection in the trend of the contours but for α of 140 km there appears to be no significant disturbance due to the load. We can therefore deduce that the minimum value of α to produce the observed deformation is between 110 and 140 km.

Isostatic Anomalies

Another method of assessing the flexural parameter is by comparing the surface gravity field with the theoretical gravity field due to the compensation for topography. The same assumptions are made regarding earth structure as earlier, with the further assumption that at depth there is a density discontinuity surface, undulations of which provide the compensation for the surface loading. Such a surface would correspond to the seismic Moho and may occur well within the boundaries of the lithosphere. Isostatic anomalies were computed for a crustal thickness of 40 km and for different values of the flexural parameter on a 30 ft × 1° grid covering the Interior Plains. This gave about 120 separate points and in Fig. 9 the variance of the isostatic anomalies about their mean is plotted against the flexural parameter. The variance is least for an α of about 120 km giving good agreement with the minimum value estimated from the loading study.

Effect of Time on the Flexural Parameter

In applying this value of the flexural parameter to the crustal loading of an ice sheet, allowance must be made for the length of time the load is acting. This is because the relaxation of stresses by creep will cause a decrease in the flexural parameter with time. The effect can be thought of as due to the time dependent strain of creep which effectively decreases the value of Young's Modulus with time (which is

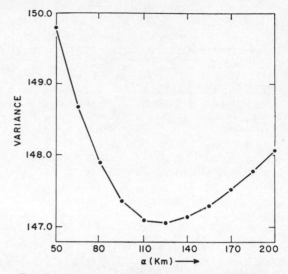

FIG. 9. Variance of isostatic anomalies about their mean computed for different values of flexural parameter. Compensation depth is 40 km. The minimum occurs about $\alpha = 120$ km.

the ratio of stress to strain) and hence lowers the effective flexural rigidity of the elastic sheet. On this basis the estimate of the flexural parameter of 180 km derived from short term loading in Fennoscandia from McConnell's (1968) values of a crustal thickness of 120 km and a Young's Modulus of 6.5×10^{11} dynes/cm² is compatible with this value of 120 km derived from the Interior Plains.

Ice Edge Profile

The calculated surface profile at the edge of a continental ice sheet loading a lithosphere with a flexural parameter of 150 km is shown in Fig. 10. The ice sheet surface profile from Weertman (1961), is approximated by an exponential curve for convenience in calculation. The radius of the ice sheet is 450 km and elevation is 1.8 km and is consequently smaller than the Laurentide Ice Sheets at their maximum. The diagram serves, however, to indicate the principal effects on the profile of the ground surface of an elastic as opposed to a hydrostatic crust. There are three important features.

(1) A forebulge is produced caused by an elastic upward bending of the lithosphere above its equilibrium position. The amplitude of the forebulge is independent of the flexural parameter and is given approximately by $H/100$ where H is the elevation in the center of the ice sheet. The distance the forebulge occurs from

8

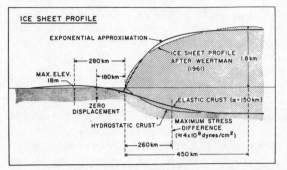

FIG. 10. Profile across an ice front. The ice sheet profile is approximated by the exponential curve. Note the amplitude and position of the forebulge, the proglacial depression, and the position of maximum stress difference.

the ice edge is dependent on the flexural parameter and is substantially independent of the load. It is given approximately by $1.9 \times \alpha$.

(2) At the ice edge the ground surface is depressed below the equilibrium position by a distance of $H/11.5$ or for a 1.8 km ice sheet by 155 m. The shape of the curve rising from there to the forebulge is characteristic and is given by the equation

$$w = \frac{H}{23} e^{-\frac{x}{\alpha}} \left(2 \cos \frac{x}{\alpha} - \sin \frac{x}{\alpha} \right)$$

where w is the deflection, H and α as before, x the distance from the ice edge.

(3) The maximum stress difference will occur at the base of the lithosphere and about 200–300 km in from the edge of the ice sheet. The magnitude of the maximum stress difference depends largely on the magnitude of the load, and is approximately given by 2×10^8 dyne/cm^2 for each 1 km of elevation of the ice sheet.

Forebulge

Evidence for the existence of a forebulge is seen in the rapid rise of sea level of Northeastern United States (Newman and March 1968) and the changes in level indicated by geodetic and tide gauge levelling in the Great Lakes Region (Moore 1948; McGinnis 1968); observations interpreted as a collapse in the forebulge after deglaciation. The amplitude of the forebulge on the Atlantic Coast exceeded 80 m (Newman and March 1968, Table I) and judging from the present rate of subsidence was of the same or greater amplitude in the

Great Lakes Region. It is unlikely that elastic bending is adequate to explain a forebulge of this size for two reasons: firstly, an 80 m forebulge requires an ice sheet of about 8 km thickness which is much greater than other estimates of ice thickness; secondly, the half wavelength of the present downwarping on the Atlantic seaboard is greater than 1000 km, yet the predicted half wavelength for an elastic lithosphere ($\alpha = 180$ km) is 560 km. The earlier idea of Daly (1940) that the peripheral bulge is caused by the lateral extrusion of mantle material from beneath the crust underlying the ice sheet would seem to provide a more adequate mechanism.

Proglacial Depression

Dynamic forces generated by movement in the mantle will be effective over a very large distance and will produce a regional warping upon which the shorter wavelength elastic bending will be superimposed. The elastic effects will dominate the crustal flexure within a few hundred kilometers of the ice edge and we may expect that warping in that region will be due, principally, to elastic bending. This point is well illustrated in the warped surface of the uppermost beach level of Lake Algonquin. Lake Algonquin, a glacially impounded lake in the vicinity of the present Great Lakes, drained in its later stages through the Fossmill Outlet in Ontario (Chapman 1954). The highest beach level therefore occurred when the ice blocked that outlet and the ice edge lay between Fossmill and the nearest upper beach level, about 70 km to the south. The uplift of upper beach levels is plotted against distance from the Fossmill outlet, in Fig. 11, and two extreme positions for the ice edge are indicated as "A" and "B". For "A", the depression of the ground surface at the ice edge was 170 m below the highest point of the profile and for "B" the depression was 240 m. These values for the depression correspond to elevations of the ice surface in the center of the ice sheet of 1.7 and 2.6 km respectively, which are not unreasonable values (Weertman 1961). Ground surface profiles of different flexural constants drawn to pass through point "A" are also shown in Fig. 11. The observed profile lies in the envelope, $\alpha = 180 \pm 30$ km. Both this value of α and the correspondence in shape

9

Contribution No. 298

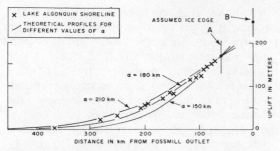

FIG. 11. Upper beach levels of Pleistocene Lake Algonquin. The distance of observations of the beach from the Fossmill outlet are plotted against their uplift relative to the horizontal region more than 500 km from the outlet. Solid lines show the predicted profile for an ice sheet the edge of which has the position and vertical displacement shown by "A". The observations lie in the envelope $\alpha = 180 \pm 30$ km.

between observed and theoretical profiles indicates a close agreement with elastic warping. Thus the profile of the upper beach levels of Lake Algonquin is consistent with the peripheral depression of an ice sheet which has the edge located at, or a short way south of the Fossmill outlet.

This model is a static one and can be contrasted with Broecker's (1966) dynamic model. In that model the upper beach level is thought to be formed during the retreat of the ice front and the uplift is a function of the rate of retreat and relaxation time of uplift. Broecker's model, however, requires a weak crust with a flexural parameter of less than 20 km.

Maximum Stress Difference

The ice edge profile demonstrates that the maximum stress difference for a 2 km thick ice sheet would be about 4×10^8 dynes/cm².

In the laboratory the ultimate strength of peridotite under extension is about 2×10^9 dynes/cm² (Griggs *et al.* 1960) but in nature this is likely to be less due to very slow strain rates (see Handin 1966, pp. 233–235), and the elastic limit will be lower still. It is possible therefore that as an ice sheet grows the maximum stress difference could increase beyond the elastic limit and anelastic effects, plastic deformation, and even fracture may occur. Once they have occurred future movement will tend to be concentrated in this zone. The zone, if it exists, will occur parallel to, and set in several hundred kilometers from the ice margin. It is possible that the elliptical trough around

the edge of the Canadian Shield, now occupied by the major lakes of Canada on the west, and the Viscount Melville Sound in the north and the St. Lawrence in the southeast, may mark this zone.

Discussion and Summary

The elastic effects of the lithosphere on the shape of an isostatic depression produced by an ice sheet can be important near the ice front, and observations of the warped Lake Algonquin shoreline agree well with the expected elastic warping of a lithosphere with a flexural parameter similar to that determined from independent gravity and structural studies in the Interior Plains of Canada. There is thus a growing body of evidence that the lithosphere of the stable platforms of North America and Fennoscandia (McConnell 1968) and even the oceanic crust in the Pacific (Gunn 1943*b*) behave as elastic sheets with a much higher flexural parameter than previously thought by Vening–Meinesz (1931).

The elastic effects of the lithosphere on the dynamics of crustal uplift are, however, two edged. On the one hand we have Niskanen's (1949) argument given earlier. On the other hand we have McConnell's (1968) argument that because of elastic stresses produced by downwarping rebound is accelerated in the return to equilibrium following deglaciation. The inference from the present study where the flexural parameter in the Interior Plains for very long term loading is only slightly smaller than short term loading due to glaciation would suggest decay of stress is very slow and that the effect of elasticity is nearer to McConnell's position than Niskanen's but that retardation may still have to be considered in the later stages of uplift.

Elastic flexure is insufficient in itself to produce a forebulge of the amplitude and extent implied by the downwarping on the Atlantic Seaboard and in East-central United States, and the dynamic effects of subcrustal flow of mantle material will have to be considered in this regard. The gravity map itself indicates that flow of matter has probably occurred as there is a significant mass deficiency in the central part of the area occupied by the Laurentide Ice Sheets. Whether the tentative indication of systematically positive anomalies in the area peripheral to the Ice Sheets is an indication of under·compensation and represents the remains of the forebulge will require further work, particularly in studies of the present direction of ground surface movement in that area. The present indication is that the inference seems to be true on the Atlantic Seaboard and eastwards to Lake Michigan but data are lacking elsewhere.

Another deduction from the gravity field is the apparently large amount of uplift that has yet to occur. This in turn suggests that our present interpretation of the decay as a simple exponential one is only partly true and the comparatively rapid decay of the past indicating a low upper mantle viscosity, is superimposed on a much slower, and, as far as amplitude of movement goes, more important, decay presumably related to some different structural effect. This is another aspect that requires further study.

Acknowledgments

It is a pleasure to acknowledge gratefully the assistance of members of the Observatories Branch and the Geological Survey of Canada in the development of this paper. In particular I would like to thank Dr. M. K. Paul and Michael Dence of Gravity Division and Dr. J. G. Fyles of Quaternary Research Division of the Geological Survey of Canada.

ANDREWS, J. T. 1968. Postglacial rebound in Arctic Canada: similarity and prediction of uplift curves. Can. J. Earth Sci. 5, pp. 39–47.

BENTLEY, C. R. 1964. The structure of Antarctica and its ice cover. Res. Geophysics II, pp. 335–389. Massachusetts Institute of Technology Press, Cambridge, Massachusetts.

BROECKER, W. S. 1966. Glacial rebound and deformation of shorelines of proglacial lakes. J. Geophys. Res. 71, pp. 4777–4782.

CHAPMAN, L. J. 1954. An outlet of Lake Algonquin at Fossmill, Ontario. Proc. Geol. Assoc. Canada, 6, pp. 61–68.

CRAIG, B. G. 1968. Late-glacial and postglacial history of the Hudson Bay region. Earth Science Symposium on Hudson Bay, Geol. Surv. Canada, 68–53.

DALY, R. A. 1940. Strength and structure of the earth. Prentice Hall, New York. 434 pp.

GRIGGS, D. T., TURNER, F. J., and HEARD, H. C. 1960. Deformation of rocks at 500–800 °C. Geol. Soc. Amer., Mem. 79, p. 39.

GUNN, R. 1943a. A quantitative evaluation of the influence of the lithosphere on anomalies of gravity. J. Franklin Inst. 236, p. 47.

——— 1943b. A quantitative study of isobaric equilibrium and gravity anomalies in the Hawaiian Islands. J. Franklin Inst. 236, p. 373.

HAMILTON, R. A., BROOKE, F. R., PEACOCK, S. D., BOWATER, S., and BULL, C. 1956. British North Greenland Expedition, 1952–1954. Scientific Results, Geog. J. 122, pp. 203–241.

HANDIN, J. 1966. Strength and ductility. Geol. Soc. Amer., Mem. 97, Editor S. P. Clark.

HERTZ, H. 1884. On the equilibrium of floating elastic plates. Weidmann's Annalen, 22, pp. 449–455.

INNES, M. J. S., GOODACRE, A. K., WESTON, A., and WEBER, J. R. 1968. Gravity and isostasy in the Hudson Bay Region. Pt. V. Science, History and Hudson Bay. Vol. 2. Editor C. S. Beals.

JEFFREYS, H. 1959. The Earth. 4th ed. Cambridge University Press, London. 420 pp.

LEE, H. A. 1959. Surficial geology of the southern District of Keewatin and the Keewatin ice divide, Northwest Territories. Geol. Surv. Can., Bull. 51.

McCONNELL, R. J., JR. 1968. Viscosity of the mantle from relaxation time spectra of isostatic adjustment. J. Geophys. Res. 73, pp. 7089–7105.

McGINNIS, L. D. 1968. Glacial crustal bending. Bull. Geol. Soc. Amer. 79, pp. 769–776.

MOORE, S. 1948. Crustal movement in the Great Lakes area. Bull. Geol. Soc. Amer. 59, pp. 697–710.

NEWMAN, W. S. and MARCH, S. 1968. Littoral of the Northeastern United States: Late Quaternary Warping. Science, 160, pp. 1110–1112.

NISKANEN, E. 1939. On the upheaval of land in Fennoscandia. Ann. Acad. Sci. Fennicae. A, LIII, No. 10.

——— 1949. On the elastic resistance of the earth's crust. Ann. Acad. Sci. Fennicae A III, Geologica–Geographica, 21.

OROWAN, E. 1965. Mechanical properties of the crust and mantle. Boeing-Scientific Research Laboratories Document D1-82-0485.

RAMBERG, H. 1967. Gravity, deformation and the earth's crust. Academic Press, London. 214 pp.

VENING-MEINESZ, F. A. 1931. Une nouvelle méthode pour la réduction isostatique regionale de l'intensité de la pesanteur. Bull. Géod., 29.

WEERTMAN, J. 1961. Equilibrium profile of ice caps. J. Glaciology, 3, pp. 953–964.

Appendix

(1) Flexural Parameter

The upward pressure caused by the elastic bending of a thin plate is $D\Delta^4 w$; where D is the flexural rigidity of the plate and is equal to $E \cdot T^3 / 12(1 - \sigma^2)$; E is Young's Modulus, T the plate thickness, σ is Poisson's ratio (Hertz 1884). If the plate is floating on a fluid of density ρ_z and has displacement 'w', and the top surface is

11   Contribution No. 298

covered by material ρ_i, then there is an additional upward pressure of $w \cdot (\rho_z - \rho_i) \cdot g$ where g is the acceleration of gravity.

Then the total upward pressure is

[1.1] $\qquad D\Delta^4 w + w \cdot (\rho_z - \rho_i) \cdot g$

In the region of a load this expression is equal to the load, away from the load this expression is zero.

Expression [1.1] is rewritten

[1.2] $\qquad \Delta^4 w + \dfrac{w \cdot (\rho_m - \rho_i) \cdot g}{D} = (\Delta^4 + 4a^4) w$

where

[1.3] $\qquad 4a^4 = \dfrac{(\rho_m - \rho_i) \cdot g}{D}$

or

[1.4] $\quad \alpha = \dfrac{1}{a} = 4\sqrt{\dfrac{ET^3}{3(\rho_m - \rho_i) g (1 - \sigma^2)}}$

If the problem is two dimensional then

[1.5] $\qquad \dfrac{d^4 w}{dx^4} + 4a^4 \cdot w = 0$

has solutions of the form

$$w = e^{-ax}(A \cos ax + B \sin ax) + e^{ax}(C \cdot \cos ax$$
$$+ D \sin ax)$$

The flexural parameter is $\alpha(\frac{1}{a})$ and has the dimensions of length.

(2) Ice Edge Profile

The profile of the ice sheet is

$$h = h_0(1 - e^{bx}), \quad x < 0$$

Therefore the load $= h \cdot \rho_i \cdot g$
where h = elevation, h_0 the elevation at $x \to \infty$, x the distance from the ice edge and b the exponential constant (for a large ice sheet $1/b \approx 120$ to 180 km).
Then

[2.1] $\dfrac{d^4 w}{dx^4} + 4a^4 \cdot w = \dfrac{h_0(1 - e^{bx}) \cdot \rho_i \cdot g}{D}$,
$$x < 0$$

and

[2.2] $\qquad \dfrac{d^4 w}{dx^4} + 4a^4 \cdot w = 0, \quad x > 0$

These equations are solved for the boundary

conditions: w and its first three differentials are continuous at $x = 0$, and w is finite at $x = \pm \infty$. Then:

(a)

[2.3] $\quad w = \dfrac{h_0 \cdot \rho_i}{\rho_m - \rho_i} \left\{ e^{ax} \left(\dfrac{\sin ax}{2} - \dfrac{4}{5} \right) + 1 \right\}$,
$$x < 0$$

[2.4] $\quad = \dfrac{h_0 \cdot \rho_i}{10 \cdot (\rho_m - \rho_i)} \{ e^{-ax}(2 \cos ax - \sin ax) \}$,
$$x > 0$$

on the assumption that $a \approx b$.

(b)

[2.5] $\dfrac{dw}{dx} = \dfrac{-ah_0 \cdot \rho_i}{\rho_m - \rho_i} \{ e^{-ax}(3 \cos ax + \sin ax) \}$,
$$x > 0$$
$$= 0 \quad \text{when} \quad ax = \tan^{-1}(-3)$$
$$= 1.89$$

Therefore the turning point of w occurs when $ax = 1.89$ ($\rho_m = 3.3$, $\rho_i = 1.0$)

(c)

[2.6] $\quad w_{max} = \dfrac{h_0}{23} \{ e^{-1.89}[2 \cos (1.89)$
$$- \sin (1.89)] \}, \quad x > 0$$
$$= \dfrac{h_0 \cdot 1.578}{23 \cdot 6.62}$$
$$= 0.0104 \cdot h_0$$

(d)

[2.7] $\quad w = 0$, when $2 \cos ax = \sin ax$, $x > 0$
$$\text{or} \qquad ax = \tan^{-1}(2)$$
$$= 1.108$$

(e)
when $\qquad x = 0$ ($\rho_m = 3.3$, $\rho_i = 1.0$)
$$w = \dfrac{h_0 \cdot \rho_i \cdot 2}{10(\rho_m - \rho_i)} = \dfrac{h_0}{11.5}$$

Note: The amplitude of the profile from the ice edge to the forebulge is independent of the flexural parameter (c and e) and is approximately $0.1\, h_0$, distance to the turning point from the ice edge is given by

$$x = 1.89 \cdot \alpha$$

Therefore the mean gradient of the proglacial

12

Contribution No. 298

depression is a measure of α/h_0 and the vertical displacement at $x = 0$ is a measure of h_0.

(3) Maximum Stress Difference
  According to Jeffreys (1959)

$$p_{11} = -Ez \frac{\partial^2 w}{\partial x^2}$$

in a thin elastic sheet where p is the stress; E is Young's Modulus, z distance from the midplane, positive downwards, w displacement of the midplane from its equilibrium position.

[3.1]   From [2.3] $\dfrac{\partial^2 w}{\partial x^2} = \dfrac{a^2 \cdot h_0 \rho_i}{(\rho_m - \rho_i)}$

$$\times e^{-ax}\left(\cos ax - \tfrac{4}{5}\right), \quad x > 0$$

and

[3.2] $\dfrac{\partial^3 w}{\partial x^3} = \dfrac{a^3 \cdot h_0 \rho_i}{(\rho_m - \rho_i)}$

$$\times e^{-ax}\left(\tfrac{4}{5} - \cos ax - \sin ax\right)$$

Therefore p_{11} is stationary when

[3.3] $\cos ax + \sin ax = \tfrac{4}{5}$

          or $ax = 1.754$

Therefore $\dfrac{\partial^2 w}{\partial x^2} = \dfrac{-a^2 \cdot h_0 \cdot \rho_i}{(\rho - \rho_i)} 0.17$

and $E = 6.5 \times 10^{11}$ dynes/cm^2,

   $z = 5 \times 10^6$ cm,

   $h_0 = h' \times 10^5$ cm (h' in km)

   $\rho_i = 1.0$ g/cm^3 $\rho_m = 3.3$ g/cm^3

   $a = 1/150$ km

   $p_{11} = 0.85 \times 10^8 \times h'$ dynes/cm^2

and $p_{33} = -h_0 \cdot \rho_i \cdot g(1 - e^{ax})$ at $ax = 1.754$

   $= -1.1 \times 10^8 \times h'$ dynes/cm^2

Therefore maximum stress difference $= 1.95 \times 10^8 \times h'$ dynes/cm^2.

13

471

Contribution No. 298

References

An asterisk indicates that the paper is included in this volume.

Anderson, D. L., and O'Connell, R. 1967: Viscosity of the earth. *Geophys. J. Roy. Astron. Soc.*, **14**, 287–295.

Andrews, J. T. 1968: Postglacial rebound in Arctic Canada: similarities and prediction of uplift curves. *Can J. Earth Sci.*, **5**, 39–47.

Andrews, J. T. 1970: A geomorphological study of post-glacial uplift with particular reference to Arctic Canada. *Inst. Brit. Geogr. Spec. Publ.*, No. 2, 156 pp.

Andrews, J. T., and Barnett, D. M. 1972: Analysis of strandline tilt directions in relation to ice centers and postglacial crustal deformation, Laurentide Ice Sheet. *Geogr. Ann.*, **54**, Ser. A, 1–11.

Andrews, V. T., Buckley, J. T., and England, J. H. 1970: Late-glacial chronology and glacio-isostatic recovery, Home Bay, east Baffin Island, Canada. *Bull. Geol. Soc. Amer.*, **81**, 1123–1148.

Bell, R. 1896: Proofs of the rising of the land around Hudson Bay. *Amer. J. Sci.*, **151**, 219–228.

Bird, J. B. 1954: Postglacial marine submergence in central Arctic Canada. *Bull. Geol. Soc. Amer.*, **65**, 457–464.

Blake, W., Jr. 1970: Studies of glacial history in Arctic Canada. I. Pumice, radiocarbon dates, and differential postglacial uplift in the eastern Queen Elizabeth Islands. *Can. J. Earth Sci.*, **7**, 634–664.

*Bloom, A. L. 1963: Late-Pleistocene fluctuations of sealevel and postglacial crustal rebound in coastal Maine. *Amer. J. Sci.*, **261**, 862–879.

Bloom, A. L. 1970: Pleistocene shorelines: a new test of isostasy. *Bull. Geol. Soc. Amer.*, **78**, 1477–1494.

Bravais, A. 1840: Sur les lignes d'ancien niveau de la mer dans le Finmark. *Compt. Rend. Acad. Sci.*, **X**, 691.

Broecker, W. S. 1962: The contribution of pressure-induced phase changes to glacial rebound. *J. Geophys. Res.*, **67**, 4837–4842.

*Broecker, W.S. 1966: Glacial rebound and the deformation of the shorelines of proglacial lakes. *J. Geophys. Res.*, **71**, 4777–4794.

Broecker, W. S. 1970: Discussion. *Can. J. Earth Sci.*, **7**, 430.

*Brotchie, J. F., and Silvester, R. 1969: On crustal flexure. *J. Geophys. Res.*, **74**, 5240–5252.

473

Bryson, R. A., Wendland, W. M., Ives. J. D., and Andrews, J. T. 1969: Radiocarbon isochrones on the disintegration of the Laurentide Ice Sheet. *Arct. Alp. Res.,* **1**, 1–14.

Canadian Journal of Earth Sciences. 1970: Papers presented at the Symposium on Recent Crustal Movements, Ottawa, Canada, March 18–19, 1969. *Can. J. Earth Sci.,* **7**, 553–734.

Cathles, L. M., II. 1971: *The viscosity of the earth's mantle.* Unpublished Ph.D. thesis, Princeton University, 328 pp.

Crary, A. P. 1971: Thickness of ice and isostatic adjustments of ice-rock interface, *in* Quam, L. O. (ed.), *Research in the Antarctic,* Amer. Ass. Adv. Sci., Washington, D.C. 341–349.

*Daly, R. A. 1934: The changing world of the ice age. Yale University Press, New Haven, 271 pp.

*De Geer, G. 1892: On Pleistocene changes of level in eastern North America. *Proc. Boston Soc. Nat. Hist.,* **25**, 454–477.

De Geer, G. 1893: Praktiskt geologiska undersökningar inöm Hallands län. *S.G.U.,* Ser. C, No. 131.

*Donner, J. J. 1969: A profile across Fennoscandia of late Weichelian and Flandrian shore lines. *Comm. Phys.-Math.,* **36**, 1–23.

*Elson, J. A. 1967: Geology of Glacial Lake Agassiz, *in* Mayer-Oakes, W. J. (ed.), *Life, Land and Water,* University of Manitoba Press, Winnipeg, 37–43, 71–80.

Fairbridge, R. W. 1961: Eustatic changes in sea level. *Phys. Chem. of the Earth,* **5**, 99–185.

*Farrand, W. R. 1962: Postglacial uplift in North America. *Amer. J. Sci.,* **260**, 181–199.

Farrand, W. R., and Gajda, R. T. 1962: Isobases on the Wisconsin marine limit in Canada. *Geogr. Bull.,* **17**, 5–22.

Feyling-Hanssen, R. W., and Olsson, I. 1959: Five radiocarbon datings of postglacial shorelines in central Spitsbergen. *Norsk Geogr. Tids.,* **17**, 122–131.

*Fillon, R. H. 1972: Possible causes of the variability of postglacial uplift in North America. *Quaternary Res.,* **1**, 522–531.

Goldthwait, J. W. 1907: The abondoned shorelines of eastern Wisconsin. *Wisconsin Geol. Nat. Hist. Sur.,* Bull. 17, Sci. Ser. 5, 134 pp.

*Goldthwait, J. W. 1908: A reconstruction of water planes of the extinct glacial lakes in the Lake Michigan Basin. *J. Geol.,* **16**, 459–476.

Gutenberg, B. 1941: Changes in sea level, postglacial uplift and mobility of the earth's interior. *Bull. Geol. Soc. Amer.,* **52**, 721–772.

Heiskanen, W. A., and Vening-Meinesz, F. A. 1958: The earth and its gravity field. Institute British Geographers, 1966: The vertical displacement of shorelines in Highland Britain. *Trans. Inst. Brit. Geogr.,* No. 39.

Ives, J. D. 1963: Determination of the marine limit in the eastern Canadian Arctic. *Geogr. Bull.,* **19**, 117–122.

*Jamieson, T. F. 1882: On the cause of the depression and re-elevation of the land during the glacial period. *Geol. Mag.,* **9**, 400–407.

Johnson, A. M. 1970: *Physical processes in geology.* Freeman, Cooper and Co., San Francisco, 577 pp.

Kane, E. K. 1856: *The second Grinnell Expedition in search of Sir John Franklin, 1853, 1854, 1855.* Vols I and II, Childs and Peterson, Philadelphia, 464 and 467 pp.

*Kupsch, W. O. 1967: Postglacial uplift—a review, *in* Mayer-Oakes, W. J. (ed.), *Life, Land and Water.* Manitoba University Press, Winnipeg, 155–186.

Lasca, N. P. 1966: Postglacial delevelling in Skeldal, Northeast Greenland. *Arctic,* **19**, 349–353.

Lee, H. A. 1962: Method of deglaciation, age of submergence and rate of uplift of west and east Hudson Bay. *Biul. Peryglac.,* **11**, 239–245.

Lidén, R. 1938: Den senkuartara standförskjutningens fölopp och kronologi i Ångermanland. *Geol. För. Stockh. Förh.,* **60**, 397–404.

Løken, O. H. 1962: The late-glacial and postglacial emergence and deglaciation of northernmost Labrador. *Geog. Bull.,* **17**, 23–56.

*Løken, O. H. 1965: Postglacial emergence at the south end of Inugsuin Fiord, Baffin Island, N.W.T. *Geogr. Bull.,* **7**, 243–258.

*McConnell, R. K., Jr. 1968: Viscosity of the mantle from relaxation time spectra of isostatic adjustment. *J. Geophys. Res.*, **73**, 7089–7105.

Mörner, N.-A. 1969: The late Quaternary history of the Kattegatt Sea and the Swedish west coast. *Sveriges Geol. Unders.*, Ser. C, Nr. 640, 487 pp.

*Mörner, N.-A. 1971: Eustatic changes during the last 20,000 years and a method of separating the isostatic and eustatic factors in an uplifted area. *Palaeogeogr. Palaeoclimat. Palaeoecol.*, **9**, 153–181.

*Nansen, F. 1921: The strandflat and isostasy. *Christiana Vidensk-Selsk.*, Skr. 1, Mat-Naturv., **11**, 350 pp.

Nansen, F. 1928: The earth's crust, its surface forms and isostatic adjustment, *Norsk Vidensk.-Acad.*, KI,I, No. 12.

*Newman, W. S., Fairbridge, R. W., and March, S. 1971: Marginal subsidence of glaciated areas: United States, Baltic and North Sea. *Études sur le quaternaire dans le monde, Paris*, 795–801.

Nikolayev, N. I. 1967: The latest stage in the evolution of Fennoscandia, the Kola Peninsula, and Karelia. *Bull. Mosk. Obschchestvo Ispytateley Pritody*, Otdel Geol., **42**, 49–68 (in Russian, English abstract in *GeoAbstracts*, 69A/918).

Olsson, I., and Blake, W., Jr. 1962: Problems of radiocarbon dating of raised beaches based on experience in Spitsbergen. *Norsk Geogr. Tids.*, **18**, 47–64.

Prest, V. K. 1969: Retreat of Wisconsin and Recent ice in North America. *Geol. Surv. Canada*, Map 1257A.

Prest, V. K. 1970. Quaternary geology of Canada. Chapter XII, in *Geology and Economic Minerals of Canada*. Econ. Geol. Rept. No. 1, 5th ed., 676–764.

Prest, V. K., Grant, D. R., and Rampton, V. N. 1968: Glacial Map of Canada. *Geol. Surv. Canada*, Map 1253A.

Robinson, H. H. 1908: Ancient waterplanes and crustal deformation. *J. Geol.*, **16**, 347–356.

Sauramo, M. 1939: The mode of land upheaval in Fennoscandia during late-Quaternary time. *Bull. Comm. Geol. Finland*, **125**, 39–64.

Schofield, J. C. 1964: Postglacial sea levels and isostatic uplift. *N. Z. J. Geol. Geophys.*, **7**, 359–370.

*Schytt, V., Hoppe, G., Blake, W., Jr., and Grosswald, M. G. 1967: The extent of the Würm glaciation in the European Arctic. *Intern. Assn. Hydrol. Sci. Publ. 79*, 207–216.

*Shaler, N. S. 1874: Preliminary report on the recent changes of level on the coast of Maine. *Mem. Boston Soc. Nat. Hist.*, **2**, 321–340.

Shepard, F. P. 1963: Thirty-five thousand years of sea level, *in Essays in Marine Geology*, University of Southern California Press, Los Angeles, 1–10.

*Sissons, J. B. 1962: A re-interpretation of the literature on late-glacial shorelines in Scotland with particular reference to the Forth area. *Trans. Edinburgh Geol. Soc.*, **19**, 83–99.

Tanner, V. 1930: Studier över Kvartarsystem et i Fennoscandia nordliga delar IV. *Bull. Comm. Geol. Finland*, **88**, Fennia 53, 600 pp.

Ten Brink, N. 1971: *Holocene delevelling and glacial history between Søndre Strømfjord and the Greenland Ice Sheet, West Greenland*. Unpublished Ph.D. thesis, University of Washington, Seattle, 191 pp.

Upham, W. 1896: The Glacial Lake Agassiz. *U.S. Geol. Surv. Monograph*, 25.

Walcott, R. I. 1970: Flexural rigidity, thickness and viscosity of the lithosphere. *J. Geophys. Res.*, **75**, 3941–3954.

*Wallcott, R. I. 1970: Isostatic response to loading of the crust in Canada. *Can. J. Earth Sci.*, **7**, 716–727.

Walcott, R. I. 1972a: Past sea levels, eustasy and deformation of the earth. *Quaternary Res.*, **2**, 1–14.

Walcott, R. I. 1972b: Late Quaternary vertical movements in eastern North America: quantitative evidence of glacio-isostatic rebound. *Rev. Geophys. Space Phys.*, **10**, 849–884.

*Washburn, A. L., and Stuiver, M. 1962: Radiocarbon-dated postglacial delevelling in Northeast Greenland and its implications. *Arctic*, **15**, 66–72.

Author Citation Index

Subject Index